NONEQUILIBRIUM STATISTICAL PHYSICS

Nonequilibrium Statistical Physics

Linear Irreversible Processes

Noëlle Pottier
Laboratoire Matière et Systèmes Complexes, CNRS, and
Université Paris Diderot – Paris 7

OXFORD
UNIVERSITY PRESS

OXFORD
UNIVERSITY PRESS

Great Clarendon Street, Oxford, OX2 6DP,
United Kingdom

Oxford University Press is a department of the University of Oxford.
It furthers the University's objective of excellence in research, scholarship,
and education by publishing worldwide. Oxford is a registered trade mark of
Oxford University Press in the UK and in certain other countries

First Edition published in 2010
First published in paperback 2014

Published in the United States of America by Oxford University Press
198 Madison Avenue, New York, NY 10016, United States of America

British Library Cataloguing in Publication Data
Data available

ISBN 978–0–19–871227–5

Preface

The subjects treated in this book belong to the extremely vast field of nonequilibrium statistical physics. We encounter, in all domains of physics, a very large variety of situations and phenomena involving systems which are not in thermodynamic equilibrium, and this, at all scales. One of the difficulties of the statistical physics of such systems lies in the fact that, contrary to equilibrium where we have a unified approach allowing us to explain the macroscopic properties of matter in terms of the microscopic interactions (J.W. Gibbs), out of equilibrium we have only a limited number of results with a general scope. The approaches used to describe the passage from the microscopic description to the macroscopic one in the case of nonequilibrium systems are varied, and they may depend on the particular system under study. It is however possible to classify these approaches into two large groups, using, in the first one, kinetic equations (L. Boltzmann), and, in the second one, the linear response theory (R. Kubo). In spite of their diversity, these methods have in common several essential fundamental features on which they rely.

One of the aims of this book is to clarify some central ideas common to these different approaches, taking various physical systems as examples. Due to the vastness of the subject, only near-equilibrium situations, in which the irreversible processes involved may be qualified as linear, are taken into consideration.

Although extremely diverse, out-of-equilibrium phenomena display very generally the essential role played by the existence in the systems under study of well-separated time scales. Most of these time scales, very short, are associated with the microscopic degrees of freedom, while other ones, in small number and much longer, are macroscopic and characterize the slow variables. The book attempts in particular to stress, for each of the approaches, the importance of the role played by the separation of time scales.

A central property common, in the linear framework, to the different approaches, is the fluctuation-dissipation theorem, which expresses the relation between the response of a system in a slightly out-of-equilibrium situation and the equilibrium fluctuations of the dynamical variables concerned. This result constitutes the cornerstone common to the different methods of nonequilibrium statistical physics in the linear range.

The elements prerequisite for the reading of this book are rather limited. It is however necessary to master the basic notions of quantum mechanics and equilibrium statistical physics. As for the mathematical techniques used in the book, they are standard. Generally speaking, the notions necessary to the understanding of each chapter are included in it or have been provided in the preceding chapters, and the calculations are explained in detail. At the end of each chapter, a list of textbooks on the subject (in alphabetic order by author) is provided. It is followed, where possible, by a list of original papers (in chronological order).

The general organization of the book is briefly described below.

Basic notions

In statistical physics, each macroscopic variable is a statistical average of the corresponding microscopic quantities. The notions of mean and of fluctuations (and, more generally, the useful definitions and results concerning random variables as well as random processes) are drawn together in Chapter 1. The most important results in view of the study of fluctuations are the central limit theorem which underpins the central role played in physics by Gaussian laws, and the Wiener–Khintchine theorem which relates the autocorrelation function and the spectral density of a stationary random process.

Thermodynamics of irreversible processes

Chapter 2 is devoted to the thermodynamics of irreversible processes. One of the macroscopic characteristics of out-of-equilibrium systems is the occurrence of irreversible processes, such as transport or relaxation phenomena. Irreversible processes have a dissipative character, which means that the systems in which they take place lose energy (this energy is transferred to the environment of the system, and it never comes back). Accordingly, an out-of-equilibrium system is a place of strictly positive entropy production. In the case of locally equilibrated systems, it is possible, relying on the slow variables and on the entropy production, to extend thermodynamics (originally limited to the study of equilibrium states) to the description of irreversible processes.

It is in systems near equilibrium that this theory is the most firmly established. Transport phenomena then obey linear phenomenological laws. The thermodynamics of irreversible processes allows us to establish some properties of the transport coefficients, among which symmetry or antisymmetry relations between kinetic coefficients (L. Onsager), as well as the Einstein relation linking the mobility and the diffusion coefficient. This relation constitutes the very first formulation of the fluctuation-dissipation theorem.

Introduction to nonequilibrium statistical physics

While it allows us to establish some properties of the transport coefficients, the thermodynamics of irreversible processes does not provide us with any recipe to calculate them explicitly.

For this purpose, it is necessary to start from a microscopic description of the out-of-equilibrium systems, and to make the link between this description and the observed properties at the macroscopic scale. Chapters 3 and 4 offer an introduction to this approach. The principal tools of the statistical physics of out-of-equilibrium classical and quantum systems are presented, in particular the evolution equations of the distribution function and of the density operator.

Kinetic approaches

Chapters 5 to 9 are devoted to the description of transport phenomena by means of irreversible kinetic equations, mainly the Boltzmann equation.

To begin with, in Chapter 5, we consider the ideal classical gas of molecules undergoing binary collisions (historically the first system to have been studied by means of a kinetic equation). The kinetic approach relies in this case on the molecular chaos hypothesis, that is, on the hypothesis of the absence of correlations between the velocities of two molecules about to collide. This assumption leads to the (irreversible) Boltzmann equation for the distribution function.

In Chapter 6, we show how the Boltzmann equation allows us, through convenient approximations, to determine the transport coefficients of the gas. In Chapter 7, we derive the hydrodynamic equations from the Boltzmann equation. Chapter 8 is devoted to the application of the Boltzmann equation to solid-state physics, where it is widely used to determine transport coefficients in semiconductors and in metals (semiclassical Bloch–Boltzmann theory of transport).

Very generally, the validity of the kinetic approaches relies on the existence in the system under study of two well-separated time scales. In dilute gases, the shorter time scale is the duration of the collision, while the longer one is the mean time interval separating two successive collisions of a given molecule. In Chapter 9, this type of approach is generalized to systems in which the interactions may be considered as local and instantaneous. Under decorrelation assumptions analogous to the molecular chaos hypothesis, the out-of-equilibrium evolution of such systems may be described by irreversible kinetic equations, generically called master equations.

Brownian motion

Chapters 10 and 11 deal with Brownian motion, that is, with the erratic motion of a particle immersed in a fluid of much lighter molecules.

Brownian motion is one of the paradigmatic problems of nonequilibrium statistical physics. It is generally studied by means of the Langevin equation, which describes the evolution of the Brownian particle velocity over time intervals intermediate between a short time scale, namely the correlation time of the random force exerted on the particle, and a long time scale, namely the relaxation time of the particle's mean velocity. We recover in this framework the Einstein relation between the mobility and the diffusion coefficient.

The theory of Brownian motion plays a role all the more important as the Brownian particle may be not a true particle, but the representation of a collective property of a macroscopic system.

We present in addition a microscopic model of the Brownian motion of a particle coupled with a thermal bath made up of an infinite ensemble of harmonic oscillators in thermal equilibrium. This model is widely used to describe the dissipative dynamics of various classical or quantum systems (A.O. Caldeira and A.J. Leggett).

Linear response theory

The linear response theory (R. Kubo) is developed in Chapters 12 to 14.

If we consider only near-equilibrium systems, the observed physical quantities depart slightly from their equilibrium values, and we expect linear deviations with respect to the perturbations driving the system from equilibrium. The linear response theory makes precise the relation between the linear response functions and the equilibrium fluctuations of the dynamical variables concerned. Once the linearity hypothesis is admitted, this theory is fully general.

To begin with, the notions of linear response functions and of equilibrium correlation functions are introduced in Chapter 12. Then, the general linear response theory establishing the relationship between these two types of quantities is developed in Chapters 13 and 14, the latter chapter concentrating on the fluctuation-dissipation theorem.

Equilibrium correlation functions thus play a central role in nonequilibrium statistical physics. Many properties of out-of-equilibrium systems, as for instance the coefficients involved in the linear phenomenological laws of transport, are determined by equilibrium correlation functions. Also, these functions provide a useful way of interpreting numerous scattering experiments involving either radiation or particles.

Transport coefficients

In Chapters 15 and 16, we show how the linear response theory allows us to obtain microscopic expressions for the transport coefficients in terms of the equilibrium correlation functions of the appropriate currents. These expressions constitute the Green–Kubo formulas.

In Chapter 15, we establish the microscopic expression for the electrical conductivity tensor in terms of the correlation functions of the relevant components of the electrical current. In the homogeneous case, we deduce from this expression the real part of the conductivity of a non-interacting electron gas in terms of matrix elements of one-particle currents (Kubo–Greenwood formula). The corrections with respect to the semiclassical conductivity deduced from the Boltzmann equation are due to quantum interference effects. We also briefly discuss the Landauer's approach of the conductance of mesoscopic systems and the one-dimensional localization phenomenon.

Chapter 16 deals with 'thermal' transport coefficients, such as the thermal conductivity or the diffusion coefficient, which cannot be directly calculated using the Kubo theory, as well as with the way they can be determined through different experiments (for instance, light scattering experiments).

To conclude this preface, I would like to emphasize that this book has not been the work of a person alone, but that it is based on teaching prepared and executed in team.

Indeed, this book stems from the lecture notes of a graduate course given during several years in the 'DEA de Physique des Solides' of the Paris/Orsay Universities. Most of the chapter supplements originate from the guided work and the associated homework problems. I thus wish to first express my thanks to Jean Klein, Sylvie Rousset, and Frédérick Bernardot, with whom I closely collaborated for several years in the context of this teaching. I am also especially grateful to Frédérick Bernardot for his careful reading of the manuscript, as well as for his precise and judicious remarks.

My thanks also go to the numerous students and readers of the lecture notes, who, by their remarks, contributed to the improvement of the manuscript.

Finally, it is a pleasure to thank Michèle Leduc, director of the series 'Savoirs actuels' (EDP Sciences/CNRS Éditions), whose constant encouragement enabled the successful production of the French version of this book, as well as Sönke Adlung, from Oxford University Press, for his steady support.

Contents

Chapter 1

Random variables and random processes

In statistical physics, we are led to consider the physical quantities characterizing the macroscopic state of a system consisting of a large number of particles as statistical averages of the corresponding microscopic quantities. The macroscopic variables, thus defined as averages, are accompanied by fluctuations due to the thermal agitation of the associated microscopic degrees of freedom. When the system under study is out of equilibrium, the temporal evolutions of the means and of the fluctuations have to be taken into account in the description and the modelization of the phenomena concerning it. Random variables and random processes are thus essential tools of nonequilibrium statistical physics.

Some fundamental notions on these topics are, for this reason, gathered in this first chapter. To begin with, we introduce the probability distributions and the moments of random variables in one or in several dimensions. We then study the distribution of the sum of two or of several independent random variables, and the central limit theorem concerning the distribution of the sum of N independent random variables in the limit $N \to \infty$. This theorem is of crucial importance in statistical physics, since it is the basis of the prominent role played there by the Gaussian laws.

Then, about random processes, we introduce the notion of stationarity, and we briefly discuss the ergodicity properties, namely, the equivalence between temporal averages and statistical averages. We consider more specifically the stationary random processes, and we present the outline of their harmonic analysis, a method well suited to the study of processes governed by linear differential equations. We derive in particular the Wiener–Khintchine theorem relating the noise spectral density and the autocorrelation function of a stationary random process.

1. Random variables, moments, and characteristic function

1.1. Definition

A *random variable*[1] is a number $X(\zeta)$ associated with each result ζ of an experiment. In that sense, it is a function whose definition domain is the set of results of the experiment. In order to define a random variable, we have to specify, on the one hand, the set of its possible values, called the *set of states*, and, on the other hand, the *probability distribution* over this set. The set of possible values may be either discrete or continuous over a given interval (or, partly discrete, partly continuous). Moreover, the set of states may be multidimensional (the random variable is then denoted vectorially by $\boldsymbol{X}$).

In the real one-dimensional case, the probability distribution of a random variable X is given by a function[2] $p(x)$ non-negative,

$$p(x) \geq 0, \tag{1.1.1}$$

and normalized such that:

$$\int_{-\infty}^{\infty} p(x)\, dx = 1. \tag{1.1.2}$$

The probability that the random variable X takes a value between x and $x + dx$ is equal to $p(x)dx$. The function $p(x)$ characterizing the probability distribution of the variable X is also called the *probability density*[3] of X. In physics, a probability density is generally a dimensioned quantity: its dimensions are the inverse of those of the random variable concerned.

The case of a variable likely to take discrete values may be treated in an analogous way by introducing delta functions in the probability density. For instance, if a random variable takes the discrete values $x_1, x_2 \ldots$ with the probabilities $p_1, p_2 \ldots$, we can formally treat it as a continuous random variable of probability density:

$$p(x) = \sum_i p_i\, \delta(x - x_i), \tag{1.1.3}$$

with:

$$p_i \geq 0, \qquad \sum_i p_i = 1. \tag{1.1.4}$$

1.2. Moments

The *average* or *expectation value* $\langle f(X)\rangle$ of any function $f(X)$ defined over the considered state space is:

$$\boxed{\langle f(X)\rangle = \int_{-\infty}^{\infty} f(x)p(x)\, dx,} \tag{1.1.5}$$

[1] We can equally say *stochastic variable*, or also *variate*: these three expressions are synonyms.

[2] A *realization*, or possible value, of X is denoted here by x. The upper case letter thus denotes the random variable, and the lower case letter one of its realizations. When no confusion will be possible between these two notions, we will employ a unique notation.

[3] For more clarity, we will sometimes denote it as $p_X(x)$.

provided that the integral on the right-hand side of equation (1.1.5) does exist.

In particular, $\mu_m = \langle X^m \rangle$ is the *moment* of order m of X. The first moment $\mu_1 = \langle X \rangle$ is the *mean value* of X. When its mean value $\langle X \rangle$ vanishes, the random variable X is said to be *centered*. The quantity:

$$\sigma^2 = \left\langle (X - \langle X \rangle)^2 \right\rangle = \mu_2 - \mu_1^2 \tag{1.1.6}$$

is the *variance* of X. It is the square of the *standard deviation* or *root-mean-square deviation* $\Delta X = \sigma$, which has the same dimensions as $\langle X \rangle$. The root-mean-square deviation σ determines the effective width of the distribution $p(x)$. The variance σ^2 is non-negative and it vanishes only when the variable X is non-random. Its two first moments are the most important characteristics of a probability distribution.

If the function $p(x)$ does not decrease sufficiently rapidly as x tends towards infinity, some of the moments of X may be not defined. An extreme example of such a behavior is given by the *Lorentz law* (or *Cauchy law*):

$$p(x) = \frac{a}{\pi} \frac{1}{(x - x_0)^2 + a^2}, \qquad a > 0, \tag{1.1.7}$$

all moments of which are diverging (we can, however, define the first moment by symmetry setting $\mu_1 = x_0$). The other moments of the Cauchy law, and thus in particular its variance, are all infinite.

1.3. Characteristic function

The *characteristic function* $G(k)$ of a random variable X is defined by:

$$\boxed{G(k) = \left\langle e^{ikX} \right\rangle = \int_{-\infty}^{\infty} e^{ikx} p(x)\, dx.} \tag{1.1.8}$$

The probability density being an integrable function of x, the characteristic function as defined by formula (1.1.8) exists for any k real. It is the Fourier transform of $p(x)$. We therefore have, inversely:

$$p(x) = \frac{1}{2\pi} \int_{-\infty}^{\infty} e^{-ikx} G(k)\, dk. \tag{1.1.9}$$

The characteristic function has the properties:

$$G(k = 0) = 1, \qquad |G(k)| \leq 1. \tag{1.1.10}$$

The characteristic function is also the *moment generating function*, in the sense that the coefficients of its Taylor expansion in powers of k are the moments μ_m:

$$G(k) = \sum_{m=0}^{\infty} \frac{(ik)^m}{m!} \mu_m. \tag{1.1.11}$$

The derivatives of $G(k)$ at $k = 0$ thus exist up to the same order as the moments. However, the generating function $G(k)$ exists even when the moments μ_m are not defined. For instance, the Cauchy law (1.1.7) has no finite moments, but its characteristic function is:

$$G(k) = \frac{a}{\pi} \int_{-\infty}^{\infty} \frac{e^{ikx}}{(x - x_0)^2 + a^2}\, dx = e^{-a|k| + ikx_0}. \tag{1.1.12}$$

The function $e^{-a|k|+ikx_0}$ is not differentiable at $k = 0$, in accordance with the fact that the moments of the Cauchy law do not exist.

2. Multivariate distributions

2.1. Joint densities, marginal densities, and conditional densities

When several random variables come into play, which is for instance the case when we consider a multidimensional random variable, it is necessary to introduce several types of probability distributions.

• *Joint probability density*

Let $\boldsymbol{X}$ be an n-dimensional random variable, of components $X_1, \ldots, X_n$. Its probability density $p_n(x_1, \ldots, x_n)$ is called the *joint probability density* of the n variables $X_1, \ldots, X_n$.

• *Marginal probability density*

Let us consider a subset of $s < n$ relevant variables $X_1, \ldots, X_s$. The probability density of these s variables, independently of the values taken by the irrelevant variables $X_{s+1}, \ldots, X_n$, is obtained by integrating over these latter variables:

$$p_s(x_1, \ldots, x_s) = \int p_n(x_1, \ldots, x_s, x_{s+1}, \ldots, x_n)\, dx_{s+1} \ldots dx_n. \tag{1.2.1}$$

The density (1.2.1) is called the *marginal probability density* of the s relevant variables.

• *Conditional probability density*

We can attribute fixed values $x_{s+1}, \ldots, x_n$ to the $n - s$ variables $X_{s+1}, \ldots, X_n$, and consider the joint probability distribution of the s variables $X_1, \ldots, X_s$. This latter distribution is called the *conditional probability density* of $X_1, \ldots, X_s$. It is denoted by $p_{s|n-s}(x_1, \ldots, x_s | x_{s+1}, \ldots, x_n)$.

2.2. Statistical independence

The joint probability density p_n is equal to the product of the marginal probability density that $X_{s+1}, \ldots, X_n$ take the values $x_{s+1}, \ldots, x_n$ and the conditional probability density that, once this realized, the variables $X_1, \ldots, X_s$ take the values $x_1, \ldots, x_s$:

$$p_n(x_1, \ldots, x_n) = p_{n-s}(x_{s+1}, \ldots, x_n) p_{s|n-s}(x_1, \ldots, x_s | x_{s+1}, \ldots, x_n). \tag{1.2.2}$$

This is the *Bayes' rule*, generally written in the form:

$$p_{s|n-s}(x_1, \ldots, x_s | x_{s+1}, \ldots, x_n) = \frac{p_n(x_1, \ldots, x_n)}{p_{n-s}(x_{s+1}, \ldots, x_n)}. \tag{1.2.3}$$

If the n variables may be divided into two subsets $X_1, \ldots, X_s$ and $X_{s+1}, \ldots, X_n$ such that p_n factorizes, that is, if we can write:

$$p_n(x_1, \ldots, x_n) = p_s(x_1, \ldots, x_s) p_{n-s}(x_{s+1}, \ldots, x_n), \tag{1.2.4}$$

these two subsets are said to be *statistically independent.*

2.3. Moments and characteristic function

The moments of a multivariate distribution are defined by:

$$\langle X_1^{m_1} \ldots X_n^{m_n} \rangle = \int p_n(x_1, \ldots, x_n) x_1^{m_1} \ldots x_n^{m_n} \, dx_1 \ldots dx_n. \tag{1.2.5}$$

The characteristic function $G_n(k_1, \ldots, k_n)$ is a function of n real variables $k_1, \ldots, k_n$ defined by:

$$G_n(k_1, \ldots, k_n) = \langle e^{i(k_1 X_1 + \cdots + k_n X_n)} \rangle. \tag{1.2.6}$$

Its Taylor expansion in powers of the variables k_i $(i = 1, \ldots, n)$ generates the moments:

$$G_n(k_1, \ldots, k_n) = \sum_{m_1, \ldots, m_n = 0}^{\infty} \frac{(ik_1)^{m_1} \ldots (ik_n)^{m_n}}{m_1! \ldots m_n!} \langle X_1^{m_1} \ldots X_n^{m_n} \rangle. \tag{1.2.7}$$

If the two subsets $X_1, \ldots, X_s$ and $X_{s+1}, \ldots, X_n$ are statistically independent, the characteristic function factorizes:

$$G_n(k_1, \ldots, k_s, k_{s+1}, \ldots, k_n) = G_s(k_1, \ldots, k_s) G_{n-s}(k_{s+1}, \ldots, k_n). \tag{1.2.8}$$

In the same way, all moments factorize:

$$\langle X_1^{m_1} \ldots X_s^{m_s} X_{s+1}^{m_{s+1}} \ldots X_n^{m_n} \rangle = \langle X_1^{m_1} \ldots X_s^{m_s} \rangle \langle X_{s+1}^{m_{s+1}} \ldots X_n^{m_n} \rangle. \tag{1.2.9}$$

2.4. Second-order moments: variances and covariances

The second-order moments are especially important in physics, where their knowledge suffices in most applications. They form a matrix $\langle X_i X_j \rangle$ of dimensions $n \times n$. We also define the *covariance matrix*, of dimensions $n \times n$ and of elements:

$$\nu_{ij} = \langle (X_i - \langle X_i \rangle)(X_j - \langle X_j \rangle) \rangle = \langle X_i X_j \rangle - \langle X_i \rangle \langle X_j \rangle. \tag{1.2.10}$$

The diagonal elements of the covariance matrix are the previously defined variances, and are thus positive, while the off-diagonal elements, called the *covariances*, are of any sign.

We can show, using the Schwarz inequality, that:

$$|\nu_{ij}|^2 \leq \sigma_i^2 \sigma_j^2, \tag{1.2.11}$$

where σ_i and σ_j denote the root-mean-square deviations of X_i and X_j. The normalized quantity:

$$\rho_{ij} = \frac{\nu_{ij}}{\sigma_i \sigma_j} = \frac{\langle X_i X_j \rangle - \langle X_i \rangle \langle X_j \rangle}{\sigma_i \sigma_j}, \tag{1.2.12}$$

bounded by -1 and $+1$, is called the *correlation coefficient* of the variables X_i and X_j. Two random variables are said to be *non-correlated* when their covariance vanishes (no hypothesis is made concerning the higher-order moments). Non-correlation is a weaker property than statistical independence.

2.5. Complex random variables

A complex random variable $Z = X + iY$ is a set of two real random variables $\{X, Y\}$. The probability density $p_Z(z)$ is simply the joint probability density of X and Y. The normalization condition is written as:

$$\int p_Z(z)\, d^2z = 1, \qquad d^2z = dx\, dy. \tag{1.2.13}$$

The definition of moments can be extended to complex random variables. If $Z_1, \ldots, Z_n$ are complex random variables, their covariance matrix is defined by:

$$\nu_{ij} = \big\langle (Z_i - \langle Z_i \rangle)(Z_j^* - \langle Z_j^* \rangle) \big\rangle = \big\langle Z_i Z_j^* \big\rangle - \big\langle Z_i \big\rangle \big\langle Z_j^* \big\rangle. \tag{1.2.14}$$

The variances $\sigma_i^2 = \langle |Z_i - \langle Z_i \rangle|^2 \rangle$ are non-negative, and the correlation coefficients ρ_{ij} are complex and of modulus smaller than 1.

3. Addition of random variables

3.1. Probability density of the sum of two independent random variables

Let X_1 and X_2 be two independent random variables of joint distribution $p_X(x_1, x_2)$. The probability that the random variable $Y = X_1 + X_2$ takes a value between y and $y + dy$ is:

$$p_Y(y)\, dy = \iint_{y < x_1 + x_2 < y + dy} p_X(x_1, x_2)\, dx_1 dx_2. \tag{1.3.1}$$

We deduce from equation (1.3.1) the expression for the density $p_Y(y)$:

$$p_Y(y) = \iint \delta(x_1 + x_2 - y) p_X(x_1, x_2)\, dx_1 dx_2 = \int p_X(x_1, y - x_1)\, dx_1. \tag{1.3.2}$$

If the variables X_1 and X_2 are independent, the density $p_X(x_1, y - x_1)$ factorizes. Equation (1.3.2) then becomes:

$$\boxed{p_Y(y) = \int p_{X_1}(x_1) p_{X_2}(y - x_1)\, dx_1.} \tag{1.3.3}$$

The probability density of the sum of two independent random variables is thus the convolution product of their individual probability densities.

We can equally derive this result by remarking that, if the variables X_1 and X_2 are independent, the characteristic function of Y factorizes:

$$G_Y(k) = \langle e^{ik(X_1+X_2)} \rangle = \langle e^{ikX_1} \rangle \langle e^{ikX_2} \rangle = G_{X_1}(k)\, G_{X_2}(k). \tag{1.3.4}$$

The result (1.3.3) follows from formula (1.3.4) by inverse Fourier transformation.

3.2. Mean and variance of the sum of two random variables

In any case, we have:

$$\langle Y \rangle = \langle X_1 \rangle + \langle X_2 \rangle. \tag{1.3.5}$$

The average of a sum is thus the sum of the averages, be the variables X_1 and X_2 correlated or not.

If X_1 and X_2 are uncorrelated, the variance of their sum is equal to the sum of their variances:

$$\sigma_Y^2 = \sigma_{X_1}^2 + \sigma_{X_2}^2. \tag{1.3.6}$$

4. Gaussian distributions

4.1. The one-variate Gaussian distribution

The most general form of the one-variate *Gaussian distribution* is:

$$p(x) = Ce^{-\frac{1}{2}Ax^2 - Bx}. \tag{1.4.1}$$

The Gaussian distribution is also called the *normal distribution*. The parameter A is a positive constant determining the width of the Gaussian. The parameter B fixes the position of its maximum. The normalization constant C is expressed in terms of A and B as:

$$C = \left(\frac{A}{2\pi}\right)^{1/2} e^{-B^2/2A}. \tag{1.4.2}$$

It is often preferable in practice to express the parameters A and B in terms of the mean $\mu_1 = -B/A$ and the variance $\sigma^2 = 1/A$. Accordingly, we write the Gaussian distribution as:

$$\boxed{p(x) = \frac{1}{\sigma\sqrt{2\pi}} \exp\left[-\frac{(x-\mu_1)^2}{2\sigma^2}\right].} \tag{1.4.3}$$

The characteristic function of the distribution (1.4.3) is:

$$G(k) = e^{i\mu_1 k - \frac{1}{2}\sigma^2 k^2}. \tag{1.4.4}$$

All the moments of the Gaussian law are finite, in accordance with the fact that the function $e^{i\mu_1 k - \frac{1}{2}\sigma^2 k^2}$ is infinitely differentiable at $k = 0$. They can be expressed in terms of the two first moments μ_1 and μ_2, or in terms of the mean μ_1 and the variance σ^2.

When $X_1, \dots, X_n$ are independent Gaussian random variables, their sum $Y = X_1 + \dots + X_n$ is, too, a Gaussian random variable. Its distribution is fully determined by the mean and the variance of Y, which are respectively equal to the sums of the mean and of the variances of the variables X_i $(i = 1, \dots, n)$.

4.2. The n-variate Gaussian distribution

The most general form of the n-variate Gaussian distribution is:

$$p_n(x_1, \dots, x_n) = C \exp\left(-\frac{1}{2} \sum_{i,j=1}^{n} A_{ij} x_i x_j - \sum_{i=1}^{n} B_i x_i \right), \tag{1.4.5}$$

where the matrix $\boldsymbol{A}$ of elements A_{ij} is a positive definite symmetric matrix of dimensions $n \times n$. In vectorial notations, we write:

$$p_n(\boldsymbol{x}) = C \exp\left(-\frac{1}{2} \boldsymbol{x}.\boldsymbol{A}.\boldsymbol{x} - \boldsymbol{B}.\boldsymbol{x} \right), \tag{1.4.6}$$

where $\boldsymbol{x}$ denotes the vector of components $x_1, \dots, x_n$, and $\boldsymbol{B}$ the vector of components $b_1, \dots, b_n$. We can obtain the normalization constant C by passing to the variables in which the matrix $\boldsymbol{A}$ is diagonal. We thus get:

$$C = (2\pi)^{-n/2} \left(\operatorname{Det} \boldsymbol{A} \right)^{1/2} \exp\left(-\frac{1}{2} \boldsymbol{B}.\boldsymbol{M}.\boldsymbol{B} \right), \tag{1.4.7}$$

where $\operatorname{Det} \boldsymbol{A}$ denotes the determinant of the matrix $\boldsymbol{A}$, and $\boldsymbol{M} = \boldsymbol{A}^{-1}$, its inverse matrix.

The characteristic function of the distribution (1.4.6) is:

$$G_n(\boldsymbol{k}) = \exp\left(-\frac{1}{2} \boldsymbol{k}.\boldsymbol{M}.\boldsymbol{k} - i\boldsymbol{k}.\boldsymbol{M}.\boldsymbol{B} \right), \tag{1.4.8}$$

where $\boldsymbol{k}$ is the vector of components $k_1, \dots, k_n$. When developing the expression (1.4.8) for $G_n(\boldsymbol{k})$ in powers of $\boldsymbol{k}$, we get the expressions for the means and for the covariances:

$$\langle X_i \rangle = -\sum_j M_{ij} B_j, \tag{1.4.9}$$

$$\nu_{ij} = \big\langle (X_i - \langle X_i \rangle)(X_j - \langle X_j \rangle) \big\rangle = M_{ij}. \tag{1.4.10}$$

The covariance matrix of the Gaussian distribution (1.4.6) is $\boldsymbol{M} = \boldsymbol{A}^{-1}$.

A multivariate Gaussian distribution is thus fully determined by the means and the covariance matrix of the variables. If the variables are uncorrelated, the matrices $\boldsymbol{A}$ and $\boldsymbol{M} = \boldsymbol{A}^{-1}$ are diagonal, and so the variables are independent. Thus, although in general non-correlation is a weaker property than statistical independence, they constitute, in the Gaussian case, equivalent properties.

4.3. The two-variate case

In the two-variate case, the covariance matrix is:

$$M = \begin{pmatrix} \sigma_1^2 & \rho_{12}\sigma_1\sigma_2 \\ \rho_{12}\sigma_1\sigma_2 & \sigma_2^2 \end{pmatrix}. \tag{1.4.11}$$

Its inverse is the matrix:

$$A = \frac{1}{1-\rho_{12}^2} \begin{pmatrix} 1/\sigma_1^2 & -\rho_{12}/\sigma_1\sigma_2 \\ -\rho_{12}/\sigma_1\sigma_2 & 1/\sigma_2^2 \end{pmatrix}, \tag{1.4.12}$$

of determinant:

$$\mathrm{Det}\, A = \frac{1}{\sigma_1^2\sigma_2^2(1-\rho_{12}^2)}. \tag{1.4.13}$$

The two-variate Gaussian distribution (of centered variates) may thus be written as:

$$p_2(x_1, x_2) = \frac{1}{2\pi\sigma_1\sigma_2(1-\rho_{12}^2)^{1/2}} \exp\left[-\frac{1}{2(1-\rho_{12}^2)}\left(\frac{x_1^2}{\sigma_1^2} - \frac{2\rho_{12}x_1x_2}{\sigma_1\sigma_2} + \frac{x_2^2}{\sigma_2^2}\right)\right]. \tag{1.4.14}$$

Formula (1.4.14) displays the fact that, when two Gaussian random variables are uncorrelated ($\rho_{12} = 0$), their joint probability distribution factorizes into the product of their individual densities. We can verify in this particular case that uncorrelated Gaussian random variables are statistically independent.

4.4. Property of the correlations

An important property of Gaussian random variables is the fact that all higher-order correlations may be expressed in terms of second-order correlations between pairs of variables. Thus, the even-order moments of a multivariate centered Gaussian distribution have the property:

$$\langle X_i X_j X_k X_l \ldots \rangle = \sum \langle X_p X_q \rangle \langle X_u X_v \rangle \ldots, \tag{1.4.15}$$

where the summation extends to all possible subdivisions in pairs of the indexes $i, j, k, l, \ldots$, while the odd-order moments all vanish.

5. The central limit theorem

5.1. Stability of the Gaussian law

Let us first consider a set of N independent random variables $X_1, \ldots, X_N$, each of them having the same Gaussian probability density $p_X(x) = (2\pi\sigma^2)^{-1/2}\exp(-x^2/2\sigma^2)$, of zero mean[4] and variance σ^2. For any N, the random variable Y_N as defined by:

$$Y_N = \frac{X_1 + \cdots + X_N}{N^{1/2}} \tag{1.5.1}$$

[4] For convenience, the variables X_i are taken here as centered, the generalization to non-centered variables being straightforward.

is, too, a Gaussian variable of zero mean and variance $\langle Y_N^2\rangle = N^{-1}\sum_{i=1}^N \langle X_i^2\rangle = \sigma^2$. The probability distribution of Y_N being the same as that of the original variables, the Gaussian law is said to be *stable* with respect to the addition of random variables.[5]

5.2. Statement and justification of the central limit theorem

The *central limit theorem*, which was established by P.-S. de Laplace in 1812, stipulates that, even when $p_X(x)$ is not a Gaussian law, but another distribution of zero mean and finite variance σ^2, the distribution of Y_N is still the Gaussian law of zero mean and variance σ^2 in the limit $N \to \infty$. This property of convergence towards the Gaussian *attraction domain* is at the origin of the prominent role played by the Gaussian distribution in statistical physics. Indeed, in many situations where a fluctuating variable Y is involved, the fluctuations are the sum of contributions coming from a large number of independent causes.[6]

To understand the origin of this property, let us consider the characteristic function corresponding to an arbitrary (centered) distribution $p_X(x)$:

$$G_X(k) = \int_{-\infty}^{\infty} e^{ikx} p_X(x)\, dx. \tag{1.5.2}$$

The individual variables being independent, the characteristic function of Y_N is:

$$G_{Y_N}(k) = \left[G_X\left(\frac{k}{N^{1/2}}\right)\right]^N. \tag{1.5.3}$$

We therefore have:

$$\log G_{Y_N}(k) = N \log G_X\left(\frac{k}{N^{1/2}}\right). \tag{1.5.4}$$

We can write, expanding[7] in powers of $k/N^{1/2}$,

$$\log G_X\left(\frac{k}{N^{1/2}}\right) = -\frac{1}{2}\sigma^2\frac{k^2}{N} + O\left(\frac{k^3}{N^{3/2}}\right), \tag{1.5.5}$$

a formula from which we deduce:[8]

$$\log G_{Y_N}(k) = -\frac{1}{2}\sigma^2 k^2 + O\left(\frac{k^3}{N^{1/2}}\right). \tag{1.5.6}$$

---

[5] Note that the variable Y_N is not equal to the sum of the original variables, but to the product of this sum and the scaling factor $c_N = N^{-1/2}$.

[6] Actually, the central limit theorem applies even when the individual laws are not identical. However, its discussion is simpler when all variables X_i are identically distributed, which we assume here to be the case.

[7] It is preferable to expand the logarithms, which vary more slowly than the functions themselves.

[8] In formula (1.5.5) (*resp.* (1.5.6)), the symbol $O(k^3/N^{3/2})$ (*resp.* $O(k^3/N^{1/2})$) stands for a quantity of order $k^3/N^{3/2}$ (*resp.* $k^3/N^{1/2}$).

As $N \to \infty$, the term $O(k^3/N^{1/2})$ approaches zero because of the factor $N^{1/2}$ in the denominator. Therefore, $G_{Y_N}(k)$ tends towards $e^{-\sigma^2 k^2/2}$, which is the characteristic function of the Gaussian distribution of zero mean and variance σ^2.

5.3. Discussion

In this discussion, we will continue to limit ourselves to the case of identically distributed variables X_i. Even in this case, the central limit theorem is actually valid under more general hypotheses than the ones we made (namely, of independent individual random variables having a distribution of finite variance).

First of all, the condition of statistical independence of the individual random variables, if it is a sufficient condition, is not a necessary one. Indeed, for the theorem to apply, these variables must not exhibit long-range correlations (that is, correlations between variables X_i and X_j with $|i-j| \gg 1$).[9] Short-range correlations may however be present without affecting the result. It is clearly more difficult to treat the case of strongly correlated individual random variables.

Secondly, the condition of finite variance of the distribution $p_X(x)$ is also a sufficient, but not necessary, condition of convergence towards the normal law. However, the distribution of the individual random variables which are summed must not be too 'broad'.[10] We identify precisely the functions $p_X(x)$ which belong to the attraction domain of the normal law by the criterion:

$$\lim_{y\to\infty} y^2 \frac{\int_{|x|>y} p_X(x)\,dx}{\int_{|x|<y} x^2 p_X(x)\,dx} = 0. \tag{1.5.7}$$

For instance, a distribution which decreases as $|x|^{-3}$ for $|x| \to \infty$ belongs to the attraction domain of the normal law, in spite of the fact that its variance is infinite. All distributions decreasing faster than $|x|^{-3}$ for $|x| \to \infty$ also belong to the domain of attraction of the normal law, which is thus extremely vast.

This is the reason why the Gaussian law is omnipresent in physical situations, the exceptions to this law (qualified as 'anomalous behaviors') being comparatively much rarer. The anomalous behaviors correspond to the fact that other laws than the Gaussian one possess the stability property with respect to the summation of individual random variables.[11] The stable laws were studied and classified by P. Lévy and A. Khintchine in 1936. They, too, possess attraction domains, which may be characterized by proper generalizations of the central limit theorem, and to which belong the individual so-called broad laws.

---

[9] This is quite apparent if we think for instance of the case where the N variables are identical.

[10] An example of *broad law* is provided by the Cauchy law (1.1.7), whose moments are all infinite. The sum of N independent variables distributed according a Cauchy law, is, too, distributed according a Cauchy law, which we can check by using the corresponding characteristic functions. However large N may be, there is no tendency towards the normal law.

[11] The sum of the individual random variables must, for each stable law, be multiplied by an appropriate scaling factor c_N. For instance, in the case of the Cauchy law, we have $c_N = N^{-1}$.

6. Random processes

When a set of random variables $X_1, X_2, \ldots$ is not countable, we cannot label the different variables using a discrete index. We therefore introduce, for this purpose, a continuous parameter t. The quantity $X(t)$ is then a *random function* of t. When t is the time, which we assume here to be the case, $X(t)$ is a *random process* (or *stochastic process*).

Taking at each instant, for the random variable X, one of its possible realizations x, we obtain a realization $x(t)$ of the process $X(t)$. Such a realization is also called a *sample*. A sample does not depend on time in a deterministic way.

6.1. Ensemble average

For each value of t, $X(t)$ is a random variable, defined over some domain, with a probability density $p(x,t)$. This density is normalized such that:

$$\int_{-\infty}^{\infty} p(x,t)\, dx = 1. \tag{1.6.1}$$

The average value of X at time t, or *one-time average*, is defined by:

$$\langle X(t) \rangle = \int_{-\infty}^{\infty} x p(x,t)\, dx. \tag{1.6.2}$$

To define the average, we can also consider the set of all realizations or samples $\{x^{(r)}(t)\}$ of the process $X(t)$. The average of X at time t may be obtained by averaging over this ensemble of realizations:

$$\langle X(t) \rangle = \lim_{N\to\infty} \frac{1}{N} \sum_{r=1}^{N} x^{(r)}(t). \tag{1.6.3}$$

Equations (1.6.2) and (1.6.3) are equivalent definitions of the *ensemble average* or *statistical average* of $X(t)$.

6.2. Joint probability densities and correlations

The probability density $p(x,t)$ that $X(t)$ takes the value x at time t is called the *one-time density*. It is sometimes denoted as $p_1(x,t)$. While it gives access to the average $\langle X(t)\rangle$, and, more generally, to the different *one-time moments* $\langle X^m(t)\rangle$, the density $p_1(x,t)$ does not however fully describe the process.

In particular, the one-time density does not yield any information about the possible correlations between $X(t_1)$ and $X(t_2)$ at different times t_1 and t_2. This information is contained in the *two-time density* $p_2(x_1,t_1;x_2,t_2)$, which is the joint probability density that $X(t)$ takes the values x_1 at time t_1 and x_2 at time t_2. From the properties of the joint probability densities, we get the consistency condition:

$$\int p_2(x_1,t_1;x_2,t_2)\, dx_2 = p_1(x_1,t_1), \tag{1.6.4}$$

which specifies that the result of the integration over x_2 must not depend on t_2. The two-time density allows us to calculate *two-time averages* such as:

$$\langle X(t_1)X(t_2)\rangle = \int x_1 x_2 p_2(x_1,t_1;x_2,t_2)\,dx_1 dx_2. \tag{1.6.5}$$

A quantity which can be calculated with the help of one- and two-time densities is the *autocorrelation function* $\kappa(t_1,t_2)$ of the process $X(t)$, as defined by:

$$\kappa(t_1,t_2) = \big\langle \big[X(t_1) - \langle X(t_1)\rangle\big]\big[X(t_2) - \langle X(t_2)\rangle\big]\big\rangle = \langle X(t_1)X(t_2)\rangle - \langle X(t_1)\rangle\langle X(t_2)\rangle. \tag{1.6.6}$$

The autocorrelation function allows us in particular to estimate the time range of the correlations.

In the same way, in order to calculate quantities such as $\langle X(t_1)\dots X(t_n)\rangle$, we have to know the *n-time density* $p_n(x_1,t_1;\dots;x_n,t_n)$. For any integer $s < n$, we have the consistency condition:

$$\int p_n(x_1,t_1;\dots;x_s,t_s;x_{s+1},t_{s+1};\dots;x_n,t_n)\,dx_{s+1}\dots dx_n = p_s(x_1,t_1;\dots;x_s,t_s). \tag{1.6.7}$$

The density p_n allows us to calculate *n-time averages*, such as:

$$\langle X(t_1)\dots X(t_n)\rangle = \int x_1 \dots x_n p_n(x_1,t_1;\dots;x_n,t_n)\,dx_1\dots dx_n. \tag{1.6.8}$$

There exists an infinite hierarchy of joint probability densities p_n. Each of them includes all the information contained in the preceding ones, and contains supplementary information. We need in principle to know all the p_n's to have at hand a full specification of the stochastic process.[12]

6.3. Multicomponent random processes

If a stochastic process consists of several components $X_1(t),\dots,X_n(t)$, it may be convenient to consider it as an n-dimensional vector $\boldsymbol{X}(t)$.

We define the *correlation matrix*, of dimensions $n \times n$ and of elements:

$$\begin{aligned}\kappa_{ij}(t_1,t_2) &= \big\langle \big[X_i(t_1) - \langle X_i(t_1)\rangle\big]\big[X_j(t_2) - \langle X_j(t_2)\rangle\big]\big\rangle \\ &= \langle X_i(t_1)X_j(t_2)\rangle - \langle X_i(t_1)\rangle\langle X_j(t_2)\rangle.\end{aligned} \tag{1.6.9}$$

The diagonal elements of the correlation matrix represent *autocorrelations*, and the off-diagonal ones *crossed correlations*.

---

[12] However, the processes whose present is not at all, or only weakly, influenced by their past history, may be characterized more simply. These processes are called *Markov processes* (see Chapter 9).

6.4. Complex random processes

A complex random process $Z(t)$ is defined as a set of two real random processes $\{X(t), Y(t)\}$. The probability density $p(z,t)$ is the joint probability density of X and Y at time t. The average of $Z(t)$ is:

$$\langle Z(t)\rangle = \int zp(z,t)\, d^2z. \tag{1.6.10}$$

For such a process, we define a complex autocorrelation function by:

$$\kappa(t_1,t_2) = \big\langle \big[Z(t_1) - \langle Z(t_1)\rangle\big]\big[Z^*(t_2) - \langle Z^*(t_2)\rangle\big]\big\rangle = \big\langle Z(t_1)Z^*(t_2)\big\rangle - \big\langle Z(t_1)\big\rangle\big\langle Z^*(t_2)\big\rangle. \tag{1.6.11}$$

6.5. Gaussian processes

A stochastic process is said to be *Gaussian* if all the joint probability densities $p_n(x_1,t_1;\ldots;x_n,t_n)$ are Gaussian distributions. A Gaussian process $X(t)$ is fully specified by the one-time average $\langle X(t)\rangle$ and the two-time average $\langle X(t_1)X(t_2)\rangle$. For instance, for a centered Gaussian process, the averages involving an even number of times may be written as:

$$\big\langle X(t_i)X(t_j)X(t_k)X(t_l)\ldots\big\rangle = \sum \big\langle X(t_p)X(t_q)\big\rangle\big\langle X(t_u)X(t_v)\big\rangle\ldots, \tag{1.6.12}$$

where the summation extends to all possible subdivisions in pairs of the indexes $i,j,k,l,\ldots$, while the averages involving an odd number of times vanish.

Gaussian processes, which are especially simple to treat, are frequently used in physics for an approximate description of random phenomena.

7. Stationarity and ergodicity

7.1. Stationarity

A stochastic process is said to be *stationary* when all the probability densities p_n are invariant under an arbitrary translation of the origin of time. In this case, the different averages are not modified by such a translation. In other words, for any n, any T, and any $t_1,\ldots,t_n$, we have the identity:

$$\big\langle X(t_1+T)\ldots X(t_n+T)\big\rangle = \big\langle X(t_1)\ldots X(t_n)\big\rangle. \tag{1.7.1}$$

The average of a centered stochastic process $X(t)$ does not depend on time. It is often convenient to deduce from $X(t)$ this constant average $\langle X\rangle$, and to work with the centered process $X(t) - \langle X\rangle$.

The autocorrelation function $\kappa(t_1,t_2)$ of a stationary stochastic process only depends on $\tau = t_1 - t_2$. In the real case, the autocorrelation function $\kappa(\tau)$ is an even function of τ, generally approaching zero as $|\tau| \to \infty$. A widespread case in physics is that of exponentially decreasing autocorrelation functions:

$$\kappa(\tau) = Ce^{-|\tau|/\tau_c}. \tag{1.7.2}$$

Such a function is negligible for $|\tau| \gg \tau_c$. The characteristic time τ_c is the *correlation time* of the random process.

In the case of a multicomponent stationary centered process, the elements of the correlation matrix may simply be written as:

$$\kappa_{ij}(\tau) = \langle X_i(\tau)X_j(0)\rangle. \tag{1.7.3}$$

For a multicomponent stationary process (centered or not), we have the property:

$$\kappa_{ij}(\tau) = \kappa_{ji}(-\tau). \tag{1.7.4}$$

7.2. Ergodicity

Let us consider a stationary random process $X(t)$, possibly complex, whose different realizations are denoted by $x^{(r)}(t)$. It frequently happens in practice that any given realization of the process contains all the statistical information available about it. The stationary random process $X(t)$ is then said to be *ergodic*.

To make this notion precise, we shall first define the *temporal average* of the stationary process $X(t)$. For this purpose, we consider a particular realization $x^{(r)}(t)$, simply denoted by $x(t)$, and a finite time interval $(t - \frac{T}{2}, t + \frac{T}{2})$, of width T. The temporal average of the process over this time interval is, by definition:

$$\overline{X(t)}^T = \frac{1}{T}\int_{t-\frac{T}{2}}^{t+\frac{T}{2}} x(t')\,dt'. \tag{1.7.5}$$

It depends both on t and on the interval width T, as well as on the particular considered realization. In the limit in which T tends towards infinity, $\overline{X(t)}^T$ becomes the temporal average of the process, denoted by $\overline{X}$:

$$\overline{X} = \lim_{T\to\infty} \overline{X(t)}^T = \lim_{T\to\infty} \frac{1}{T}\int_{t-\frac{T}{2}}^{t+\frac{T}{2}} x(t')\,dt'. \tag{1.7.6}$$

The temporal average $\overline{X}$ depends neither on t nor on T any longer, but it may a priori still depend on the considered realization. If this is not the case, it then coincides with the ensemble average or statistical average $\langle X\rangle$, and the process $X(t)$ is said to be *ergodic in the mean.*[13]

More generally, a random stationary process $X(t)$ is said to be *ergodic in the full sense* if there is an identity between the temporal average and the statistical average, not only as far as $X(t)$ is concerned, but also for all the products of the type $X(t_1)X^*(t_2)\ldots$ which serve to define the higher-order correlation functions.[14]

---

[13] It can be shown that a sufficient condition for a stationary stochastic process to be ergodic in the mean is that the correlations decrease sufficiently rapidly at large times for the autocorrelation function to be integrable. Then every particular realization of the process contains enough statistical information for the temporal average to coincide with the ensemble average.

[14] Criteria may also be established for the ergodicity of higher-order correlation functions.

8. Random processes in physics: the example of Brownian motion

8.1. Brownian motion

In 1827, the botanist R. Brown discovered under the microscope the irregular motion of small pollen particles in suspension in water. He also observed that small mineral particles may undergo similar erratic motions, which precludes to attribute to a possibly 'vital force' specific to biological objects this incessant and disordered motion. A. Einstein gave in 1905 the first clear theoretical explanation of this physical phenomenon. Direct experimental verifications of the Einstein theory were carried out in 1908 by J. Perrin.[15]

Fluctuation phenomena such as those evident in Brownian motion are universally widespread. For instance, the thermal agitation of electrons in a conductor in thermodynamic equilibrium gives rise to fluctuations of the electric current which passes through it, and of the potential difference between its extremities. These fluctuations constitute the *thermal noise*, studied in 1928 by J.B. Johnson and H. Nyquist.[16] Generally speaking, all experimentally observed quantities, or *macroscopic variables*, are accompanied by fluctuations due to the thermal agitation of the associated microscopic degrees of freedom.

In most cases, the fluctuations of macroscopic variables are extremely small with respect to their average values, and they may be neglected. However, since the fluctuations reflect the motions at the microscopic scale in the system under consideration, their analysis is important for its study.

8.2. Moving to a continuous-time description

Brownian motion also played an important historical role in mathematics. It was to represent the displacement of a Brownian particle that a stochastic process was constructed for the first time (N. Wiener, 1923).

In order to be able to modelize a physical phenomenon by a random process, it is necessary to pass from the discrete-time description imposed by the experiment to a continuous-time one. Suppose for instance that we observe under the microscope a Brownian particle during a time interval $0 \leq t \leq T$, and that we record the projection of the particle's position on an axis Ox as a function of time. Repeating N times the observations as time evolves, we get N values of the particle's coordinate, $x(t_1), \ldots, x(t_N)$. In contrast to what happens in mechanics, it is impossible to make deterministic predictions, and we have to adopt a probabilistic point of view. The value $x(t)$ of the Brownian particle's coordinate at time t is a realization of a random variable, and each of the observed series $\{x(t_j)\}$ is a sample of a statistical ensemble. If we could proceed to a continuous observation, we would get a random function of time or stochastic process $X(t)$ with t as a continuous parameter. In practice, we make the observations at discrete times $t_1 < \ldots < t_N$, and we thus obtain a set of N numbers, $x(t_1), \ldots, x(t_N)$. The mathematical description by a continuous-time process is

[15] Brownian motion will be studied in Chapters 10 and 11.

[16] See Supplement 10C.

obtained by taking the limit of very large N, and intervals between the observation times getting smaller and smaller.

9. Harmonic analysis of stationary random processes

Let $X(t)$ be a stationary stochastic process. The *harmonic analysis* of this process consists in studying the properties of the Fourier series of $X(t)$, or those of its Fourier transform. This analysis is especially useful in linear problems.[17]

It has however to be carried out with some caution, since any given realization $x(t)$ of the process is, a priori, neither a periodic function expandable in Fourier series, nor a function integrable or square-integrable possessing a well-defined Fourier transform.

9.1. Fourier transform of a stationary process

A realization $x(t)$ of the stationary process $X(t)$ does not vanish as $t \to \pm\infty$. The function $x(t)$ is thus neither integrable nor square-integrable, and its Fourier transform does not exist in the ordinary sense. We can however define a Fourier transform of $x(t)$ in the following way. We consider a large interval of finite width T of the time axis. The process under consideration being stationary, this time interval may be taken starting from any origin. We generally choose the origin $t = 0$. We define the Fourier transform $x(\omega)$ of the function $x_T(t)$, equal to $x(t)$ over the interval $(0, T)$ and vanishing outside this interval:

$$x(\omega) = \int_{-\infty}^{\infty} x_T(t) e^{i\omega t}\, dt = \int_0^T x(t) e^{i\omega t}\, dt. \tag{1.9.1}$$

Inversely, we have:

$$x_T(t) = \frac{1}{2\pi} \int_{-\infty}^{\infty} x(\omega) e^{-i\omega t}\, d\omega. \tag{1.9.2}$$

The limit $T \to \infty$ will be taken at the end of the calculation.

As for the stochastic process $X(t)$ itself, we sometimes write symbolically[18] (it then being understood that the above described procedure has been used),

$$X(\omega) = \int_{-\infty}^{\infty} X(t) e^{i\omega t}\, dt, \tag{1.9.3}$$

and, inversely:

$$X(t) = \frac{1}{2\pi} \int_{-\infty}^{\infty} X(\omega) e^{-i\omega t}\, d\omega. \tag{1.9.4}$$

---

[17] An illustration of the efficiency of harmonic analysis in linear problems will be given in Chapter 10, with the harmonic analysis of the Langevin equation for Brownian motion.

[18] For the sake of simplicity, we use the same notation $x(.)$ for the function $x(t)$ and its Fourier transform $x(\omega)$, as well as the same notation $X(.)$ for the process $X(t)$ and its Fourier transform $X(\omega)$.

9.2. Fourier series of a stationary process

A realization $x(t)$ of the stationary process $X(t)$ is not a periodic function. We can nevertheless define its Fourier series. To this end, we consider this function over a large interval of finite width T of the time axis, taken starting from any origin (we can choose, as previously, the origin $t = 0$). For fixed T, it is possible to expand in Fourier series the function obtained by periodizing $x(t)$ (that is, by repeating this function in an identical manner over each interval of width T of the time axis). This expansion coincides with $x(t)$ over the interval $0 \leq t \leq T$:

$$x(t) = \sum_{n=-\infty}^{\infty} a_n e^{-i\omega_n t}, \qquad 0 \leq t \leq T. \tag{1.9.5}$$

The angular frequencies ω_n and the Fourier coefficients a_n are given by the usual formulas:

$$\omega_n = \frac{2\pi n}{T}, \qquad a_n = \frac{1}{T}\int_0^T x(t)e^{i\omega_n t}\,dt, \qquad n = 0, \pm 1, \pm 2, \ldots. \tag{1.9.6}$$

The limit $T \to \infty$ will be taken at the end of the calculations.

As for the (stationary) stochastic process, we write, in a symbolic way:

$$X(t) = \sum_{n=-\infty}^{\infty} A_n e^{-i\omega_n t}, \qquad 0 \leq t \leq T. \tag{1.9.7}$$

The Fourier coefficient a_n is a realization of the random variable A_n defined by:

$$A_n = \frac{1}{T}\int_0^T X(t)e^{i\omega_n t}\,dt. \tag{1.9.8}$$

The value of T being fixed, we have the relation:

$$A_n = \frac{1}{T}X(\omega_n) \tag{1.9.9}$$

between the Fourier coefficient A_n and the Fourier transform $X(\omega_n)$.

9.3. Consequences of the stationarity

Let us now examine the consequences of stationarity on the coefficients of the Fourier series expansion.

• *One-time averages*

The process under consideration being stationary, $\langle X(t)\rangle = \langle X\rangle$ is a constant. The average of A_n being given by:

$$\langle A_n\rangle = \frac{1}{T}\int_0^T \langle X(t)\rangle e^{i\omega_n t}\,dt, \tag{1.9.10}$$

we have:

$$\begin{cases} \langle A_n \rangle = 0, & n \neq 0 \\ \langle A_0 \rangle = \dfrac{1}{T} \displaystyle\int_0^T \langle X(t) \rangle \, dt = \langle X \rangle. \end{cases} \tag{1.9.11}$$

A realization a_0 of A_0 is the temporal average of a realization $x(t)$ of the process $X(t)$ over the interval $(0, T)$:

$$a_0 = \overline{x(t)}^T = \frac{1}{T} \int_0^T x(t) \, dt. \tag{1.9.12}$$

If the process $X(t)$ is ergodic in the mean, which we assume to be the case,[19] then:

$$\lim_{T \to \infty} \overline{x(t)}^T = \langle X \rangle. \tag{1.9.13}$$

All realizations a_0 of A_0 being then equal to $\langle X \rangle$, A_0 is a non-random variable. We can therefore focus the interest on the centered process $X(t) - \langle X \rangle$ and assume, without loss of generality, that we have:

$$\langle A_n \rangle = 0, \qquad n = 0, \pm 1, \pm 2, \ldots. \tag{1.9.14}$$

• *Two-time averages*

For a stationary process, the two-time averages only depend on the difference of the two times involved. The autocorrelation function $\kappa(\tau) = \langle X^*(t) X(t+\tau) \rangle$ of the process $X(t)$, assumed to be centered, may be written, using the series expansion (1.9.7), as:

$$\kappa(\tau) = \sum_{n=-\infty}^{\infty} \sum_{n'=-\infty}^{\infty} \langle A_n A_{n'}^* \rangle e^{-i(\omega_n - \omega_{n'})t} e^{-i\omega_n \tau}. \tag{1.9.15}$$

This function having to be independent of t for any τ, we get:

$$\langle A_n A_{n'}^* \rangle = \langle |A_n|^2 \rangle \delta_{n,n'}. \tag{1.9.16}$$

Thus, there is no correlation between two Fourier coefficients of unequal angular frequencies.

10. The Wiener–Khintchine theorem

10.1. Spectral density of a stationary centered process

Let us consider a centered stationary stochastic process $X(t)$ characterized by real realizations $x(t)$. The Fourier coefficients of $x(t)$ take the form:

$$a_n = a'_n + i a''_n, \qquad a_{-n} = a_n^* = a'_n - i a''_n. \tag{1.10.1}$$

[19] More precisely, we assume that the autocorrelation function $\kappa(\tau)$ is integrable, which is a sufficient condition to ensure ergodicity in the mean.

The mean square of the Fourier component A_n of $X(t)$ is:

$$\langle |A_n|^2 \rangle = \langle {A'_n}^2 \rangle + \langle {A''_n}^2 \rangle, \tag{1.10.2}$$

where A'_n and A''_n denote the random variables of respective realizations a'_n and a''_n.

When a convenient filter is used to select the angular frequencies belonging to the interval $(\omega, \omega + \Delta\omega)$, the mean observable intensity is:

$$\sigma(\omega)\Delta\omega = \sum_{\omega_n \text{ in } (\omega, \omega+\Delta\omega)} \langle |A_n|^2 \rangle. \tag{1.10.3}$$

The right-hand side of equation (1.10.3) involves a sum over all angular frequencies included in the considered band of width $\Delta\omega$. The number of modes of this type is:

$$\frac{\Delta\omega}{2\pi/T} = \frac{T}{2\pi}\Delta\omega. \tag{1.10.4}$$

In the limit $T \to \infty$, we can write, provided that $\langle |A_n|^2 \rangle$ is a continuous function of the angular frequency:

$$\sigma(\omega) = \lim_{T\to\infty} \frac{T}{2\pi} \langle |A_n|^2 \rangle. \tag{1.10.5}$$

Rather than $\sigma(\omega)$, we generally use the quantity:

$$S(\omega) = 2\pi\sigma(\omega) = \lim_{T\to\infty} T \langle |A_n|^2 \rangle, \tag{1.10.6}$$

called the *spectral density* or the *noise spectrum*[20] of the process $X(t)$. Using the relation (1.9.9), we can also express $S(\omega)$ as a function of the squared modulus of the Fourier transform $X(\omega)$:

$$\boxed{S(\omega) = \lim_{T\to\infty} \frac{1}{T} \langle |X(\omega)|^2 \rangle.} \tag{1.10.7}$$

The introduction of the spectral density allows us to make explicit the continuous limit of equation (1.9.16) displaying the fact that there is no correlation between Fourier coefficients of unequal angular frequencies. Indeed, in this limit:

$$\boxed{\langle X(\omega) X^*(\omega') \rangle = 2\pi\delta(\omega - \omega') S(\omega).} \tag{1.10.8}$$

Formula (1.10.8) has been established assuming that $S(\omega)$ is a continuous function of ω. This relation may also be viewed as defining the spectral density.

[20] It is equally called the *power spectrum*.

10.2. The Wiener–Khintchine theorem

Let us come back to the expression (1.9.15) for the autocorrelation function $\kappa(\tau)$, assumed to be integrable, of a centered stationary process ergodic in the mean. On account of the decorrelation property (1.9.16), we have:

$$\kappa(\tau) = \sum_{n=-\infty}^{\infty} \langle |A_n|^2 \rangle e^{-i\omega_n \tau}. \tag{1.10.9}$$

In the limit $T \to \infty$, the discrete summation in formula (1.10.9) is replaced by an integration and we can write, the relation (1.9.9) being taken into account:

$$\kappa(\tau) = \lim_{T\to\infty} \frac{1}{2\pi T} \int_{-\infty}^{\infty} \langle |X(\omega)|^2 \rangle e^{-i\omega\tau}\, d\omega. \tag{1.10.10}$$

The autocorrelation function $\kappa(\tau)$ thus appears as the Fourier transform of the spectral density $S(\omega)$:

$$\boxed{\kappa(\tau) = \frac{1}{2\pi} \int_{-\infty}^{\infty} S(\omega) e^{-i\omega\tau}\, d\omega.} \tag{1.10.11}$$

Inversely, we have:

$$\boxed{S(\omega) = \int_{-\infty}^{\infty} \kappa(\tau) e^{i\omega\tau}\, d\tau.} \tag{1.10.12}$$

Equations (1.10.11) and (1.10.12) constitute the *Wiener–Khintchine theorem*, demonstrated by N. Wiener in 1930 and A. Khintchine in 1934, which states that the autocorrelation function and the spectral density of a stationary stochastic process form a Fourier transform pair.[21] Both quantities contain the same information on the stochastic process under consideration.

10.3. Generalization to a non-centered process

The previous demonstration concerns a centered process $X(t)$, with an autocorrelation function $\kappa(\tau) = \langle X^*(t)X(t+\tau)\rangle$ integrable and thus having a continuous Fourier transform $S(\omega)$.

In the case of a non-centered process $X(t)$, the autocorrelation function is defined by $\kappa(\tau) = \langle [X^*(t) - \langle X^*\rangle][X(t+\tau) - \langle X\rangle]\rangle$. The function $\kappa(\tau)$, which we still assume to be integrable, is nothing but the autocorrelation function $\langle Y^*(t)Y(t+\tau)\rangle$ of the centered process $Y(t) = X(t) - \langle X\rangle$ associated with the fluctuations of $X(t)$ around its mean value $\langle X\rangle$. In the form of the relations (1.10.11) and (1.10.12), the Wiener–Khintchine theorem applies to the centered process $Y(t)$. In these formulas, $\kappa(\tau)$ may

[21] An alternative derivation, shedding more light on some of the validity conditions of the Wiener–Khintchine theorem, is proposed in the appendix at the end of this chapter.

be interpreted, either as the autocorrelation function of $Y(t)$ or as that of $X(t)$, while $S(\omega)$ is the spectral density of $Y(t)$, that is, of the fluctuations of $X(t)$. For a non-centered stationary stochastic process, the Wiener–Khintchine theorem thus expresses the fact that the autocorrelation function and the fluctuations spectral density form a Fourier transform pair. The spectral density of the non-centered process $X(t)$ of mean $\langle X\rangle$ is not a continuous function of ω. It is the sum of the spectral density $S(\omega)$ of the fluctuations of $X(t)$ and the singular term $2\pi|\langle X\rangle|^2\delta(\omega)$. Its inverse Fourier transform is the function $\langle X^*(t)X(t+\tau)\rangle = \kappa(\tau) + |\langle X\rangle|^2$.

10.4. An example

Let us consider again the example (1.7.2) of an autocorrelation function $\kappa(\tau)$ decreasing exponentially with a time constant τ_c, the parameter C being chosen such that $\int_{-\infty}^{\infty}\kappa(\tau)\,d\tau = 1$:

$$\kappa(\tau) = \frac{1}{2\tau_c}e^{-|\tau|/\tau_c}. \qquad (1.10.13)$$

This cusped autocorrelation function, non-differentiable at the origin, corresponds by Fourier transformation to a Lorentzian spectral density:

$$S(\omega) = \frac{\omega_c^2}{\omega_c^2+\omega^2}, \qquad \omega_c = \tau_c^{-1}. \qquad (1.10.14)$$

In the limit $\tau_c \to 0$, $\kappa(\tau)$ tends towards the function $\delta(\tau)$. Correlatively, $S(\omega)$ tends towards a constant equal to 1. We are in the *white noise* limit.

Appendix

1A. An alternative derivation of the Wiener–Khintchine theorem

The demonstration which follows sheds more light on the validity conditions of the theorem. We have:

$$\langle |A_n|^2 \rangle = \frac{1}{T^2}\int_0^T\int_0^T \langle X(t)X(t')\rangle e^{i\omega_n(t-t')}\,dt dt'. \qquad \text{(1A.1)}$$

The autocorrelation function appearing in the integrand depends only on the time difference $\tau = t - t'$. The integration over t and t' takes place in the square $0 \le t \le T$, $0 \le t' \le T$ (Fig. 1.1).

When $t > t'$, we take as new integration variables[22] $t - t' = \tau$ and t'. The integration over t' is done from 0 to $T - \tau$, which yields a contribution to $\langle |A_n|^2 \rangle$ equal to:

$$\frac{1}{T^2}\int_0^T (T-\tau)\kappa(\tau)e^{i\omega_n\tau}\,d\tau. \qquad \text{(1A.2)}$$

In the same way, the integration of the part $t < t'$ yields a contribution to $\langle |A_n|^2 \rangle$ equal to:

$$\frac{1}{T^2}\int_0^T (T-\tau)\kappa(-\tau)e^{-i\omega_n\tau}\,d\tau, \qquad \text{(1A.3)}$$

that is, to:

$$\frac{1}{T^2}\int_{-T}^0 (T+\tau)\kappa(\tau)e^{i\omega_n\tau}\,d\tau. \qquad \text{(1A.4)}$$

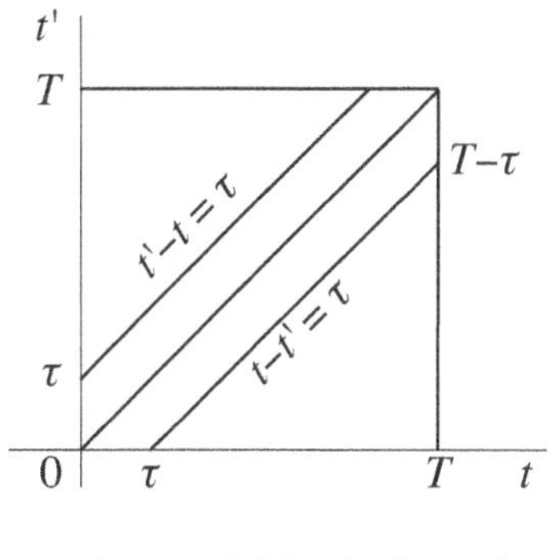

Fig. 1.1 Changing the integration variables in formula (1A.1).

22 The corresponding Jacobian is equal to 1.

Adding the contributions (1A.2) and (1A.4) to $\langle |A_n|^2 \rangle$, gives:

$$\langle |A_n|^2 \rangle = \frac{1}{T} \int_{-T}^{T} \left(1 - \frac{|\tau|}{T}\right) \kappa(\tau) e^{i\omega_n \tau} \, d\tau. \tag{1A.5}$$

Importing the result (1A.5) into the expression (1.10.6) for the spectral density, gives:

$$S(\omega) = \lim_{T \to \infty} \int_{-T}^{T} \left(1 - \frac{|\tau|}{T}\right) \kappa(\tau) e^{i\omega \tau} \, d\tau. \tag{1A.6}$$

In the limit $T \to \infty$, we get:

$$\boxed{S(\omega) = \int_{-\infty}^{\infty} \kappa(\tau) e^{i\omega \tau} \, d\tau.} \tag{1A.7}$$

We also have, inversely:

$$\boxed{\kappa(\tau) = \frac{1}{2\pi} \int_{-\infty}^{\infty} S(\omega) e^{-i\omega \tau} \, d\omega.} \tag{1A.8}$$

The couple of Fourier relations (1A.7) and (1A.8) constitute the Wiener–Khintchine theorem.

Given the inequality:

$$\int_{-T}^{T} \left(1 - \frac{|\tau|}{T}\right) |\kappa(\tau)| \, d\tau \leq \int_{-T}^{T} |\kappa(\tau)| \, d\tau, \tag{1A.9}$$

a sufficient condition for the Wiener–Khintchine theorem to be valid is that the integral $\int_{-\infty}^{\infty} |\kappa(\tau)| \, d\tau$ be convergent.[23]

[23] We recover here a sufficient condition for the process to be ergodic in the mean.

Bibliography

S. CHANDRASEKHAR, Stochastic problems in physics and astronomy, Rev. Mod. Phys. **15**, 1 (1943). Reprinted in *Selected papers on noise and stochastic processes* (N. WAX editor), Dover Publications, New York, 2003.

W. FELLER, *An introduction to probability theory and its applications*, Vol. 1, Wiley, New York, third edition, 1968; Vol. 2, Wiley, New York, second edition, 1971.

C.W. GARDINER, *Handbook of stochastic methods*, Springer-Verlag, Berlin, third edition, 2004.

B.V. GNEDENKO and A.N. KOLMOGOROV, *Limit distributions for sums of independent random variables*, Addison-Wesley, Cambridge (Mass.), 1954.

N.G. VAN KAMPEN, *Stochastic processes in physics and chemistry*, North-Holland, Amsterdam, third edition, 2007.

R. KUBO, M. TODA, and N. HASHITSUME, *Statistical physics II: nonequilibrium statistical mechanics*, Springer-Verlag, Berlin, second edition, 1991.

L. MANDEL and E. WOLF, *Optical coherence and quantum optics*, Cambridge University Press, Cambridge, 1995.

R.M. MAZO, *Brownian motion: fluctuations, dynamics, and applications*, Oxford University Press, Oxford, 2002.

A. PAPOULIS, *Probability, random variables, and stochastic processes*, McGraw-Hill, New York, 1984.

F. REIF, *Fundamentals of statistical and thermal physics*, McGraw-Hill, New York, 1965.

References

N. WIENER, Generalized harmonic analysis, *Acta Math.* **55**, 117 (1930).

A. KHINTCHINE, Korrelationstheorie der stationären stochastichen Prozesse, *Math. Ann.* **109**, 604 (1934).

P. LÉVY, *Théorie de l'addition des variables aléatoires*, Gauthier-Villars, Paris, 1954. Reprinted, Éditions J. Gabay, Paris, 2003.

M.C. WANG and G.E. UHLENBECK, On the theory of the Brownian motion II, *Rev. Mod. Phys.* **17**, 323 (1945). Reprinted in *Selected papers on noise and stochastic processes* (N. WAX editor), Dover Publications, New York, 2003.

J.-P. BOUCHAUD and A. GEORGES, Anomalous diffusion in disordered media: statistical mechanisms, models and physical applications, *Phys. Rep.* **195**, 127 (1990).

Chapter 2

Linear thermodynamics of irreversible processes

The thermodynamics of irreversible processes is a macroscopic theory dealing with states and processes in out-of-equilibrium systems. It allows us, at least when the deviations from equilibrium are not too important, to have to hand a unified treatment of transport phenomena and relaxation processes. This theory thus goes beyond the framework of equilibrium thermodynamics, where only 'reversible' processes, in the course of which the system is supposed to be, at any time, in an equilibrium state, can be taken into account. Out of equilibrium, irreversible processes play an essential role. For instance, a system maintained out of equilibrium by applied constraints reacts in being subjected to fluxes: the system is the place of transport phenomena. Or, after having been submitted to applied constraints, a system returns to equilibrium once these constraints are suppressed: this is a relaxation process.

Not far from equilibrium, transport phenomena are governed by linear phenomenological laws. For instance, if we apply an electric field to a gas of charged particles, there is an electric current which, if the field is not too high, depends linearly on it: this is Ohm's law (G. Ohm, 1822). There are many other examples of linear laws of transport. We could quote the viscous flow law (I. Newton, 1687), the thermal conduction law (J. Fourier, 1811), the law of diffusion of matter (A. Fick, 1855) ... These laws involve experimentally measured transport coefficients.

The general theoretical description of these phenomena requires the introduction of a matrix of kinetic coefficients related to the transport coefficients. The symmetry properties of the system allow us to reduce the number of independent coefficients: this is Curie's principle (P. Curie, 1894). The general theory made great progress owing to L. Onsager (1931), who established symmetry or antisymmetry relations, called the reciprocity relations, between the kinetic coefficients. Finally, the thermodynamics of irreversible processes took its current form after the work of several physicists, among whom I. Prigogine who established, in 1945, the minimum entropy production theorem characterizing stationary out-of-equilibrium states.

1. A few reminders of equilibrium thermodynamics

1.1. The maximum entropy principle

In its modern formulation, due to H.B. Callen (1960), equilibrium thermodynamics is deduced from a postulate concerning the existence and the maximum property of entropy. This postulate is equivalent to the usual principles of thermodynamics. The 'à la Callen' formulation turns out, however, to be better suited than the traditional one to the extension of thermodynamics to the description of irreversible processes.

The statement of the postulate is as follows: "The equilibrium states of a system are parametrized on the macroscopic scale by a set of extensive variables X_i. For each equilibrium state, we can define a quantity, called the entropy and denoted by S, which is a positive, continuous and differentiable function of the X_i variables:

$$\boxed{S = S(X_i).} \tag{2.1.1}$$

This function possesses the additivity (or extensivity) property: the entropy of a system made up of several subsystems is the sum of the entropies of the constituent subsystems. In an isolated composite system, the lifting of one or of several constraints allows for exchanges of extensive quantities between different subsystems. These exchanges result in modifications of the X_i variables relative to the subsystems. These modifications must be compatible with the remaining constraints, as well as with the conservation laws. The values taken by the extensive variables in the equilibrium state reached after the lifting of one or of several internal constraints correspond to the maximum of the entropy compatible with the remaining constraints."

1.2. Conserved variables, conjugate intensive variables, and equations of state

One of the extensive X_i variables is the internal energy E. For a fluid system, the volume V and the number of molecules N are also extensive variables. In the case of a fluid mixture, we have to include, among the extensive variables, the numbers of molecules N_i of the different constituents.

When a system is composed of several subsystems, the X_i's are the set of variables such as the energy, the volume, the numbers of molecules ... relative to each subsystem. The index i then denotes both the nature of the variable and the subsystem. These variables are *conserved variables*: any change of one of them is compensated by an opposite change of the sum of the corresponding variables relative to the other subsystems. In the case of a fluid mixture, this is the case for E, V, and (in the absence of chemical reactions) for the N_i's.

More generally, in a continuous medium, the X_i's refer to each volume element, and the index i represents both the nature of the extensive variable and the coordinates of the point of space. We then have to introduce, in place of the extensive variables themselves, their densities per unit volume. In the case of a fluid, we thus introduce the energy per unit volume $\varepsilon(\boldsymbol{r})$ and the number of particles per unit volume $n(\boldsymbol{r})$.

The fundamental relation (2.1.1) contains all the thermodynamic information about the system. The differential of the entropy:

$$dS = \sum_i \frac{\partial S}{\partial X_i} dX_i, \tag{2.1.2}$$

may be written as:

$$dS = \sum_i F_i dX_i \tag{2.1.3}$$

(*Gibbs relation*). The F_i are the intensive variables conjugate to the conserved quantities X_i. The relations:

$$F_i = \frac{\partial S}{\partial X_i} \tag{2.1.4}$$

are the *equations of state* of the system. For a fluid mixture at equilibrium in the absence of chemical reactions, formula (2.1.1) takes the form:

$$S = S(E, V, N_i). \tag{2.1.5}$$

The corresponding Gibbs relation is:

$$dS = \frac{1}{T}dE + \frac{\mathcal{P}}{T}dV - \sum_i \frac{\mu_i}{T}dN_i, \tag{2.1.6}$$

where T and $\mathcal{P}$ denote the temperature and the pressure, and μ_i the chemical potential per molecule of the constituent i. The intensive variables conjugate to E, V, and N_i are respectively $1/T$, $\mathcal{P}/T$, and $-\mu_i/T$.

2. Description of irreversible processes: affinities and fluxes

Real, irreversible, physical processes always bring into play out-of-equilibrium states. We shall now introduce the *generalized forces* or *affinities* which produce irreversible processes inside a system, as well as the *fluxes* representing the *response* of the system to these affinities. Affinities and fluxes are the fundamental quantities in the description of irreversible processes. To define them, we must have at our disposal a convenient parametrization of the out-of-equilibrium system. This can be achieved in terms of the *slow variables*.

2.1. Slow variables

If an isolated system consisting of several subsystems is prepared in an initial equilibrium state, and if, after the lifting of some constraints, some extensive X_i variables relative to the subsystems may evolve by internal relaxation, the global system reaches a final equilibrium state maximizing the entropy $S(X_i)$.

If the X_i's are slow variables (the other variables, microscopic, relaxing very rapidly over the time scales characteristic of the evolution of the X_i's), we can legitimately consider the system as being at any time in an equilibrium state parametrized by the X_i's. We then defines an *instantaneous entropy* $S(X_i)$ at each step of the relaxation of the X_i's. This separation of the variables into two groups implies a separation of time scales between the microscopic rapid variables and the slow ones. The existence of well-separated time scales may schematically be represented by the inequality:

$$\tau_c \ll \tau_r, \tag{2.2.1}$$

in which τ_c symbolizes the characteristic evolution times of the microscopic variables and τ_r the relaxation times of the slow variables.

2.2. Affinities and fluxes in a discrete system composed of two subsystems

The most interesting irreversible processes take place in continuous media, for example the conduction of heat in a bar in which there is a uniform temperature gradient. It is however simpler, to begin with, to deal with the case of a discrete system composed of two subsystems likely to exchange a globally conserved extensive quantity X_i.

Let us therefore consider an isolated system composed of two weakly coupled subsystems, and an extensive thermodynamic quantity, globally conserved, taking the values X_i and X_i' in each of the subsystems. Since the ensemble is isolated and the coupling is weak, we have:

$$X_i + X_i' = X_i^{(0)} = \text{Const.} \tag{2.2.2}$$

For the global system, both X_i and X_i' are internal variables. The total entropy $S^{(0)}$ is the sum of the entropies of each of the subsystems:

$$S^{(0)} = S(X_i) + S'(X_i'). \tag{2.2.3}$$

The equilibrium values of X_i and X_i' are determined by the maximum of the total entropy (2.2.3), that is, by the condition:

$$\left.\frac{\partial S^{(0)}}{\partial X_i}\right|_{X_i^{(0)}} = \left.\frac{\partial (S+S')}{\partial X_i}\right|_{X_i^{(0)}} = \frac{\partial S}{\partial X_i} - \frac{\partial S'}{\partial X_i'} = F_i - F_i' = 0. \tag{2.2.4}$$

Therefore, if the quantity:

$$\boxed{\mathcal{F}_i = F_i - F_i'} \tag{2.2.5}$$

vanishes, the system is at equilibrium. Otherwise, an irreversible process takes place and brings the system back into equilibrium. The quantity $\mathcal{F}_i$ thus acts as a generalized force governing the process of return towards equilibrium. Generalized forces are also called affinities.

Let us take the example of two fluid systems separated by a diathermal[1] wall, and let X_i be the energy. Its associate or *conjugate* affinity is $(1/T) - (1/T')$, where T and

[1] A diathermal wall is a wall allowing for heat exchanges.

T' are the temperatures of the subsystems. If this affinity vanishes, no heat exchange can take place across the diathermal wall. Conversely, if this affinity is non-zero, a heat exchange will take place between the two subsystems. Similarly, if the wall is movable and if X_i represents the volume, its conjugate affinity is $(\mathcal{P}/T) - (\mathcal{P}'/T')$, where $\mathcal{P}$ and $\mathcal{P}'$ are the pressures of the subsystems. If this affinity is non-zero, there will be a volume exchange between the two subsystems. If the wall is permeable and if X_i represents a number of molecules, its conjugate affinity is $(\mu'/T') - (\mu/T)$, where μ and μ' are the chemical potentials per molecule of the subsystems. If this affinity is non-zero, an exchange of molecules will take place between the two subsystems.

We characterize the response of the system to the applied force by the rate of change of the extensive quantity X_i. The flux J_i is defined by:

$$\boxed{J_i = \frac{dX_i}{dt}.} \tag{2.2.6}$$

In the above example, each flux vanishes if its conjugate affinity vanishes. Inversely, a non-zero affinity leads to a non-zero conjugate flux. Generally speaking, it is the relation between fluxes and affinities which characterizes the changes arising in the course of irreversible processes.

In the discrete case described here, affinities as well as fluxes have a scalar character.

2.3. The entropy production

In each particular situation, we have to properly identify affinities and fluxes. To this end, the most convenient method consists in considering the rate of change of the total entropy $S^{(0)}$ in the course of an irreversible process. The total entropy being a function of the extensive X_i variables, its rate of change, called the *entropy production*, is of the form:

$$\frac{dS^{(0)}}{dt} = \sum_i \frac{\partial S^{(0)}}{\partial X_i}\frac{dX_i}{dt}, \tag{2.2.7}$$

that is:

$$\boxed{\frac{dS^{(0)}}{dt} = \sum_i \mathcal{F}_i J_i.} \tag{2.2.8}$$

The entropy production has a bilinear structure: it is the sum of the products of each flux and its conjugate affinity.

This property of the entropy production is not restricted to the case of a discrete system, but it generalizes to a continuous medium.[2]

[2] See Subsection 4.4.

3. The local equilibrium hypothesis

3.1. Formulation

The thermodynamics of irreversible processes deals with large systems, considered as continuous media supposed, at any time, to be in *local equilibrium.* In other words, we imagine that, at any time, the system (which, as a whole, is out of equilibrium) may be divided into cells[3] of intermediate size, small enough for the thermodynamic quantities to vary only slightly inside, but large enough so as to be treated as thermodynamic subsystems in contact with their environment. It is then possible to define local thermodynamic quantities (uniform inside each cell, but varying from one cell to another).

3.2. Local entropy

The local entropy per unit volume is denoted by $\sigma(\boldsymbol{r})$. As for the other extensive quantities X_i, we introduce their local densities per unit volume[4] $\xi_i(\boldsymbol{r})$ (for instance, the mass per unit volume $\rho(\boldsymbol{r})$). For practical reasons, we also introduce the local densities per unit mass of the entropy and of the X_i variables, which we respectively denote by $s(\boldsymbol{r})$ and $x_i(\boldsymbol{r})$. The local densities per unit volume and those per unit mass of each quantity are linked together through formulas involving $\rho(\boldsymbol{r})$:

$$\sigma(\boldsymbol{r}) = \rho(\boldsymbol{r})s(\boldsymbol{r}), \qquad \xi_i(\boldsymbol{r}) = \rho(\boldsymbol{r})x_i(\boldsymbol{r}). \tag{2.3.1}$$

According to the *local equilibrium hypothesis*, the local entropy is, at any time, the same function of the local thermodynamic quantities as is the equilibrium entropy of the global ones. In other words, according to this hypothesis, the Gibbs relation is equally valid locally. We write accordingly:

$$dS = \int d\boldsymbol{r} \sum_i F_i(\boldsymbol{r}) d\xi_i(\boldsymbol{r}), \tag{2.3.2}$$

and:

$$d(\rho s) = \sum_i F_i d(\rho x_i). \tag{2.3.3}$$

The local intensive variables $F_i(\boldsymbol{r})$ are thus defined at any time as functional derivatives of S:

$$F_i(\boldsymbol{r}) = \frac{\partial \sigma(\boldsymbol{r})}{\partial \xi_i(\boldsymbol{r})} = \frac{\delta S}{\delta \xi_i(\boldsymbol{r})}\cdot \tag{2.3.4}$$

It is in this way, for instance, that we define in a fluid mixture a local temperature $T(\boldsymbol{r})$, a local pressure $\mathcal{P}(\boldsymbol{r})$, and a local chemical potential $\mu_i(\boldsymbol{r})$ for the molecules of chemical species i. The relations (2.3.4) are the *local equations of state*. They do not

---

[3] We adopt here the *Eulerian* description, in which the cells are fixed volume elements. Each of these cells constitutes an open system, exchanging, in particular, particles with the surrounding cells.

[4] The dependence with respect to the space coordinates being here explicit, the index i refers only to the nature of the considered variable.

depend on the presence of gradients: indeed, the local Gibbs relation (2.3.2) having the same form as the equilibrium Gibbs relation (2.1.3), the local equations of state (2.3.4) have the same form as the equilibrium equations of state (2.1.4).

In the case of a fluid, the mechanism ensuring the tendency towards an equilibrium can be described in terms of collisions between molecules. Local equilibrium can be achieved only if these collisions are frequent enough. Conversely, local equilibrium may be not realized in a very rarefied gas in which collisions are not sufficiently numerous to thermalize the molecules.[5]

3.3. General validity criterion

The subdivision of a system into local equilibrium cells may be carried out over a physical basis if there exists an intrinsic length λ such that the number $N_\lambda = (N/V)\lambda^3$ of particles inside a cell of volume λ^3 has only small relative fluctuations ($\delta N_\lambda/N_\lambda \ll 1$). In this case, we can treat the local equilibrium cell of size λ as a thermodynamic system. Such an intrinsic length scale may be defined in gases as the typical distance covered by a molecule between two successive collisions. This distance is called the *mean free path*[6] and is generally denoted by ℓ. We can then choose[7] $\lambda = \ell$.

The local equilibrium cells are open to the transportation of energy and matter. For local equilibrium to be maintained in the presence of applied gradients, the relative changes of the thermodynamic variables induced by these gradients inside each cell must not exceed the equilibrium fluctuations. A general criterion for local equilibrium may thus be the following: if a local thermodynamic variable Π undergoes equilibrium fluctuations $\delta\Pi_{\text{eq}}$ in a cell of size λ, and if an external gradient produces a change $\lambda|\nabla\Pi| = \Delta\Pi$ of this variable over a distance λ, local equilibrium for the variable Π may be considered as maintained if we have:[8]

$$\frac{\Delta\Pi}{\Pi} < \frac{\delta\Pi_{\text{eq}}}{\Pi} \ll 1. \tag{2.3.5}$$

[5] Such a gas is called a *Knudsen gas*.

[6] See Chapters 5 and 7.

[7] For instance, in the case of argon at room temperature and at atmospheric pressure, the density is $n \sim 3 \times 10^{19}$ cm^{-3} and the mean free path $\ell \sim 10^{-5}$ cm, so that the number of molecules in a local equilibrium cell is $N_\lambda = n\ell^3 \sim 3 \times 10^4$. This number is sufficiently large for allowing us to neglect fluctuations and to use thermodynamics: the relative fluctuations $\delta N_\lambda/N_\lambda$ are of order $N_\lambda^{-1/2}$ (see Supplement 2A), that is, 5×10^{-3}.

[8] Let us examine for instance the criterion (2.3.5), as applied to the temperature, in the case of argon at room temperature and at atmospheric pressure. It yields the order of magnitude of the maximum acceptable temperature gradient for the gas to remain in local equilibrium:

$$\frac{\ell|\nabla T|}{T} < \frac{\delta T_{\text{eq}}}{T} \sim N_\lambda^{-1/2} \sim 5 \times 10^{-3}.$$

Taking again the mean free path value as quoted in Note 7, we find that the local equilibrium hypothesis for the temperature remains valid as long as the temperature gradient does not exceed 10^5 K.cm^{-1} (which in fact confers a large domain of validity to this hypothesis).

4. Affinities and fluxes in a continuous medium in local equilibrium

4.1. Balance equations: global balance and local balance

Let us consider a macroscopic system consisting of a continuous medium confined in a volume V limited by a closed surface Σ. Each extensive thermodynamic quantity[9] (conserved or not) $A(t)$ may be written in the form:

$$A(t) = \int_V \rho(\boldsymbol{r}, t) a(\boldsymbol{r}, t) \, d\boldsymbol{r}, \tag{2.4.1}$$

where $a(\boldsymbol{r}, t)$ represents the density per unit mass of $A(t)$.

More precisely, we are interested here in the balance equations in a fluid medium in the absence of chemical reactions.[10] In the case of mass, energy, and entropy, $A(t)$ has a scalar character, while in the case of momentum, $A(t)$ has a vectorial character. For a scalar quantity,[11] the *global balance equation* reads:

$$\frac{dA(t)}{dt} + \int_\Sigma (\boldsymbol{J}_A . \boldsymbol{\nu}) \, d\Sigma = \int_V \sigma_A \, d\boldsymbol{r}, \tag{2.4.2}$$

where $\boldsymbol{\nu}$ represents the unit vector normal to the surface Σ and oriented towards the exterior of the volume V. In equation (2.4.2), σ_A is the *source density* (or, if the case arises, the *sink density*) of A, and $\boldsymbol{J}_A$ is the *current density*, or *flux density*,[12] of A. Using the Green formula, we rewrite equation (2.4.2) in the equivalent form:

$$\frac{dA(t)}{dt} + \int_V (\nabla . \boldsymbol{J}_A) \, d\boldsymbol{r} = \int_V \sigma_A \, d\boldsymbol{r}. \tag{2.4.3}$$

Equation (2.4.3) being valid for any volume V, we get from it the *local balance equation*:

$$\frac{\partial(\rho a)}{\partial t} + \nabla . \boldsymbol{J}_A = \sigma_A. \tag{2.4.4}$$

4.2. Local balance of the conserved extensive quantities

For the extensive conserved quantities X_i, for which there exists, by definition, neither source nor sink ($\sigma_{X_i} = 0$), the local balance equation (2.4.4) takes the form of a conservation equation, known as the *continuity equation*:

$$\frac{\partial(\rho x_i)}{\partial t} + \nabla . \boldsymbol{J}_i = 0. \tag{2.4.5}$$

In equation (2.4.5), $\boldsymbol{J}_i$ denotes the flux of X_i.

---

[9] The notion of extensive quantity is, however, defined unambiguously only in the absence of long-range forces such as the electric or magnetic forces in a polarized or magnetized medium. In the presence of such forces, the energy indeed acquires a dependence with respect to the shape of the system (which corresponds to the existence of a depolarizing field). Problems of this type also appear in elastic media. Generally speaking, the existence of long-range forces prevents us from defining a local density.

[10] See also Chapter 7.

[11] For a vectorial quantity such as momentum, the form taken by the balance equation (which, in the absence of applied external forces, is a conservation equation) is more complicated. In particular, the corresponding density of flux is a tensor (see Chapter 7 on this subject).

[12] No ambiguity being possible, the normal use is to drop the word 'density', and to speak of current or of flux.

4.3. Local balance of the entropy

The entropy is a non-conserved extensive quantity. Its global balance equation:

$$\frac{dS}{dt} + \int_\Sigma (\boldsymbol{J_S}.\boldsymbol{\nu})\, d\Sigma = \int_V \sigma_S\, d\boldsymbol{r}, \tag{2.4.6}$$

has the form:

$$\frac{dS}{dt} = \frac{dS_{\text{exch}}}{dt} + \frac{dS_{\text{int}}}{dt}. \tag{2.4.7}$$

The term $dS_{\text{exch}}/dt = -\int_\Sigma \boldsymbol{J_S}.\boldsymbol{\nu}\, d\Sigma$ is the contribution to dS/dt due to the exchange of entropy between the system and its environment. The term $dS_{\text{int}}/dt = \int_V \sigma_S\, d\boldsymbol{r}$, called the entropy production, is related to changes internal to the system. The entropy production is actually the rate of change of the entropy of the global isolated ensemble consisting of the system of interest and its environment.[13] It is therefore a non-negative quantity. Irreversible phenomena contribute to a strictly positive entropy production. They are also called *dissipative phenomena*.

The local balance equation associated with equation (2.4.6) is:

$$\frac{\partial(\rho s)}{\partial t} + \nabla.\boldsymbol{J}_S = \sigma_S. \tag{2.4.8}$$

In equation (2.4.8), $\boldsymbol{J}_S$ and $\sigma_S \geq 0$ are respectively the entropy flux and the entropy source.

4.4. Affinities, fluxes, and the entropy source in a continuous medium

In order to identify affinities and fluxes, it is necessary to have at one's disposal an explicit expression for the entropy source. We shall study here the particular case of a continuous medium in which the only fluxes coming into play, produced by appropriate forces, are the energy flux and the particle fluxes of each constituent. This simplified approach, although it does not allow us to treat problems such as viscous dissipation, is interesting in that it makes it easy to show the structure of the expression for σ_S.

According to the local equilibrium hypothesis, the local entropy is the same function of the local thermodynamic quantities as is the equilibrium entropy of the global ones. Its differential is thus of the form (2.3.3), which suggests defining by analogy the entropy flux $\boldsymbol{J}_S$ as:

$$\boldsymbol{J}_S = \sum_i F_i \boldsymbol{J}_i, \tag{2.4.9}$$

where $\boldsymbol{J}_i$ is the flux of the extensive conserved quantity X_i. The first term involved in the formula (2.4.8) for σ_S is, according to equation (2.3.3):

$$\frac{\partial(\rho s)}{\partial t} = \sum_i F_i \frac{\partial(\rho x_i)}{\partial t}. \tag{2.4.10}$$

[13] The environment is a thermostat or a reservoir with which the system of interest is in contact. It is considered as being very little perturbated by the processes in which it is involved. In other words, all the processes concerning the environment are supposed to be reversible, without entropy production.

The second term may be calculated as follows:

$$\nabla.\boldsymbol{J}_S = \nabla.(\sum_i F_i \boldsymbol{J}_i) = \sum_i \nabla F_i.\boldsymbol{J}_i + \sum_i F_i \nabla.\boldsymbol{J}_i. \tag{2.4.11}$$

The entropy source σ_S (formula (2.4.8)) then reads:

$$\sigma_S = \sum_i F_i \frac{\partial(\rho x_i)}{\partial t} + \sum_i \nabla F_i.\boldsymbol{J}_i + \sum_i F_i \nabla.\boldsymbol{J}_i. \tag{2.4.12}$$

According to the continuity equations (2.4.5) for the extensive conserved variables, the first and third terms of the expression (2.4.12) for σ_S cancel each other. We are thus left with:

$$\sigma_S = \sum_i \nabla F_i.\boldsymbol{J}_i. \tag{2.4.13}$$

Introducing then the affinity $\mathcal{F}_i$ as defined by:

$$\boxed{\mathcal{F}_i = \nabla F_i,} \tag{2.4.14}$$

we finally write the entropy source in the form:

$$\boxed{\sigma_S = \textstyle\sum_i \mathcal{F}_i.\boldsymbol{J}_i.} \tag{2.4.15}$$

In the same way as in a discrete system, the entropy source has the bilinear structure of a sum of the products of each flux and its conjugate affinity. The fluxes are the current densities $\boldsymbol{J}_i$ of the conserved extensive variables (and not their time derivatives as it is the case in a discrete system). As for the affinities $\mathcal{F}_i$, they are the gradients of the local intensive variables (and not, as in the discrete case, the differences between the values taken by these variables in two neighboring subsystems). More precisely, we have:

$$\sigma_S = \boldsymbol{J}_E.\nabla(\frac{1}{T}) + \sum_i \boldsymbol{J}_{N_i}.\nabla(-\frac{\mu_i}{T}), \tag{2.4.16}$$

where $\boldsymbol{J}_E$ denotes the energy flux and $\boldsymbol{J}_{N_i}$ the particle flux of the constituent i. The expression for σ_S is often rewritten with the aid of the entropy flux. This flux is related to $\boldsymbol{J}_E$ and to the $\boldsymbol{J}_{N_i}$'s by formula (2.4.9), which can be written more explicitly as:

$$\boldsymbol{J}_S = \frac{1}{T}\boldsymbol{J}_E - \sum_i \frac{\mu_i}{T}\boldsymbol{J}_{N_i}. \tag{2.4.17}$$

This gives:[14]

$$\sigma_S = -\frac{1}{T}\boldsymbol{J}_S.\nabla T - \frac{1}{T}\sum_i \boldsymbol{J}_{N_i}.\nabla\mu_i. \tag{2.4.18}$$

[14] This displays the fact that affinities and fluxes are not defined in a unique way.

Let us now come back to the formula (2.4.15) for σ_S. Established in the particular case of a fluid in which the only fluxes coming into play are the energy flux and the particle fluxes of each constituent, this expression does not constitute the most general form of the entropy source in a continuous medium. Indeed, various types of irreversible processes, of various tensorial characters, may occur in such a medium. Chemical reactions are scalar processes, heat and matter transport are vectorial processes, whereas the viscous transport is a tensorial process of order 2. The bilinear structure of the source of entropy is however fully general.

5. Linear response

In a continuous medium in local equilibrium, the fluxes at a given point of space and at a given time are determined by the values of the affinities at this point and at this time:[15]

$$J_i(\boldsymbol{r},t) = J_i\big[\mathcal{F}_1(\boldsymbol{r},t), \mathcal{F}_2(\boldsymbol{r},t), \ldots\big]. \tag{2.5.1}$$

The response of the system is local and instantaneous, which implies that the quantities concerned vary sufficiently slowly in space and in time. Besides, formula (2.5.1) also expresses the fact that each flux may depend, not only on its conjugate affinity ('direct effect'), but also on the other affinities ('indirect effects').

Near equilibrium, this formula may be written in the form of a Taylor series expansion. Since the fluxes vanish if the affinities vanish, this expansion possesses no zeroth-order term. When the deviations from equilibrium are small,[16] the expansion may be limited to first order:

$$\boxed{J_i(\boldsymbol{r},t) = \textstyle\sum_k L_{ik}\mathcal{F}_k(\boldsymbol{r},t).} \tag{2.5.2}$$

The quantities:

$$L_{ik} = \frac{\partial J_i}{\partial \mathcal{F}_k}, \tag{2.5.3}$$

called the *kinetic coefficients*, are determined by the equilibrium values of the intensive variables F_i (such as the temperature or the pressure), and they do not depend on the constraints maintaining the system out of equilibrium (such as pressure or temperature gradients):

$$L_{ik} = L_{ik}(F_1, F_2, \ldots). \tag{2.5.4}$$

The *matrix of kinetic coefficients* $\boldsymbol{L}$ of elements L_{ik} characterizes the *linear response* of the system.

---

[15] In this section, for the sake of simplicity, the symbols J_i and $\mathcal{F}_i$ are used to represent scalar fluxes and affinities, and components of vectorial and/or tensorial ones as well. The index i thus denotes here both the nature of the conserved variable and, when necessary, the relevant component.

[16] The domain of validity of the linearity hypothesis must be studied in each particular situation. It is not necessarily identical to the domain of validity of local equilibrium.

In the linear regime, taking into account the form (2.5.2) of the fluxes, we write the entropy source, which is the sum of the products of each flux by its conjugate affinity, as:

$$\sigma_S = \sum_{ik} L_{ik}\mathcal{F}_i\mathcal{F}_k. \tag{2.5.5}$$

Since σ_S is non-negative, the elements of the matrix $\boldsymbol{L}$ obey the inequalities:

$$L_{ii} \geq 0, \qquad L_{ii}L_{kk} \geq \frac{1}{4}(L_{ik} + L_{ki})^2. \tag{2.5.6}$$

Furthermore, only the symmetric part $\boldsymbol{L}^{(s)}$ of the matrix $\boldsymbol{L}$ contributes to the entropy production:

$$\sigma_S = \sum_{ik} L_{ik}\mathcal{F}_i\mathcal{F}_k = \sum_{ik} L_{ik}^{(s)}\mathcal{F}_i\mathcal{F}_k. \tag{2.5.7}$$

In the case of effects implying rapid variations in space and time, local equilibrium may be not realized. Then, the linear response relations do not take the form (2.5.2), but involve *non-local* and *retarded* kinetic coefficients (which signifies that the fluxes at a given point of space and at a given time depend on the affinities at other points of space and at times before the time of interest). Such effects lie beyond the framework of the linear thermodynamics of irreversible processes. They will be studied in statistical physics using the linear response theory.[17]

6. A few simple examples of transport coefficients

The variables in terms of which we usually express experimental results are not those involved in the general theoretical equations formulated in terms of affinities and fluxes. For instance, the heat current produced by a temperature inhomogeneity is usually expressed as a function of ∇T, and not as a function of the affinity $\nabla(1/T)$. Therefore, instead of the kinetic coefficients L_{ik}, we rather use in practice *transport coefficients* (which are related to the kinetic coefficients).

As simple examples, we shall present the transport coefficients involved in the Ohm's, Fick's, and Fourier's laws, and discuss their relations with the corresponding kinetic coefficients.[18]

6.1. Electrical conductivity

An electric field $\boldsymbol{E}$ is applied to a gas of charged particles, or to the electrons of a metal, or to the charge carriers of a semiconductor. We assume, for the sake of simplicity, that there is only one type of charge carrier.[19] The system is assumed to be macroscopically

[17] See Chapters 15 and 16.

[18] On that subject see also Chapter 6, where these transport coefficients, as well as the kinetic coefficients, are calculated for a *Lorentz gas* of charged particles (that is, for a gas of charged particles undergoing collisions with fixed scattering centers).

[19] In a metal or in a highly doped n-type semiconductor, the carriers are electrons with charge e ($e < 0$), whereas in a highly doped p-type semiconductor, they are holes with charge $-e$. In an intrinsic semiconductor, we would have to add the contributions from the different types of carrier, electrons and holes.

neutral (the carrier density is thus uniform) and maintained at uniform temperature. In these conditions, the medium is subjected to an electric current, whose density $\boldsymbol{J}$ depends linearly on the field $\boldsymbol{E}$ provided that this field is not too intense. This is *Ohm's law*,

$$\boxed{\boldsymbol{J} = \underline{\sigma}.\boldsymbol{E},} \tag{2.6.1}$$

where $\underline{\sigma}$ is the *electrical conductivity tensor* of components $\sigma_{\alpha\beta}$.

For carriers with charge q, the electric current density $\boldsymbol{J}$ is related to the particle flux $\boldsymbol{J}_N$ through $\boldsymbol{J} = q\boldsymbol{J}_N$. In the presence of an electric field $\boldsymbol{E} = -\nabla\phi$, the conjugate affinity of $\boldsymbol{J}_N$ is not $\nabla(-\mu/T)$ but $\nabla(-\overline{\mu}/T)$, where $\overline{\mu} = \mu + q\phi$ is the *electrochemical potential.*[20]

To simplify, let us suppose that the medium is isotropic. The tensor $\underline{\sigma}$ is then proportional to the unit matrix: $\sigma_{\alpha\beta} = \sigma\delta_{\alpha\beta}$ (the same is true in a crystalline medium of cubic symmetry). The scalar σ is the *electrical conductivity*. Ohm's law (2.6.1) simply becomes:

$$\boldsymbol{J} = \sigma\boldsymbol{E}, \tag{2.6.2}$$

whereas the formula (2.5.2) of the general theory reads:

$$\boldsymbol{J}_N = L_{NN}\nabla(-\frac{\overline{\mu}}{T}). \tag{2.6.3}$$

At uniform temperature and uniform carrier density, the gradient of electrochemical potential is $\nabla\overline{\mu} = -q\boldsymbol{E}$. Comparison of formulas (2.6.2) and (2.6.3) thus allows us to relate σ to L_{NN}:

$$\sigma = \frac{q^2}{T}L_{NN}. \tag{2.6.4}$$

6.2. Diffusion coefficient

We consider particles dissolved or suspended in a fluid,[21] in the absence of convective motion. If, because of an external perturbation or of a spontaneous fluctuation, their density is not uniform, there appears a particle flux tending to restore the spatial homogeneity of the density. This is the matter diffusion phenomenon. At uniform temperature, the particle flux $\boldsymbol{J}_N$ depends linearly on the density gradient ∇n, provided that this gradient is not too high. This is *Fick's law*,

$$\boxed{\boldsymbol{J}_N = -\underline{D}.\nabla n,} \tag{2.6.5}$$

where $\underline{D}$ is the *diffusion tensor* of components $D_{\alpha\beta}$.

---

[20] Indeed, it is then necessary to include, in the Helmholtz free energy F, the electrostatic energy of the charge carriers, equal to $q\phi$ per particle. By taking the derivative of F with respect to the number of particles, we do not get the chemical potential μ, but instead the electrochemical potential $\overline{\mu} = \mu + q\phi$.

[21] The mixture under consideration is supposed to be a binary one. We limit ourselves here to the case where the dissolved or suspended molecules are in relatively small number as compared to the other molecules. Then the binary mixture, dilute, consists of a solute in a solvent.

In a medium either isotropic or of cubic symmetry, the tensor $\underline{D}$ is proportional to the unit matrix: $D_{\alpha\beta} = D\delta_{\alpha\beta}$. The scalar D is the *diffusion coefficient*. Fick's law (2.6.5) simply becomes:

$$\boldsymbol{J}_N = -D\nabla n, \tag{2.6.6}$$

whereas the formula (2.5.2) of the general theory reads:

$$\boldsymbol{J}_N = L_{NN}\nabla(-\frac{\mu}{T}). \tag{2.6.7}$$

In the case of a dilute mixture in mechanical equilibrium at uniform temperature, the chemical potential gradient only depends on the density gradient of the solute:

$$\nabla\mu = \frac{\partial\mu}{\partial n}\bigg|_T \nabla n. \tag{2.6.8}$$

Comparison of formulas (2.6.6) and (2.6.7) then allows us to relate D to L_{NN}:

$$D = \frac{1}{T}\frac{\partial\mu}{\partial n}\bigg|_T L_{NN}. \tag{2.6.9}$$

6.3. The Einstein relation

Formula (2.6.9) also applies to the diffusion coefficient of a gas of charged particles. In this context, by eliminating L_{NN} between formulas (2.6.4) and (2.6.9), we get a relation between D and σ:

$$D = \sigma\frac{1}{q^2}\frac{\partial\mu}{\partial n}\bigg|_T. \tag{2.6.10}$$

We usually introduce the *drift mobility* μ_D of the charge carriers, defined as the ratio, in a stationary regime, between their mean velocity and the electric field. This microscopic quantity is directly related to the electrical conductivity, a macroscopic quantity:

$$\sigma = nq\mu_D. \tag{2.6.11}$$

Formula (2.6.10) takes the form of a relation between D and μ_D:

$$\boxed{\frac{D}{\mu_D} = \frac{n}{q}\frac{\partial\mu}{\partial n}\bigg|_T.} \tag{2.6.12}$$

The existence of such a relation leads us to think that, in the vicinity of equilibrium, the linear response to an external perturbation (as characterized here by the mobility), on the one hand, and the equilibrium fluctuations (characterized by the diffusion coefficient), on the other hand, involve the same microscopic mechanisms. We observe here, in a particular case, a general behavior of near-equilibrium systems.[22]

[22] See Chapters 13 and 14.

In order to confer a more explicit character to formula (2.6.12), we have to compute the thermodynamic derivative $(\partial\mu/\partial n)_T$, which we will do in both particular cases of a non-degenerate gas (for instance the gas of charge carriers in a non-degenerate semiconductor) and of a highly degenerate gas (such as the electron gas in a metal). In both examples, we will consider, for the sake of simplicity, the charge carriers as non-interacting.

• *Non-degenerate gas of charge carriers*

The carriers are considered as forming an ideal classical gas. Whatever the system dimensionality, we have:

$$\left.\frac{\partial\mu}{\partial n}\right|_T = \frac{kT}{n} \tag{2.6.13}$$

(k denotes the Boltzmann constant). Formula (2.6.12) reads, in this case:

$$\boxed{\frac{D}{\mu_D} = \frac{kT}{q}.} \tag{2.6.14}$$

Formula (2.6.14) is the *Einstein relation.*[23]

• *Fermion gas at* $T = 0$

Consider an ideal gas of fermions of mass m. At zero temperature, its chemical potential μ is nothing but the Fermi energy $\varepsilon_F = \hbar^2 k_F^2/2m$. The Fermi wave vector k_F is determined by the gas density n, that is (at three dimensions), $k_F = (3\pi^2 n)^{1/3}$. We have:

$$\left.\frac{\partial\mu}{\partial n}\right|_{T=0} = \frac{1}{2n(\varepsilon_F)}, \tag{2.6.15}$$

where $n(\varepsilon_F) = \pi^{-2}2^{-1/2}(m/\hbar^2)^{3/2}\varepsilon_F^{1/2}$ is the density of states in energy per unit volume and per spin direction at the Fermi level. Formula (2.6.12) reads, in this case:

$$\frac{D}{\mu_D} = \frac{2}{3}\frac{\varepsilon_F}{q}, \tag{2.6.16}$$

or, in terms of the Fermi temperature[24] T_F:

$$\boxed{\frac{D}{\mu_D} = \frac{kT_F}{q}.} \tag{2.6.17}$$

[23] The Einstein relation will be studied anew in several different contexts later in the book (see Chapter 6, Chapter 10 devoted to Brownian motion, and Chapter 16).

[24] In three dimensions, the Fermi temperature is defined by $\varepsilon_F = \frac{3}{2}kT_F$. Note that relation (2.6.17) is formally similar to the Einstein relation (2.6.14), provided that T be replaced by T_F.

6.4. Thermal conductivity of an insulating solid

A temperature gradient ∇T is applied to an insulating solid. It results in an energy flux $\boldsymbol{J}_E$, which depends linearly on the temperature gradient, if this gradient is not too high. This is *Fourier's law*,

$$\boxed{\boldsymbol{J}_E = -\underline{\kappa}.\nabla T,} \tag{2.6.18}$$

where $\underline{\kappa}$ is the *thermal conductivity tensor* of components $\kappa_{\alpha\beta}$. The material being an insulator, heat conduction takes place solely via phonons (the quasi-particles associated with the lattice vibration modes). Since the number of phonons is not conserved, their chemical potential vanishes. This is the reason why the energy flux does not contain a term[25] in $\nabla(-\mu/T)$.

If the medium is either isotropic or of cubic symmetry, the tensor $\underline{\kappa}$ is proportional to the unit matrix: $\kappa_{\alpha\beta} = \kappa\delta_{\alpha\beta}$. The scalar κ is the *thermal conductivity* of the insulating solid. Fourier's law (2.6.18) reduces to:

$$\boldsymbol{J}_E = -\kappa\nabla T, \tag{2.6.19}$$

whereas the formula (2.5.2) of the general theory reads:

$$\boldsymbol{J}_E = L_{EE}\nabla\left(\frac{1}{T}\right). \tag{2.6.20}$$

The comparison between formulas (2.6.19) and (2.6.20) allows us to relate κ to L_{EE}:

$$\kappa = \frac{1}{T^2}L_{EE}. \tag{2.6.21}$$

7. Curie's principle

In each of the above examples, a given affinity solely produces a flux of its conjugate quantity; this is a direct effect. In other cases an affinity produces additional fluxes of quantities which are not conjugate to it; such effects are qualified as indirect. In complex systems, transport phenomena are thus frequently coupled. For instance, in a fluid mixture, a temperature gradient produces a heat flux and a diffusion flux as well. This phenomenon is called the *thermodiffusion*.[26] Coupled transport phenomena are also involved in *thermoelectric effects*, which occur in metals and in semiconductors.[27]

In most systems, the components of the different fluxes do not actually depend on all the affinities. This is a consequence of the *symmetry principle* or *Curie's principle* (P. Curie, 1894). This principle is traditionally stated as follows: "When some causes

[25] The situation is different in a conductor, where thermal conduction is mainly due to the charge carriers (see Supplement 2B).

[26] See Supplement 2C.

[27] See Supplement 2B.

produce some effects, the symmetry elements of the causes must also be found in the effects they produce. When some effects display some dissymmetry, this dissymmetry must also be found in the causes which produced them." We mean here by 'cause' a physical system with its environment (external fields, constraint fields ...), and by 'effect' a physical property.

However, Curie's principle applies in this simple form only if the considered effect is unique. If this is not the case, an effect may have fewer symmetry elements than its cause (in other words, a given solution of a problem may have fewer symmetry elements than the data of this problem). In the linear range, the simple formulation of Curie's principle is sufficient.

Curie's principle has several consequences for the linear thermodynamics of irreversible processes, in isotropic media and in anisotropic ones as well.

- *Isotropic media*

In an isotropic medium, fluxes and affinities whose tensorial order differ by one unit (for instance, scalar and vector) cannot be coupled. Indeed, if in the coupling relation (2.5.2) the difference of tensorial orders between fluxes and affinities were odd, the kinetic coefficients[28] $L_{ik}^{\alpha\beta}$ would be the components of an odd-order tensor. Such a tensor cannot remain invariant with respect to rotations. Isotropic systems may therefore be characterized only by kinetic coefficients either scalar or of even tensorial order. In an isotropic medium, the transportation of scalar quantities such as the particle number or the energy involve tensors such as $L_{NN}^{\alpha\beta}$ or $L_{EE}^{\alpha\beta}$, proportional to the unit matrix: $L_{NN}^{\alpha\beta} = L_{NN}\delta_{\alpha\beta}$, $L_{EE}^{\alpha\beta} = L_{EE}\delta_{\alpha\beta}$. The same is true for the electrical conductivity, diffusion, and thermal conductivity tensors, which are respectively of the form $\sigma_{\alpha\beta} = \sigma\delta_{\alpha,\beta}$, $D_{\alpha\beta} = D\delta_{\alpha,\beta}$, and $\kappa_{\alpha\beta} = \kappa\delta_{\alpha,\beta}$.

- *Anisotropic media*

In an anisotropic medium, for instance a crystalline medium invariant with respect to some transformations of a symmetry group, the consideration of symmetry properties allows us to reduce the number of independent kinetic coefficients. Thus, in a cubic crystal, the tensors $L_{NN}^{\alpha\beta}$ and $L_{EE}^{\alpha\beta}$ are proportional to the unit matrix, as in an isotropic medium. The same is true of the tensors of electrical conductivity, diffusion, and thermal conductivity.

8. The reciprocity relations

In 1931, L. Onsager put forward the idea according to which symmetry or antisymmetry relations between kinetic coefficients, called the *reciprocity relations*, must exist in all thermodynamic systems in which transport and relaxation phenomena are properly described by linear laws. The Onsager reciprocity relations allow us in practice to reduce the number of experiments necessary to measure the transport coefficients. These relations concern the off-diagonal elements of the matrix $\boldsymbol{L}$, that is, the kinetic coefficients describing indirect effects. Fundamentally, they come from the reversibility

---

[28] The indexes i, k ... denote here the nature of the conserved variable, and the exponents α, β... the relevant components.

of the microscopic equations of motion, that is, from their invariance under time-reversal[29] (provided however that, if the case arises, we also change the sign of the magnetic field and/or of the rotation vector).

In order to be able to formulate the reciprocity relations resulting from this invariance, we first have to make sure that the chosen affinities and fluxes possess a well-defined parity under time-reversal. Generally speaking, the entropy being independent of the direction of time, the entropy source σ_S is an odd function of t. From the general structure of the expression for σ_S (formula (2.4.15)), it results that, if affinities and fluxes possess a well-defined parity, each affinity has a parity opposite to that of its conjugate flux. Quantities such as the energy density or the particle density are invariant under time-reversal. They are said to be even, or that their *signature* under time-reversal is $\epsilon_i = +1$. The fluxes of these quantities are odd. Conversely, a quantity such as the momentum density changes sign under time-reversal. It is an odd quantity, of signature $\epsilon_i = -1$. Its flux (tensorial) is even.

8.1. The Onsager relations

For extensive conserved variables of densities x_i and x_k invariant under time-reversal, the Onsager relations are symmetry relations expressing the identity of the kinetic coefficients L_{ik} and L_{ki}:

$$\boxed{L_{ik} = L_{ki}.} \tag{2.8.1}$$

Let us consider the example of a conductor with one type of charge carrier, submitted to a temperature gradient and an electrochemical potential gradient. These gradients give rise to a particle flux and an energy flux. The general formulas (2.5.2) take the form:[30]

$$\begin{cases} \boldsymbol{J}_N = L_{NN}\nabla(-\dfrac{\overline{\mu}}{T}) + L_{NE}\nabla(\dfrac{1}{T}) \\ \\ \boldsymbol{J}_E = L_{EN}\nabla(-\dfrac{\overline{\mu}}{T}) + L_{EE}\nabla(\dfrac{1}{T}). \end{cases} \tag{2.8.2}$$

Both particle and energy densities being invariant under time-reversal, the corresponding Onsager reciprocity relation reads:[31]

$$L_{EN} = L_{NE}. \tag{2.8.3}$$

---

[29] In classical mechanics, time-reversal changes the signs of the velocities, without modifying the positions. In quantum mechanics, time-reversal associates with a wave function its complex conjugate quantity.

[30] For the sake of simplicity, the conductor is assumed to be isotropic.

[31] The relation (2.8.3) was established as early as 1854 by W. Thomson (later Lord Kelvin) in the framework of the study of the thermoelectric effects related to the simultaneous presence of a particle flux and an energy flux in conductors (see Supplement 2B).

8.2. The Onsager–Casimir relations

The reciprocity relations took their final form in 1945 owing to H.B.G. Casimir. For conserved extensive variables of densities x_i and x_k, and signatures ϵ_i and ϵ_k, the reciprocity relations, called *the Onsager–Casimir relations*, read:

$$\boxed{L_{ik} = \epsilon_i \epsilon_k L_{ki}.} \tag{2.8.4}$$

According to the sign of the product $\epsilon_i \epsilon_k$, the Onsager–Casimir relations express the symmetry or the antisymmetry of the kinetic coefficients L_{ik} and L_{ki}. In the case $\epsilon_i \epsilon_k = 1$, the coefficient L_{ik} corresponds to a coupling of processes with the same parity, and the reciprocity relation is a symmetry relation ($L_{ik} = L_{ki}$). When $\epsilon_i \epsilon_k = -1$, L_{ik} corresponds to a coupling between processes with different parities, and the reciprocity relation is an antisymmetry relation ($L_{ik} = -L_{ki}$).

8.3. Generalization

When irreversible processes take place in the presence of a magnetic field $\boldsymbol{H}$ and/or in a system rotating at the angular frequency $\boldsymbol{\Omega}$, the Onsager–Casimir relations read:

$$\boxed{L_{ik}(\boldsymbol{H}, \boldsymbol{\Omega}) = \epsilon_i \epsilon_k L_{ki}(-\boldsymbol{H}, -\boldsymbol{\Omega}).} \tag{2.8.5}$$

They relate the kinetic coefficients of two physical systems which differ from one another by the change of sign of the parameters $\boldsymbol{H}$ and/or $\boldsymbol{\Omega}$. If there is neither magnetic field nor rotation, the Onsager–Casimir relations, which are then given by formula (2.8.4), simply express the symmetry or the antisymmetry of the kinetic coefficients of a given system.

9. Justification of the reciprocity relations

Demonstration of the reciprocity relations can only be carried out by making use of the link with a microscopic description of the out-of-equilibrium system (thus, in statistical physics). We shall limit ourselves here to giving a justification of these relations based on an hypothesis proposed by L. Onsager about the evolution of thermodynamic fluctuations.[32]

9.1. Equilibrium fluctuations

Let us consider an isolated ensemble composed of a system, itself either isolated or coupled to a thermostat or a reservoir with which it is likely to exchange heat and, possibly, particles. The probability of a fluctuation in which the entropy and the extensive variables of the system undergo fluctuations δS and δX_i around their equilibrium

[32] This is the fluctuations *regression hypothesis*, which can be demonstrated in the framework of the linear response theory (see Chapters 13 and 14).

values may be obtained with the aid of the theory of thermodynamic fluctuations,[33] developed in particular by A. Einstein in 1910.

To calculate the probability of a fluctuation in an isolated system at equilibrium, Einstein proposed to make use of the fundamental Boltzmann formula for entropy. The entropy of an isolated system is proportional to the logarithm of the number of accessible microscopic states (the proportionality relation involves the Boltzmann constant k). By 'inverting' this formula, we get the probability of a fluctuation. If the system is not isolated, the entropy variation to consider is that of the global isolated ensemble composed of the system and its environment, that is, the entropy variation δS_{int} related to the changes internal to the system. The probability of such a fluctuation is thus given by the *Einstein formula*:

$$\boxed{w \sim \exp\left(\frac{\delta S_{\text{int}}}{k}\right).} \tag{2.9.1}$$

The entropy variation related to the changes internal to the system is of the form $\delta S_{\text{int}} = \delta S - \delta S_{\text{exch}}$, where δS is the variation of the entropy of the system and $\delta S_{\text{exch}} = \sum_i F_i \delta X_i$, the entropy exchange between the system and its environment. The quantity F_i is the equilibrium value, defined by the environment properties, of the intensive variable conjugate of X_i. The probability of a fluctuation characterized by the variations δS and δX_i of the entropy and of the other extensive variables thus reads:

$$\boxed{w \sim \exp\left(\frac{\delta S}{k} - \frac{1}{k}\sum_i F_i \delta X_i\right).} \tag{2.9.2}$$

9.2. A property of the correlations

We shall prove that there are no correlations between the fluctuations of an extensive quantity X_i and those of the intensive quantities other than its conjugate variable F_i. More precisely, we shall establish the identity:

$$\boxed{\langle \delta X_i \mathcal{F}_j \rangle = -k\delta_{ij},} \tag{2.9.3}$$

where the generalized forces $\mathcal{F}_j = \delta F_j$ are the fluctuations of the intensive variables F_j. In formula (2.9.3), the symbol $\langle \ldots \rangle$ denotes the average calculated using the distribution w.

We get from formula (2.9.2):

$$\frac{\partial w}{\partial F_i} = -\frac{1}{k} w\, \delta X_i. \tag{2.9.4}$$

[33] See Supplement 2A.

The average value $\langle \delta X_i \mathcal{F}_j \rangle$, defined by:

$$\langle \delta X_i \mathcal{F}_j \rangle = \int w \, \delta X_i \delta F_j \prod_l d(\delta X_l), \tag{2.9.5}$$

can be written, in virtue of formula (2.9.4):

$$\langle \delta X_i \mathcal{F}_j \rangle = -k \int \frac{\partial w}{\partial F_i} \delta F_j \prod_l d(\delta X_l). \tag{2.9.6}$$

Now we have the identity:

$$\frac{\partial}{\partial F_i} \int w \delta F_j \prod_l d(\delta X_l) = \int w \frac{\partial \delta F_j}{\partial F_i} \prod_l d(\delta X_l) + \int \frac{\partial w}{\partial F_i} \delta F_j \prod_l d(\delta X_l), \tag{2.9.7}$$

whose right-hand side vanishes, since it corresponds to the derivative of the mean value of a fluctuation, a quantity null by definition. Since F_i represents the equilibrium value of the intensive variable conjugate of X_i, we have $\partial \delta F_j / \partial F_i = -\delta_{ij}$. This gives:

$$\int \frac{\partial w}{\partial F_i} \delta F_j \prod_l d(\delta X_l) = \delta_{ij}. \tag{2.9.8}$$

Importing this result into formula (2.9.6), we demonstrate the property (2.9.3).

9.3. Regression of the fluctuations: justification of the reciprocity relations

Consider, for the sake of simplicity, in a system with neither magnetic field nor rotation, extensive quantities X_i and X_k of densities invariant under time-reversal ($\epsilon_i = \epsilon_k = +1$). The equilibrium correlation function $\langle \delta X_i \delta X_k(\tau) \rangle$ has, due to time-reversal invariance, the property:

$$\langle \delta X_i \delta X_k(\tau) \rangle = \langle \delta X_i \delta X_k(-\tau) \rangle. \tag{2.9.9}$$

from which we deduce, using the stationarity property, the equality:

$$\langle \delta X_i \delta X_k(\tau) \rangle = \langle \delta X_i(\tau) \delta X_k \rangle. \tag{2.9.10}$$

Taking the derivative with respect to τ of both sides of equation (2.9.10), gives:

$$\langle \delta X_i \delta \dot{X}_k \rangle = \langle \delta \dot{X}_i \delta X_k \rangle. \tag{2.9.11}$$

According to the Onsager *regression hypothesis*,[34] fluctuations relax by following the same laws as the macroscopic quantities:

$$\delta \dot{X}_i = \sum_j L_{ij} \mathcal{F}_j, \qquad \delta \dot{X}_k = \sum_j L_{kj} \mathcal{F}_j. \tag{2.9.12}$$

[34] This hypothesis constitutes one of the first statements of the *fluctuation-dissipation theorem* (see Chapter 14). It will be verified in Chapter 10 for the particular case of the equilibrium fluctuations of a Brownian particle's velocity.

We deduce from equations (2.9.11) and (2.9.12) the relation:

$$\sum_j L_{kj}\langle\delta X_i \mathcal{F}_j\rangle = \sum_j L_{ij}\langle\delta X_k \mathcal{F}_j\rangle. \tag{2.9.13}$$

In virtue of the property (2.9.3), the identity (2.9.13) reduces to the Onsager relation (2.8.1) relative to L_{ik} and L_{ki}. It is a symmetry relation, in accordance with the fact that X_i and X_k have been assumed to be invariant under time-reversal.

9.4. Discussion

The previous arguments rely upon the fluctuation regression hypothesis (formula (2.9.12)). However, to show that the fluctuations relax according to the same laws as the averages, we have to study them at the microscopic level. In a strictly thermodynamic framework, it is therefore impossible to demonstrate the regression hypothesis. This is the reason why the arguments presented in this chapter do not constitute a genuine proof. They nevertheless have the interest of shedding light on the fundamental link between the reciprocity relations and the reversibility of the microscopic equations of motion.

10. The minimum entropy production theorem

For systems submitted to time-independent external constraints, some out-of-equilibrium states play a particularly important role, analogous in a certain sense to the role played in thermodynamics by equilibrium states. These states are the *stationary states* characterized, by definition, by time-independent variables X_i (state variables). As shown by I. Prigogine in 1945, the out-of-equilibrium stationary states correspond to a minimum of the entropy production. The *minimum entropy production theorem* is however only valid under relatively restrictive hypotheses.

In a stationary state, the state variables are time-independent. In the domain of validity of the local equilibrium hypothesis, the same is true of the entropy (function of the state variables). We therefore have:

$$\frac{dS}{dt} = \frac{dS_{\text{exch}}}{dt} + \frac{dS_{\text{int}}}{dt} = 0. \tag{2.10.1}$$

In formula (2.10.1), dS_{exch}/dt represents the rate of change of the entropy exchange between the system and its environment, and $P_S = dS_{\text{int}}/dt$, the entropy production related to changes internal to the system. The entropy production is non-negative:

$$\frac{dS_{\text{int}}}{dt} = \int_V \sigma_S \, dV \geq 0. \tag{2.10.2}$$

We deduce from equations (2.10.1) and (2.10.2) the inequality:

$$\frac{dS_{\text{exch}}}{dt} \leq 0. \tag{2.10.3}$$

Thus, to maintain a system in an out-of-equilibrium stationary state, it is necessary to continually transfer entropy from this system to the outside medium.

In order to characterize the way in which the system evolves towards such a stationary state, we calculate the time-derivative of the entropy production. We can show that, if the kinetic coefficients are constant and obey the reciprocity relations, and if the boundary conditions imposed on the system are time-independent, the rate of change of the entropy production obeys the inequality:

$$\frac{dP_S}{dt} \leq 0. \tag{2.10.4}$$

The entropy production P_S being non-negative, the inequality (2.10.4) shows that the rate of change of the entropy production always eventually vanishes. The system then finds itself in a stationary state characterized by the smallest entropy production compatible with the imposed boundary conditions. This is the statement of the minimum entropy production theorem.[35]

[35] We do not provide here a general demonstration of this theorem. An illustration of it will be provided in connection with the thermoelectric effects (see Supplement 2B).

Bibliography

R. Balian, *From microphysics to macrophysics*, Vol. 1, Springer-Verlag, Berlin, 1991; Vol. 2, Springer-Verlag, Berlin, 1992.

H.B. Callen, *Thermodynamics and an introduction to thermostatistics*, Wiley, New York, second edition, 1985.

S.R. de Groot and P. Mazur, *Non-equilibrium thermodynamics*, North-Holland, Amsterdam, 1962. Reprinted, Dover Publications, New York, 1984.

R. Haase, *Thermodynamics of irreversible processes*, Addison Wesley, Reading, 1969. Reprinted, Dover Publications, New York, 1990.

H.J. Kreuzer, *Nonequilibrium thermodynamics and its statistical foundations*, Clarendon Press, Oxford, 1981.

L.D. Landau and E.M. Lifshitz, *Statistical physics*, Butterworth-Heinemann, Oxford, third edition, 1980.

D.A. McQuarrie, *Statistical mechanics*, University Science Books, Sausalito, second edition, 2000.

G. Nicolis and I. Prigogine, *Self-organization in nonequilibrium systems*, Wiley, New York, 1977.

References

A. Einstein, Theorie der Opaleszenz von homogenen Flüssigkeiten und Flüssigkeitsgemischen in der Nähe des kritischen Zustandes, *Annalen der Physik* **33**, 1275 (1910).

L. Onsager, Reciprocal relations in irreversible processes. I., *Phys. Rev.* **37**, 405 (1931); II., *Phys. Rev.* **38**, 2265 (1931).

H.B.G. Casimir, On Onsager's principle of microscopic reversibility, *Rev. Mod. Phys.* **17**, 343 (1945).

Supplement 2A

Thermodynamic fluctuations

1. The fluctuations

The theory of fluctuations at thermodynamic equilibrium was established by J.W. Gibbs in 1902 and A. Einstein in 1910. The fluctuations are viewed as random variables, and we have to determine their probability distribution in order to be able to calculate their different moments. To begin with, it will be shown that the definition of the equilibrium fluctuations cannot be carried out in the same way, depending on whether we are concerned with the fluctuations of extensive quantities or of intensive ones.

1.1. Fluctuations of extensive quantities

It is for those extensive quantities, which, besides a thermodynamic meaning, have a mechanical definition, for instance the internal energy E as well as, in the case of a fluid system, the volume V and the number of molecules N, that the notion of fluctuations can be introduced most naturally.

This notion is, however, more tricky to define for the entropy, since entropy is an extensive quantity with no mechanical definition. To define the entropy fluctuations, we can start from the 'à la Callen' formulation, in which the entropy of a system is a function $S(X_i)$ of the extensive variables X_i parametrizing its equilibrium states. We define the entropy fluctuation as the variation of the function $S(X_i)$ ensuing from the fluctuations of the X_i's:

$$\delta S = \sum_i \frac{\partial S}{\partial X_i} \delta X_i. \tag{2A.1.1}$$

Formula (2A.1.1) is formally similar to the Gibbs relation between the differential of the entropy and the differentials of the X_i variables.

1.2. Fluctuations of intensive quantities

For intensive variables such as temperature, pressure, chemical potential . . ., the notion of fluctuations seems a priori to be much less clear. At first sight, it even seems in principle impossible to define, since in statistical physics these quantities appear as Lagrange multipliers associated with the extensive variables that the system is likely to exchange with a reservoir. In this sense, the intensive variables are fixed and do not fluctuate. It is however sometimes necessary to be able to give a sense to their fluctuations.[1]

In this respect, we have to consider these fluctuations from a slightly different point of view, corresponding to a more physical approach. For instance, a sufficiently sensitive and small thermometer, introduced into a system at equilibrium, gives an indication of temperature which fluctuates with time: such a thermometer in fact measures the instantaneous energy of its immediate environment. From this point of view, the fluctuations of intensive quantities (in this example, the temperature) are actually both a measure and the result of the fluctuations of the extensive ones (here, the energy).

Before defining the probability of a fluctuation characterized by the variations δS and δX_i of the entropy and of the other extensive variables, let us come back in more detail to the general maximum entropy principle, as applied to the global isolated ensemble composed of the system and the environment with which it is in contact.

2. Consequences of the maximum entropy principle

Generally speaking, a fluctuation of the entropy of a system in contact with an environment is of the form:

$$\delta S = \delta S_{\text{exch}} + \delta S_{\text{int}}, \tag{2A.2.1}$$

where δS_{exch} comes out from the exchange of entropy between the system and its environment, and δS_{int} from the entropy variation related to changes internal to the system. Interestingly, the fluctuation $\delta S_{\text{int}} = \delta S - \delta S_{\text{exch}}$ represents the fluctuation of the entropy of the global isolated ensemble composed of the system and its environment. In virtue of the maximum entropy principle applied to this isolated ensemble at equilibrium, we have:

$$\delta S_{\text{int}} \leq 0. \tag{2A.2.2}$$

The inequality (2A.2.2) applies to any thermodynamic system, whatever the equilibrium conditions under which it is placed. It includes all the situations for which the second principle of thermodynamics is usually expressed in terms of thermodynamic potentials. Let us now check this latter statement using a few examples.

- *Isolated system*

For an isolated system, we have $\delta S_{\text{exch}} = 0$, since there is no interaction with an environment. We thus have $\delta S = \delta S_{\text{int}}$. The second principle for an isolated system implies $\delta S \leq 0$, that is, $\delta S_{\text{int}} \leq 0$.

[1] It is for instance the case in the study of the validity of the local equilibrium criterion for an intensive quantity Π such as the temperature or the pressure: the variation $\Delta\Pi$ produced inside a local equilibrium cell by an imposed gradient must indeed remain smaller than the equilibrium fluctuation $\delta\Pi_{\text{eq}}$.

• *Closed system likely to exchange heat with a thermostat*

For a *closed* system not exchanging particles with the outside but likely to exchange heat with a thermostat at temperature T, we have $\delta S_{\text{exch}} = \delta Q/T$. Let us take the example of a fluid with a fixed number of molecules and at constant volume, in contact with a thermostat at temperature T. The variation of the Helmholtz free energy $F = E - TS$ reads:

$$\delta F = T\delta S_{\text{exch}} - T(\delta S_{\text{exch}} + \delta S_{\text{int}}) = -T\delta S_{\text{int}}. \tag{2A.2.3}$$

In this case, the second principle, therefore, reads $\delta S_{\text{int}} \leq 0$ or, equivalently, $\delta F \geq 0$. Similarly, in the case of a fluid with a fixed number of molecules, at constant pressure $\mathcal{P}$ and in contact with a thermostat at temperature T, the variation of the Gibbs free energy $G = E + \mathcal{P}V - TS$ reads:

$$\delta G = T\delta S_{\text{exch}} - \mathcal{P}\delta V + \mathcal{P}\delta V - T(\delta S_{\text{exch}} + \delta S_{\text{int}}) = -T\delta S_{\text{int}}. \tag{2A.2.4}$$

The second principle thus reads equally $\delta S_{\text{int}} \leq 0$ or $\delta G \geq 0$.

In the same way, we can treat other situations, for instance the case of an *open* system likely to exchange heat as well as particles with a reservoir. We retrieve, in all cases, the inequality (2A.2.2). This fully general formulation highlights the essential role played by the entropy fluctuation δS_{int} related to changes internal to the system.

3. Probability of a fluctuation: the Einstein formula

For an isolated system at equilibrium, the Einstein formula is the 'inverse' of the fundamental Boltzmann formula for the entropy. It gives the probability w of a fluctuation in the course of which the entropy undergoes a fluctuation δS:

$$w \sim \exp\left(\frac{\delta S}{k}\right). \tag{2A.3.1}$$

For a non-isolated system, the entropy fluctuation to take into account is not δS, but instead $\delta S_{\text{int}} = \delta S - \sum_i F_i \delta X_i$ (F_i denotes the equilibrium value, defined by the thermostat or the reservoir, of the intensive variable conjugate of X_i). The probability of a fluctuation characterized by the variations δS and δX_i of the entropy and of the other extensive variables is thus:

$$\boxed{w \sim \exp\left(\frac{\delta S}{k} - \frac{1}{k}\sum_i F_i \delta X_i\right).} \tag{2A.3.2}$$

The form (2A.3.2) of the distribution w is fully general. It allows us to compute the different moments of the fluctuations of the thermodynamic quantities around their equilibrium values. Using this expression for w, we can in particular demonstrate the identity:

$$\langle \delta X_i \delta F_j \rangle = -k\delta_{ij}, \tag{2A.3.3}$$

which expresses the absence of correlations between the fluctuations of an extensive quantity and those of the intensive quantities other than its conjugate variable.

4. Equilibrium fluctuations in a fluid of N molecules

4.1. The distribution w_N

Let us consider a fluid consisting of one kind of molecules of fixed number N, in equilibrium at constant temperature T and pressure $\mathcal{P}$. On account of formulas (2A.3.2) and (2A.2.4), the probability distribution of the fluctuations, denoted w_N, is given by:

$$w_N \sim \exp\left(-\frac{\delta G}{kT}\right). \tag{2A.4.1}$$

The variation of the Gibbs free energy in the course of a fluctuation characterized by variations δE, δS, δV of energy, entropy, and volume is $\delta G = \delta E - T\delta S + \mathcal{P}\delta V$. We thus have:

$$w_N \sim \exp\left(-\frac{\delta E - T\delta S + \mathcal{P}\delta V}{kT}\right). \tag{2A.4.2}$$

4.2. Small fluctuations case

We shall now determine an approximate form of w_N, valid only for small fluctuations, but more convenient to use in this case than the general expression (2A.4.2).

For small fluctuations, we can expand δE in terms of δS and δV. At leading order, the argument of the exponential in formula (2A.4.2) vanishes, which corresponds to the fact that w_N is maximal at equilibrium. At next order, we get:

$$\delta E - T\delta S + \mathcal{P}\delta V = \frac{1}{2}\left[\frac{\partial^2 E}{\partial S^2}(\delta S)^2 + 2\frac{\partial^2 E}{\partial S\partial V}\delta S\delta V + \frac{\partial^2 E}{\partial V^2}(\delta V)^2\right]. \tag{2A.4.3}$$

The temperature and the pressure of the fluid being respectively given by:

$$T = \left.\frac{\partial E}{\partial S}\right|_V, \qquad \mathcal{P} = -\left.\frac{\partial E}{\partial V}\right|_S, \tag{2A.4.4}$$

we define the fluctuations of these intensive quantities as resulting from those of the extensive quantities S and V:

$$\delta T = \frac{\partial^2 E}{\partial S^2}\delta S + \frac{\partial^2 E}{\partial S\partial V}\delta V, \qquad \delta\mathcal{P} = -\left(\frac{\partial^2 E}{\partial S\partial V}\delta S + \frac{\partial^2 E}{\partial V^2}\delta V\right). \tag{2A.4.5}$$

Thus, formula (2A.4.3) can also be written as:

$$\delta E - T\delta S + \mathcal{P}\delta V = \frac{1}{2}(\delta T\delta S - \delta\mathcal{P}\delta V). \tag{2A.4.6}$$

Importing the expression (2A.4.6) for $\delta E - T\delta S + \mathcal{P}\delta V$ into formula (2A.4.2), we obtain the approximate form of w_N valid for small fluctuations:

$$\boxed{w_N \sim \exp\left(\frac{\delta\mathcal{P}\delta V - \delta T\delta S}{2kT}\right).} \tag{2A.4.7}$$

Formula (2A.4.7) gives the probability distribution of a fluctuation characterized by small variations $\delta\mathcal{P}$, δV, δT, δS of pressure, volume, temperature, and entropy. It allows us to obtain the moments of the equilibrium fluctuations of the various thermodynamic quantities for a fluid with a fixed number of molecules,[2] and this in an exact manner up to order two.

In order to compute these moments, we have to choose a pair of thermodynamic variables, one of a mechanical nature, the other of a thermal one, in terms of which we will express w_N.

- *Variables V and T*

This choice of variables having been made, we express the fluctuations of entropy and pressure in terms of those of volume and temperature. We thus write:[3]

$$\delta S = \frac{Nc_v}{T}\delta T + \frac{\partial\mathcal{P}}{\partial T}\bigg|_V \delta V, \qquad \delta\mathcal{P} = \frac{\partial\mathcal{P}}{\partial T}\bigg|_V \delta T + \frac{\partial\mathcal{P}}{\partial V}\bigg|_T \delta V, \tag{2A.4.8}$$

where c_v denotes the specific heat at constant volume per molecule. Importing the expressions (2A.4.8) for δS and $\delta\mathcal{P}$ into formula (2A.4.7), gives:

$$w_N \sim \exp\left[-\frac{Nc_v}{2kT^2}(\delta T)^2 + \frac{1}{2kT}\frac{\partial\mathcal{P}}{\partial V}\bigg|_T (\delta V)^2\right]. \tag{2A.4.9}$$

Thus, the probability of a small fluctuation is approximately a Gaussian distribution[4] of the two variables δT and δV. The temperature and volume fluctuations are not correlated:

$$\boxed{\langle \delta T \delta V\rangle = 0.} \tag{2A.4.10}$$

The mean square fluctuations of temperature and volume are given by:

$$\boxed{\langle(\delta T)^2\rangle = \frac{kT^2}{Nc_v}, \qquad \langle(\delta V)^2\rangle = kTV\chi_T.} \tag{2A.4.11}$$

---

[2] If the number of molecules of the fluid is allowed to vary, we can write an analogous approximate formula for the probability w of a fluctuation,

$$w \sim \exp\left(\frac{\delta\mathcal{P}\delta V - \delta T\delta S - \delta\mu\delta N}{2kT}\right),$$

where μ denotes the chemical potential per molecule. The physical conditions being different, the moments computed with the aid of w are not necessarily identical to those obtained with the aid of w_N.

[3] In the expression for δS, we have taken into account the Maxwell relation $(\partial S/\partial V)_T = (\partial\mathcal{P}/\partial T)_V$.

[4] This approximate form of w_N allows us to compute exactly the second-order moments of the fluctuations, but not their higher-order moments.

In formula (2A.4.11), $\chi_T = -(1/V)(\partial V/\partial \mathcal{P})_T$ is the isothermal compressibility of the fluid.

Using δT_{eq} and δV_{eq} to denote the average equilibrium fluctuations of temperature and volume,[5] we have:

$$\frac{\delta T_{\text{eq}}}{T} = \left(\frac{k}{Nc_v}\right)^{1/2}, \qquad \frac{\delta V_{\text{eq}}}{V} = \left(\frac{kT\chi_T}{V}\right)^{1/2}. \tag{2A.4.12}$$

In practice, rather than in the volume fluctuations, we are interested in the fluctuations of the density $n = N/V$. The number of molecules being fixed, we have, using δn_{eq} to denote the average equilibrium fluctuation of the density:

$$\frac{\delta n_{\text{eq}}}{n} = \frac{\delta V_{\text{eq}}}{V} = \left(\frac{kT\chi_T}{V}\right)^{1/2}. \tag{2A.4.13}$$

We deduce from equation (2A.4.13) the relative fluctuation of the number of molecules $N_\lambda = nV_\lambda$ in a local equilibrium cell of fixed volume V_λ within a fluid in local equilibrium:

$$\frac{\delta N_\lambda}{N_\lambda} = \left(\frac{nkT\chi_T}{N_\lambda}\right)^{1/2}. \tag{2A.4.14}$$

- *Variables $\mathcal{P}$ and S*

We express the fluctuations of volume and temperature in terms of those of pressure and entropy. We thus write:[6]

$$\delta V = \frac{\partial T}{\partial \mathcal{P}}\bigg|_S \delta S + \frac{\partial V}{\partial \mathcal{P}}\bigg|_S \delta\mathcal{P}, \qquad \delta T = \frac{T}{Nc_p}\delta S + \frac{\partial T}{\partial \mathcal{P}}\bigg|_S \delta\mathcal{P}, \tag{2A.4.15}$$

where c_p denotes the specific heat at constant pressure per molecule. Importing the expressions (2A.4.15) for δV and δT into formula (2A.4.7), gives:

$$w_N \sim \exp\left[\frac{1}{2kT}\frac{\partial V}{\partial \mathcal{P}}\bigg|_S (\delta\mathcal{P})^2 - \frac{1}{2kNc_p}(\delta S)^2\right]. \tag{2A.4.16}$$

The distribution w_N now appears as a Gaussian distribution of the two variables $\delta\mathcal{P}$ and δS. The pressure and entropy fluctuations are not correlated:

$$\boxed{\langle \delta\mathcal{P}\delta S\rangle = 0.} \tag{2A.4.17}$$

---

[5] The root-mean-square deviations or average equilibrium fluctuations δT_{eq} and δV_{eq} are defined by the formulas:

$$(\delta T_{\text{eq}})^2 = \langle(\delta T)^2\rangle, \qquad (\delta V_{\text{eq}})^2 = \langle(\delta V)^2\rangle.$$

[6] In the expression for δV, we have taken into account the Maxwell relation $(\partial V/\partial S)_T = (\partial T/\partial \mathcal{P})_S$.

The mean square fluctuations of pressure and entropy are given by:[7]

$$\langle(\delta\mathcal{P})^2\rangle = \frac{kTc_p}{Vc_v\chi_T}, \qquad \langle(\delta S)^2\rangle = kNc_p. \tag{2A.4.18}$$

If we denote by $\delta\mathcal{P}_{\text{eq}}$ the average equilibrium fluctuation of pressure,[8] we have:

$$\frac{\delta\mathcal{P}_{\text{eq}}}{\mathcal{P}} = \left(\frac{kTc_p}{\mathcal{P}^2Vc_v\chi_T}\right)^{1/2}. \tag{2A.4.19}$$

Formulas (2A.4.12) and (2A.4.19) show that, in a fluid of N molecules, the average relative fluctuations of temperature and pressure are of order $N^{-1/2}$. This result plays an important role in the discussion of the local equilibrium criteria for temperature and pressure.

[7] We have made use of the relation between the isentropic and isothermal compressibilities:

$$-\frac{1}{V}\frac{\partial V}{\partial\mathcal{P}}\bigg|_S = \frac{c_v}{c_p}\chi_T.$$

[8] The root-mean-square deviation or average equilibrium fluctuation $\delta\mathcal{P}_{\text{eq}}$ is defined by the formula:

$$(\delta\mathcal{P}_{\text{eq}})^2 = \langle(\delta\mathcal{P})^2\rangle.$$

Bibliography

H.B. CALLEN, *Thermodynamics and an introduction to thermostatistics*, Wiley, New York, second edition, 1985.

D.L. GOODSTEIN, *States of matter*, Prentice-Hall, Englewood Cliffs, 1975. Reprinted, Dover Publications, New York, 2002.

H.J. KREUZER, *Nonequilibrium thermodynamics and its statistical foundations*, Clarendon Press, Oxford, 1981.

L.D. LANDAU and E.M. LIFSHITZ, *Statistical physics*, Butterworth-Heinemann, Oxford, third edition, 1980.

G. NICOLIS and I. PRIGOGINE, *Self-organization in nonequilibrium systems*, Wiley, New York, 1977.

References

A. EINSTEIN, Theorie der Opaleszenz von homogenen Flüssigkeiten und Flüssigkeitsgemischen in der Nähe des kritischen Zustandes, *Annalen der Physik* **33**, 1275 (1910).

H.B. CALLEN, Thermodynamic fluctuations, *Non-equilibrium thermodynamics, variational techniques, and stability* (R.J. DONNELLY, R. HERMAN, and I. PRIGOGINE editors), The University of Chicago Press, Chicago, 1965.

Supplement 2B

Thermoelectric effects

1. Introduction

The thermoelectric effects are phenomena associated with the simultaneous presence of an electric current and a heat flux¡ in a system, in practice a metal or a semiconductor. These effects were studied in 1854 by W. Thomson (later Lord Kelvin).

We assume, for the sake of simplicity, that there exists in the conductor only one type of carrier, with charge q. The system is assumed to be macroscopically neutral (the carrier density is thus uniform). The temperature $T(\boldsymbol{r})$ and the electrostatic potential $\phi(\boldsymbol{r})$ may vary from one point to another inside the sample, but they remain constant over the course of time. Since the density is uniform, the local chemical potential depends on $\boldsymbol{r}$ solely via the local temperature: it is accordingly denoted by $\mu[T(\boldsymbol{r})]$. The local electrochemical potential is:

$$\overline{\mu}(\boldsymbol{r}) = \mu\big[T(\boldsymbol{r})\big] + q\phi(\boldsymbol{r}). \tag{2B.1.1}$$

The inhomogeneities of temperature and electrochemical potential give rise to a particle flux $\boldsymbol{J}_N$ and an energy flux $\boldsymbol{J}_E$. The conjugate affinities of $\boldsymbol{J}_N$ and $\boldsymbol{J}_E$ are respectively $\nabla(-\overline{\mu}/T)$ and $\nabla(1/T)$. The linear response relations between the fluxes and the affinities are of the form:[1]

$$\begin{cases} \boldsymbol{J}_N = L_{NN}\nabla(-\dfrac{\overline{\mu}}{T}) + L_{NE}\nabla(\dfrac{1}{T}) \\ \boldsymbol{J}_E = L_{EN}\nabla(-\dfrac{\overline{\mu}}{T}) + L_{EE}\nabla(\dfrac{1}{T}). \end{cases} \tag{2B.1.2}$$

The Onsager reciprocity relation is a symmetry relation: $L_{EN} = L_{NE}$. The equations (2B.1.2) determine the transport properties of the conductor in a linear regime. The entropy source σ_S reads:

$$\sigma_S = \boldsymbol{J}_E.\nabla(\frac{1}{T}) + \boldsymbol{J}_N.\nabla(-\frac{\overline{\mu}}{T}). \tag{2B.1.3}$$

[1] For the sake of simplicity, we consider a medium in which the fluxes as well as the affinities are parallel to the same given direction. It may be either an isotropic medium or a one-dimensional medium (fluxes produced in wires or in bars), a geometry well adapted to the analysis of thermoelectric effects. The kinetic coefficients are thus scalars.

2. The entropy source

However, in practice, rather than the energy flux $\boldsymbol{J}_E$, we are interested in the heat flux $\boldsymbol{J}_Q^* = T\boldsymbol{J}_S$, where $\boldsymbol{J}_S$ denotes the entropy flux, defined as $T\boldsymbol{J}_S = \boldsymbol{J}_E - \overline{\mu}\boldsymbol{J}_N$. We have:

$$\boldsymbol{J}_Q^* = \boldsymbol{J}_E - \overline{\mu}\boldsymbol{J}_N. \tag{2B.2.1}$$

The flux $\boldsymbol{J}_Q^*$ contains, besides the heat conduction term denoted usually by $\boldsymbol{J}_Q$, a term describing the heat transport due to convection. Both fluxes $\boldsymbol{J}_Q^*$ and $\boldsymbol{J}_Q$ are related through:

$$\boldsymbol{J}_Q^* = \boldsymbol{J}_Q + Ts_p\boldsymbol{J}_N, \tag{2B.2.2}$$

where s_p denotes the entropy per particle.[2] The term $Ts_p\boldsymbol{J}_N$ represents the convective contribution to $\boldsymbol{J}_Q^*$.

It is possible to rewrite the entropy source choosing fluxes $\boldsymbol{J}_N$ and $\boldsymbol{J}_Q^*$, in place of $\boldsymbol{J}_N$ and $\boldsymbol{J}_E$. We obtain:

$$\sigma_S = \boldsymbol{J}_Q^*.\nabla(\frac{1}{T}) - \boldsymbol{J}_N.\frac{1}{T}\nabla\overline{\mu}. \tag{2B.2.3}$$

The conjugate affinities of $\boldsymbol{J}_N$ and $\boldsymbol{J}_Q^*$ are respectively $-(1/T)\nabla\overline{\mu}$ and $\nabla(1/T)$. In the linear regime, we write response relations of the form:

$$\begin{cases} \boldsymbol{J}_N = -L_{11}\dfrac{1}{T}\nabla\overline{\mu} + L_{12}\nabla(\dfrac{1}{T}) \\ \\ \boldsymbol{J}_Q^* = -L_{21}\dfrac{1}{T}\nabla\overline{\mu} + L_{22}\nabla(\dfrac{1}{T}). \end{cases} \tag{2B.2.4}$$

Formulas (2B.2.4) must be supplemented by the Onsager symmetry relation $L_{12} = L_{21}$.

Before analyzing the various thermoelectric effects, let us compare both families of kinetic coefficients appearing in the systems of equations (2B.1.2) and (2B.2.4). Replacing $\boldsymbol{J}_Q^*$ in the second of equations (2B.2.4) by its definition (2B.2.1), we can show that the two families of kinetic coefficients are related through the formulas:

$$\begin{cases} L_{11} = L_{NN} \\ L_{12} = L_{NE} - \overline{\mu}L_{NN}, \qquad L_{21} = L_{EN} - \overline{\mu}L_{NN} \\ L_{22} = L_{EE} - \overline{\mu}(L_{EN} + L_{NE}) + \overline{\mu}^2 L_{NN}, \end{cases} \tag{2B.2.5}$$

and, inversely:

$$\begin{cases} L_{NN} = L_{11} \\ L_{NE} = L_{12} + \overline{\mu}L_{11}, \qquad L_{EN} = L_{21} + \overline{\mu}L_{11} \\ L_{EE} = L_{22} + \overline{\mu}(L_{12} + L_{21}) + \overline{\mu}^2 L_{11}. \end{cases} \tag{2B.2.6}$$

---

[2] The relation (2B.2.2) between the fluxes $\boldsymbol{J}_Q^*$ and $\boldsymbol{J}_Q$ will be established in Subsection 6.3.

3. Isothermal electrical conduction

3.1. Ohm's law

When the temperature is uniform, the electric current density is:

$$\boldsymbol{J} = q\boldsymbol{J}_N = -qL_{11}\frac{1}{T}\nabla\overline{\mu}, \tag{2B.3.1}$$

with $\nabla\overline{\mu} = -q\boldsymbol{E}$ ($\boldsymbol{E} = -\nabla\phi(\boldsymbol{r})$ is the electric field). Relation (2B.3.1) is nothing but Ohm's law $\boldsymbol{J} = \sigma\boldsymbol{E}$. The electrical conductivity σ is thus related to L_{11}:

$$\sigma = \frac{q^2}{T}L_{11}. \tag{2B.3.2}$$

We can also write, in virtue of the identity between L_{11} and L_{NN} as displayed by formulas (2B.2.5) and (2B.2.6):

$$\sigma = \frac{q^2}{T}L_{NN}. \tag{2B.3.3}$$

3.2. The Joule effect

The *Joule effect* is connected with the passage of an electric current in a conductor placed under conditions of uniform electric field, uniform carrier density, and uniform temperature. If the temperature is maintained constant during the course of time, the energy provided to the system is released in the form of heat in the external medium. The Joule power dissipated per unit volume, $(dQ/dt)_{\text{Joule}}$, is given by the energy balance:

$$\left.\frac{dQ}{dt}\right|_{\text{Joule}} = -\nabla.\boldsymbol{J}_E = -\nabla.(\boldsymbol{J}_Q^* + \overline{\mu}\boldsymbol{J}_N). \tag{2B.3.4}$$

The temperature being uniform and the medium macroscopically neutral, we have $\nabla.\boldsymbol{J}_Q^* = 0$ and $\nabla.\boldsymbol{J}_N = 0$. From the relation:

$$\nabla\overline{\mu} = -\frac{q^2}{\sigma}\boldsymbol{J}_N, \tag{2B.3.5}$$

we obtain the expression for the Joule power dissipated per unit volume:

$$\boxed{\left.\frac{dQ}{dt}\right|_{\text{Joule}} = \frac{1}{\sigma}\boldsymbol{J}^2.} \tag{2B.3.6}$$

The divergence of the entropy flux vanishes ($\nabla.\boldsymbol{J}_S = 0$), and the entropy source (2B.2.3) reads:

$$\sigma_S = \frac{1}{\sigma T}\boldsymbol{J}^2. \tag{2B.3.7}$$

The local balance equation for the entropy is thus:

$$\frac{\partial(\rho s)}{\partial t} = \frac{1}{\sigma T}\boldsymbol{J}^2. \tag{2B.3.8}$$

This entropy production is at the root of the Joule effect, and it expresses the irreversibility of the charge transport. It is characterized by the electrical conductivity σ, a positive quantity. The electrical conductivity is a transport coefficient or *dissipative coefficient* (and not a thermodynamic equilibrium quantity).

4. Open-circuit thermal conduction

Consider an open circuit, in which consequently $\boldsymbol{J} = q\boldsymbol{J}_N = 0$. In the presence of a temperature gradient ∇T, there is an electrochemical potential gradient given, according to the first of equations (2B.2.4), by:

$$\nabla\overline{\mu} = -\frac{1}{T}\frac{L_{12}}{L_{11}}\nabla T. \qquad \text{(2B.4.1)}$$

Since there is no particle flux, both fluxes $\boldsymbol{J}_Q^*$ and $\boldsymbol{J}_Q$ are identical. Importing the expression (2B.4.1) for $\nabla\overline{\mu}$ into the second of equations (2B.2.4), we obtain Fourier's law of heat conduction,

$$\boldsymbol{J}_Q^* = \boldsymbol{J}_Q = -\kappa\nabla T, \qquad \text{(2B.4.2)}$$

with:

$$\kappa = \frac{1}{T^2}\frac{L_{11}L_{22} - L_{12}L_{21}}{L_{11}}. \qquad \text{(2B.4.3)}$$

The coefficient κ, a positive quantity, is the open-circuit thermal conductivity. Using formulas (2B.2.5), we can also write κ in the form:[3]

$$\kappa = \frac{1}{T^2}\frac{L_{EE}L_{NN} - L_{NE}L_{EN}}{L_{NN}}. \qquad \text{(2B.4.4)}$$

5. The Seebeck effect

In an open circuit, a temperature gradient ∇T is accompanied by an electrochemical potential gradient $\nabla\overline{\mu}$ given by formula (2B.4.1). This is the *Seebeck effect*, discovered by T. Seebeck in 1821. We usually set:

$$\frac{1}{q}\nabla\overline{\mu} = -\eta\nabla T, \qquad \text{(2B.5.1)}$$

where η denotes the *thermoelectric power* or *Seebeck coefficient* of the material. Comparing equations (2B.4.1) and (2B.5.1) gives:

$$\eta = \frac{1}{qT}\frac{L_{12}}{L_{11}}. \qquad \text{(2B.5.2)}$$

The Seebeck effect exists only if L_{12} is non-zero. In that sense it is an indirect effect. We can also write:

$$\eta = \frac{1}{qT}\left(\frac{L_{NE}}{L_{NN}} - \overline{\mu}\right). \qquad \text{(2B.5.3)}$$

[3] In a conductor, thermal conduction is mainly due to charge carriers, that is, to particles whose number is conserved. The principal mechanism coming into play is thus different from the one at work in an insulating solid (in this latter case, heat transport is due solely to phonons, whose number is not conserved, and we have $\kappa = L_{EE}/T^2$).

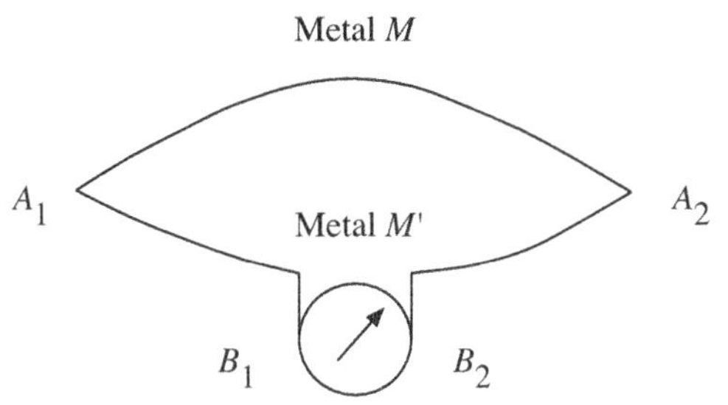

Fig. 2B.1 Schematic drawing of a thermocouple.

To observe the Seebeck effect, we realize the thermocouple represented in Fig. 2B.1. It is a circuit made of two different conductors, namely, two wires of thermopowers η and η', welded together at points A_1 and A_2. The welds are brought up to unequal temperatures T_1 and T_2. The circuit being open, we measure between points B_1 and B_2 the potential difference:

$$\phi_2 - \phi_1 = \frac{1}{q} \int_{B_1}^{B_2} \nabla\overline{\mu}.\boldsymbol{dl}, \tag{2B.5.4}$$

the integral being taken along the circuit $B_1A_1A_2B_2$. Taking into account formula (2B.5.1), we get:

$$\boxed{\phi_2 - \phi_1 = \int_{T_1}^{T_2} (\eta' - \eta)\, dT.} \tag{2B.5.5}$$

The potential difference $\phi_2 - \phi_1$ is an electrochemical potential difference.[4]

6. The Peltier effect

The *Peltier effect*, discovered in 1834, is the indirect effect inverse of the Seebeck effect, in the sense that it involves the kinetic coefficient L_{21} (or L_{EN}), whereas the Seebeck effect involves L_{12} (or L_{NE}).

6.1. Description

At uniform temperature, an electric current of density $\boldsymbol{J} = q\boldsymbol{J}_N$ is accompanied by a heat flux:

$$\boldsymbol{J}_Q^* = \pi \boldsymbol{J}, \tag{2B.6.1}$$

[4] This potential difference is measured with the aid of a highly resistive voltmeter, so that the current in the circuit is negligible. Once calibrated, the device allows us to measure the temperature of A_2, that of A_1 being fixed at a reference temperature T_1. The measurement is independent of the ambient temperature.

where π denotes the *Peltier coefficient*. We have, according to equations (2B.2.4),

$$\pi = \frac{1}{q}\frac{L_{21}}{L_{11}}, \tag{2B.6.2}$$

or:

$$\pi = \frac{1}{q}\left(\frac{L_{EN}}{L_{NN}} - \overline{\mu}\right). \tag{2B.6.3}$$

To observe the Peltier effect, we consider an isothermal junction between two different conductors M and M', through which an electric current of density $\boldsymbol{J}$ flows. The junction receives a heat flux $\pi\boldsymbol{J}$ from the side of M, and loses a heat flux $\pi'\boldsymbol{J}$ to the side of M'. This results in an absorption or a release of heat at the interface, $(dQ/dt)_{\text{Peltier}}$, equal, per unit time and unit area, to:

$$\boxed{\left.\frac{dQ}{dt}\right|_{\text{Peltier}} = (\pi - \pi')J.} \tag{2B.6.4}$$

The Peltier effect is linear in $\boldsymbol{J}$; depending on the direction of the electric current, an absorption or a release of heat[5] takes place at the junction.

6.2. The second Kelvin relation

Comparing formulas (2B.5.2) for the thermoelectric power and (2B.6.2) for the Peltier coefficient, and taking into account the Onsager relation $L_{12} = L_{21}$, we obtain the *second Kelvin relation*, demonstrated on an empirical basis in 1854:

$$\boxed{\pi = \eta T.} \tag{2B.6.5}$$

Thus, the four kinetic coefficients associated with the thermoelectric effects may be expressed with the aid of three independent experimentally measurable dissipative coefficients, such as σ, κ, and η.

6.3. Relation between the heat fluxes $\boldsymbol{J}_Q$ and $\boldsymbol{J}_Q^*$

Eliminating $\nabla\overline{\mu}$ between the expressions (2B.2.4) for $\boldsymbol{J}_N$ and $\boldsymbol{J}_Q^*$, we get, taking into account the second Kelvin relation (2B.6.5):

$$\boldsymbol{J}_Q^* = -\kappa\nabla T + \eta q T\boldsymbol{J}_N. \tag{2B.6.6}$$

As displayed by formula (2B.6.6), the flux $\boldsymbol{J}_Q^* = T\boldsymbol{J}_S$ actually appears as the sum of two terms. The first, $\boldsymbol{J}_Q = -\kappa\nabla T$, corresponds to heat transport due to thermal conduction in the presence of the imposed temperature gradient. The second, $\eta qT\boldsymbol{J}_N$, results from convective heat transport due to the drift of electrical charges. Everything thus happens as if each charge carrier carried with it an entropy $s_p = \eta q$ (formula (2B.2.2)).

---

[5] Refrigerating or temperature regulation devices based on the Peltier effect have been built since the 1960s. They make use of semiconducting materials.

7. The Thomson effect

When an electric current flows in a conducting bar of inhomogeneous temperature $T(\boldsymbol{r})$, the amount of heat exchanged between the bar and the external medium is the sum of three contributions, corresponding to thermal conduction, Joule effect, and a supplementary effect, the *Thomson effect*.

In the presence of a uniform electric current of density $\boldsymbol{J} = q\boldsymbol{J}_N$, the energy flux flowing in the conducting bar is, according to formulas (2B.2.1) and (2B.6.6):

$$\boldsymbol{J}_E = \boldsymbol{J}_Q^* + \overline{\mu}\boldsymbol{J}_N = -\kappa\nabla T + (\overline{\mu} + \eta qT)\boldsymbol{J}_N. \tag{2B.7.1}$$

The power exchanged with the outside per unit volume, given by the energy balance:

$$\frac{dQ}{dt} = -\nabla.\boldsymbol{J}_E = \nabla.(\kappa\nabla T) - \nabla.\big[(\overline{\mu} + \eta qT)\boldsymbol{J}_N\big], \tag{2B.7.2}$$

reads, since the flux $\boldsymbol{J}_N$ is uniform:

$$\frac{dQ}{dt} = \nabla.(\kappa\nabla T) - \boldsymbol{J}_N.\nabla(\overline{\mu} + \eta qT). \tag{2B.7.3}$$

The expression for $\nabla\overline{\mu}$ is obtained from the first of equations (2B.2.4), and formulas (2B.3.2) and (2B.5.2):

$$\nabla\overline{\mu} = -\eta q\nabla T - \frac{q^2}{\sigma}\boldsymbol{J}_N. \tag{2B.7.4}$$

We thus have:

$$\frac{dQ}{dt} = \nabla.(\kappa\nabla T) - qT\boldsymbol{J}_N.\nabla\eta + \frac{q^2}{\sigma}\boldsymbol{J}_N^2, \tag{2B.7.5}$$

that is, the thermopower being a function of the local temperature:

$$\frac{dQ}{dt} = \nabla.(\kappa\nabla T) - T\frac{d\eta}{dT}\boldsymbol{J}.\nabla T + \frac{1}{\sigma}\boldsymbol{J}^2. \tag{2B.7.6}$$

In the expression (2B.7.6) for dQ/dt, the first term represents the power exchanged with the outside due to thermal conduction. The third term represents the power dissipated by the Joule effect. As for the second term, it corresponds to the Thomson effect: it represents the power exchanged with the external medium when an electric current of density $\boldsymbol{J}$ goes through the temperature gradient ∇T. The *Thomson coefficient* α is defined as the Thomson power per unit electric current density and unit temperature:

$$\boxed{\left.\frac{dQ}{dt}\right|_{\text{Thomson}} = -\alpha\boldsymbol{J}.\nabla T.} \tag{2B.7.7}$$

In contrast to the Joule effect, the Thomson effect is linear in $\boldsymbol{J}$; depending on the current direction, there is an absorption or a release of heat. According to formula (2B.7.6), the Thomson coefficient is related to the Seebeck coefficient:

$$\alpha = T\frac{d\eta}{dT}. \tag{2B.7.8}$$

We also have, on account of formula (2B.6.5):

$$\boxed{\alpha = \frac{d\pi}{dT} - \eta.} \tag{2B.7.9}$$

Formula (2B.7.9) is the *first Kelvin relation*. It can also be established directly with the aid of the first law of thermodynamics.

Purely thermodynamic considerations do not allow us to go further. To obtain explicit expressions for the three dissipative coefficients σ, κ, and η, we have to turn to a microscopic theory such as the Boltzmann transport equation[6] or the Kubo linear response theory.[7]

8. An illustration of the minimum entropy production theorem

The entropy source corresponding to thermoelectric effects is given, for instance, by formula (2B.2.3). The fluxes $\boldsymbol{J}_N$ and $\boldsymbol{J}_Q^*$ are related to the affinities $-(1/T)\nabla\overline{\mu}$ and $\nabla(1/T)$ through the linear laws (2B.2.4). Using these formulas, as well as the symmetry relation $L_{12} = L_{21}$, we rewrite σ_S in the form:

$$\sigma_S = L_{11}\frac{1}{T^2}(\nabla\overline{\mu})^2 - 2L_{12}\frac{1}{T}\nabla\overline{\mu}\cdot\nabla(\frac{1}{T}) + L_{22}\Big[\nabla(\frac{1}{T})\Big]^2. \tag{2B.8.1}$$

Let us consider as an example the stationary state with no particle current. In this state, we have:

$$\boldsymbol{J}_N = -L_{11}\frac{1}{T}\nabla\overline{\mu} + L_{12}\nabla(\frac{1}{T}) = 0. \tag{2B.8.2}$$

It is possible to retrieve equation (2B.8.2) as resulting from the condition according to which σ_S should be minimum at fixed temperature gradient. This latter condition reads:

$$\frac{\partial\sigma_S}{\partial(\nabla\overline{\mu}/T)}\bigg|_{\nabla(1/T)} = 0. \tag{2B.8.3}$$

Since we have:

$$\frac{\partial\sigma_s}{\partial(\nabla\overline{\mu}/T)}\bigg|_{\nabla(1/T)} = -2\Big[-L_{11}\frac{1}{T}\nabla\overline{\mu} + L_{12}\nabla(\frac{1}{T})\Big] = -2\boldsymbol{J}_N, \tag{2B.8.4}$$

both conditions $\boldsymbol{J}_N = 0$ and $\partial\sigma_S/\partial(\nabla\overline{\mu}/T)|_{\nabla(1/T)} = 0$ are indeed equivalent. The stationary state with no particle current is thus, at fixed ∇T, a state of minimum entropy production. At fixed temperature gradient, the system establishes an electrochemical potential gradient such that no particle transport takes place. With respect to this parameter, the entropy production has an extremum. This extremum can only be a minimum. Indeed the entropy source which, in a linear system, is a quadratic form of the type $\sigma_S = \sum_{ik} L_{ik}\mathcal{F}_i\mathcal{F}_k$, must be non-negative.

It can be shown that this stationary state is stable with respect to small local perturbations.

---

[6] See Chapter 8 and Supplement 8B.

[7] See Chapters 15 and 16.

Bibliography

R. BALIAN, *From microphysics to macrophysics*, Vol. 2, Springer-Verlag, Berlin, 1992.

H.B. CALLEN, *Thermodynamics and an introduction to thermostatistics*, Wiley, New York, second edition, 1985.

S.R. DE GROOT and P. MAZUR, *Non-equilibrium thermodynamics*, North-Holland, Amsterdam, 1962. Reprinted, Dover Publications, New York, 1984.

R. HAASE, *Thermodynamics of irreversible processes*, Addison Wesley, Reading, 1969. Reprinted, Dover Publications, New York, 1990.

H.J. KREUZER, *Nonequilibrium thermodynamics and its statistical foundations*, Clarendon Press, Oxford, 1981.

Supplement 2C

Thermodiffusion in a fluid mixture

1. Introduction

When put in the presence of a temperature gradient, the constituents of a uniform mixture manifest a tendency towards separation. This indirect phenomenon is called the *thermodiffusion*, or the *Soret effect*. It takes place in gases and in liquids, and also in solids. The inverse effect of the Soret effect is called the *Dufour effect*. This relates to the fact that a concentration gradient imposed to a mixture, besides the diffusion current which tends to re-establish homogeneity, also produces a heat flux.

We shall describe here these effects in the simple case of a binary fluid mixture in the absence of external forces, viscous phenomena, and chemical reactions.

2. Diffusive fluxes in a binary mixture

We consider a fluid mixture, liquid or gaseous, made up of two constituents A and B. Since there are no external forces, the pressure is uniform when local equilibrium is attained. We denote by n_A and n_B the numbers of molecules of species A and B per unit volume of the fluid, and by c the concentration in A defined by:

$$c = \frac{n_A}{n_A + n_B} = \frac{n_A}{n}. \tag{2C.2.1}$$

The concentration varies over the course of time and also from one point of space to another, because of diffusion and of the fluid macroscopic motion. We denote by $\boldsymbol{u}_A$ and $\boldsymbol{u}_B$ the local mean velocities of the constituents A and B, and by $\boldsymbol{u}$ the local barycentric velocity defined by:

$$\boldsymbol{u} = c\boldsymbol{u}_A + (1-c)\boldsymbol{u}_B. \tag{2C.2.2}$$

We also introduce the diffusive fluxes of A and B with respect to the local barycentric velocity:

$$\boldsymbol{J}_A^{\text{diff}} = nc(\boldsymbol{u}_A - \boldsymbol{u}), \qquad \boldsymbol{J}_B^{\text{diff}} = n(1-c)(\boldsymbol{u}_B - \boldsymbol{u}). \tag{2C.2.3}$$

These fluxes are not independent. Indeed we have:

$$\boldsymbol{J}_A^{\text{diff}} = -\boldsymbol{J}_B^{\text{diff}}. \tag{2C.2.4}$$

If the fluid consists of molecules all of the same species, there is no diffusion flux. If for instance $c = 0$, there are solely B molecules in the mixture. Then we have $\boldsymbol{u}_B = \boldsymbol{u}$. The diffusive fluxes thus vanish: $\boldsymbol{J}_A^{\text{diff}} = -\boldsymbol{J}_B^{\text{diff}} = 0$. A similar reasoning holds in the case $c = 1$.

3. The entropy source

The entropy source in the fluid mixture is expressed as:

$$\sigma_S = \boldsymbol{J}_E.\nabla\Big(\frac{1}{T}\Big) + \boldsymbol{J}_A.\nabla\Big(-\frac{\mu_A}{T}\Big) + \boldsymbol{J}_B.\nabla\Big(-\frac{\mu_B}{T}\Big), \tag{2C.3.1}$$

where $\boldsymbol{J}_E$ is the energy flux and $\boldsymbol{J}_k$ the flux of the constituent k ($k = A$ or $k = B$). In formula (2C.3.1), T denotes the local temperature and μ_k the local chemical potential per molecule of the constituent k.

The flux of A is the sum of a convective flux and the diffusive flux $\boldsymbol{J}_A^{\text{diff}}$:

$$\boldsymbol{J}_A = nc\boldsymbol{u} + \boldsymbol{J}_A^{\text{diff}}. \tag{2C.3.2}$$

In the same way, we have:

$$\boldsymbol{J}_B = n(1-c)\boldsymbol{u} + \boldsymbol{J}_B^{\text{diff}}. \tag{2C.3.3}$$

Using the Gibbs–Helmholtz relation:

$$\nabla\Big(\frac{\mu_k}{T}\Big) = \frac{1}{T}\nabla\mu_k\big|_T + h_k\nabla\Big(\frac{1}{T}\Big), \tag{2C.3.4}$$

where $\nabla\mu_k|_T$ denotes the isothermal gradient of μ_k, and h_k the enthalpy per molecule of the constituent k, we obtain:

$$\sigma_S = \boldsymbol{J}_E.\nabla\Big(\frac{1}{T}\Big) - n\boldsymbol{u}.\Big(c\nabla\mu_A\big|_T + (1-c)\nabla\mu_B\big|_T + \big[ch_A + (1-c)h_B\big]\nabla\Big(\frac{1}{T}\Big)\Big) + \boldsymbol{J}_A^{\text{diff}}.\nabla\Big(-\frac{\mu_A}{T}\Big) + \boldsymbol{J}_B^{\text{diff}}.\nabla\Big(-\frac{\mu_B}{T}\Big). \tag{2C.3.5}$$

The pressure being uniform, the Gibbs–Duhem relation for the mixture reads:

$$c\nabla\mu_A\big|_T + (1-c)\nabla\mu_B\big|_T = 0. \tag{2C.3.6}$$

We thus get:

$$\sigma_S = (\boldsymbol{J}_E - nh\boldsymbol{u}).\nabla\Big(\frac{1}{T}\Big) + \boldsymbol{J}_A^{\text{diff}}.\nabla\Big(-\frac{\mu_A}{T}\Big) + \boldsymbol{J}_B^{\text{diff}}.\nabla\Big(-\frac{\mu_B}{T}\Big), \tag{2C.3.7}$$

where:

$$h = ch_A + (1 - c)h_B \tag{2C.3.8}$$

is the enthalpy per molecule of the mixture. Introducing the heat flux[1] $\boldsymbol{J}_Q = \boldsymbol{J}_E - nh\boldsymbol{u}$, we obtain an expression for the entropy source involving the fluxes $\boldsymbol{J}_Q$, $\boldsymbol{J}_A^{\text{diff}}$, and $\boldsymbol{J}_B^{\text{diff}}$:

$$\sigma_S = \boldsymbol{J}_Q.\nabla\left(\frac{1}{T}\right) + \boldsymbol{J}_A^{\text{diff}}.\nabla\left(-\frac{\mu_A}{T}\right) + \boldsymbol{J}_B^{\text{diff}}.\nabla\left(-\frac{\mu_B}{T}\right). \tag{2C.3.9}$$

Introducing the *diffusion flux* $\boldsymbol{J}^{\text{diff}} = \boldsymbol{J}_A^{\text{diff}} = -\boldsymbol{J}_B^{\text{diff}}$ and the *chemical potential of the mixture* $\mu = \mu_A - \mu_B$, we can rewrite σ_S as:

$$\sigma_S = \boldsymbol{J}_Q.\nabla\left(\frac{1}{T}\right) + \boldsymbol{J}^{\text{diff}}.\nabla\left(-\frac{\mu}{T}\right), \tag{2C.3.10}$$

or, equivalently, as:[2]

$$\sigma_S = \left(\boldsymbol{J}_Q - \mu\boldsymbol{J}^{\text{diff}}\right).\nabla\left(\frac{1}{T}\right) - \boldsymbol{J}^{\text{diff}}.\frac{1}{T}\nabla\mu. \tag{2C.3.11}$$

4. Linear relations between fluxes and affinities

The expression (2C.3.11) for σ_S involves the fluxes $\boldsymbol{J}_Q - \mu\boldsymbol{J}^{\text{diff}}$ and $\boldsymbol{J}^{\text{diff}}$, whose conjugate affinities are respectively $\nabla(1/T)$ and $-(1/T)\nabla\mu$. The linear relations between these fluxes and these affinities read:

$$\begin{cases} \boldsymbol{J}^{\text{diff}} = -L_{11}\dfrac{1}{T}\nabla\mu + L_{12}\nabla\left(\dfrac{1}{T}\right) \\[2ex] \boldsymbol{J}_Q - \mu\boldsymbol{J}^{\text{diff}} = -L_{21}\dfrac{1}{T}\nabla\mu + L_{22}\nabla\left(\dfrac{1}{T}\right), \end{cases} \tag{2C.4.1}$$

---

[1] We have:

$$\boldsymbol{J}_Q = \boldsymbol{J}_E - h_A nc\boldsymbol{u} - h_B n(1-c)\boldsymbol{u}.$$

Besides, setting $\boldsymbol{J}_Q^* = T\boldsymbol{J}_S$, we have:

$$\boldsymbol{J}_Q^* = \boldsymbol{J}_E - \mu_A\boldsymbol{J}_A - \mu_B\boldsymbol{J}_B.$$

The heat flux due to thermal conduction, denoted here by $\boldsymbol{J}_Q^{\text{cond}}$, is defined by:

$$\boldsymbol{J}_Q^{\text{cond}} = \boldsymbol{J}_Q^* - Ts_A\boldsymbol{J}_A - Ts_B\boldsymbol{J}_B,$$

where s_k is the entropy per molecule of the constituent k. We have the relation:

$$\boldsymbol{J}_Q^{\text{cond}} = \boldsymbol{J}_Q - h_A\boldsymbol{J}_A^{\text{diff}} - h_B\boldsymbol{J}_B^{\text{diff}}.$$

The difference between $\boldsymbol{J}_Q$ and $\boldsymbol{J}_Q^{\text{cond}}$ represents that part of the heat transport which is due to matter diffusion.

[2] In a pure fluid, the diffusive fluxes vanish, and the entropy source reduces to its thermal part (we do not take into account here the viscous phenomena):

$$\sigma_S = \boldsymbol{J}_Q.\nabla\left(\frac{1}{T}\right).$$

This expression for σ_S will be encountered in Chapter 16, in which we calculate microscopically the thermal conductivity of a fluid.

where the kinetic coefficients verify the Onsager symmetry relation $L_{12} = L_{21}$.

4.1. Thermal conductivity of the mixture

Eliminating $\nabla\mu$ between the two equations (2C.4.1), we obtain an expression for $\boldsymbol{J}_Q$ in terms of $\boldsymbol{J}^{\text{diff}}$ and $\nabla(1/T)$:

$$\boldsymbol{J}_Q = \Big(\mu + \frac{L_{21}}{L_{11}}\Big)\boldsymbol{J}^{\text{diff}} + \frac{L_{11}L_{22} - L_{12}L_{21}}{L_{11}}\nabla\Big(\frac{1}{T}\Big). \tag{2C.4.2}$$

When the diffusion flux vanishes ($\boldsymbol{J}^{\text{diff}} = 0$), there is solely thermoconduction ($\boldsymbol{J}_Q = -\kappa\nabla T$). This allows us to deduce the expression for the thermal conductivity of the fluid mixture in terms of the kinetic coefficients:

$$\kappa = \frac{1}{T^2}\frac{L_{11}L_{22} - L_{12}L_{21}}{L_{11}}. \tag{2C.4.3}$$

4.2. Diffusion and thermodiffusion coefficients

The chemical potential μ of a binary fluid mixture depends a priori on the pressure, the temperature, and the concentration. Since the pressure is uniform, we simply have:

$$\nabla\mu = \frac{\partial\mu}{\partial T}\bigg|_{p,c}\nabla T + \frac{\partial\mu}{\partial c}\bigg|_{p,T}\nabla c. \tag{2C.4.4}$$

Introducing the expression (2C.4.4) for $\nabla\mu$ into equations (2C.4.1) and taking into account the Onsager symmetry relation, we obtain for the diffusion flux and the heat flux the expressions:

$$\begin{cases} \boldsymbol{J}^{\text{diff}} = -nD\Big(\nabla c + \dfrac{k_T}{T}\nabla T\Big) \\[2ex] \boldsymbol{J}_Q = \Big(\mu + k_T\dfrac{\partial\mu}{\partial c}\bigg|_{p,T} - T\dfrac{\partial\mu}{\partial T}\bigg|_{p,c}\Big)\boldsymbol{J}^{\text{diff}} - \kappa\nabla T. \end{cases} \tag{2C.4.5}$$

Comparing equations (2C.4.1) and (2C.4.5), we obtain by identification the expressions for D and k_T in terms of the kinetic coefficients and the thermodynamic derivatives. The diffusion coefficient is given by:

$$D = \frac{L_{11}}{n}\frac{1}{T}\frac{\partial\mu}{\partial c}\bigg|_{p,T}. \tag{2C.4.6}$$

The quantity $D_T = k_T D$ is called the *thermodiffusion coefficient*. We have:

$$D_T = \frac{T}{n}\Big(\frac{L_{11}}{T}\frac{\partial\mu}{\partial T}\bigg|_{p,c} + \frac{L_{12}}{T^2}\Big). \tag{2C.4.7}$$

The ratio $k_T = D_T/D$ is termed the *Soret coefficient*.

4.3. Positivity of the entropy source

Coming back to formula (2C.3.11), and using expressions (2C.4.1) for fluxes and formula (2C.4.3) for the thermal conductivity, as well as the Onsager relation $L_{12} = L_{21}$, we get:

$$\sigma_S = \frac{\kappa(\nabla T)^2}{T^2} + \frac{(\boldsymbol{J}^{\text{diff}})^2}{L_{11}}. \tag{2C.4.8}$$

Each term of the entropy source must be separately non-negative. The positivity of the first term implies $\kappa > 0$. As for the second term, its positivity implies $L_{11} > 0$, and therefore, since the thermodynamic derivative $(\partial\mu/\partial c)_{p,T}$ is always positive, $D > 0$.

5. The Soret and Dufour effects

The diffusion flux due to the temperature gradient is determined by the thermodiffusion coefficient D_T. This effect was studied in gases in 1879 by C. Soret. The expression (2C.4.5) for $\boldsymbol{J}^{\text{diff}}$ displays the fact that, in the absence of a diffusion flux, the existence of a temperature gradient leads to the appearance of a concentration gradient. Both gradients are related through:

$$\nabla c = -\frac{k_T}{T}\nabla T. \tag{2C.5.1}$$

In order to have evidence of the Soret effect in a gaseous mixture of average concentration in A equal to c_0, we place it in a closed box between two horizontal plates. We denote by T_{bottom} the temperature of the bottom plate, and by T_{top} that of the top one (we take $T_{\text{top}} > T_{\text{bottom}}$ in order to prevent the appearance of convection currents). We assume that the Soret coefficient has the form:

$$k_T = \alpha c(1 - c) \tag{2C.5.2}$$

(an expression consistent with the fact that no thermodiffusion can take place in a pure fluid). Making the approximation:

$$k_T \simeq \alpha c_0(1 - c_0), \tag{2C.5.3}$$

we obtain for the concentration difference between the top and the bottom of the box the expression:

$$c_{\text{top}} - c_{\text{bottom}} = -\alpha c_0(1 - c_0)\log\frac{T_{\text{top}}}{T_{\text{bottom}}}. \tag{2C.5.4}$$

A substance with a positive (*resp.* negative) Soret coefficient k_T tends to diffuse towards the coldest (*resp.* hottest) region and to accumulate there.[3]

The inverse effect of the Soret effect, that is, the existence of a heat current due to a concentration gradient, was studied in 1872 by L. Dufour.

[3] The Soret effect is thus the basis of a method of separation of isotopic gaseous mixtures (however, in liquid phase, the order of magnitude of the effect is too small to allow us to consider its practical application).

Bibliography

L.D. LANDAU and E.M. LIFSHITZ, *Fluid mechanics*, Butterworth-Heinemann, Oxford, second edition, 1987.

References

L. DUFOUR, Über die Diffusion der Gase durch poröse Wände und die sie begleitenden Temperaturveränderungen, *Annalen der Physik* **28**, 490 (1873).

C. SORET, Propagation de la chaleur dans les cristaux, *Comptes rendus des séances de la Société de Physique et d'Histoire Naturelle de Genève* **10**, 15 (1893); *Archives des Sciences Physiques et Naturelles (Genève)* **29**, 4 (1893).

Chapter 3

Statistical description of out-of-equilibrium systems

In most problems discussed in this book, the material systems under consideration, be they gaseous, liquid, or solid, are composed of a very large number of particles. When such a system is in equilibrium (or in local equilibrium), its macroscopic state may be parametrized by a limited number of extensive variables (or by their local densities). On the other hand, its microscopic state is not known exactly: the number of microscopic degrees of freedom coming into play is indeed so huge that it is completely impossible to have at one's disposal a full knowledge of the system at this level.

Such a detailed knowledge is in fact not necessary to describe the system in a macroscopic way. In statistical physics, we define the variables parametrizing the macroscopic state of a system of a very large number of particles as statistical averages of the variables associated with the microscopic states compatible with the macroscopic state under consideration. These microscopic variables are functions of the coordinates and momenta of all the particles of the system. In order to determine their averages, we introduce, for a classical system, the phase space distribution function or, for a quantum system, the density operator.

Out of equilibrium, the averages defining the macroscopic variables are functions of time. One of the most used procedures to determine their temporal dependence consists in studying the evolution of the phase space distribution function or of the density operator. The notions of phase space distribution function and density operator, as well as their temporal evolution, are therefore the subject of this introductory chapter to out-of-equilibrium statistical physics.

1. The phase space distribution function

1.1. The phase space

Let us consider a classical system of N point particles in the three-dimensional space. This system may be a gas or a liquid, in which case the translational motions of the molecules are accurately described by the laws of classical mechanics. The number N is assumed to be fixed (the system is closed). A *microscopic state* of the system is specified by the $3N$ spatial coordinates $q_1, \ldots, q_{3N}$ of the particles and the $3N$ components $p_1, \ldots, p_{3N}$ of their momenta.

We introduce the *phase space* of the system, defined as a $6N$-dimensional space with coordinates $q_1, \ldots, q_{3N}$, $p_1, \ldots, p_{3N}$. A microscopic state of the system is thus specified by a point in the phase space. As time evolves, the point representative of the state of the system moves in the phase space in a way determined by the microscopic equations of motion.[1]

The notion of phase space can be extended to the case in which other degrees of freedom than the translational ones are likely to be treated classically.[2] A system with s classical degrees of freedom is thus described by s configurational variables or generalized coordinates q_i, and s generalized momenta p_i. The microscopic state of the system is then represented by a point in a $2s$-dimensional phase space.

The dynamical variables or microscopic observables of the system are functions of the $6N$ (or, more generally, the $2s$) coordinates and momenta (possibly generalized).

1.2. The Gibbs ensemble

Generally speaking, when N or s is very large,[3] to refer to a system in a given macroscopic state amounts referring to an extremely large number of microscopic states,[4] or, adopting a formulation proposed by J.W. Gibbs in 1902, to a collection of an extremely large number of identical systems, finding themselves in different microscopic states but in the same macroscopic state. This collection of systems constitutes a statistical ensemble called the *Gibbs ensemble*.

To define the macroscopic variables, we can, according to Gibbs' idea, build an extremely large number $\mathcal{N}$ of copies of the system, all these copies satisfying the imposed values of the macroscopic constraints, and study the properties of the ensemble of phase space points corresponding to the different microscopic states compatible with these constraints. The macroscopic variables are then defined as ensemble averages of the corresponding microscopic variables.

---

[1] The trajectory of the representative point in phase space is either a closed curve or a curve which never intersects itself.

[2] This may for instance be the case of the rotational degrees of freedom of polyatomic molecules.

[3] For macroscopic systems, N and s are generally of the order of Avogadro's number.

[4] For instance, a macroscopic state of an isolated homogeneous fluid of N molecules in a box is defined by macroscopic properties such as the energy and the volume. An extremely large number of ways to distribute these molecules in space, as well as to distribute between them the total energy, is compatible with these macroscopic data.

1.3. Phase space distribution function

The number $\mathcal{N}$ of systems of the Gibbs ensemble being extremely large, the corresponding representative points are dense in the phase space. We are thus led to introduce their density, called the *phase space distribution function*. Coming back to the example of the classical fluid of N point particles, the phase space distribution function is a function f of the coordinates $q_1, \dots, q_{3N}, p_1, \dots, p_{3N}$, as well as, when out-of-equilibrium, of time t. By definition, the quantity:

$$f(q_1, \dots, q_{3N}, p_1, \dots, p_{3N}, t)\, dq_1 \dots dq_{3N} dp_1 \dots dp_{3N} \tag{3.1.1}$$

is the probability for the representative point of the microscopic state of the fluid to be found at time t in the volume element $dq_1 \dots dq_{3N} dp_1 \dots dp_{3N}$ centered at the point of coordinates $q_1, \dots, q_{3N}, p_1, \dots, p_{3N}$. This probability may also be denoted for short by $f(q,p,t)\, dqdp$. The phase space distribution function is a non-negative quantity. A priori, it must be normalized in such a way that:

$$\int f(q,p,t)\, dqdp = 1, \tag{3.1.2}$$

the integration being carried out over the whole phase space.

Once the distribution function is known, we calculate the ensemble average $\langle A(t) \rangle$ of a microscopic variable $A(q,p)$ as the phase space integral of $A(q,p)$ weighted by $f(q,p,t)$:

$$\boxed{\langle A(t) \rangle = \frac{\int A(q,p) f(q,p,t)\, dqdp}{\int f(q,p,t)\, dqdp}.} \tag{3.1.3}$$

The expectation value $\langle A(t) \rangle$ represents a macroscopic variable. It remains unchanged if the normalization of $f(q,p,t)$ is modified. In practice, we frequently use, instead of condition (3.1.2), the normalization:

$$\frac{1}{N! h^{3N}} \int f(q,p,t)\, dqdp = 1, \tag{3.1.4}$$

where h is the Planck constant.[5]

1.4. Evolution of $f(q,p,t)$: the Liouville equation

Let us consider a classical isolated system of N particles, described by a time-independent Hamiltonian $H(q_i, p_i)$. The microscopic equations of motion are *Hamilton's equations:*

$$\dot{q}_i = \frac{\partial H}{\partial p_i}, \qquad \dot{p}_i = -\frac{\partial H}{\partial q_i}, \qquad i = 1, \dots, 3N. \tag{3.1.5}$$

[5] This latter normalization choice allows us to obtain classical statistical mechanics as a limiting case of quantum statistical mechanics. The factor $1/N!$ is associated with the indistinguishability of particles, and the factor $1/h^{3N}$ allows us to define the volume element in phase space $dqdp/N!h^{3N}$ and the distribution function $f(q,p,t)$ as dimensionless quantities.

Starting from equations (3.1.5), it is possible to establish the evolution equation of the distribution function $f(q, p, t)$. For this purpose, we study the temporal evolution of the number n of representative points contained in an arbitrary volume $\mathcal{V}$ of the phase space. This number is given by the integral:[6]

$$n = \mathcal{N} \int_{\mathcal{V}} f(q, p, t) \, dq dp. \tag{3.1.6}$$

The rate of change of n is:

$$\frac{dn}{dt} = \mathcal{N} \int_{\mathcal{V}} \frac{\partial f}{\partial t} \, dq dp. \tag{3.1.7}$$

Now, the representative points are neither created nor deleted. Therefore, the rate of change of n is equal to the flux of representative points across the surface Σ enclosing the volume $\mathcal{V}$. We thus have:

$$\frac{dn}{dt} = -\mathcal{N} \int_{\Sigma} (f\boldsymbol{u}.\boldsymbol{\nu}) \, d\Sigma, \tag{3.1.8}$$

where $\boldsymbol{u}$ represents the $6N$-components vector $\dot{q}_1, \ldots, \dot{q}_{3N}, \dot{p}_1, \ldots, \dot{p}_{3N}$, and $\boldsymbol{\nu}$ the unit vector normal to Σ and oriented towards the exterior of $\mathcal{V}$. The surface integral on the right-hand side of equation (3.1.8) may be transformed into a volume integral:

$$\frac{dn}{dt} = -\mathcal{N} \int_{\mathcal{V}} \nabla.(f\boldsymbol{u}) \, dq dp. \tag{3.1.9}$$

The expressions (3.1.7) and (3.1.9) for dn/dt must be identical whatever the volume $\mathcal{V}$. We deduce from this identity the local conservation equation of the number of representative points in the phase space:

$$\frac{\partial f}{\partial t} + \nabla.(f\boldsymbol{u}) = 0. \tag{3.1.10}$$

In order to take into account the microscopic equations of motion, we make explicit $\nabla.(f\boldsymbol{u})$:

$$\nabla.(f\boldsymbol{u}) = \sum_{i=1}^{3N} \left(\frac{\partial f}{\partial q_i} \dot{q}_i + \frac{\partial f}{\partial p_i} \dot{p}_i \right) + \sum_{i=1}^{3N} \left(\frac{\partial \dot{q}_i}{\partial q_i} + \frac{\partial \dot{p}_i}{\partial p_i} \right) f. \tag{3.1.11}$$

The second sum appearing on the right-hand side of equation (3.1.11) vanishes as a consequence of equations (3.1.5). We are thus left with:

$$\nabla.(f\boldsymbol{u}) = \sum_{i=1}^{3N} \left(\frac{\partial f}{\partial q_i} \dot{q}_i + \frac{\partial f}{\partial p_i} \dot{p}_i \right). \tag{3.1.12}$$

[6] Here, we use the normalization (3.1.2).

Using once again Hamilton's equations, we can write the local conservation equation (3.1.10) in the form of the *Liouville equation*:

$$\frac{\partial f}{\partial t} + \sum_{i=1}^{3N} \left(\frac{\partial H}{\partial p_i} \frac{\partial f}{\partial q_i} - \frac{\partial H}{\partial q_i} \frac{\partial f}{\partial p_i} \right) = 0. \tag{3.1.13}$$

The sum appearing on the left-hand side of equation (3.1.13) is the *Poisson bracket* of H and f, which we denote by[7] $\{H, f\}$. We thus have:

$$\frac{\partial f}{\partial t} + \{H, f\} = 0. \tag{3.1.14}$$

Equation (3.1.14) may formally be rewritten as:

$$\frac{\partial f}{\partial t} = -i\mathcal{L}f. \tag{3.1.15}$$

In equation (3.1.15), $\mathcal{L}$ denotes the *Liouville operator* defined by:

$$\mathcal{L}\bullet = -i\{H, \bullet\}, \tag{3.1.16}$$

where the symbol $\bullet$ stands for any function of the generalized coordinates and momenta. The Hamiltonian H being time-independent, the formal integration of the evolution equation (3.1.15) yields, for the distribution function at time t, the expression:

$$f(q, p, t) = e^{-i\mathcal{L}t} f(q, p). \tag{3.1.17}$$

In formula (3.1.17), $f(q, p)$ represents the distribution function at a given time chosen as origin.

1.5. An equivalent form of the Liouville equation

Coming back to the local conservation equation (3.1.10), we can also write it as:

$$\frac{\partial f}{\partial t} + \boldsymbol{u}.\nabla f + f \nabla.\boldsymbol{u} = 0. \tag{3.1.18}$$

---

[7] The sign convention adopted in the definition of the Poisson bracket is not universal. Here, we use the convention:

$$\{\alpha, \beta\} = \sum_{i=1}^{3N} \left(\frac{\partial \alpha}{\partial p_i} \frac{\partial \beta}{\partial q_i} - \frac{\partial \alpha}{\partial q_i} \frac{\partial \beta}{\partial p_i} \right).$$

On account of Hamilton's equations, we have $\nabla.\boldsymbol{u} = 0$. Introducing the *material derivative* or *hydrodynamic derivative*, defined by:[8]

$$\frac{d}{dt} = \frac{\partial}{\partial t} + \boldsymbol{u}.\nabla, \tag{3.1.19}$$

we obtain once again, from equation (3.1.18), the Liouville equation, now written in the form:

$$\boxed{\frac{df}{dt} = 0.} \tag{3.1.20}$$

Equation (3.1.20) can be interpreted as the fact that the set of the representative points in the phase space behaves like an incompressible fluid (that is, like a fluid whose density does not change when we follow the flow over the course of time).

2. The density operator

In quantum statistical physics, it is the *density operator* $\rho(t)$ which plays the role of the distribution function $f(q,p,t)$ in classical statistical physics. However, $\rho(t)$ carries a richer information than does $f(q,p,t)$, since it makes precise the quantum correlations existing between dynamical variables, correlations which are absent classically.

2.1. Pure states and statistical mixtures of states

Let us consider an isolated quantum system, described by a time-independent Hamiltonian H. A microscopic state, also called a *pure state*, is a state of the form:

$$|\phi\rangle = \sum_n c_n |\phi_n\rangle, \tag{3.2.1}$$

where the $\{|\phi_n\rangle\}$ denote the eigenstates of a complete orthonormal base. The complex numbers c_n are the coefficients of the expansion of the pure state $|\phi\rangle$ over the states $|\phi_n\rangle$. The expectation value of a microscopic observable A in the pure state $|\phi\rangle$ is:

$$\langle A\rangle = \langle\phi|A|\phi\rangle = \sum_{n,m} \langle\phi_n|A|\phi_m\rangle c_n^* c_m. \tag{3.2.2}$$

In statistical physics, we aim to describe statistical ensembles of systems finding themselves in a given macroscopic state defined by the specification of macroscopic variables. A macroscopic state does not correspond to a well-defined microscopic state, but to a set of microscopic states $\{|\psi_i\rangle\}$. An occurrence probability p_i is associated

[8] Formula (3.1.19) establishes the link, as far the derivation with respect to time is concerned, between the *Lagrangian* description, in which the derivative d/dt takes the fluid motion into account, and the *Eulerian* description, in which the derivative $\partial/\partial t$ is calculated at a fixed point in space.

with each of these states. This is the reason why a macroscopic state is also called a *statistical mixture*. The probabilities p_i are non-negative and normalized:

$$p_i \geq 0, \qquad \sum_i p_i = 1. \tag{3.2.3}$$

The expectation value of a physical quantity A in the considered macroscopic state is:

$$\langle A \rangle = \sum_i p_i \langle \psi_i | A | \psi_i \rangle. \tag{3.2.4}$$

2.2. Definition of the density operator

The states $\{|\psi_i\rangle\}$ of the statistical mixture decompose over the orthonormal base $\{|\phi_n\rangle\}$:

$$|\psi_i\rangle = \sum_n c_{ni} |\phi_n\rangle. \tag{3.2.5}$$

Making use of expansion (3.2.5) in formula (3.2.4), we get:

$$\langle A \rangle = \sum_i p_i \sum_{n,m} \langle \phi_n | A | \phi_m \rangle c_{ni}^* c_{mi}. \tag{3.2.6}$$

We then introduce the density operator ρ defined by:

$$\boxed{\rho = \textstyle\sum_i |\psi_i\rangle p_i \langle \psi_i|.} \tag{3.2.7}$$

The density operator is Hermitean and positive.[9] Over the orthonormal base $\{|\phi_n\rangle\}$, its matrix elements are:

$$\rho_{mn} = \langle \phi_m | \rho | \phi_n \rangle = \sum_i p_i c_{ni}^* c_{mi}. \tag{3.2.8}$$

It is thus characterized by the *density matrix* of elements ρ_{mn}. The normalization condition (3.2.3) implies that the trace of the density operator equals unity:

$$\mathrm{Tr}\,\rho = \sum_n \langle \phi_n | \rho | \phi_n \rangle = \sum_i p_i \sum_n |c_{ni}|^2 = \sum_i p_i = 1. \tag{3.2.9}$$

All the information about the macroscopic state of the system is contained in the density operator. The diagonal element $\rho_{nn} = \langle \phi_n | \rho | \phi_n \rangle = \sum_i p_i |c_{ni}|^2$ represents the average probability of finding the system in the state $|\phi_n\rangle$. This is the reason why the diagonal elements of the density matrix are called the *populations*. The off-diagonal elements are called the *coherences*.

---

[9] The positivity of ρ signifies that, for any state $|\phi\rangle$, we have $\langle \phi | \rho | \phi \rangle \geq 0$.

We have the relation:

$$\langle A \rangle = \sum_{n,m} \langle \phi_m | \rho | \phi_n \rangle \langle \phi_n | A | \phi_m \rangle, \tag{3.2.10}$$

that is:

$$\boxed{\langle A \rangle = \mathrm{Tr}(\rho A).} \tag{3.2.11}$$

2.3. Evolution of $\rho(t)$: the Liouville–von Neumann equation

The system being isolated, each state $|\psi_i(t)\rangle$ of the statistical mixture evolves according to the Schrödinger equation:

$$i\hbar \frac{d|\psi_i(t)\rangle}{dt} = H|\psi_i(t)\rangle. \tag{3.2.12}$$

The density operator defined by formula (3.2.7) thus evolves according to the *Liouville–von Neumann equation*:

$$\frac{d\rho}{dt} = -\frac{i}{\hbar}[H, \rho]. \tag{3.2.13}$$

Equation (3.2.13) may formally be rewritten as:

$$\boxed{\frac{d\rho}{dt} = -i\mathcal{L}\rho.} \tag{3.2.14}$$

The symbol $\mathcal{L}$ denotes here the quantum *Liouville* operator,[10] defined by:

$$\mathcal{L}\bullet = \frac{1}{\hbar}[H, \bullet], \tag{3.2.15}$$

where the symbol $\bullet$ stands for any operator.

The Hamiltonian being time-independent, the formal integration of the evolution equation (3.2.14) yields, for the density operator at time t, the expression:

$$\rho(t) = e^{-i\mathcal{L}t}\rho. \tag{3.2.16}$$

In formula (3.2.16), ρ represents the density operator at a given time chosen as origin.[11] We can also write, equivalently:

$$\rho(t) = e^{-iHt/\hbar}\rho e^{iHt/\hbar}. \tag{3.2.17}$$

If the $\{|\phi_n\rangle\}$ are the eigenstates of H, of energies ε_n, we have:

$$\begin{cases} \rho_{nn}(t) = \rho_{nn} \\ \rho_{mn}(t) = \rho_{mn} e^{-i\omega_{mn} t}, \qquad m \neq n, \end{cases} \tag{3.2.18}$$

with $\omega_{mn} = (\varepsilon_m - \varepsilon_n)/\hbar$. The populations do not depend on time and the coherences oscillate at the Bohr angular frequencies ω_{mn}.

---

[10] The Liouville operator does not act in the state space, but in the operator space: this is why it is sometimes referred to as a *superoperator*.

[11] As displayed by formulas (3.2.16) and (3.1.17), the use of the Liouville operator leads to formally analogous expressions for $\rho(t)$ and $f(q, p, t)$.

3. Systems at equilibrium

The most often encountered equilibrium situations are either that of a closed system likely to exchange heat with a thermostat (canonical equilibrium) or that of an open system likely to exchange heat and particles with a reservoir (grand canonical equilibrium). Let us make precise the form of the density operator (*resp.* of the classical distribution function) in both situations.

3.1. Closed system: canonical equilibrium

The density operator (*resp.* the distribution function) of a system of Hamiltonian H, in equilibrium with a thermostat at temperature T, is the canonical density operator (*resp.* the canonical distribution function):

$$\boxed{\rho \text{ (resp. } f) = \frac{1}{Z} e^{-\beta H}, \qquad \beta = (kT)^{-1}.} \tag{3.3.1}$$

The partition function Z is expressed as a discrete sum over the states of the system (*resp.* as an integral over the phase space of the system).

- *Quantum case*

We have:

$$Z = \mathrm{Tr}(e^{-\beta H}), \tag{3.3.2}$$

where the trace concerns all the states of the system.

- *Classical case*

The trace is replaced by a phase space integral. The partition function of a classical system of N particles in a three-dimensional space reads for instance:

$$Z_N = \frac{1}{N! h^{3N}} \int e^{-\beta H} \, dq dp. \tag{3.3.3}$$

3.2. Open system: grand canonical equilibrium

The density operator (*resp.* the distribution function) of a system of Hamiltonian H, in equilibrium with a reservoir of temperature T and chemical potential μ, is the grand canonical density operator (*resp.* the grand canonical distribution function):

$$\boxed{\rho \text{ (resp. } f) = \frac{1}{\Xi} e^{-\beta (H - \mu N)},} \tag{3.3.4}$$

where Ξ stands for the grand partition function.

• *Quantum case*

For a quantum system, we have:

$$\Xi = \mathrm{Tr}[e^{-\beta(H-\mu N)}], \tag{3.3.5}$$

where the trace concerns all states with all possible values of the number N of particles.

• *Classical case*

In order to obtain the grand partition function of a classical system, we can for instance use the general relation:

$$\Xi = \sum_{N=0}^{\infty} e^{\beta\mu N} Z_N, \tag{3.3.6}$$

in which Z_N represents the partition function of a classical system of N identical particles (formula (3.3.3)).

4. Evolution of the macroscopic variables: classical case

Let $A(q,p)$ be a microscopic dynamical variable of the system. In order to study the temporal evolution of the macroscopic variable associated with $\langle A(t)\rangle$, we can adopt one of two approaches, which differ from each other by the way the temporal evolutions are taken into account, but which, in the end, lead to identical results.

4.1. The two approaches

• *First approach*

We consider that the variable $A(q,p)$ does not depend on time, and that its distribution function $f(q,p,t)$ evolves with time according to the Liouville equation (3.1.15). The expectation value $\langle A(t)\rangle$ given by formula (3.1.3) reads, with the aid of the normalization (3.1.4),

$$\langle A(t)\rangle = \frac{1}{N!h^{3N}} \int A(q,p) f(q,p,t)\, dqdp, \tag{3.4.1}$$

that is, using the expression (3.1.17) for $f(q,p,t)$:

$$\boxed{\langle A(t)\rangle = \frac{1}{N!h^{3N}} \int A(q,p) e^{-i\mathcal{L}t} f(q,p)\, dqdp.} \tag{3.4.2}$$

• *Second approach*

This relies on the fact that the generalized coordinates and momenta evolve with time according to Hamilton's equations (3.1.5). We thus consider that the dynamical variable $A[q(t),p(t)] = A(t)$ itself evolves with time. We have:

$$\frac{dA}{dt} = \sum_{i=1}^{3N} \left(\frac{\partial A}{\partial q_i}\dot{q}_i + \frac{\partial A}{\partial p_i}\dot{p}_i \right), \tag{3.4.3}$$

that is, taking into account Hamilton's equations:

$$\frac{dA}{dt} = \sum_{i=1}^{3N} \left(\frac{\partial A}{\partial q_i} \frac{\partial H}{\partial p_i} - \frac{\partial A}{\partial p_i} \frac{\partial H}{\partial q_i} \right). \tag{3.4.4}$$

The sum on the right-hand side of equation (3.4.4) is the Poisson bracket of H and A. Equation (3.4.4) thus reads:

$$\frac{dA}{dt} = \{H, A\}, \tag{3.4.5}$$

that is, in terms of the Liouville operator (3.1.16):

$$\boxed{\frac{dA}{dt} = i\mathcal{L}A.} \tag{3.4.6}$$

The Hamiltonian being time-independent, the formal integration of equation (3.4.6) yields the expression:

$$A(t) = e^{i\mathcal{L}t} A(q, p). \tag{3.4.7}$$

In this approach, the expectation value $\langle A(t) \rangle$ reads:

$$\langle A(t) \rangle = \frac{1}{N! h^{3N}} \int A(t) f(q, p)\, dq dp, \tag{3.4.8}$$

that is, using formula (3.4.7):

$$\boxed{\langle A(t) \rangle = \frac{1}{N! h^{3N}} \int e^{i\mathcal{L}t} A(q, p) f(q, p)\, dq dp.} \tag{3.4.9}$$

4.2. Equivalence of the two approaches

The two approaches described above are respectively the classical analogs of the points of view known in quantum mechanics as the Schrödinger and Heisenberg pictures. As in quantum mechanics,[12] these two approaches are equivalent, in the sense that they yield the same results for the expectation values of physical quantities.

To establish this equivalence in the classical case, the simplest method is to compare the expressions for $d\langle A(t) \rangle / dt$ in the first approach,

$$\frac{d\langle A(t) \rangle}{dt} = \frac{1}{N! h^{3N}} (-i) \int A\mathcal{L}f\, dq dp, \tag{3.4.10}$$

[12] The proof of the equivalence in the quantum case will be carried out in the next section.

and in the second one:

$$\frac{d\langle A(t)\rangle}{dt} = \frac{1}{N!h^{3N}} i \int \mathcal{L}Af\, dqdp. \tag{3.4.11}$$

To demonstrate the identity of the expressions (3.4.10) and (3.4.11), it is enough to demonstrate the following property of the Liouville operator:

$$-i \int A\mathcal{L}f\, dqdp = i \int \mathcal{L}Af\, dqdp, \tag{3.4.12}$$

a property which reads, in a more explicit way:

$$-\int A\{H,f\}\, dqdp = \int \{H,A\}f\, dqdp. \tag{3.4.13}$$

Formula (3.4.13) can be proved using integrations by parts and the fact that the distribution function f vanishes for $q_i = \pm\infty$ or $p_i = \pm\infty$. It follows from property (3.4.12) that the expressions (3.4.10) and (3.4.11) for $d\langle A(t)\rangle/dt$ are identical, hence the identity of the expressions (3.4.2) and (3.4.9) for $\langle A(t)\rangle$. This latter property is expressed through the equality:

$$\int A(q,p)e^{-i\mathcal{L}t}f(q,p)\, dqdp = \int e^{i\mathcal{L}t}A(q,p)f(q,p)\, dqdp, \tag{3.4.14}$$

which allows us to transfer the time dependence of the distribution function to the dynamical variables, and vice versa.

5. Evolution of the macroscopic variables: quantum case

Consider an observable A of a quantum system. To determine the temporal evolution of the macroscopic variable associated with the expectation value $\langle A(t)\rangle$, we can adopt either the Schrödinger picture or the Heisenberg one.

5.1. The two pictures

• *The Schrödinger picture*

The density operator evolves with time according to the Liouville–von Neumann equation (3.2.13). The expectation value $\langle A(t)\rangle$, given by:

$$\langle A(t)\rangle = \operatorname{Tr}\big[\rho(t)A\big], \tag{3.5.1}$$

reads, according to the expression (3.2.17) for $\rho(t)$:

$$\boxed{\langle A(t)\rangle = \operatorname{Tr}\big(e^{-iHt/\hbar}\,\rho\, e^{iHt/\hbar}\,A\big).} \tag{3.5.2}$$

• *The Heisenberg picture*

The observables evolve with time. The evolution equation of an observable A is the Heisenberg equation,

$$i\hbar\frac{dA}{dt} = [A, H], \tag{3.5.3}$$

which also reads, in terms of the Liouville operator $\mathcal{L}$ defined by equation (3.2.15):

$$\boxed{\frac{dA}{dt} = i\mathcal{L}A.} \tag{3.5.4}$$

The Hamiltonian being time-independent, the formal integration of equation (3.5.4) yields the expression:[13]

$$A(t) = e^{i\mathcal{L}t}A, \tag{3.5.5}$$

which may also be written as:

$$A(t) = e^{iHt/\hbar}\,A\,e^{-iHt/\hbar}. \tag{3.5.6}$$

The expectation value $\langle A(t)\rangle$ is expressed as:

$$\langle A(t)\rangle = \mathrm{Tr}\big[\rho A(t)\big], \tag{3.5.7}$$

that is, according to expression (3.5.6) for $A(t)$:

$$\boxed{\langle A(t)\rangle = \mathrm{Tr}\big(\rho\, e^{iHt/\hbar}\,A\,e^{-iHt/\hbar}\big).} \tag{3.5.8}$$

5.2. Equivalence between both pictures

The Schrödinger picture and the Heisenberg one are equivalent: they do indeed yield the same results for the expectation values of physical quantities as displayed by the identity between expressions (3.5.2) and (3.5.8) for $\langle A(t)\rangle$ (the proof of the equivalence relies on the invariance of the trace under a circular permutation of the operators).

---

[13] Once again, the use of the Liouville operator leads to a formal analogy between the quantum formula (3.5.5) and the classical formula (3.4.7).

Bibliography

R. Balescu, *Statistical dynamics. Matter out of equilibrium*, Imperial College Press, London, 1997.

R. Balian, *From microphysics to macrophysics*, Vol. 1, Springer-Verlag, Berlin, 1991.

C. Cohen-Tannoudji, B. Diu, and F. Laloë, *Quantum mechanics*, Vol. 1, Hermann and Wiley, Paris, second edition, 1977.

H. Grabert, *Projection operator techniques in nonequilibrium statistical mechanics*, Springer Tracts in Modern Physics **95**, Springer-Verlag, Berlin, 1982.

R. Kubo, M. Toda, and N. Hashitsume, *Statistical physics II: nonequilibrium statistical mechanics*, Springer-Verlag, Berlin, second edition, 1991.

L.D. Landau and E.M. Lifshitz, *Mechanics*, Butterworth-Heinemann, Oxford, third edition, 1976.

D.A. McQuarrie, *Statistical mechanics*, University Science Books, Sausalito, second edition, 2000.

M. Toda, R. Kubo, and N. Saitô, *Statistical physics I: equilibrium statistical mechanics*, Springer-Verlag, Berlin, second edition, 1992.

D. Zubarev, V. Morozov, and G. Röpke, *Statistical mechanics of nonequilibrium processes*, Vol. 1: *Basic concepts, kinetic theory*, Akademie Verlag, Berlin, 1996.

Chapter 4

Classical systems: reduced distribution functions

In order to determine the distribution function of a classical system such as a gas or a liquid, it is necessary to know its Hamiltonian. The structure of this depends on the nature of the interactions between the particles, as well as on the possible presence of external fields. In this chapter, we focus on a system of particles whose interactions are limited to pair interactions. Even in this case, solving the Liouville equation constitutes an impracticable task, mainly because the distribution function concerns the whole set of interacting particles.

However, the dynamical variables of practical interest, for instance the kinetic energy or the potential energy of the system, are sums of quantities each involving the coordinates and the momenta of a very small number of particles. This is why we are led to introduce reduced distribution functions (concerning a limited number of particles), knowledge of which suffices to determine the expectation values of the relevant physical dynamical variables. Among these distribution functions, the one-particle distribution function plays the most important role. Its evolution may be deduced from a hierarchy of coupled equations successively involving the one-particle, two-particle ... distribution functions.

At this stage, we are therefore led to devise approximation schemes enabling us to obtain from this hierarchy a closed evolution equation for the one-particle distribution function. Several approximate evolution equations have thus been introduced, each of them suited to a specific physical context. As an example, we present in this chapter the Vlasov equation which describes the beginning of the evolution of an ionized gas from an out-of-equilibrium initial state.

1. Systems of classical particles with pair interactions

Consider a system of $N \gg 1$ classical identical particles in interaction, for instance molecules, atoms, or ions. These particles, considered as point particles, are assumed to be confined inside a box of given volume.

The form of the Hamiltonian of the system depends on the type of interactions between particles as well as on the possible presence of applied external fields.

• *Particles with pair interactions*

For a system of N identical point particles of mass m in the absence of external fields, it is possible, in most practically interesting cases, to write the Hamiltonian as a sum of kinetic energy terms and pair interaction potential energy terms:[1]

$$H_N = \sum_{i=1}^{N} \frac{\boldsymbol{p}_i^2}{2m} + \sum_{\substack{i,j=1 \\ i<j}}^{N} u\big(|\boldsymbol{r}_i - \boldsymbol{r}_j|\big). \tag{4.1.1}$$

In formula (4.1.1), each pair interaction term is assumed to depend only on the distance between the two particles i and j of the considered pair.[2] Interactions involving more than two particles are not taken into account.

• *Particles with pair interactions in the presence of a scalar potential*

In the presence of a scalar potential $\phi(\boldsymbol{r})$ associated with either a gravitational field or an external electric field, the Hamiltonian is:

$$H_N = \sum_{i=1}^{N} \frac{\boldsymbol{p}_i^2}{2m} + \sum_{i=1}^{N} \phi(\boldsymbol{r}_i) + \sum_{\substack{i,j=1 \\ i<j}}^{N} u\big(|\boldsymbol{r}_i - \boldsymbol{r}_j|\big). \tag{4.1.2}$$

• *Particles with pair interactions in the presence of a scalar potential and of a vector potential*

If the particles carry a charge q and are submitted to an external electromagnetic field deriving from a scalar potential $\phi(\boldsymbol{r}, t)$ and a vector potential $\boldsymbol{A}(\boldsymbol{r}, t)$, the Hamiltonian becomes:

$$H_N = \sum_{i=1}^{N} \frac{\big[\boldsymbol{p}_i - q\boldsymbol{A}(\boldsymbol{r}_i, t)/c\big]^2}{2m} + q\sum_{i=1}^{N} \phi(\boldsymbol{r}_i, t) + \sum_{\substack{i,j=1 \\ i<j}}^{N} u\big(|\boldsymbol{r}_i - \boldsymbol{r}_j|\big), \tag{4.1.3}$$

where c is the velocity of light.

---

[1] The pair interactions are interactions between the particles taken two by two.

[2] Some most used pair interaction potentials are described in an appendix at the end of this chapter.

2. The Liouville equation

2.1. The phase space distribution function

As a general rule, for a classical system described by generalized coordinates q_i and generalized momenta p_i, the knowledge of the phase space distribution function $f(q, p, t)$ enables us to calculate the expectation value $\langle A(t)\rangle$ of a dynamical variable $A(q, p)$ (the ensemble average $\langle A(t)\rangle$ is expressed as the phase space integral of the concerned dynamical variable weighted by the distribution function).

For a system of N identical and indistinguishable point particles of coordinates $\boldsymbol{r}_i$ and momenta $\boldsymbol{p}_i$ $(i = 1, \ldots, N)$, the distribution function $f(\boldsymbol{r}_1, \boldsymbol{p}_1, \ldots, \boldsymbol{r}_N, \boldsymbol{p}_N, t)$ is not modified when we interchange two particles. It is thus invariant under any permutation of the indexes i. We will use the following normalization:

$$\frac{1}{N!h^{3N}}\int f(\boldsymbol{r}_1, \boldsymbol{p}_1, \ldots, \boldsymbol{r}_N, \boldsymbol{p}_N, t)\, d\boldsymbol{r}_1 d\boldsymbol{p}_1 \ldots d\boldsymbol{r}_N d\boldsymbol{p}_N = 1. \tag{4.2.1}$$

Note that the distribution function itself may be written in the form of an average in the phase space:

$$f(\boldsymbol{r}'_1, \boldsymbol{p}'_1, \ldots, \boldsymbol{r}'_N, \boldsymbol{p}'_N, t) = N!h^{3N}\langle\prod_{i=1}^{N}\delta(\boldsymbol{r}_i - \boldsymbol{r}'_i)\delta(\boldsymbol{p}_i - \boldsymbol{p}'_i)\rangle. \tag{4.2.2}$$

The evolution of f is governed by the Liouville equation $df/dt = 0$, which we will make explicit on account of the specific form of the Hamiltonian H_N.

2.2. General structure of the Liouville equation

The Liouville equation reads:

$$\frac{\partial f}{\partial t} + \sum_{i=1}^{N}\dot{\boldsymbol{r}}_i.\nabla_{\boldsymbol{r}_i} f + \sum_{i=1}^{N}\dot{\boldsymbol{p}}_i.\nabla_{\boldsymbol{p}_i} f = 0. \tag{4.2.3}$$

In equation (4.2.3), $\nabla_{\boldsymbol{r}_i}$ (*resp.* $\nabla_{\boldsymbol{p}_i}$) denotes the gradient with respect to the coordinate (*resp.* momentum) variable of the particle i. While $\dot{\boldsymbol{r}}_i$ is clearly just the velocity $\boldsymbol{v}_i$, it is on the other hand necessary to make explicit $\dot{\boldsymbol{p}}_i$. We have, in a general way,

$$\frac{d\boldsymbol{p}_i}{dt} = \boldsymbol{F}_i, \tag{4.2.4}$$

where $\boldsymbol{F}_i$ represents the force exerted on the particle i. This force is the sum of the force $\boldsymbol{X}_i$ due to the scalar and vector potentials possibly present,[3] and of the forces $\boldsymbol{X}_{ij}$ due to the other particles:

$$\boldsymbol{F}_i = \boldsymbol{X}_i + \sum_{\substack{j=1\\ j\neq i}}^{N}\boldsymbol{X}_{ij}. \tag{4.2.5}$$

[3] The calculation of the force $\boldsymbol{X}_i$ exerted on a particle in the presence of a scalar potential and of a vector potential describing an external electromagnetic field is carried out in an appendix at the end of this chapter.

The force $\boldsymbol{X}_{ij}$ derives from the corresponding pair interaction potential:

$$\boldsymbol{X}_{ij} = -\nabla_{\boldsymbol{r}_i} u\big(|\boldsymbol{r}_i - \boldsymbol{r}_j|\big). \tag{4.2.6}$$

We will now introduce the expression for $\dot{\boldsymbol{p}}_i$ in equation (4.2.3), first in the case without interactions ($\boldsymbol{F}_i = \boldsymbol{X}_i$), then in the case of pair interactions ($\boldsymbol{F}_i = \boldsymbol{X}_i + \sum_{\substack{j=1\\ j\neq i}}^{N} \boldsymbol{X}_{ij}$).

2.3. Non-interacting particles

When the particles do not interact, the Hamiltonian H_N decomposes into a sum of one-particle terms (the form of these terms depends on the physics of the problem, that is, in particular, on the presence of external fields). The force $\boldsymbol{F}_i$ reduces to the force $\boldsymbol{X}_i$. Thus, it only depends on the parameters of the particle i. The left-hand side of the Liouville equation (4.2.3) is therefore composed of sums of one-particle terms. This equation reads:

$$\boxed{\frac{\partial f}{\partial t} + \sum_{i=1}^{N} \boldsymbol{v}_i . \nabla_{\boldsymbol{r}_i} f + \sum_{i=1}^{N} \boldsymbol{X}_i . \nabla_{\boldsymbol{p}_i} f = 0.} \tag{4.2.7}$$

2.4. Particles in the presence of pair interactions

In the presence of pair interactions, it is no longer possible to decompose H_N into a sum of one-particle terms. The force $\boldsymbol{F}_i$ is the sum of $\boldsymbol{X}_i$ and the forces $\boldsymbol{X}_{ij}$ (formula (4.2.5)). The Liouville equation now reads:

$$\boxed{\frac{\partial f}{\partial t} + \sum_{i=1}^{N} \boldsymbol{v}_i . \nabla_{\boldsymbol{r}_i} f + \sum_{i=1}^{N} \boldsymbol{F}_i . \nabla_{\boldsymbol{p}_i} f = 0.} \tag{4.2.8}$$

The similarity between equations (4.2.7) and (4.2.8) is only an apparent one. Indeed, in contrast to $\boldsymbol{X}_i$, the force $\boldsymbol{F}_i$ involves the parameters of all particles j interacting with the particle i.

For a system with $N \gg 1$ particles, it is an impossible task to directly tackle the resolution of such an equation. This is the reason why we establish a formalism relying on a hierarchy of coupled equations for *reduced* distribution functions, that is, for distribution functions concerning only a limited number of particles. This approach makes it possible, through some approximations, to eliminate the irrelevant information in a simpler way than by working directly on the Liouville equation (4.2.8) obeyed by the full distribution function.

3. Reduced distribution functions: the BBGKY hierarchy

For the sake of simplicity, it will be assumed from now on that the particles in pair interaction evolve either in the absence of a potential or in the presence of a scalar potential $\phi(\boldsymbol{r})$ corresponding for instance to a gravitational field. The Hamiltonian H_N is therefore of the form (4.1.1) or (4.1.2). Since there is no vector potential, $\dot{\boldsymbol{r}}_i = \boldsymbol{v}_i = \boldsymbol{p}_i/m$ is independent of $\boldsymbol{r}_i$. In the same way, $\dot{\boldsymbol{p}}_i = -\nabla\phi(\boldsymbol{r}_i) - \sum_{\substack{j=1\\ j\neq i}}^{N} \nabla_{\boldsymbol{r}_i} u(|\boldsymbol{r}_i - \boldsymbol{r}_j|)$ is independent of $\boldsymbol{p}_i$.

3.1. Reduced distribution functions

The dynamical variables of practical interest, as for instance the kinetic energy or the potential energy of the system, are sums of functions each involving the coordinates and the momenta of a very small number of particles.[4] This is, in particular, the case of the Hamiltonian H_N, whose expectation value reads:

$$\langle H_N\rangle = \langle E_c(t)\rangle + \langle E_p^{\rm grav}(t)\rangle + \langle E_p^{\rm int}(t)\rangle, \tag{4.3.1}$$

with:

$$\langle E_c(t) + E_p^{\rm grav}(t)\rangle = \frac{1}{N!h^{3N}}\int\Big[\sum_{i=1}^{N}\frac{\boldsymbol{p}_i^2}{2m} + \phi(\boldsymbol{r}_i)\Big] f(\boldsymbol{r}_1,\boldsymbol{p}_1,\ldots,\boldsymbol{r}_N,\boldsymbol{p}_N,t)\,d\tau_N \tag{4.3.2}$$

and:

$$\langle E_p^{\rm int}(t)\rangle = \frac{1}{N!h^{3N}}\int\Big[\sum_{\substack{i,j=1\\ i<j}}^{N} u(|\boldsymbol{r}_i-\boldsymbol{r}_j|)\Big] f(\boldsymbol{r}_1,\boldsymbol{p}_1,\ldots,\boldsymbol{r}_N,\boldsymbol{p}_N,t)\,d\tau_N. \tag{4.3.3}$$

In formulas (4.3.2) and (4.3.3), $\langle E_c(t)\rangle$, $\langle E_p^{\rm grav}(t)\rangle$, and $\langle E_p^{\rm int}(t)\rangle$ denote respectively the expectation values of the kinetic energy, the gravitational potential energy, and the interaction potential energy. Also, we have introduced the condensed notation:

$$d\tau_N = \prod_{i=1}^{N} d\Omega_i, \qquad d\Omega_i = d\boldsymbol{r}_i d\boldsymbol{p}_i. \tag{4.3.4}$$

In order to calculate averages of the type (4.3.2) or (4.3.3), it is not necessary to know the full distribution function $f(\boldsymbol{r}_1,\boldsymbol{p}_1,\ldots,\boldsymbol{r}_N,\boldsymbol{p}_N,t)$ (this latter function containing, besides the relevant information, information about the correlations between three, four, or more, particles). It is enough to know the one-particle and two-particle reduced

---

[4] The total kinetic energy is the sum of the individual kinetic energies $\boldsymbol{p}_i^2/2m$ (one-particle dynamical variables). In the same way, the potential energy in a gravitational field described by the potential $\phi(\boldsymbol{r})$ is the sum of the individual potential energies $\phi(\boldsymbol{r}_i)$. The interaction potential energy is the sum of the pair interaction potential energies (two-particle dynamical variables).

distribution functions, respectively denoted by $f^{(1)}(\boldsymbol{r}_1,\boldsymbol{p}_1,t)$ and $f^{(2)}(\boldsymbol{r}_1,\boldsymbol{p}_1,\boldsymbol{r}_2,\boldsymbol{p}_2,t)$, and defined by:

$$\begin{cases} f^{(1)}(\boldsymbol{r}_1,\boldsymbol{p}_1,t) = \dfrac{1}{(N-1)!h^{3N}} \displaystyle\int f(\boldsymbol{r}_1,\boldsymbol{p}_1,\ldots,\boldsymbol{r}_N,\boldsymbol{p}_N,t)\,d\Omega_2\ldots d\Omega_N \\ f^{(2)}(\boldsymbol{r}_1,\boldsymbol{p}_1,\boldsymbol{r}_2,\boldsymbol{p}_2,t) = \dfrac{1}{(N-2)!h^{3N}} \displaystyle\int f(\boldsymbol{r}_1,\boldsymbol{p}_1,\ldots,\boldsymbol{r}_N,\boldsymbol{p}_N,t)\,d\Omega_3\ldots d\Omega_N. \end{cases} \tag{4.3.5}$$

The functions $f^{(1)}$ and $f^{(2)}$ obey the relations:

$$\int f^{(1)}\,d\Omega_1 = N, \qquad \int f^{(2)}\,d\Omega_1 d\Omega_2 = N(N-1), \qquad \int f^{(2)}\,d\Omega_2 = (N-1)f^{(1)}. \tag{4.3.6}$$

In terms of $f^{(1)}$ and $f^{(2)}$, $\langle H_N\rangle$ reads:

$$\langle H_N\rangle = \int\left[\frac{p_1^2}{2m}+\phi(\boldsymbol{r}_1)\right]f^{(1)}(\boldsymbol{r}_1,\boldsymbol{p}_1,t)\,d\Omega_1 + \frac{1}{2}\int u(|\boldsymbol{r}_1-\boldsymbol{r}_2|)\,f^{(2)}(\boldsymbol{r}_1,\boldsymbol{p}_1,\boldsymbol{r}_2,\boldsymbol{p}_2,t)\,d\Omega_1 d\Omega_2. \tag{4.3.7}$$

In the same way, and more generally, we define reduced distribution functions concerning any number n $(1 \leq n < N)$ of particles:

$$f^{(n)}(\boldsymbol{r}_1,\boldsymbol{p}_1,\ldots,\boldsymbol{r}_n,\boldsymbol{p}_n,t) = \frac{1}{(N-n)!h^{3N}}\int f(\boldsymbol{r}_1,\boldsymbol{p}_1,\ldots,\boldsymbol{r}_N,\boldsymbol{p}_N,t)\,d\Omega_{n+1}\ldots d\Omega_N. \tag{4.3.8}$$

The n-particle reduced distribution function is proportional to the probability density of finding n particles at $\boldsymbol{r}_1,\boldsymbol{p}_1,\ldots,\boldsymbol{r}_n,\boldsymbol{p}_n$ at time t, whatever the positions and momenta of the $(N-n)$ other particles.

The reduced distribution function $f^{(1)}$ may be interpreted as the distribution function in a one-particle (and six-dimensional) phase space. Indeed, the average number of particles in a volume element $d\boldsymbol{r}d\boldsymbol{p}$ around the point $(\boldsymbol{r},\boldsymbol{p})$ of this phase space is given by:

$$\Big\langle\sum_i \delta(\boldsymbol{r}-\boldsymbol{r}_i)\delta(\boldsymbol{p}-\boldsymbol{p}_i)\Big\rangle\,d\boldsymbol{r}d\boldsymbol{p} = f^{(1)}(\boldsymbol{r},\boldsymbol{p})\,d\boldsymbol{r}d\boldsymbol{p}. \tag{4.3.9}$$

Among the reduced distribution functions, the most useful in practice are the one-particle and two-particle distribution functions. We now want to establish their evolution equations in the presence of pair interactions. To begin with, we will address the simpler case of non-interacting particles.

3.2. Evolution of $f^{(1)}$: the case without interactions

The left-hand side of the Liouville equation (4.2.7) involves sums of one-particle terms. Integrating this equation over $d\Omega_2\ldots d\Omega_N$ in the whole phase space, we obtain a closed evolution equation for $f^{(1)}$. The calculation relies on the fact that the distribution function $f(q,p,t)$ vanishes for $q_i = \pm\infty$ or $p_i = \pm\infty$. All terms in which the presence of a derivative allows us to reduce by one the number of successive integrations thus lead to a vanishing result.[5] We show in this way that, in the absence of interactions between the particles, $f^{(1)}$ obeys a closed evolution equation:

$$\boxed{\frac{\partial f^{(1)}}{\partial t} + \boldsymbol{v}_1.\nabla_{\boldsymbol{r}_1} f^{(1)} + \boldsymbol{X}_1.\nabla_{\boldsymbol{p}_1} f^{(1)} = 0.} \tag{4.3.10}$$

The term $\boldsymbol{v}_1.\nabla_{\boldsymbol{r}_1} f^{(1)}$ is called the *drift term*, and the term $\boldsymbol{X}_1.\nabla_{\boldsymbol{p}_1} f^{(1)}$ the *driving term*. Equation (4.3.10) is a simple conservation equation. It is identical to the Liouville equation for the distribution function of a system which would reduce to a unique particle evolving in a six-dimensional phase space.[6]

3.3. Evolution of $f^{(1)}$ in the presence of pair interactions: the BBGKY hierarchy

We now start from the Liouville equation (4.2.8). The pair interactions provide the following contribution to $\partial f^{(1)}/\partial t$:

$$\frac{1}{(N-1)!h^{3N}} \int \sum_{j=2}^{N} \boldsymbol{X}_{1j}.\nabla_{\boldsymbol{p}_1} f \, d\Omega_2 \ldots d\Omega_N. \tag{4.3.11}$$

The contribution (4.3.11) is a sum of $(N-1)$ terms corresponding to the different values taken by the index j. These terms are all equal, since, due to the indistinguishability of the particles, the distribution function f is symmetric with respect to its different arguments $\boldsymbol{r}_i, \boldsymbol{p}_i$. The definition of $f^{(2)}$ (equation (4.3.5)) being taken into account, the contribution of any of these $(N-1)$ terms, after integrating over $d\Omega_3 \ldots d\Omega_N$, reads:

$$\frac{1}{N-1} \int \boldsymbol{X}_{12}.\nabla_{\boldsymbol{p}_1} f^{(2)} \, d\Omega_2. \tag{4.3.12}$$

The one-particle distribution function thus obeys, in the presence of pair interactions, an evolution equation involving the two-particle distribution function:

$$\boxed{\frac{\partial f^{(1)}}{\partial t} + \boldsymbol{v}_1.\nabla_{\boldsymbol{r}_1} f^{(1)} + \boldsymbol{X}_1.\nabla_{\boldsymbol{p}_1} f^{(1)} + \int \boldsymbol{X}_{12}.\nabla_{\boldsymbol{p}_1} f^{(2)} \, d\Omega_2 = 0.} \tag{4.3.13}$$

---

[5] The calculation is fairly easy to do in the case considered here, in which, in the absence of a vector potential, $\dot{\boldsymbol{r}}_i$ is independent of $\boldsymbol{r}_i$ and $\dot{\boldsymbol{p}}_i$ is independent of $\boldsymbol{p}_i$. The result may be extended to the case where a vector potential is present. It is then necessary to take into account the appropriate Hamilton's equations (these equations are given in an appendix at the end of this chapter).

[6] Equation (4.3.10) for the one-particle distribution function was introduced in stellar dynamics by J.H. Jeans in 1915 under the name of the *collisionless Boltzmann equation* or the *Liouville equation*.

The drift term and the driving term are the same as in equation (4.3.10) for the case without interactions. The existence of pair interactions leads to the presence of an extra term, whose expression involves the two-particle distribution function.

We can, in a similar way, derive the evolution equation of $f^{(2)}$. In the presence of pair interactions, this equation involves $f^{(3)}$:

$$\frac{\partial f^{(2)}}{\partial t} + \boldsymbol{v}_1.\nabla_{\boldsymbol{r}_1} f^{(2)} + \boldsymbol{v}_2.\nabla_{\boldsymbol{r}_2} f^{(2)} + (\boldsymbol{X}_1 + \boldsymbol{X}_{12}).\nabla_{\boldsymbol{p}_1} f^{(2)} + (\boldsymbol{X}_2 + \boldsymbol{X}_{21}).\nabla_{\boldsymbol{p}_2} f^{(2)} + \int (\boldsymbol{X}_{13}.\nabla_{\boldsymbol{p}_1} f^{(3)} + \boldsymbol{X}_{23}.\nabla_{\boldsymbol{p}_2} f^{(3)})\, d\Omega_3 = 0. \tag{4.3.14}$$

In the same way, the evolution equation of $f^{(n)}$ involves $f^{(n+1)}$. We thus get, step by step, a chain of coupled equations, the last of which, for $n = N$, is nothing but the Liouville equation for the full distribution function $f^{(N)} = f$. None of these equations, except for the last one, is closed: we have indeed to know $f^{(n+1)}$ in order to be able to determine the evolution of $f^{(n)}$ (for $n < N$).

This set of equations, the index n varying from 1 to N, constitutes the *BBGKY hierarchy*, from the names of the different physicists who established it independently: N.N. Bogoliubov (1946), M. Born (1946), M.S. Green (1946), J.G. Kirkwood (1946), and J. Yvon (1935). This system of chain equations, of a hierarchical structure, involving distribution functions with a reduced but increasing number of particles, is exact.[7] Considered as a whole, it is equivalent to the Liouville equation for the full distribution function. As long as no approximation is made, the BBGKY hierarchy is valid whatever the particle density (provided, however, that we remain in the framework of pair interactions). Its major interest is to allow us to proceed to approximations leading to truncating the hierarchy at some level (in practice, after $f^{(1)}$).

An example of such an approximation is the *mean field approximation*, in which the correlations between particles are neglected. This approximation allows us to obtain an evolution equation for the one-particle distribution function applicable to the plasmas (ionized gases): this is the *Vlasov equation*.

4. The Vlasov equation

4.1. Mean field approximation

Let us come back to the BBGKY hierarchy, and assume that we can neglect the correlations in phase space and write in an approximate way:

$$f^{(2)}(\boldsymbol{r}_1, \boldsymbol{p}_1, \boldsymbol{r}_2, \boldsymbol{p}_2, t) \simeq f^{(1)}(\boldsymbol{r}_1, \boldsymbol{p}_1, t) f^{(1)}(\boldsymbol{r}_2, \boldsymbol{p}_2, t). \tag{4.4.1}$$

Equation (4.3.13) then becomes a closed equation for $f^{(1)}$:

$$\frac{\partial f^{(1)}}{\partial t} + \boldsymbol{v}_1.\nabla_{\boldsymbol{r}_1} f^{(1)} + \left[\boldsymbol{X}_1 + \int \boldsymbol{X}_{12} f^{(1)}(\boldsymbol{r}_2, \boldsymbol{p}_2, t)\, d\Omega_2\right].\nabla_{\boldsymbol{p}_1} f^{(1)} = 0. \tag{4.4.2}$$

[7] It is valid even in the presence of scalar and vector potentials corresponding to applied external fields.

The existence of pair interactions leads to an extra driving term: the force produced on average by the other particles, namely, the force $\int \boldsymbol{X}_{12} f^{(1)}(\boldsymbol{r}_2, \boldsymbol{p}_2, t)\, d\Omega_2$, must be added to the external force $\boldsymbol{X}_1$.

The approximation made in discarding the correlations in phase space belongs to the general class of mean field approximations. The mean field, which here is the force $\int \boldsymbol{X}_{12} f^{(1)}(\boldsymbol{r}_2, \boldsymbol{p}_2, t)\, d\Omega_2$, itself depends on the distribution function which we aim to determine. Equation (4.4.2), generalized to a mixture of two types of particles (electrons and ions), was proposed by A.A. Vlasov in 1938 for studying the evolution of an out-of-equilibrium plasma.[8] A plasma is a medium in which the Coulomb interactions between the charged particles play an essential role. Owing to the long range of these interactions, the mean potential acting on a given particle is actually produced by a very large number of other particles. The main effect of the Coulomb interactions is thus the collective mean field effect.

4.2. Time-reversal invariance

At the microscopic level, the Hamiltonian H_N only depends on the generalized coordinates q_i and generalized momenta p_i. Hamilton's equations are invariant under a change of sign of time:[9] $t \to -t$. This is also the case of the Liouville equation for f, as well as of the BBGKY hierarchy equations, as long as no approximation likely to break this invariance is made.

The decorrelation approximation (4.4.1) preserves the time-reversal invariance. As a consequence, the Vlasov equation, like the Liouville equation, is invariant under time-reversal: if we change t into $-t$ in equation (4.4.2), we have to simultaneously change $\boldsymbol{p}$ into $-\boldsymbol{p}$, and we note that $f^{(1)}(\boldsymbol{r}, -\boldsymbol{p}, -t)$ obeys the same equation as $f^{(1)}(\boldsymbol{r}, \boldsymbol{p}, t)$. The Vlasov equation in fact describes the beginning of the evolution of an ionized gas from an initial out-of-equilibrium state, but it cannot describe the irreversible process which eventually leads this plasma towards equilibrium once the external constraints have been removed.

5. Gauge invariance

Consider a system of charged particles in the presence of a scalar potential and a vector potential describing an external electromagnetic field. The Liouville equation and the BBGKY hierarchy have been obtained in the framework of the Hamiltonian

---

[8] A substance whose temperature is elevated successively passes through liquid, then gaseous, phases. When the temperature further increases, the translational kinetic energy of each particle becomes higher than the ionization energy of its constituent atoms. The collisions between the molecules then put the electrons in free states and leave positively charged ions, the gas remaining globally neutral. This state of matter is called a plasma. A plasma is thus composed of positively and negatively charged particles in non-bounded states. These particles interact via Coulomb interactions. To describe a plasma, we have to introduce a couple of one-particle distribution functions, one for the electrons, the other for the ions. We can write, for these two distribution functions, Vlasov equations generalizing equation (4.4.2). Some phenomena occurring in out-of-equilibrium plasmas are properly described by these equations. We can quote in particular the *Landau damping* of plasma waves (see Supplement 6A).

[9] In the presence of a vector potential, we also have to change the sign of the latter.

formalism in which the dynamical variables are expressed in terms of the generalized coordinates and momenta, and where we make use of the potentials, and not of the fields. As a consequence, these evolution equations are not gauge-invariant.

To display the gauge invariance, it is necessary to work with the aid of gauge-independent quantities, the dynamical variables being expressed in terms of the coordinates and the velocities, and the use of the electric and magnetic fields replacing that of the potentials. We will illustrate this point on the simple example of the Liouville equation for the one-particle distribution function[10] (the case without interactions). This equation reads, in the Hamiltonian formalism,

$$\frac{\partial f(\boldsymbol{r},\boldsymbol{p},t)}{\partial t}+\boldsymbol{v}.\nabla_{\boldsymbol{r}}f(\boldsymbol{r},\boldsymbol{p},t)+\boldsymbol{X}.\nabla_{\boldsymbol{p}}f(\boldsymbol{r},\boldsymbol{p},t)=0, \tag{4.5.1}$$

where $\boldsymbol{X}$ denotes the force due to the scalar and vector potentials.[11] In order to respect gauge invariance, we must, in place of equation (4.5.1), write the evolution equation of the one-particle distribution function $F(\boldsymbol{r},\boldsymbol{v},t)$ of gauge-independent arguments $\boldsymbol{r}$ and $\boldsymbol{v}$. This latter function is defined through the identity:

$$F(\boldsymbol{r},\boldsymbol{v},t)\,d\boldsymbol{r}d\boldsymbol{v}=f(\boldsymbol{r},\boldsymbol{p},t)\,d\boldsymbol{r}d\boldsymbol{p}. \tag{4.5.2}$$

Since $\boldsymbol{p}=m\boldsymbol{v}+q\boldsymbol{A}(\boldsymbol{r},t)/c$, we have:

$$F(\boldsymbol{r},\boldsymbol{v},t)=m^3f(\boldsymbol{r},\boldsymbol{p},t). \tag{4.5.3}$$

We can deduce,[12] from equation (4.5.1) for $f(\boldsymbol{r},\boldsymbol{p},t)$, the evolution equation of $F(\boldsymbol{r},\boldsymbol{v},t)$:

$$\frac{\partial F(\boldsymbol{r},\boldsymbol{v},t)}{\partial t}+\boldsymbol{v}.\nabla_{\boldsymbol{r}}F(\boldsymbol{r},\boldsymbol{v},t)+\frac{1}{m}q\big[\boldsymbol{E}(\boldsymbol{r},t)+\frac{1}{c}\boldsymbol{v}\times\boldsymbol{H}(\boldsymbol{r},t)\big].\nabla_{\boldsymbol{v}}F(\boldsymbol{r},\boldsymbol{v},t)=0. \tag{4.5.4}$$

In the gauge-invariant equation (4.5.4), the driving force is the Lorentz force $q\big[\boldsymbol{E}(\boldsymbol{r},t)+\boldsymbol{v}\times\boldsymbol{H}(\boldsymbol{r},t)/c\big]$.

This type of analysis may be extended to the evolution in the presence of pair interactions.[13]

[10] For the sake of simplicity, this distribution function is denoted here by f (instead of $f^{(1)}$).

[11] The expression for $\boldsymbol{X}$ in terms of the scalar and vector potentials is given in an appendix at the end of this chapter.

[12] The details of the calculation are given in the appendix.

[13] The gauge-invariant evolution equation, extended to the case with pair interactions, plays an important role in electronic transport in the presence of a magnetic field (see Chapters 6 and 8).

Appendices

4A. Pair interaction potentials

The form of the interaction potential between two point particles depends in particular on the nature, charged or neutral, of the particles.

4A.1. Interaction between two charged particles

• *Coulomb potential*

The electrostatic interaction between two particles of charge q separated by a distance r is described by the Coulomb potential[14] $u_C(r)$:

$$u_C(r) = \frac{q^2}{r}. \tag{4A.1}$$

As with any function decreasing as a power law, the Coulomb potential lacks a parameter characterizing its range, which may therefore be considered as infinite.

• *Screened Coulomb potential*

In a gas of charged particles, the Coulomb interaction between two particles is actually screened by the other particles. At equilibrium, the resulting effective interaction is described by the screened Coulomb potential[15] $u_{C,\mathrm{scr}}(r)$:

$$u_{C,\mathrm{scr}}(r) = \frac{q^2 e^{-k_0 r}}{r}. \tag{4A.2}$$

The range of $u_{C,\mathrm{scr}}(r)$ is characterized by the screening length k_0^{-1}, whose expression depends on the character, degenerate or not, of the gas.

In a classical gas of particles of density n in equilibrium at temperature T, the screening length is the *Debye length*: $k_0^{-1} = (kT/4\pi nq^2)^{1/2}$. In a quantum gas at $T = 0$, for instance a gas of electrons with charge $q = e$, the screening length is the *Thomas–Fermi length*: $k_0^{-1} = (kT_F/4\pi ne^2)^{1/2}$, where T_F denotes the Fermi temperature.[16]

---

[14] We are using Gauss units.

[15] This potential also appears in nuclear physics, where it is known as the *Yukawa potential*.

[16] Note that the expression for the Thomas–Fermi length is formally similar to that for the Debye length, the temperature T being replaced by the Fermi temperature T_F.

4A.2. Interaction between two neutral particles

The two particles represent schematically two neutral atoms. The interaction between two neutral atoms is repulsive at short distances (because of the Pauli principle, the electronic clouds do not interpenetrate), and attractive at large distances (this is the van der Waals interaction between the induced dipolar moments). This interaction is generally modelized either by the *Lennard–Jones potential* which conveniently describes its characteristics over the whole set of values of r or, in a more schematic way, by the *hard spheres potential.*

• *Lennard–Jones potential*

The Lennard–Jones potential $u_{LJ}(r)$ is:

$$u_{LJ}(r) = u_0\left[\left(\frac{r_0}{r}\right)^{12} - 2\left(\frac{r_0}{r}\right)^{6}\right]. \tag{4A.3}$$

It depends on two parameters, the maximal strength u_0 of the attractive interaction and the position r_0 of the minimum (which defines the range of the interaction).

• *Hard spheres potential*

We sometimes use, instead of the Lennard–Jones potential, a simplified model, the hard spheres potential $u_{HS}(r)$:

$$u_{HS}(r) = \begin{cases} \infty, & r < r_0 \\ 0, & r > r_0. \end{cases} \tag{4A.4}$$

The hard spheres model represents spherical atoms of radius $r_0/2$ moving freely in space, except for collisions with other atoms at distance r_0. The hard spheres potential has no parameter characterizing its strength, the latter being from construction either vanishing (for $r > r_0$) or infinite (for $r < r_0$).

4B. Hamilton's equations for a charged particle

The Hamiltonian of a particle of mass m and charge q in the presence of a scalar potential $\phi(\boldsymbol{r}, t)$ and of a vector potential $\boldsymbol{A}(\boldsymbol{r}, t)$ is:

$$H_1(\boldsymbol{r}, \boldsymbol{p}) = \frac{[\boldsymbol{p} - q\boldsymbol{A}(\boldsymbol{r}, t)/c]^2}{2m} + q\phi(\boldsymbol{r}, t). \tag{4B.1}$$

The electric and magnetic fields derive from the scalar and vector potentials:

$$\begin{cases} \boldsymbol{E}(\boldsymbol{r}, t) = -\nabla\phi(\boldsymbol{r}, t) - \dfrac{1}{c}\dfrac{\partial \boldsymbol{A}(\boldsymbol{r}, t)}{\partial t} \\ \boldsymbol{H}(\boldsymbol{r}, t) = \nabla \times \boldsymbol{A}(\boldsymbol{r}, t). \end{cases} \tag{4B.2}$$

The set $\{\phi(\boldsymbol{r}, t),\ \boldsymbol{A}(\boldsymbol{r}, t)\}$ constitutes a gauge. The gauge is not unique, in the sense that the potentials $\phi'(\boldsymbol{r}, t)$ and $\boldsymbol{A}'(\boldsymbol{r}, t)$, defined by the formulas:

$$\begin{cases} \phi'(\boldsymbol{r}, t) = \phi(\boldsymbol{r}, t) - \dfrac{1}{c}\dfrac{\partial \chi(\boldsymbol{r}, t)}{\partial t} \\ \boldsymbol{A}'(\boldsymbol{r}, t) = \boldsymbol{A}(\boldsymbol{r}, t) + \nabla\chi(\boldsymbol{r}, t), \end{cases} \tag{4B.3}$$

where $\chi(\boldsymbol{r},t)$ is any function of $\boldsymbol{r}$ and t, describe the same electromagnetic field as $\phi(\boldsymbol{r},t)$ and $\boldsymbol{A}(\boldsymbol{r},t)$.

The evolution equation of the particle's position operator $\boldsymbol{r}$ is the Hamilton's equation $d\boldsymbol{r}/dt = \nabla_{\boldsymbol{p}} H_1$, which reads, in virtue of formula (4B.1):

$$\frac{d\boldsymbol{r}}{dt} = \frac{\boldsymbol{p} - q\boldsymbol{A}(\boldsymbol{r},t)/c}{m}. \tag{4B.4}$$

Thus, in the presence of a vector potential $\boldsymbol{A}(\boldsymbol{r},t)$, the relation between the momentum $\boldsymbol{p}$ and the velocity $\boldsymbol{v} = d\boldsymbol{r}/dt$ of a particle of mass m and charge q is:

$$\boldsymbol{p} = m\boldsymbol{v} + \frac{q\boldsymbol{A}(\boldsymbol{r},t)}{c}. \tag{4B.5}$$

The kinetic momentum $m\boldsymbol{v}$ is a true, gauge-independent, physical quantity. Equation (4B.5) shows that on the other hand the momentum $\boldsymbol{p}$ depends on the gauge choice.

In the presence of electric and magnetic fields, we have:

$$m\frac{d\boldsymbol{v}}{dt} = q\big[\boldsymbol{E}(\boldsymbol{r},t) + \frac{1}{c}\boldsymbol{v}\times\boldsymbol{H}(\boldsymbol{r},t)\big]. \tag{4B.6}$$

The expression for the Lorentz force on the right-hand side of equation (4B.6) involves the fields $\boldsymbol{E}(\boldsymbol{r},t)$ and $\boldsymbol{H}(\boldsymbol{r},t)$. The Lorentz force is thus gauge-invariant. The evolution equation of the momentum operator $\boldsymbol{p}$ is the Hamilton's equation $d\boldsymbol{p}/dt = -\nabla_{\boldsymbol{r}} H_1$, from which we get, for instance, using formula (4B.1), the evolution equation of p_x:[17]

$$\frac{dp_x}{dt} = -q\frac{\partial\phi(\boldsymbol{r},t)}{\partial x} + \frac{q}{c}\boldsymbol{v}.\frac{\partial\boldsymbol{A}(\boldsymbol{r},t)}{\partial x}. \tag{4B.7}$$

We finally write:

$$\frac{d\boldsymbol{p}}{dt} = \boldsymbol{X}, \tag{4B.8}$$

where:

$$\boldsymbol{X} = -q\nabla\phi(\boldsymbol{r},t) + \frac{q}{c}\boldsymbol{v}\times\big[\nabla\times\boldsymbol{A}(\boldsymbol{r},t)\big] + \frac{q}{c}(\boldsymbol{v}.\nabla)\boldsymbol{A}(\boldsymbol{r},t) \tag{4B.9}$$

represents the force due to the scalar and vector potentials.

---

[17] Formula (4B.7) can also be derived in the following way. We deduce from equation (4B.5) the equality:

$$\frac{d\boldsymbol{p}}{dt} = m\frac{d\boldsymbol{v}}{dt} + \frac{q}{c}\frac{d\boldsymbol{A}(\boldsymbol{r},t)}{dt},$$

which also reads:

$$\frac{d\boldsymbol{p}}{dt} = m\frac{d\boldsymbol{v}}{dt} + \frac{q}{c}\frac{\partial\boldsymbol{A}(\boldsymbol{r},t)}{\partial t} + \frac{q}{c}(\boldsymbol{v}.\nabla)\boldsymbol{A}(\boldsymbol{r},t).$$

Making use of the expression (4B.6) for $m d\boldsymbol{v}/dt$, we get:

$$\frac{d\boldsymbol{p}}{dt} = q\Big[\boldsymbol{E}(\boldsymbol{r},t) + \frac{1}{c}\boldsymbol{v}\times\boldsymbol{H}(\boldsymbol{r},t)\Big] + \frac{q}{c}\frac{\partial\boldsymbol{A}(\boldsymbol{r},t)}{\partial t} + \frac{q}{c}(\boldsymbol{v}.\nabla)\boldsymbol{A}(\boldsymbol{r},t),$$

that is:

$$\frac{d\boldsymbol{p}}{dt} = -q\nabla\phi(\boldsymbol{r},t) + \frac{q}{c}\boldsymbol{v}\times\Big[\nabla\times\boldsymbol{A}(\boldsymbol{r},t)\Big] + \frac{q}{c}(\boldsymbol{v}.\nabla)\boldsymbol{A}(\boldsymbol{r},t).$$

We thus recover expression (4B.7) for dp_x/dt.

4C. Gauge invariance of the Liouville equation

To deduce from the Liouville equation (4.5.1) for $f(\boldsymbol{r},\boldsymbol{p},t)$ the gauge-invariant Liouville equation (4.5.4) obeyed by $F(\boldsymbol{r},\boldsymbol{v},t)$, we first multiply equation (4.5.1) by m^3:

$$m^3\frac{\partial f(\boldsymbol{r},\boldsymbol{p},t)}{\partial t}+m^3\boldsymbol{v}.\nabla_{\boldsymbol{r}}f(\boldsymbol{r},\boldsymbol{p},t)+m^3\dot{\boldsymbol{p}}.\nabla_{\boldsymbol{p}}f(\boldsymbol{r},\boldsymbol{p},t)=0. \tag{4C.1}$$

We then successively compute the three terms on the left-hand side of equation (4C.1).

- *Term* $m^3\partial f/\partial t$

From the equality:

$$m^3 f(\boldsymbol{r},\boldsymbol{p},t)=F\left[\boldsymbol{r},\frac{\boldsymbol{p}-q\boldsymbol{A}(\boldsymbol{r},t)/c}{m},t\right], \tag{4C.2}$$

we deduce:

$$m^3\frac{\partial f}{\partial t}=\frac{\partial F}{\partial t}-\frac{1}{m}\frac{q}{c}\frac{\partial \boldsymbol{A}(\boldsymbol{r},t)}{\partial t}\cdot\nabla_{\boldsymbol{v}}F. \tag{4C.3}$$

- *Term* $m^3\boldsymbol{v}.\nabla_{\boldsymbol{r}}f$

We have the relation:

$$m^3\frac{\partial f}{\partial x}=\frac{\partial F}{\partial x}-\frac{1}{m}\frac{q}{c}\left(\frac{\partial F}{\partial v_x}\frac{\partial A_x}{\partial x}+\frac{\partial F}{\partial v_y}\frac{\partial A_y}{\partial x}+\frac{\partial F}{\partial v_z}\frac{\partial A_z}{\partial x}\right), \tag{4C.4}$$

as well as analogous formulas for $\partial f/\partial y$ and $\partial f/\partial z$. The quantity $m^3\boldsymbol{v}.\nabla_{\boldsymbol{r}}f$ can thus be expressed in terms of F:

$$\begin{aligned}\boldsymbol{v}.\nabla_{\boldsymbol{r}}F-\frac{1}{m}\frac{q}{c}\Bigg\{v_x&\left(\frac{\partial F}{\partial v_x}\frac{\partial A_x}{\partial x}+\frac{\partial F}{\partial v_y}\frac{\partial A_y}{\partial x}+\frac{\partial F}{\partial v_z}\frac{\partial A_z}{\partial x}\right)\\&+v_y\left(\frac{\partial F}{\partial v_x}\frac{\partial A_x}{\partial y}+\frac{\partial F}{\partial v_y}\frac{\partial A_y}{\partial y}+\frac{\partial F}{\partial v_z}\frac{\partial A_z}{\partial y}\right)\\&+v_z\left(\frac{\partial F}{\partial v_x}\frac{\partial A_x}{\partial z}+\frac{\partial F}{\partial v_y}\frac{\partial A_y}{\partial z}+\frac{\partial F}{\partial v_z}\frac{\partial A_z}{\partial z}\right)\Bigg\}.\end{aligned} \tag{4C.5}$$

- *Term* $m^3\dot{\boldsymbol{p}}.\nabla_{\boldsymbol{p}}f(\boldsymbol{r},\boldsymbol{p},t)$

We have:

$$m^3\nabla_{\boldsymbol{p}}f(\boldsymbol{r},\boldsymbol{p},t)=\frac{1}{m}\nabla_{\boldsymbol{v}}F(\boldsymbol{r},\boldsymbol{v},t). \tag{4C.6}$$

As the quantity $\dot{\boldsymbol{p}}$, given by the formula:

$$\dot{\boldsymbol{p}}=q\big[\boldsymbol{E}(\boldsymbol{r},t)+\frac{1}{c}\boldsymbol{v}\times\boldsymbol{H}(\boldsymbol{r},t)\big]+\frac{q}{c}\frac{\partial \boldsymbol{A}(\boldsymbol{r},t)}{\partial t}+\frac{q}{c}(\boldsymbol{v}.\nabla)\boldsymbol{A}(\boldsymbol{r},t), \tag{4C.7}$$

is a sum of three terms, the quantity $m^3\dot{\boldsymbol{p}}.\nabla_{\boldsymbol{p}}f(\boldsymbol{r},\boldsymbol{p},t)$ also appears as a sum of three terms.

The first one involves the Lorentz force. It reads:

$$\frac{1}{m}q\big[\boldsymbol{E}(\boldsymbol{r},t)+\frac{1}{c}\boldsymbol{v}\times\boldsymbol{H}(\boldsymbol{r},t)\big].\nabla_{\boldsymbol{v}}F. \tag{4C.8}$$

The second one, given by:

$$\frac{1}{m}\frac{q}{c}\frac{\partial \boldsymbol{A}(\boldsymbol{r},t)}{\partial t}\cdot\nabla_{\boldsymbol{v}}F, \tag{4C.9}$$

cancels the opposite term appearing in the contribution (4C.3). The third one, given by:

$$\frac{1}{m}\frac{q}{c}\big[(\boldsymbol{v}.\nabla)\boldsymbol{A}(\boldsymbol{r},t)\big].\nabla_{\boldsymbol{v}}F, \tag{4C.10}$$

also reads:

$$\begin{aligned}\frac{1}{m}\frac{q}{c}\bigg\{\bigg(v_x\frac{\partial A_x}{\partial x}+v_y\frac{\partial A_x}{\partial y}+v_z\frac{\partial A_x}{\partial z}\bigg)\frac{\partial F}{\partial v_x}\\ +\bigg(v_x\frac{\partial A_y}{\partial x}+v_y\frac{\partial A_y}{\partial y}+v_z\frac{\partial A_y}{\partial z}\bigg)\frac{\partial F}{\partial v_y}\\ +\bigg(v_x\frac{\partial A_z}{\partial x}+v_y\frac{\partial A_z}{\partial y}+v_z\frac{\partial A_z}{\partial z}\bigg)\frac{\partial F}{\partial v_z}\bigg\}.\end{aligned} \tag{4C.11}$$

It cancels the term between brackets in expression (4C.5).

Combining all contributions, and taking into account the above-mentioned compensations, we obtain for the distribution function $F(\boldsymbol{r},\boldsymbol{v},t)$ the gauge-invariant Liouville equation (4.5.4), in which the driving force is the Lorentz force.

Bibliography

R. Balescu, *Statistical dynamics. Matter out of equilibrium*, Imperial College Press, London, 1997.

R. Balian, *From microphysics to macrophysics*, Vol. 1, Springer-Verlag, Berlin, 1991; Vol. 2, Springer-Verlag, Berlin, 1992.

C. Cohen-Tannoudji, B. Diu, and F. Laloë, *Quantum mechanics*, Vol. 1, Hermann and Wiley, Paris, second edition, 1977.

D.L. Goodstein, *States of matter*, Prentice-Hall, Englewood Cliffs, 1975. Reprinted, Dover Publications, New York, 2002.

K. Huang, *Statistical mechanics*, Wiley, New York, second edition, 1987.

H.J. Kreuzer, *Nonequilibrium thermodynamics and its statistical foundations*, Clarendon Press, Oxford, 1981.

R. Kubo, M. Toda, and N. Hashitsume, *Statistical physics II: nonequilibrium statistical mechanics*, Springer-Verlag, Berlin, second edition, 1991.

E.M. Lifshitz and L.P. Pitaevskii, *Physical kinetics*, Butterworth-Heinemann, Oxford, 1981.

J.A. McLennan, *Introduction to nonequilibrium statistical mechanics*, Prentice-Hall, Englewood Cliffs, 1989.

D.A. McQuarrie, *Statistical mechanics*, University Science Books, Sausalito, second edition, 2000.

M. Toda, R. Kubo, and N. Saitô, *Statistical physics I: equilibrium statistical mechanics*, Springer-Verlag, Berlin, second edition, 1992.

D. Zubarev, V. Morozov, and G. Röpke, *Statistical mechanics of nonequilibrium processes*, Vol. 1: *Basic concepts, kinetic theory*, Akademie Verlag, Berlin, 1996.

Chapter 5

The Boltzmann equation

The statistical evolution of a classical system of N particles with pair interactions can in principle be studied by means of the BBGKY hierarchy for the reduced distribution functions. If no approximation is made, the evolution equation of the one-particle distribution function involves the two-particle distribution function, and so on. The study of the evolution of the one-particle distribution function is thus tricky. However, in some cases, it is possible, through proper approximations, to obtain a closed evolution equation for this distribution function. If the effect of collisions leading to an irreversible evolution is taken into account by these approximations, the evolution equation thus obtained for the one-particle distribution function belongs to the class of kinetic equations. There are several types of such equations, each of them relating to a particular physical system placed in given conditions.

Historically, the first systems to be studied by means of a kinetic equation were the dilute classical gases of molecules undergoing binary collisions. In this case, the kinetic approach relies on the 'molecular chaos hypothesis', according to which, in the description of a collision, the possible correlations between the velocities of the two colliding molecules prior to their collision may be neglected. The evolution of the one-particle distribution function is then governed by the Boltzmann equation (L. Boltzmann, 1872). A theorem expressing the irreversibility of the evolution, namely the H-theorem, is closely associated with this equation. This approach of the irreversible evolution of a dilute classical gas turned out to be extremely fruitful. It rendered possible in particular the computation of the thermal conductivity and the viscosity coefficient of a gas in terms of microscopic quantities such as the collision time.

As a general rule, the validity of the kinetic approaches relies on the existence within the system of two well-separated time scales. In the case of dilute classical gases to which the Boltzmann equation applies, the shorter time scale is the duration of a collision, whereas the longer one is the mean time interval separating two successive collisions of a given molecule.

1. Statistical description of dilute classical gases

1.1. Dilute classical gas

The classical kinetic theory deals with a dilute gas of N identical molecules of mass m confined inside a box of volume V. The gas is considered as ideal, which means that the potential energy of interaction between the molecules is negligible relative to their kinetic energy. At a fixed temperature, this approximation is all the better for the gas being more dilute. The dilution can be expressed through the inequality:

$$r_0 \ll d, \tag{5.1.1}$$

where r_0 denotes the range of the intermolecular forces and $d \sim n^{-1/3}$ the mean distance between the molecules ($n = N/V$ is the gas density). Thus, the molecules in a dilute gas are most of the time free and independent. However, the collisions, which redistribute energy between the molecules, play an essential role in the evolution of the gas towards equilibrium. We will study this effect, taking into account binary collisions only.[1]

Besides, the temperature is assumed sufficiently high and the gas density low enough for the molecules to be conveniently represented by localized wave packets, whose dimensions, as measured by the thermal wavelength $\lambda = h(2\pi mkT)^{-1/2}$, are small compared to the mean intermolecular distance:

$$\lambda \ll d. \tag{5.1.2}$$

Any molecule of the gas may then be considered as a classical particle with well-defined position and momentum. The molecules are treated as indistinguishable particles.

1.2. Role of the one-particle distribution function

The gas, as modelized by a system of N classical indistinguishable point particles, is described by a Hamiltonian depending on the coordinates and momenta of all the particles. The phase space being $6N$-dimensional, the distribution function of the system depends, besides time, on $6N$ variables coordinates and momenta.

In the case of a dilute gas, it is however not necessary to know the full distribution function in order to describe most macroscopic properties. Indeed, a given molecule does not interact with more than one other molecule at the same time, and it moves freely between two successive collisions. The duration τ_0 of any collision is much smaller than the *collision time* τ, defined as the mean time interval separating two successive collisions of a given molecule: in a dilute gas, a molecule, for most of the time, does not interact with others. The macroscopic properties of a dilute gas can thus be obtained from the one-particle distribution function, which depends, besides time, on six variables (coordinates and momenta).

[1] Collisions involving more than two molecules will be neglected, which is justified in a dilute gas.

2. Time and length scales

2.1. Time scales

To derive an evolution equation for the one-particle distribution function f in a dilute gas amounts to obtaining an expression for df/dt appropriate to the physics of the problem. Note that, in the quantity written in the mathematical form of the derivative df/dt, the symbol dt does not stand for an infinitesimal time interval, but rather for a finite time interval Δt during which a variation Δf (represented by df) of the distribution function is produced. This variation is due in particular to the collisions taking place during Δt. This time interval has to be compared to the various time scales characteristic of the evolution of the gas.

The shortest of these time scales is the duration τ_0 of a collision, much smaller than the collision time τ: $\tau_0 \ll \tau$. The collision time is itself very small compared with the relaxation time τ_r towards a local equilibrium, since such an equilibrium results from numerous collisions:[2] $\tau \ll \tau_r$. The final time scale, denoted τ_{eq}, is much longer than τ_r, and is characteristic of the macroscopic evolution of the gas towards global thermodynamic equilibrium: $\tau_r \ll \tau_{\text{eq}}$.

The *Boltzmann equation* is a closed equation for the one-particle distribution function in a dilute classical gas. It describes the evolution of f over a time interval Δt intermediate between τ_0 and τ_r. In other words, for such a time interval, the Boltzmann equation enables us to compute $f(t + \Delta t)$ given $f(t)$. This stage of the evolution of an initially out-of-equilibrium gas, as characterized by the inequalities:

$$\tau_0 \ll \Delta t \ll \tau_r, \tag{5.2.1}$$

is called the *kinetic* stage.[3]

2.2. Length scales

Similarly, it exists within the gas several characteristic length scales, in particular the range r_0 of the intermolecular forces and the mean free path ℓ. The mean free path is defined as the length scale associated with the collision time τ: it is the mean distance covered by a molecule between two successive collisions. We can take as an estimate:[4]

$$\ell \sim \frac{1}{n r_0^2} \sim \frac{d^3}{r_0^2}. \tag{5.2.2}$$

The gas being dilute, we have $r_0 \ll d$ and, consequently, $d \ll \ell$. We have thus also $r_0 \ll \ell$. Finally, we have to take into account a final length scale, macroscopic and denoted L, characteristic of the linear dimensions of the box containing the gas.

---

[2] However, we often take for the relaxation time the estimate $\tau_r \sim \tau$ (see Chapter 6 and Chapter 7).

[3] The kinetic stage is followed by a *hydrodynamic* stage, corresponding to the evolution of f over a much longer time interval Δt ($\tau_r \ll \Delta t \ll \tau_{\text{eq}}$). In the case of a dilute gas in local equilibrium, the *hydrodynamic equations* can be derived from the Boltzmann equation (see Chapter 7).

[4] See Chapter 7.

The distances Δl which come into play in the Boltzmann equation[5] are intermediate between r_0 and ℓ:

$$r_0 \ll \Delta l \ll \ell. \qquad (5.2.3)$$

3. Notations and definitions

3.1. Notations

In order to treat systems of charged particles in the presence of external electric and magnetic fields, it is convenient to use the distribution function $F(\boldsymbol{r}, \boldsymbol{v}, t)$ of gauge-independent arguments $\boldsymbol{r}$ and $\boldsymbol{v}$.

However, it is better to use in practice, instead of $F(\boldsymbol{r}, \boldsymbol{v}, t)$, the distribution function having as arguments the position $\boldsymbol{r}$ and the kinetic momentum $m\boldsymbol{v}$ (generally denoted for short by $\boldsymbol{p}$). This latter distribution function, denoted by $f(\boldsymbol{r}, \boldsymbol{p}, t)$, is related to $F(\boldsymbol{r}, \boldsymbol{v}, t)$ through the equality:

$$f(\boldsymbol{r}, \boldsymbol{p}, t) = \frac{1}{m^3} F(\boldsymbol{r}, \boldsymbol{v}, t), \qquad (5.3.1)$$

with[6] $\boldsymbol{p} = m\boldsymbol{v}$. The quantity $f(\boldsymbol{r}, \boldsymbol{p}, t)\, d\boldsymbol{r} d\boldsymbol{p}$ represents the mean number of molecules[7] which, at time t, are in the phase space volume element $d\boldsymbol{r} d\boldsymbol{p}$ about the point $(\boldsymbol{r}, \boldsymbol{p})$.

Over a scale Δt, the collisions are considered as instantaneous. As for the linear dimensions Δl of the spatial volume element $d\boldsymbol{r}$, they are considered as much larger than r_0. The collisions which take place in this volume element modify the kinetic momentum of the molecules concerned, but they leave them inside this volume element. Over a scale Δl, the collisions are thus treated as local.

3.2. Local density and local mean velocity

The one-particle distribution function allows us to compute the local density and the local mean velocity of the gas molecules.

• *Local density*

The integral of f over the kinetic momentum is the *local density* $n(\boldsymbol{r}, t)$:

$$\boxed{n(\boldsymbol{r}, t) = \int f(\boldsymbol{r}, \boldsymbol{p}, t)\, d\boldsymbol{p}.} \qquad (5.3.2)$$

[5] For longer distances Δl ($\ell \ll \Delta l \ll L$), the evolution of the gas is described by the hydrodynamic equations (see Chapter 7).

[6] Note that $\boldsymbol{p}$ designates from now on, not the momentum, but the kinetic momentum.

[7] This definition implies that there must be a sufficient number of molecules in the spatial volume element $d\boldsymbol{r}$ or, otherwise stated, that the inequality $d \ll \Delta l$ (more restrictive than the inequality $r_0 \ll \Delta l$) be verified.

The six-dimensional space spanned by the vector $(\boldsymbol{r}, \boldsymbol{p})$ is traditionally called the *μ-space*.[8] A point in this space represents a state of a molecule of the gas. At any time, the state of a gas of N molecules is represented by N points in the μ-space. We have:

$$N = \int n(\boldsymbol{r}, t)\, d\boldsymbol{r} = \int f(\boldsymbol{r}, \boldsymbol{p}, t)\, d\boldsymbol{r} d\boldsymbol{p}. \tag{5.3.3}$$

• *Local mean velocity*

We also define the *mean local velocity* $\boldsymbol{u}(\boldsymbol{r}, t)$:

$$\boxed{\boldsymbol{u}(\boldsymbol{r}, t) = \langle \boldsymbol{v} \rangle = \frac{\int f(\boldsymbol{r}, \boldsymbol{p}, t)\boldsymbol{v}\, d\boldsymbol{p}}{\int f(\boldsymbol{r}, \boldsymbol{p}, t)\, d\boldsymbol{p}} = \frac{1}{n(\boldsymbol{r}, t)} \int f(\boldsymbol{r}, \boldsymbol{p}, t)\boldsymbol{v}\, d\boldsymbol{p}.} \tag{5.3.4}$$

4. Evolution of the distribution function

We can derive the evolution equation of $f(\boldsymbol{r}, \boldsymbol{p}, t)$ using balance arguments.[9] The gas is enclosed inside a box of volume V. An external field, either gravitational or electromagnetic, may possibly act on the molecules. The motion of the molecules between collisions is described by classical mechanics.

4.1. Evolution in the absence of collisions

In the absence of collisions, the force exerted on a particle reduces to the external force $\boldsymbol{F}$. In the case of a particle with charge q in the presence of external electric and magnetic fields $\boldsymbol{E}(\boldsymbol{r}, t)$ and $\boldsymbol{H}(\boldsymbol{r}, t)$, $\boldsymbol{F}$ is the Lorentz force $q\big[\boldsymbol{E}(\boldsymbol{r}, t) + \boldsymbol{v} \times \boldsymbol{H}(\boldsymbol{r}, t)/c\big]$. The one-particle distribution function obeys the evolution equation:

$$\frac{\partial f}{\partial t} + \boldsymbol{v}.\nabla_{\boldsymbol{r}} f + \boldsymbol{F}.\nabla_{\boldsymbol{p}} f = 0. \tag{5.4.1}$$

Equation (5.4.1), identical to the Liouville equation of a system reduced to a unique particle, expresses the density conservation in the μ-space.

4.2. The effect of collisions

In the presence of collisions, the μ-space density is not conserved, and equation (5.4.1) has to be modified accordingly. We then formally write the evolution equation of the distribution function as:

$$\boxed{\frac{\partial f}{\partial t} + \boldsymbol{v}.\nabla_{\boldsymbol{r}} f + \boldsymbol{F}.\nabla_{\boldsymbol{p}} f = \left(\frac{\partial f}{\partial t}\right)_{\text{coll}},} \tag{5.4.2}$$

---

[8] This terminology is due to P. and T. Ehrenfest (1911), who proposed to call *γ-space* the $6N$-dimensional phase space associated with the gas of N molecules and *μ-space* the six-dimensional phase space corresponding to an individual molecule.

[9] The only interactions considered here being pair interactions, it is also possible to derive the Boltzmann equation from the BBGKY hierarchy by truncating it in an appropriate way.

where the right-hand side or *collision term* $(\partial f/\partial t)_{\text{coll}}$ represents the rate of change of f due to collisions.

We usually write the collision term in a form displaying its balance structure,

$$\left(\frac{\partial f}{\partial t}\right)_{\text{coll}} = \left(\frac{\partial f}{\partial t}\right)^{(+)}_{\text{coll}} - \left(\frac{\partial f}{\partial t}\right)^{(-)}_{\text{coll}}, \tag{5.4.3}$$

where $(\partial f/\partial t)^{(+)}_{\text{coll}}$ is called the *entering collision term* and $(\partial f/\partial t)^{(-)}_{\text{coll}}$ the *leaving collision term*. The quantity $(\partial f/\partial t)^{(-)}_{\text{coll}}\, d\boldsymbol{r}d\boldsymbol{p}dt$ (*resp.* $(\partial f/\partial t)^{(+)}_{\text{coll}}\, d\boldsymbol{r}d\boldsymbol{p}dt$) represents the mean number of molecules undergoing a collision between times t and $t+dt$, one of these two molecules finding itself, before (*resp.* after) the collision, in the volume element $d\boldsymbol{r}d\boldsymbol{p}$ around the point $(\boldsymbol{r},\boldsymbol{p})$ of the μ-space.

We now have to specify the form of the entering and leaving collision terms in the case of a dilute gas of molecules undergoing binary collisions.

5. Binary collisions

The collision term $(\partial f/\partial t)_{\text{coll}}$ contains all the physics of the collisions. We will make its expression precise under the following assumptions:

(i) Only binary collisions are taken into account, which is justified provided that the gas is dilute enough.

(ii) The molecules are assumed to be devoid of internal structure. In other words, they are supposed to be monoatomic (this implies the elasticity of collisions, any energy transfer towards internal degrees of freedom being precluded).

(iii) The collisions are considered as local and instantaneous (in particular, any possible effect of the external force $\boldsymbol{F}$ on the collisions is neglected).

(iv) Finally, in the description of a collision between two molecules, any possible correlations between their velocities prior to the collision are neglected. This latter approximation is traditionally called the *molecular chaos hypothesis*. It was formulated by L. Boltzmann in 1872 under the name of *Stosszahlansatz*. This assumption, which is justified provided that the gas density is sufficiently low, plays a crucial role in the derivation of an irreversible kinetic equation.

5.1. Description of collisions in classical mechanics

Consider a collision between two molecules, denoting by $\boldsymbol{p}$ and $\boldsymbol{p}_1$ the kinetic momenta of both molecules before the collision, and by $\boldsymbol{p}'$ and $\boldsymbol{p}'_1$ their kinetic momenta after the collision. The corresponding kinetic energies are respectively denoted by ε, ε_1, and ε', ε'_1. The collision being considered as local and instantaneous, the total kinetic momentum is conserved:

$$\boldsymbol{p} + \boldsymbol{p}_1 = \boldsymbol{p}' + \boldsymbol{p}'_1. \tag{5.5.1}$$

Moreover, the collision being assumed elastic, the total kinetic energy is conserved:

$$\varepsilon + \varepsilon_1 = \varepsilon' + \varepsilon'_1. \tag{5.5.2}$$

Introducing the total kinetic momentum before the collision $\boldsymbol{\Pi} = \boldsymbol{p} + \boldsymbol{p}_1$ and the relative[10] kinetic momentum before the collision $\boldsymbol{\pi} = \frac{1}{2}(\boldsymbol{p}_1 - \boldsymbol{p})$, as well as the corresponding quantities $\boldsymbol{\Pi}'$ and $\boldsymbol{\pi}'$ after the collision, we can rewrite the conservation equations (5.5.1) and (5.5.2) in the equivalent form:

$$\boldsymbol{\Pi} = \boldsymbol{\Pi}' \tag{5.5.3}$$

and:

$$|\boldsymbol{\pi}| = |\boldsymbol{\pi}'|. \tag{5.5.4}$$

As it is elastic, the collision produces a rotation of $\boldsymbol{\pi}$ which brings it on $\boldsymbol{\pi}'$ without changing its modulus. The collision is fully determined by the knowledge of $\boldsymbol{\Pi}$ and $\boldsymbol{\pi}$, as well as of the angles (θ, ϕ), called the *scattering angles*, of $\boldsymbol{\pi}'$ with respect to $\boldsymbol{\pi}$.

The problem is equivalent to that of the scattering of a molecule by a fictitious fixed force center, represented by the point O in Fig. 5.1. The molecule comes near O with a kinetic momentum $\boldsymbol{\pi}$ and an *impact parameter* b. As $|\boldsymbol{\pi}'| = |\boldsymbol{\pi}|$, the final state is made precise by the knowledge of both scattering angles θ and ϕ, collectively denoted by Ω.

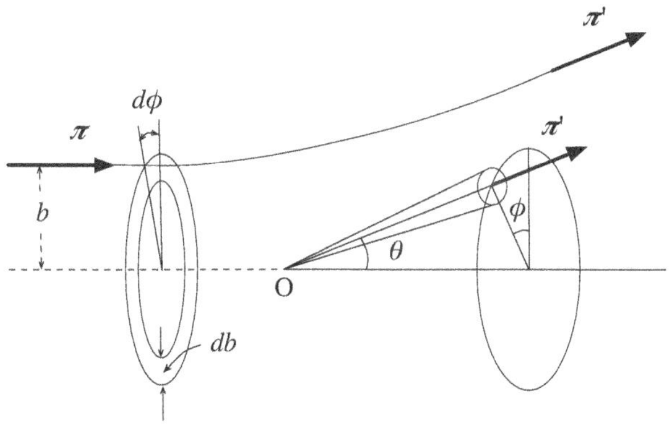

Fig. 5.1 Scattering of a molecule by a fixed force center O.

The sole knowledge of the initial kinetic momenta $\boldsymbol{p}$ and $\boldsymbol{p}_1$ does not suffice to fully determine the collision, since the impact parameter is not known precisely. The knowledge of $\boldsymbol{p}$ and $\boldsymbol{p}_1$ defines a class of collisions with various impact parameters, and thus various scattering angles. We generally describe this class of collisions by imagining a beam of particles with initial kinetic momentum $\boldsymbol{\pi}$, uniformly spread in space, incident on the force center O. The incident flux is defined as the number of molecules per unit time crossing the unit area perpendicular to the incident beam. By definition of the *differential scattering cross-section* $\sigma(\Omega)$, the number of molecules per unit time deflected in a direction pertaining to the solid angle element $d\Omega$ is equal to the product of the incident flux and $\sigma(\Omega)d\Omega$.

[10] The mass associated with the 'relative particle' is the *reduced mass*, which here equals $m/2$ since both particles are identical.

In this classical description of the collision, we write:

$$\sigma(\Omega)\, d\Omega = b\, dbd\phi. \tag{5.5.5}$$

The differential cross-section $\sigma(\Omega)$ can be measured directly. It can also be computed provided that the pair interaction potential is known.[11]

5.2. Properties of the cross-section

We shall not here carry out detailed cross-section calculations. We will only quote some general properties of $\sigma(\Omega)$, valid whatever the particular form of the pair interaction potential. Let us set, for the collision $\{\boldsymbol{p}, \boldsymbol{p}_1\} \to \{\boldsymbol{p}', \boldsymbol{p}'_1\}$:

$$\sigma(\Omega) = \sigma(\boldsymbol{p}, \boldsymbol{p}_1 | \boldsymbol{p}', \boldsymbol{p}'_1). \tag{5.5.6}$$

The interactions between the molecules being of electromagnetic origin, the microscopic equations of motion and, consequently, the scattering cross-section possess the following invariance properties:

(i) time-reversal invariance ($t \to -t$):

$$\sigma(\boldsymbol{p}, \boldsymbol{p}_1 | \boldsymbol{p}', \boldsymbol{p}'_1) = \sigma(-\boldsymbol{p}', -\boldsymbol{p}'_1 | -\boldsymbol{p}, -\boldsymbol{p}_1) \tag{5.5.7}$$

(ii) space inversion invariance ($\boldsymbol{r} \to -\boldsymbol{r}$):

$$\sigma(\boldsymbol{p}, \boldsymbol{p}_1 | \boldsymbol{p}', \boldsymbol{p}'_1) = \sigma(-\boldsymbol{p}, -\boldsymbol{p}_1 | -\boldsymbol{p}', -\boldsymbol{p}'_1) \tag{5.5.8}$$

In the following, we will be interested in the *inverse* collision $\{\boldsymbol{p}', \boldsymbol{p}'_1\} \to \{\boldsymbol{p}, \boldsymbol{p}_1\}$, obtained from the original collision $\{\boldsymbol{p}, \boldsymbol{p}_1\} \to \{\boldsymbol{p}', \boldsymbol{p}'_1\}$ by interchanging the initial and final states. The cross-section for the inverse collision is $\sigma(\boldsymbol{p}', \boldsymbol{p}'_1 | \boldsymbol{p}, \boldsymbol{p}_1)$. Making successive use of the time-reversal invariance (formula (5.5.7)) and of the space inversion invariance (formula (5.5.8)), we obtain the *microreversibility* relation:

$$\boxed{\sigma(\boldsymbol{p}', \boldsymbol{p}'_1 | \boldsymbol{p}, \boldsymbol{p}_1) = \sigma(\boldsymbol{p}, \boldsymbol{p}_1 | \boldsymbol{p}', \boldsymbol{p}'_1).} \tag{5.5.9}$$

---

[11] In contrast to the notion of impact parameter, the notion of scattering cross-section keeps a sense in quantum mechanics. The calculation of $\sigma(\Omega)$ for binary collisions in a dilute gas must be carried out quantum-mechanically (and not by using the classical formula (5.5.5)). Indeed, although the molecules may be considered as classical particles between collisions (the thermal wavelength λ being very small compared to the mean distance d between molecules), this is not the case over spatial scales comparable to the range of the scattering potential (we cannot say $\lambda \ll r_0$). Over such spatial scales, the notion of trajectory loses its significance and the analysis in terms of impact parameter is not correct.

6. The Boltzmann equation

The evolution equation of the distribution function $f(\boldsymbol{r},\boldsymbol{p},t)$ in the presence of collisions is of the form:

$$\frac{\partial f(\boldsymbol{r},\boldsymbol{p},t)}{\partial t}+\boldsymbol{v}.\nabla_{\boldsymbol{r}}f(\boldsymbol{r},\boldsymbol{p},t)+\boldsymbol{F}.\nabla_{\boldsymbol{p}}f(\boldsymbol{r},\boldsymbol{p},t)=\left(\frac{\partial f}{\partial t}\right)^{(+)}_{\text{coll}}-\left(\frac{\partial f}{\partial t}\right)^{(-)}_{\text{coll}}. \qquad (5.6.1)$$

We are looking for an explicit expression for the entering and leaving collision terms $(\partial f/\partial t)^{(-)}_{\text{coll}}$ and $(\partial f/\partial t)^{(+)}_{\text{coll}}$, the molecular chaos hypothesis being taken into account. According to this assumption, in a spatial volume element $d\boldsymbol{r}$ centered at $\boldsymbol{r}$, the mean number of pairs of molecules with kinetic momenta in the elements $d\boldsymbol{p}$ centered at $\boldsymbol{p}$ and $d\boldsymbol{p}_1$ centered at $\boldsymbol{p}_1$ can be written, at time t, in the factorized form:

$$\big[f(\boldsymbol{r},\boldsymbol{p},t)\,d\boldsymbol{r}d\boldsymbol{p}\big]\times\big[f(\boldsymbol{r},\boldsymbol{p}_1,t)\,d\boldsymbol{r}d\boldsymbol{p}_1\big]. \qquad (5.6.2)$$

6.1. Leaving collision term

The rate of decrease of $f(\boldsymbol{r},\boldsymbol{p},t)$ due to collisions may be obtained by first focusing on a given molecule situated in the spatial volume element $d\boldsymbol{r}$ centered at $\boldsymbol{r}$ and having its kinetic momentum in the element $d\boldsymbol{p}$ centered at $\boldsymbol{p}$. In the spatial volume element we are considering, there are molecules of kinetic momentum $\boldsymbol{p}_1$ which form a beam of molecules incident on the molecule of interest. The corresponding incident flux is:

$$f(\boldsymbol{r},\boldsymbol{p}_1,t)\,d\boldsymbol{p}_1|\boldsymbol{v}-\boldsymbol{v}_1|. \qquad (5.6.3)$$

The number of collisions of the type $\{\boldsymbol{p},\boldsymbol{p}_1\}\rightarrow\{\boldsymbol{p}',\boldsymbol{p}'_1\}$ taking place on the molecule of interest between times t and $t+dt$ is:

$$f(\boldsymbol{r},\boldsymbol{p}_1,t)\,d\boldsymbol{p}_1|\boldsymbol{v}-\boldsymbol{v}_1|\sigma(\Omega)\,d\Omega dt, \qquad (5.6.4)$$

where $\sigma(\Omega)$ stands for $\sigma(\boldsymbol{p},\boldsymbol{p}_1|\boldsymbol{p}',\boldsymbol{p}'_1)$. We get, taking into account all the molecules of the considered type (in number $f(\boldsymbol{r},\boldsymbol{p},t)\,d\boldsymbol{r}d\boldsymbol{p}$), and integrating over $\boldsymbol{p}_1$:

$$\left(\frac{\partial f}{\partial t}\right)^{(-)}_{\text{coll}}d\boldsymbol{r}d\boldsymbol{p}dt=f(\boldsymbol{r},\boldsymbol{p},t)\,d\boldsymbol{r}d\boldsymbol{p}\int d\boldsymbol{p}_1\int d\Omega\,\sigma(\Omega)\,dt\,|\boldsymbol{v}-\boldsymbol{v}_1|f(\boldsymbol{r},\boldsymbol{p}_1,t). \qquad (5.6.5)$$

Hence the expression for the leaving collision term is:

$$\left(\frac{\partial f}{\partial t}\right)^{(-)}_{\text{coll}}=f(\boldsymbol{r},\boldsymbol{p},t)\int d\boldsymbol{p}_1\int d\Omega\,\sigma(\Omega)|\boldsymbol{v}-\boldsymbol{v}_1|f(\boldsymbol{r},\boldsymbol{p}_1,t). \qquad (5.6.6)$$

6.2. Entering collision term

In order to compute the rate of increase of $f(\boldsymbol{r},\boldsymbol{p},t)$ due to collisions, we focus on the inverse collisions of the type $\{\boldsymbol{p}',\boldsymbol{p}'_1\}\rightarrow\{\boldsymbol{p},\boldsymbol{p}_1\}$. We thus consider a given molecule of kinetic momentum $\boldsymbol{p}'$ and an incident beam of molecules of kinetic momentum $\boldsymbol{p}'_1$. The corresponding incident flux is:

$$f(\boldsymbol{r},\boldsymbol{p}'_1,t)\,d\boldsymbol{p}'_1|\boldsymbol{v}'-\boldsymbol{v}'_1|. \tag{5.6.7}$$

The number of such collisions taking place on the considered molecule between times t and $t+dt$ is:

$$f(\boldsymbol{r},\boldsymbol{p}'_1,t)\,d\boldsymbol{p}'_1|\boldsymbol{v}'-\boldsymbol{v}'_1|\sigma'(\Omega)\,d\Omega dt, \tag{5.6.8}$$

where $\sigma'(\Omega)$ stands for $\sigma(\boldsymbol{p}',\boldsymbol{p}'_1|\boldsymbol{p},\boldsymbol{p}_1)$. The rate of change $(\partial f/\partial t)^{(+)}_{\text{coll}}$ thus verifies the equality:

$$\left(\frac{\partial f}{\partial t}\right)^{(+)}_{\text{coll}} d\boldsymbol{r}d\boldsymbol{p}dt = \int d\boldsymbol{p}'_1\int d\Omega\,\sigma'(\Omega)\,dt\,|\boldsymbol{v}'-\boldsymbol{v}'_1|f(\boldsymbol{r},\boldsymbol{p}',t)\,d\boldsymbol{r}d\boldsymbol{p}'\,f(\boldsymbol{r},\boldsymbol{p}'_1,t), \tag{5.6.9}$$

obtained by taking into account all the molecules of the considered type (in number $f(\boldsymbol{r},\boldsymbol{p}',t)\,d\boldsymbol{r}d\boldsymbol{p}'$), and integrating over $\boldsymbol{p}'_1$. From equation (5.6.9), it follows:

$$\left(\frac{\partial f}{\partial t}\right)^{(+)}_{\text{coll}} d\boldsymbol{p} = \int d\boldsymbol{p}'_1\int d\Omega\,\sigma'(\Omega)|\boldsymbol{v}'-\boldsymbol{v}'_1|f(\boldsymbol{r},\boldsymbol{p}',t)\,d\boldsymbol{p}'\,f(\boldsymbol{r},\boldsymbol{p}'_1,t). \tag{5.6.10}$$

The differential cross-sections $\sigma(\Omega)$ and $\sigma'(\Omega)$, which refer to collisions inverse of each other, are equal. Since we also have the equalities:

$$|\boldsymbol{v}-\boldsymbol{v}_1| = |\boldsymbol{v}'-\boldsymbol{v}'_1|, \tag{5.6.11}$$

and, at given scattering angles,

$$d\boldsymbol{p}d\boldsymbol{p}_1 = d\boldsymbol{p}'d\boldsymbol{p}'_1, \tag{5.6.12}$$

we get from formula (5.6.10) the expression for $(\partial f/\partial t)^{(+)}_{\text{coll}}$:

$$\left(\frac{\partial f}{\partial t}\right)^{(+)}_{\text{coll}} = \int d\boldsymbol{p}_1\int d\Omega\,\sigma(\Omega)|\boldsymbol{v}-\boldsymbol{v}_1|f(\boldsymbol{r},\boldsymbol{p}',t)f(\boldsymbol{r},\boldsymbol{p}'_1,t). \tag{5.6.13}$$

In formula (5.6.13), $\boldsymbol{p}$ is fixed, while $\boldsymbol{p}'$ and $\boldsymbol{p}'_1$ are functions of $\boldsymbol{p}$, $\boldsymbol{p}_1$ and Ω.

6.3. The Boltzmann equation

Combining the results (5.6.6) and (5.6.13), we obtain the following expression, called the *collision integral*, for the collision term of equation (5.6.1):

$$\left(\frac{\partial f}{\partial t}\right)_{\text{coll}} = \int d\boldsymbol{p}_1\int d\Omega\,\sigma(\Omega)|\boldsymbol{v}-\boldsymbol{v}_1|(f'f'_1-ff_1). \tag{5.6.14}$$

In formula (5.6.14), $\sigma(\Omega)$ represents the differential scattering cross-section for the collision $\{\boldsymbol{p}, \boldsymbol{p}_1\} \to \{\boldsymbol{p}', \boldsymbol{p}'_1\}$ as well as for the inverse collision $\{\boldsymbol{p}', \boldsymbol{p}'_1\} \to \{\boldsymbol{p}, \boldsymbol{p}_1\}$. We have introduced the following condensed notations:

$$\begin{cases} f = f(\boldsymbol{r}, \boldsymbol{p}, t) \\ f_1 = f(\boldsymbol{r}, \boldsymbol{p}_1, t) \\ f' = f(\boldsymbol{r}, \boldsymbol{p}', t) \\ f'_1 = f(\boldsymbol{r}, \boldsymbol{p}'_1, t). \end{cases} \tag{5.6.15}$$

Importing this form of the collision term into the evolution equation (5.6.1), we obtain the Boltzmann equation (1872):

$$\boxed{\frac{\partial f}{\partial t} + \boldsymbol{v}.\nabla_{\boldsymbol{r}} f + \boldsymbol{F}.\nabla_{\boldsymbol{p}} f = \int d\boldsymbol{p}_1 \int d\Omega\, \sigma(\Omega) |\boldsymbol{v} - \boldsymbol{v}_1| \big(f' f'_1 - f f_1\big).} \tag{5.6.16}$$

Equation (5.6.16) is a non-linear integro-differential equation for the one-particle distribution function in a dilute classical gas.[12] The problem of the kinetic theory of dilute classical gases amounts to that of the resolution of this equation.

It may be convenient to rewrite the Boltzmann equation in a form displaying explicitly the conservation laws of kinetic momentum and kinetic energy in each collision:

$$\begin{aligned} \frac{\partial f}{\partial t} + \boldsymbol{v}.\nabla_{\boldsymbol{r}} f + \boldsymbol{F}.\nabla_{\boldsymbol{p}} f = \frac{1}{2} \int d\boldsymbol{p}_1 d\boldsymbol{p}' d\boldsymbol{p}'_1\, W(\boldsymbol{p}', \boldsymbol{p}'_1 | \boldsymbol{p}, \boldsymbol{p}_1) \delta(\boldsymbol{p} + \boldsymbol{p}_1 - \boldsymbol{p}' - \boldsymbol{p}'_1) \\ \times \delta(\varepsilon + \varepsilon_1 - \varepsilon' - \varepsilon'_1) \big(f' f'_1 - f f_1\big). \end{aligned} \tag{5.6.17}$$

In equation (5.6.17), the quantity $W(\boldsymbol{p}', \boldsymbol{p}'_1 | \boldsymbol{p}, \boldsymbol{p}_1) = 4\sigma(\Omega)/m^2$ is proportional to the probability per unit time for two molecules of kinetic momenta $\boldsymbol{p}$ and $\boldsymbol{p}_1$ having, after their collision, kinetic momenta $\boldsymbol{p}'$ and $\boldsymbol{p}'_1$ (the factor 1/2 takes into account the indistinguishability of the molecules with kinetic momenta $\boldsymbol{p}'$ and $\boldsymbol{p}'_1$).

6.4. Generalization to a mixture

The Boltzmann equation may be generalized to a gas consisting of a mixture of molecules a and b of unequal masses m_a and m_b undergoing binary collisions.

For this purpose, we have to introduce a couple of one-particle distribution functions, $f_a(\boldsymbol{r}, \boldsymbol{p}, t)$ and $f_b(\boldsymbol{r}, \boldsymbol{p}, t)$, associated respectively with the molecules of type a and type b, as well as the differential cross-sections $\sigma_{aa}(\Omega)$, $\sigma_{bb}(\Omega)$, $\sigma_{ab}(\Omega)$, corresponding

[12] The quadratic character of the collision integral corresponds to the fact that it describes binary collisions.

to the collisions aa, bb, and ab (or ba). Arguments similar to those previously developed lead to Boltzmann equations for each of the distribution functions f_a and f_b. These equations may be written, using condensed notations of the type of those of formula (5.6.15),

$$\frac{\partial f_a}{\partial t} + \boldsymbol{v}.\nabla_{\boldsymbol{r}} f_a + \boldsymbol{F}.\nabla_{\boldsymbol{p}} f_a = \int d\boldsymbol{p}_1 \int d\Omega\, \sigma_{aa}(\Omega)|\boldsymbol{v}-\boldsymbol{v}_1|\big(f'_a f'_{a1} - f_a f_{a1}\big) + \int d\boldsymbol{p}_1 \int d\Omega\, \sigma_{ab}(\Omega)|\boldsymbol{v}-\boldsymbol{v}_1|\big(f'_a f'_{b1} - f_a f_{b1}\big), \tag{5.6.18}$$

and:

$$\frac{\partial f_b}{\partial t} + \boldsymbol{v}.\nabla_{\boldsymbol{r}} f_b + \boldsymbol{F}.\nabla_{\boldsymbol{p}} f_b = \int d\boldsymbol{p}_1 \int d\Omega\, \sigma_{bb}(\Omega)|\boldsymbol{v}-\boldsymbol{v}_1|\big(f'_b f'_{b1} - f_b f_{b1}\big) + \int d\boldsymbol{p}_1 \int d\Omega\, \sigma_{ab}(\Omega)|\boldsymbol{v}-\boldsymbol{v}_1|\big(f'_b f'_{a1} - f_b f_{a1}\big). \tag{5.6.19}$$

In the particular case in which the b molecules are much heavier than the a molecules ($m_b \gg m_a$) and where the collisions involving only a molecules may be neglected, the system reduces to a gas of non-interacting particles undergoing collisions with scattering centers which may be considered as fixed. Such a system is known as a *Lorentz gas*.[13]

7. Irreversibility

The Boltzmann equation is not time-reversal invariant. If we change t into $-t$ in equation (5.6.16), we must simultaneously change the sign of the velocities. The left-hand side then changes sign, whereas the collision integral on the right-hand side is not modified. Thus, $f(\boldsymbol{r}, -\boldsymbol{p}, -t)$ does not obey the same equation as $f(\boldsymbol{r}, \boldsymbol{p}, t)$. In other words, the dynamics described by the Boltzmann equation is irreversible: the Boltzmann equation belongs to the class of *kinetic equations*.

The determinant assumption in this respect in the derivation of the Boltzmann equation is the molecular chaos hypothesis. This assumption, which amounts to neglecting any correlation between the velocities of two molecules preparing to collide, means that no molecule undergoing a collision carries information about its previous encounters. The memory of the dynamic correlations due to the foregoing collisions is thus lost before a new one takes place. This implies in particular that there are two well-separated time scales, namely the duration of a collision, on the one hand, and

[13] The evolution equation of the one-particle distribution function of the Lorentz gas is derived in Supplement 5A via a direct calculation. It is also possible to deduce it from the Boltzmann equation (5.6.18) for f_a by a limiting procedure: the collision term in σ_{aa} must be dropped, and the term in σ_{ab} evaluated by setting $f_b(\boldsymbol{r}, \boldsymbol{p}, t)\, d\boldsymbol{r} d\boldsymbol{p} = n_i d\boldsymbol{r}$, n_i being the spatial density of the fixed scattering centers.

the mean time interval separating two successive collisions of a given molecule, on the other hand. This separation of time scales is expressed through the inequality:

$$\tau_0 \ll \tau. \tag{5.7.1}$$

The molecular chaos hypothesis introduces a clear-cut distinction between the event 'before a collision' and the event 'after a collision'. This distinction is at the source of irreversibility in the Boltzmann equation.[14]

It is worthwhile to underline the fundamental difference between the Boltzmann equation for dilute classical gases and the Vlasov equation for plasmas. The Vlasov equation is applicable to a transitory stage of the system's evolution: it is reversible and does not belong to the class of kinetic equations (it is sometimes qualified as 'collisionless'). As for the Boltzmann equation, it describes a statistical evolution in which the collisions are taken into account: it is an irreversible evolution equation. The information about the two-particle dynamics intervenes in the Boltzmann equation via a quantity of a statistical nature, namely the scattering cross-section.

8. The H-theorem

8.1. The Boltzmann's *H*-functional

The Boltzmann's *H-theorem* concerns the evolution of the entropy of an out-of-equilibrium dilute classical gas, such as it can be deduced from the Boltzmann equation. The functional $H(t)$ was defined by Boltzmann as:

$$\boxed{H(t) = \int f(\boldsymbol{r}, \boldsymbol{p}, t) \log f(\boldsymbol{r}, \boldsymbol{p}, t)\, d\boldsymbol{r} d\boldsymbol{p},} \tag{5.8.1}$$

where the one-particle distribution function $f(\boldsymbol{r}, \boldsymbol{p}, t)$ obeys the Boltzmann equation.

We associate with $H(t)$ an entropy $S_B(t)$, the *Boltzmann entropy*, defined by:

$$S_B(t) = k \int f(\boldsymbol{r}, \boldsymbol{p}, t) \Big[1 - \log h^3 f(\boldsymbol{r}, \boldsymbol{p}, t)\Big]\, d\boldsymbol{r} d\boldsymbol{p}. \tag{5.8.2}$$

The Boltzmann entropy is associated with a state of the gas characterized by the distribution function $f(\boldsymbol{r}, \boldsymbol{p}, t)$ with $f(\boldsymbol{r}, \boldsymbol{p}, t)$ a solution of the Boltzmann equation. It generally differs from the thermodynamic equilibrium entropy S and from the entropy $S(t)$ in a local equilibrium state (except when the gas finds itself either in thermodynamic equilibrium or in a local equilibrium state).

The Boltzmann entropy and the $H(t)$ functional are equal up to an additive constant:

$$\boxed{S_B(t) = -kH(t) + \text{Cste}.} \tag{5.8.3}$$

[14] Some questions linked to the understanding of the way in which irreversibility is taken into account in the Boltzmann equation were the subject of strong controversies in the nineteenth century. The objections to Boltzmann's ideas were formulated in two paradoxes, which are presented and discussed in Supplement 5B.

The Boltzmann's H-theorem, according to which $H(t)$ is a decreasing function of t (and $S_B(t)$ an increasing function of t) in an isolated system, highlights the irreversible character of the evolution as described by the Boltzmann equation.

8.2. Derivation of the H-theorem

In order to demonstrate the H-theorem, we consider the local density $\eta(\boldsymbol{r}, t)$ of the quantity $H(t)$, defined by:

$$\eta(\boldsymbol{r}, t) = \int f(\boldsymbol{r}, \boldsymbol{p}, t) \log f(\boldsymbol{r}, \boldsymbol{p}, t) \, d\boldsymbol{p}. \tag{5.8.4}$$

We have:

$$\frac{\partial \eta}{\partial t} = \int \frac{\partial}{\partial t}(f \log f) \, d\boldsymbol{p}, \tag{5.8.5}$$

that is:

$$\frac{\partial \eta}{\partial t} = \int (1 + \log f) \frac{\partial f}{\partial t} \, d\boldsymbol{p}. \tag{5.8.6}$$

To evaluate expression (5.8.6) for $\partial \eta / \partial t$, multiply both sides of the Boltzmann equation (5.6.16) by $(1 + \log f)$, and integrate over $\boldsymbol{p}$. The derivative $\partial \eta / \partial t$ appears as the sum of three contributions, respectively due to the drift term, the driving term, and the collision term.

The contribution of the drift term to $\partial \eta / \partial t$ is given by:

$$-\int (1 + \log f) \boldsymbol{v}.\nabla_{\boldsymbol{r}} f \, d\boldsymbol{p} = -\nabla_{\boldsymbol{r}}. \int \frac{\boldsymbol{p}}{m} f \log f \, d\boldsymbol{p}. \tag{5.8.7}$$

The contribution (5.8.7) may be expressed as $-\nabla_{\boldsymbol{r}}.\boldsymbol{J}_H$, where:

$$\boldsymbol{J}_H(\boldsymbol{r}, t) = \int f \log f \, \frac{\boldsymbol{p}}{m} \, d\boldsymbol{p}, \tag{5.8.8}$$

is the flux of $H(t)$.

When the external force does not depend on the velocity, we can write the contribution of the driving term as:

$$-\int (1 + \log f) \boldsymbol{F}.\nabla_{\boldsymbol{p}} f \, d\boldsymbol{p} = -\boldsymbol{F}. \int \nabla_{\boldsymbol{p}}(f \log f) \, d\boldsymbol{p}. \tag{5.8.9}$$

The distribution function vanishes when the kinetic momentum becomes infinite: the contribution (5.8.9) thus vanishes.[15]

We can therefore write the following local balance equation:[16]

$$\frac{\partial \eta}{\partial t} + \nabla.\boldsymbol{J}_H = \sigma_H, \tag{5.8.10}$$

---

[15] This result can be extended to the case where there is an external magnetic field.

[16] Since there is no ambiguity, $\nabla_{\boldsymbol{r}}$ is simply denoted by ∇.

in which the source term:

$$\sigma_H = \int (1 + \log f)(f'f_1' - ff_1)\sigma(\Omega)|\boldsymbol{v} - \boldsymbol{v}_1|\, d\boldsymbol{p}d\boldsymbol{p}_1 d\Omega \tag{5.8.11}$$

is the contribution of the collision term to $\partial\eta/\partial t$. The associate global balance equation reads:

$$\frac{dH(t)}{dt} + \int_V (\nabla.\boldsymbol{J}_H)\, d\boldsymbol{r} = \int_V \sigma_H\, d\boldsymbol{r}. \tag{5.8.12}$$

To study σ_H, we first note that this quantity remains invariant under the permutation of the kinetic momenta $\boldsymbol{p}$ and $\boldsymbol{p}_1$ (or, accordingly, of the velocities $\boldsymbol{v}$ and $\boldsymbol{v}_1$). We can therefore write:

$$\sigma_H = \int (1 + \log f_1)(f'f_1' - ff_1)\sigma(\Omega)|\boldsymbol{v} - \boldsymbol{v}_1|\, d\boldsymbol{p}d\boldsymbol{p}_1 d\Omega. \tag{5.8.13}$$

We can take for σ_H the half-sum of the expressions (5.8.11) and (5.8.13), that is:

$$\sigma_H = \frac{1}{2}\int \Big[2 + \log(ff_1)\Big](f'f_1' - ff_1)\sigma(\Omega)|\boldsymbol{v} - \boldsymbol{v}_1|\, d\boldsymbol{p}d\boldsymbol{p}_1 d\Omega. \tag{5.8.14}$$

We then carry out the change of variables $(\boldsymbol{p}, \boldsymbol{p}_1) \to (\boldsymbol{p}', \boldsymbol{p}_1')$, which amounts to considering the collision $\{\boldsymbol{p}', \boldsymbol{p}_1'\} \to \{\boldsymbol{p}, \boldsymbol{p}_1\}$, inverse of the collision $\{\boldsymbol{p}, \boldsymbol{p}_1\} \to \{\boldsymbol{p}', \boldsymbol{p}_1'\}$. The cross-sections relative to both collisions are identical. Also, we have the equalities:

$$|\boldsymbol{v} - \boldsymbol{v}_1| = |\boldsymbol{v}' - \boldsymbol{v}_1'|, \tag{5.8.15}$$

and, at given scattering angles:

$$d\boldsymbol{p}d\boldsymbol{p}_1 = d\boldsymbol{p}'d\boldsymbol{p}_1'. \tag{5.8.16}$$

We therefore have:

$$\sigma_H = \frac{1}{2}\int \Big[2 + \log(f'f_1')\Big](ff_1 - f'f_1')\sigma(\Omega)|\boldsymbol{v} - \boldsymbol{v}_1|\, d\boldsymbol{p}d\boldsymbol{p}_1 d\Omega. \tag{5.8.17}$$

Taking the half-sum of the expressions (5.8.14) and (5.8.17) for σ_H, we finally obtain:

$$\sigma_H = \frac{1}{4}\int \log\left(\frac{ff_1}{f'f_1'}\right)(f'f_1' - ff_1)\sigma(\Omega)|\boldsymbol{v} - \boldsymbol{v}_1|\, d\boldsymbol{p}d\boldsymbol{p}_1 d\Omega. \tag{5.8.18}$$

The quantity $\log(ff_1/f'f_1')(f'f_1' - ff_1)$ being only negative or vanishing, the same property holds for σ_H.

Let us now come back to the global balance equation (5.8.12). For a gas contained in a box of volume V, the integral $\int_V (\nabla.\boldsymbol{J}_H) d\boldsymbol{r}$ vanishes and $dH(t)/dt$ reduces to $\int_V \sigma_H d\boldsymbol{r}$. From there comes the H-theorem, established in 1872 by L. Boltzmann:

$$\boxed{\frac{dH(t)}{dt} \leq 0.} \tag{5.8.19}$$

On account of formula (5.8.3), this result can also be written as:

$$\boxed{\frac{dS_B(t)}{dt} \geq 0.} \tag{5.8.20}$$

The Boltzmann entropy cannot decrease over the course of time. It can only either increase or remain stationary.

8.3. Discussion

The physical origin of the increase of the Boltzmann entropy has to be found in the collisions between the gas molecules. The increase of $S_B(t)$ results from the way in which the colllisions are taken into account in the Boltzmann equation via the molecular chaos hypothesis.

The role of collisions as a source of entropy finds its origin in molecular chaos, which constantly destroys information. After a collision between two molecules, their positions and velocities are actually correlated.[17] However, each molecule will then undergo collisions with other molecules. After a number of collisions, the correlations between the two molecules first considered will fade away. This introduces an irreversibility in time.

9. Equilibrium distributions

The definition of $S_B(t)$ allows us to show that it is a bounded quantity. Besides, we have $dS_B(t)/dt \geq 0$. The Boltzmann entropy must thus tend to a limit as $t \to \infty$. In this limit, we have:

$$\frac{dS_B(t)}{dt} = 0. \tag{5.9.1}$$

Equation (5.9.1) admits several types of solutions, the *global macroscopic equilibrium* distribution, on the one hand, and *local equilibrium distributions*, on the other hand.

- *Global equilibrium distributions*

The global equilibrium distribution f_0 is the time-independent solution of the Boltzmann equation ($\partial f_0/\partial t = 0$). The global macroscopic equilibrium is the thermodynamic equilibrium. The gas at equilibrium possesses well-defined values of the density n of molecules, of their mean velocity $\boldsymbol{u}$, and of the temperature T.

- *Local equilibrium distributions*

Equation (5.9.1) also admits solutions which do not obey the Boltzmann equation. These distributions, denoted $f^{(0)}(\boldsymbol{r}, \boldsymbol{p}, t)$, are local equilibrium distributions. They correspond to well-defined values of the local density $n(\boldsymbol{r}, t)$ of molecules, of their local mean velocity $\boldsymbol{u}(\boldsymbol{r}, t)$, and of the local temperature $T(\boldsymbol{r}, t)$.

---

[17] Indeed, were their velocities reversed, both molecules would have to undergo a collision.

The time scales necessary for the establishment of these two types of equilibrium are not the same. The collisions begin by locally uniformizing the density, the mean velocity, and the temperature, but inhomogeneities continue to be present on a larger scale. In the absence of external constraints, these inhomogeneities eventually disappear. The collision term alone controls the rapid relaxation towards a local equilibrium, whereas the combined effect of the collision term, on the one hand, and of the drift and driving terms, on the other hand, controls the slow relaxation towards global equilibrium.

10. Global equilibrium

10.1. Determination of the global equilibrium distribution

To begin with, we will assume that there is no external force ($\boldsymbol{F} = 0$), in which case the system is spatially homogeneous. The distribution function, which then does not depend on $\boldsymbol{r}$, may simply be denoted by $f(\boldsymbol{p}, t)$.

According to the H-theorem, we have, for any arbitrary initial condition:

$$\lim_{t \to \infty} f(\boldsymbol{p}, t) = f_0(\boldsymbol{p}). \tag{5.10.1}$$

In the limit $t \to \infty$, we have $dH/dt = 0$. Clearly, from expression (5.8.18) for σ_H in which the integrand has a constant sign, it follows that this integrand must itself vanish in this limit. The distribution f_0 must thus verify the condition:

$$f_0(\boldsymbol{p}) f_0(\boldsymbol{p}_1) = f_0(\boldsymbol{p}') f_0(\boldsymbol{p}'_1), \tag{5.10.2}$$

that is:

$$\log f_0(\boldsymbol{p}) + \log f_0(\boldsymbol{p}_1) = \log f_0(\boldsymbol{p}') + \log f_0(\boldsymbol{p}'_1). \tag{5.10.3}$$

Equation (5.10.3) is a conservation law relative to the collision $\{\boldsymbol{p}, \boldsymbol{p}_1\} \to \{\boldsymbol{p}', \boldsymbol{p}'_1\}$. It admits solutions of the form:

$$\log f_0(\boldsymbol{p}) = \chi(\boldsymbol{p}), \tag{5.10.4}$$

where $\chi(\boldsymbol{p})$ is a *collisional invariant*.[18] Equation (5.10.3) being linear, its most general solution is a linear combination of the independent collisional invariants (namely, the mass, the three components of the kinetic momentum, and the kinetic energy), which can be formulated in the following way:

$$\log f_0(\boldsymbol{p}) = -A|\boldsymbol{p} - m\boldsymbol{u}|^2 + \log C. \tag{5.10.5}$$

The global equilibrium distribution is thus of the form:

$$f_0(\boldsymbol{p}) = Ce^{-A|\boldsymbol{p} - m\boldsymbol{u}|^2}. \tag{5.10.6}$$

[18] A collisional invariant is a quantity which is conserved in a collision:

$$\chi(\boldsymbol{p}) + \chi(\boldsymbol{p}_1) = \chi(\boldsymbol{p}') + \chi(\boldsymbol{p}'_1).$$

The parameters C and A as well as the three components of $\boldsymbol{u}$ must be determined in terms of the equilibrium properties of the system.

10.2. Parameters of the equilibrium distribution

To obtain them, we compute the total number of molecules of the gas, as well as the mean kinetic momentum and the mean kinetic energy of a molecule in the gas at equilibrium described by the distribution function (5.10.6).

• *Total number of molecules*

From:

$$N = V \int f_0(\boldsymbol{p})\, d\boldsymbol{p}. \tag{5.10.7}$$

the density of molecules is $n = C(\pi/A)^{3/2}$.

• *Mean kinetic momentum of a molecule*

From the formula:

$$\langle \boldsymbol{p} \rangle = \frac{1}{n} \int f_0(\boldsymbol{p}) \boldsymbol{p}\, d\boldsymbol{p}, \tag{5.10.8}$$

we deduce $\langle \boldsymbol{p} \rangle = m\boldsymbol{u}$. In the absence of any global translation motion of the gas, we thus have $\boldsymbol{u} = 0$.

• *Mean kinetic energy*

The mean kinetic energy of a molecule is:[19]

$$\langle \varepsilon \rangle = \frac{1}{n} \int f_0(\boldsymbol{p}) \frac{p^2}{2m}\, d\boldsymbol{p}. \tag{5.10.9}$$

We get, for $\boldsymbol{u} = 0$, $\langle \varepsilon \rangle = 3(4Am)^{-1}$.

The expressions for A and C in terms of n and $\langle \varepsilon \rangle$ being taken into account, the equilibrium distribution (5.10.6) reads:

$$f_0(\boldsymbol{p}) = n\Big(\frac{3}{4\pi\langle \varepsilon \rangle m}\Big)^{3/2} \exp\Big(-\frac{3}{4\langle \varepsilon \rangle m} p^2\Big). \tag{5.10.10}$$

The global equilibrium described here is simply the thermodynamic equilibrium. The mean kinetic energy of a molecule is thus equal to its equipartition value $3kT/2$. We therefore have:

$$\boxed{f_0(\boldsymbol{p}) = n(2\pi mkT)^{-3/2} \exp\Big(-\frac{p^2}{2mkT}\Big).} \tag{5.10.11}$$

The distribution $f_0(\boldsymbol{p})$, called the *Maxwell–Boltzmann distribution*, not only makes the collision integral (5.6.14) vanish, but furthermore is a solution of the Boltzmann

[19] In formula (5.10.9), we set $p = |\boldsymbol{p}|$.

equation (formulas (5.6.16) or (5.6.17)). When expressed in terms of the velocity, it identifies with the *Maxwell velocities distribution function*:

$$F_0(\boldsymbol{v}) = n\left(\frac{m}{2\pi kT}\right)^{3/2} \exp\left(-\frac{mv^2}{2kT}\right). \tag{5.10.12}$$

10.3. Equilibrium in the presence of an external force

In the presence of an external force $\boldsymbol{F}(\boldsymbol{r})$ deriving from a scalar potential $\phi(\boldsymbol{r})$, the global equilibrium distribution also depends on $\boldsymbol{r}$. It reads:

$$f_0(\boldsymbol{r},\boldsymbol{p}) = V\,\frac{\exp\left[-\phi(\boldsymbol{r})/kT\right]}{\int \exp\left[-\phi(\boldsymbol{r})/kT\right] d\boldsymbol{r}}\,f_0(\boldsymbol{p}), \tag{5.10.13}$$

where $f_0(\boldsymbol{p})$ is given by formula (5.10.11) (provided that there is no global translational motion). The distribution (5.10.13) actually makes the integrand in the expression (5.8.18) for σ_H vanish: indeed, $\log f_0(\boldsymbol{r},\boldsymbol{p})$ is a collisional invariant, since $\log f_0(\boldsymbol{p})$ is such an invariant and the collisions are considered as local. This distribution also makes the left-hand side of the Boltzmann equation (formulas (5.6.16) or (5.6.17)) vanish. It is thus actually the global equilibrium distribution.

10.4. Entropy at thermodynamic equilibrium

The thermodynamic equilibrium entropy is calculated with the aid of the formula:

$$S = k\int f_0(\boldsymbol{p})\left[1 - \log h^3 f_0(\boldsymbol{p})\right] d\boldsymbol{r}d\boldsymbol{p}, \tag{5.10.14}$$

deduced from the general formula (5.8.2) for $S_B(t)$. From the expression (5.10.11) for $f_0(\boldsymbol{p})$, we get:

$$S = Nk\left(\log\frac{V}{N\lambda^3} + \frac{5}{2}\right). \tag{5.10.15}$$

Formula (5.10.15) is the *Sackur–Tetrode formula* for the equilibrium entropy of an ideal classical gas.

11. Local equilibrium

11.1. Local equilibrium distributions

The collision integral of the Boltzmann equation vanishes for the global equilibrium distribution $f_0(\boldsymbol{p})$. Besides, it also vanishes for any local equilibrium distribution of the form:

$$f^{(0)}(\boldsymbol{r},\boldsymbol{p},t) = n(\boldsymbol{r},t)\left[2\pi mkT(\boldsymbol{r},t)\right]^{-3/2} \exp\left[-\frac{|\boldsymbol{p} - m\boldsymbol{u}(\boldsymbol{r},t)|^2}{2mkT(\boldsymbol{r},t)}\right]. \tag{5.11.1}$$

The distributions of the form (5.11.1), called *local Maxwell–Boltzmann distributions*, are parametrized by the local density $n(\boldsymbol{r},t)$ of molecules, their local mean velocity $\boldsymbol{u}(\boldsymbol{r},t)$, and the local temperature $T(\boldsymbol{r},t)$, all these quantities being slowly varying functions of $\boldsymbol{r}$ and t.

A local Maxwell–Boltzmann distribution $f^{(0)}$ is not a solution of the Boltzmann equation (formulas (5.6.16) or (5.6.17)). We have:

$$\left(\frac{\partial f^{(0)}}{\partial t}\right)_{\text{coll}} = 0, \tag{5.11.2}$$

but:

$$\left(\frac{\partial}{\partial t} + \boldsymbol{v}.\nabla_{\boldsymbol{r}} + \boldsymbol{F}.\nabla_{\boldsymbol{p}}\right) f^{(0)}(\boldsymbol{r},\boldsymbol{p},t) \neq 0. \tag{5.11.3}$$

11.2. The local equilibrium entropy

We associate with any local equilibrium distribution $f^{(0)}(\boldsymbol{r},\boldsymbol{p},t)$ a local equilibrium entropy $S(t)$, defined by:

$$S(t) = k\int f^{(0)}(\boldsymbol{r},\boldsymbol{p},t)\Big[1 - \log h^3 f^{(0)}(\boldsymbol{r},\boldsymbol{p},t)\Big]\, d\boldsymbol{r}d\boldsymbol{p}. \tag{5.11.4}$$

The entropy $S(t)$ is solely defined in a local equilibrium regime, that is, for times t much longer than the relaxation time τ_r. Since the distribution $f^{(0)}$ is not a solution of the Boltzmann equation, there is no 'H-theorem' which would state that $S(t)$ should increase over the course of time. This is nevertheless the case for times $t \gg \tau_r$, since the entropy production within the gas in local equilibrium is necessarily non-negative.

It can be shown that the Boltzmann entropy, the local equilibrium entropy, and the thermodynamic equilibrium entropy obey the inequalities:

$$\boxed{S_B(t) \leq S(t) \leq S,} \tag{5.11.5}$$

which express the fact that the three descriptions of the gas, first, in terms of the one-particle distribution function (Boltzmann description, entropy $S_B(t)$), then in terms of the hydrodynamic variables[20] (local equilibrium description, entropy $S(t)$), and, finally, in terms of the thermodynamic variables (global equilibrium description, entropy S) are less and less detailed.

[20] These variables are the local density $n(\boldsymbol{r},t)$, the local mean velocity $\boldsymbol{u}(\boldsymbol{r},t)$, and the local temperature $T(\boldsymbol{r},t)$.

Bibliography

R. BALESCU, *Statistical dynamics. Matter out of equilibrium*, Imperial College Press, London, 1997.

R. BALIAN, *From microphysics to macrophysics*, Vol. 1, Springer-Verlag, Berlin, 1991; Vol. 2, Springer-Verlag, Berlin, 1992.

C. CERCIGNANI, *Ludwig Boltzmann, the man who trusted atoms*, Oxford University Press, Oxford, 1998.

K. HUANG, *Statistical mechanics*, Wiley, New York, second edition, 1987.

H.J. KREUZER, *Nonequilibrium thermodynamics and its statistical foundations*, Clarendon Press, Oxford, 1981.

R. KUBO, M. TODA, and N. HASHITSUME, *Statistical physics II: nonequilibrium statistical mechanics*, Springer-Verlag, Berlin, second edition, 1991.

L.D. LANDAU and E.M. LIFSHITZ, *Mechanics*, Butterworth-Heinemann, Oxford, third edition, 1976.

E.M. LIFSHITZ and L.P. PITAEVSKII, *Physical kinetics*, Butterworth-Heinemann, Oxford, 1981.

J.A. MCLENNAN, *Introduction to nonequilibrium statistical mechanics*, Prentice-Hall, Englewood Cliffs, 1989.

D.A. MCQUARRIE, *Statistical mechanics*, University Science Books, Sausalito, second edition, 2000.

F. REIF, *Fundamentals of statistical and thermal physics*, McGraw-Hill, New York, 1965.

D. ZUBAREV, V. MOROZOV, and G. RÖPKE, *Statistical mechanics of nonequilibrium processes,* Vol. 1: *Basic concepts, kinetic theory*, Akademie Verlag, Berlin, 1996.

References

L. BOLTZMANN, Weitere Studien über das Wärmegleichgewicht unter Gasmolekülen, *Sitzungsberichte der Akademie der Wissenschaften*, Wien II, **66**, 275 (1872).

P. EHRENFEST und T. EHRENFEST, Mechanik der aus sehr zahlreichen diskreten Teilen bestehenden Systeme, *Enzyklopädie der mathematischen Wissenschaften*, Vol. 4, Leipzig, 1911.

Supplement 5A

The Lorentz gas

1. Gas in the presence of fixed scattering centers

The model of a gas of particles evolving in the presence of fixed scattering centers was initially introduced by H.A. Lorentz in 1905 in order to describe the electron gas in metals. The Lorentz gas is a set of classical non-interacting particles undergoing collisions with scattering centers considered as exterior to the system of particles. Whereas it cannot in fact concern the electron gas in metals because of the degenerate character of the latter, the Lorentz gas model conveniently applies to electrons in non-degenerate semiconductors.[1] In this case, the scattering centers are the impurities and the defects with which the electrons collide.

In the Lorentz model, the scattering centers are much heavier than the gas particles. They may for this reason be considered as fixed. Besides, their internal structure is not taken into account. The gas particles being, too, assumed to be devoid of internal structure, their collisions with the scattering centers are thus considered as elastic.

Finally, the scattering centers are assumed to be randomly distributed in space, which insures that successive collision processes will not give rise to coherence effects.

2. Time scales

The quantity $f(\boldsymbol{r}, \boldsymbol{p}, t)\, d\boldsymbol{r}d\boldsymbol{p}$ represents the mean number of particles of the Lorentz gas which, at time t, are in the phase space volume element $d\boldsymbol{r}d\boldsymbol{p}$ about the point $(\boldsymbol{r}, \boldsymbol{p})$.

The discussion about the time and length characteristic scales in the Lorentz gas is similar to that concerning dilute classical gases of particles undergoing binary collisions. In particular, the duration τ_0 of any collision is assumed to be much smaller than the collision time τ. Like the Boltzmann equation, the kinetic equation of the Lorentz gas describes the evolution of the one-particle distribution function over a time interval Δt intermediate between the duration τ_0 of a collision and the relaxation time $\tau_r \sim \tau$ towards a local equilibrium.

---

[1] The Lorentz gas model also applies to holes. Electrons and holes constitute approximately two independent gases of charge carriers. We will only be interested here in the electron distribution function (similar considerations can be made for holes).

3. Collisions with the fixed scatterers

The hypotheses concerning the collisions are as follows:

(i) Only the collisions of the particles of the Lorentz gas with the fixed scatterers are taken into account, the interactions between the gas particles themselves being neglected.

(ii) The scattering centers as well as the gas particles are assumed to be devoid of internal structure, which implies the elasticity of collisions.

(iii) The collisions are considered as local and instantaneous.

(iv) The scattering centers are assumed to be randomly distributed in space. This last hypothesis is fundamental. It is indeed this assumption which leads to an irreversible kinetic equation.

The collision with a fixed scattering center of a particle of kinetic momentum $\boldsymbol{p}$ modifies this kinetic momentum, which becomes $\boldsymbol{p}'$. The corresponding differential scattering cross-section is denoted by $\sigma(\boldsymbol{p}|\boldsymbol{p}')$. Due to the elasticity hypothesis, the kinetic momentum modulus is conserved in the collision:[2]

$$|\boldsymbol{p}| = |\boldsymbol{p}'|. \tag{5A.3.1}$$

4. Kinetic equation of the Lorentz gas

Our aim is to obtain the evolution equation of the one-particle distribution function $f(\boldsymbol{r}, \boldsymbol{p}, t)$ of the Lorentz gas. The issue is analogous to that of a dilute classical gas of molecules undergoing binary collisions: in the presence of collisions, the density $f(\boldsymbol{r}, \boldsymbol{p}, t)$ in the μ-space is not conserved. We formally write its evolution equation as:

$$\boxed{\frac{\partial f}{\partial t} + \boldsymbol{v}.\nabla_{\boldsymbol{r}} f + \boldsymbol{F}.\nabla_{\boldsymbol{p}} f = \left(\frac{\partial f}{\partial t}\right)_{\text{coll}},} \tag{5A.4.1}$$

where $\boldsymbol{F}$ denotes a possibly applied external force. As in the case of the Boltzmann equation, the collision term $(\partial f/\partial t)_{\text{coll}}$, which represents the rate of change of f due to collisions, may conveniently be written in a form displaying its balance structure:

$$\left(\frac{\partial f}{\partial t}\right)_{\text{coll}} = \left(\frac{\partial f}{\partial t}\right)^{(+)}_{\text{coll}} - \left(\frac{\partial f}{\partial t}\right)^{(-)}_{\text{coll}}. \tag{5A.4.2}$$

The quantity $(\partial f/\partial t)^{(-)}_{\text{coll}}\, d\boldsymbol{r}d\boldsymbol{p}dt$ (*resp.* $(\partial f/\partial t)^{(+)}_{\text{coll}}\, d\boldsymbol{r}d\boldsymbol{p}dt$) represents the mean number of particles undergoing a collision between times t and $t+dt$ and finding themselves, before (*resp.* after) the collision, in the volume element $d\boldsymbol{r}d\boldsymbol{p}$ around the point $(\boldsymbol{r}, \boldsymbol{p})$ of the μ-space.

---

[2] More generally, any function of the kinetic momentum modulus is conserved in the collision. The mass and the kinetic energy are collisional invariants of the Lorentz model. On the other hand, the kinetic momentum itself is not conserved.

We are looking for an explicit expression for the collision term (5A.4.2), on account of the hypotheses made about the collisions, and in particular of the fact that the scattering centers are assumed to be randomly distributed in space. Their spatial density is denoted by n_i.

4.1. Leaving collision term

We first consider a given scattering center situated in the spatial volume element $d\boldsymbol{r}$ centered at $\boldsymbol{r}$. In this volume element, there are particles of kinetic momentum in the element $d\boldsymbol{p}$ centered at $\boldsymbol{p}$ forming a beam of particles falling upon the considered scatterer. The corresponding incident flux is:

$$f(\boldsymbol{r},\boldsymbol{p},t)\,d\boldsymbol{p}|\boldsymbol{v}|. \tag{5A.4.3}$$

The number of collisions of the type $\{\boldsymbol{p}\}\to\{\boldsymbol{p}'\}$ taking place on this scattering center between times t and $t+dt$ is:

$$f(\boldsymbol{r},\boldsymbol{p},t)\,d\boldsymbol{p}|\boldsymbol{v}|\sigma(\boldsymbol{p}|\boldsymbol{p}')\,d\Omega'dt, \tag{5A.4.4}$$

where Ω' marks the direction of $\boldsymbol{p}'$. We obtain, by summing the contributions of the $n_i d\boldsymbol{r}$ scattering centers present in the volume element $d\boldsymbol{r}$ and integrating over the direction of $\boldsymbol{p}'$:

$$\left(\frac{\partial f}{\partial t}\right)^{(-)}_{\text{coll}} d\boldsymbol{r}d\boldsymbol{p}dt = n_i\,d\boldsymbol{r}\,|\boldsymbol{v}|f(\boldsymbol{r},\boldsymbol{p},t)\,d\boldsymbol{p}dt\int d\Omega'\,\sigma(\boldsymbol{p}|\boldsymbol{p}'). \tag{5A.4.5}$$

Hence the expression for the leaving collision term is:

$$\left(\frac{\partial f}{\partial t}\right)^{(-)}_{\text{coll}} = n_i|\boldsymbol{v}|f(\boldsymbol{r},\boldsymbol{p},t)\int d\Omega'\,\sigma(\boldsymbol{p}|\boldsymbol{p}'). \tag{5A.4.6}$$

4.2. Entering collision term

The term $(\partial f/\partial t)^{(+)}_{\text{coll}}$ may be calculated in an analogous way. We consider, for a given scattering center, the inverse collisions of the type $\{\boldsymbol{p}'\}\to\{\boldsymbol{p}\}$, the kinetic momentum $\boldsymbol{p}$ being given inside an element $d\boldsymbol{p}$. The incident flux, on a given scatterer, of particles of kinetic momentum $\boldsymbol{p}'$ is:

$$f(\boldsymbol{r},\boldsymbol{p}',t)\,d\boldsymbol{p}'|\boldsymbol{v}'|. \tag{5A.4.7}$$

The number of collisions of this type taking place on the considered scatterer between times t and $t+dt$ is:

$$f(\boldsymbol{r},\boldsymbol{p}',t)\,d\boldsymbol{p}'|\boldsymbol{v}'|\sigma(\boldsymbol{p}'|\boldsymbol{p})\,d\Omega dt, \tag{5A.4.8}$$

where $d\Omega$ is the solid angle element around $\boldsymbol{p}$.

We can write:

$$d\boldsymbol{p} = p^2\,dpd\Omega, \qquad d\boldsymbol{p}' = p'^2\,dp'd\Omega'. \tag{5A.4.9}$$

The collisions being elastic, we have the equalities:

$$|\boldsymbol{p}| = |\boldsymbol{p}'|, \qquad d|\boldsymbol{p}| = d|\boldsymbol{p}'|. \tag{5A.4.10}$$

The expression (5A.4.8) thus also reads:

$$f(\boldsymbol{r}, \boldsymbol{p}', t)\, p^2\, dp d\Omega'\, |\boldsymbol{v}'| \sigma(\boldsymbol{p}'|\boldsymbol{p})\, d\Omega dt. \tag{5A.4.11}$$

We obtain the rate $(\partial f/\partial t)_{\text{coll}}^{(+)}$ by summing the contributions of the $n_i d\boldsymbol{r}$ scattering centers present in the volume element $d\boldsymbol{r}$ and integrating over the direction of $\boldsymbol{p}'$:

$$\left(\frac{\partial f}{\partial t}\right)_{\text{coll}}^{(+)} d\boldsymbol{r} d\boldsymbol{p} dt = n_i\, d\boldsymbol{r}\, |\boldsymbol{v}|\, d\boldsymbol{p} dt \int d\Omega'\, \sigma(\boldsymbol{p}'|\boldsymbol{p}) f(\boldsymbol{r}, \boldsymbol{p}', t). \tag{5A.4.12}$$

From equation (5A.4.12), it follows:

$$\left(\frac{\partial f}{\partial t}\right)_{\text{coll}}^{(+)} = n_i |\boldsymbol{v}| \int d\Omega'\, \sigma(\boldsymbol{p}'|\boldsymbol{p}) f(\boldsymbol{r}, \boldsymbol{p}', t). \tag{5A.4.13}$$

4.3. The kinetic equation of the Lorentz gas

Combining results (5A.4.6) and (5A.4.13), and taking into account the identity of the cross-sections $\sigma(\boldsymbol{p}|\boldsymbol{p}')$ and $\sigma(\boldsymbol{p}'|\boldsymbol{p})$, gives the expression for the collision integral of the Lorentz gas:

$$\left(\frac{\partial f}{\partial t}\right)_{\text{coll}} = n_i |\boldsymbol{v}| \int d\Omega'\, \sigma(\boldsymbol{p}|\boldsymbol{p}') \big[f(\boldsymbol{r}, \boldsymbol{p}', t) - f(\boldsymbol{r}, \boldsymbol{p}, t)\big]. \tag{5A.4.14}$$

Importing this form of the collision term into the evolution equation (5A.4.1), gives the kinetic equation of the Lorentz gas:

$$\boxed{\frac{\partial f}{\partial t} + \boldsymbol{v}.\nabla_{\boldsymbol{r}} f + \boldsymbol{F}.\nabla_{\boldsymbol{p}} f = n_i |\boldsymbol{v}| \int d\Omega'\, \sigma(\boldsymbol{p}|\boldsymbol{p}') \big[f(\boldsymbol{r}, \boldsymbol{p}', t) - f(\boldsymbol{r}, \boldsymbol{p}, t)\big].} \tag{5A.4.15}$$

Equation (5A.4.15) is a linear partial differential equation for the one-particle distribution function. Its linearity comes from the fact that, since the scattering centers are considered as fixed and randomly distributed in space, they come into play in the evolution equation solely via their density n_i. Equation (5A.4.15) is not time-reversal invariant. It thus describes an irreversible dynamics.[3]

The kinetic equation of the Lorentz gas is commonly used in the semiclassical description of electronic transport in solids.[4]

---

[3] We can derive an H-theorem for the Lorentz gas.

[4] See Chapter 8 and Supplement 8A.

Bibliography

R. BALIAN, *From microphysics to macrophysics*, Vol. 2, Springer-Verlag, Berlin, 1992.

J.A. MCLENNAN, *Introduction to nonequilibrium statistical mechanics*, Prentice-Hall, Englewood Cliffs, 1989.

References

H.A. LORENTZ, *Arch. Neerl.* **10**, 336 (1905). Reprinted in *Collected Papers*, Martinus Nijhoff, Vol. 3, The Hague, 1936.

Supplement 5B

The irreversibility paradoxes

1. The paradoxes

The H-theorem led at the end of the nineteenth century to a somewhat confused situation. In spite of the fact that the Boltzmann equation had successfully been applied to the description of numerous physical phenomena, Boltzmann's ideas encountered strong opposition, emanating from physicists as well as from mathematicians. Their objections were formulated as two paradoxes, the *time-reversal paradox* (or *Loschmidt's paradox*) and the *recurrence paradox* (or *Zermelo's paradox*).

Both paradoxes are enunciated starting from an insufficiently precise formulation of the H-theorem, according to which the derivative dH/dt would be negative or vanishing at any time.[1]

2. The time-reversal paradox

2.1. Loschmidt's argumentation

The time-reversal paradox consists of a set of arguments developed first by W. Thomson (later Lord Kelvin) in 1874, then put forward by J. Loschmidt in 1876 as constituting an objection to the Boltzmann equation.

These arguments may be summarized in the following way. The microscopic equations of motion are invariant under time-reversal. In mechanics, there is no preferred direction of time. The H-theorem indicates that the Boltzmann equation implies a particular direction of time. Now, given at a certain time a state of the gas with certain molecular velocities, there is, at the same time, another possible state of the gas in which the molecules have the same positions and reverse velocities. The evolution of this second state towards the past is identical to the evolution of the first one towards the future. Thus, if the function $H(t)$ decreases in the first case, it must necessarily increase in the second one, which, according to Loschmidt, would contradict the H-theorem.

---

[1] The correct formulation of the H-theorem is as follows: if, at a given time t, the state of the gas verifies the molecular chaos assumption, then, at time $t + \epsilon$ ($\epsilon \to 0^+$), we have $dH/dt \leq 0$. The necessary and sufficient condition for dH/dt to vanish is that the distribution $f(\boldsymbol{r}, \boldsymbol{p}, t)$ be a Maxwell–Boltzmann distribution (either of local or of global equilibrium).

2.2. Boltzmann's answer

Boltzmann responded to Loschmidt's objection by highlighting the role of the initial conditions, as well as the statistical nature of the functional $H(t)$. For some very peculiar initial conditions,[2] $H(t)$ may indeed increase over the course of time. But there are infinitely more initial states from which $H(t)$ decreases. The function $H(t)$ is actually a quantity of a statistical nature. For an initially out-of-equilibrium system, $H(t)$ decreases on average over the course of time towards its equilibrium value, but fluctuations are always likely to arise.

Nowadays, it is possible to compute numerically the evolution of $H(t)$ in a gas. It is observed that, indeed, for an initially out-of-equilibrium gas, this function decreases on average, but that there are fluctuations around this average behavior.

3. The recurrence paradox

3.1. Zermelo's argumentation

Zermelo's paradox is based on a theorem of classical mechanics established by H. Poincaré in 1889, the *recurrence theorem*. According to this result, any mechanical system of fixed total energy contained in a finite volume returns arbitrarily near its initial state after some time, and this for almost all initial states. The time necessary for this return is called the *recurrence time*, and the corresponding cycle, the *Poincaré cycle*.

The mathematician E. Zermelo developed in 1896 an argumentation according to which the recurrence theorem would render any mechanical model such as the kinetic theory incompatible with the second law of thermodynamics, which would thus lead us to reject the kinetic theory. Otherwise stated, Zermelo thought he saw a contradiction between Boltzmann's H-theorem and the Poincaré recurrence theorem. This contradiction may be formulated in the following way: how could $H(t)$ evolve towards an equilibrium value and remain at this value (H-theorem) since, according to classical mechanics, the system must return towards its initial state (recurrence theorem)?

3.2. Boltzmann's answer

Boltzmann responded that the recurrence theorem does not contradict the H-theorem, but is instead compatible with it. For the macroscopic physical systems dealt with by the H-theorem, the recurrence times are indeed excessively large. A rough estimate shows that the duration of a Poincaré cycle of a system of N particles is of order e^N. For a macroscopic system for which $N \approx 10^{23}$, the duration of a Poincaré cycle is of order $10^{10^{23}}$ (seconds, or any other time unit). Such a time is clearly devoid of any physical significance.

Thus, the very concept of irreversibility is linked to the length of the recurrence time. For a system initially in a state characterized by a very large recurrence time, the evolution process appears de facto as irreversible.

---

[2] Such initial conditions, although extremely improbable, may for instance be obtained by reversing the velocities of all molecules in an equilibrium state attained through an evolution of the gas from an out-of-equilibrium state.

Bibliography

R. BALIAN, *From microphysics to macrophysics*, Vol. 1, Springer-Verlag, Berlin, 1991; Vol. 2, Springer-Verlag, Berlin, 1992.

S.G. BRUSH, *The kind of motion we call heat. A history of the kinetic theory of gases in the 19th century*, Vol. 1: *Physics and the atomists*, North-Holland, Amsterdam, 1976; Vol. 2: *Statistical physics and irreversible processes*, North-Holland, Amsterdam, 1976.

C. CERCIGNANI, *Ludwig Boltzmann, the man who trusted atoms*, Oxford University Press, Oxford, 1998.

K. HUANG, *Statistical mechanics*, Wiley, New York, second edition, 1987.

R. ZWANZIG, *Nonequilibrium statistical mechanics*, Oxford University Press, Oxford, 2001.

References

J. LOSCHMIDT, Über den Zustand des Wärmegleichgewichtes eines Systems von Körpern mit Rücksicht auf die Schwerkraft, *Wiener Berichte* **73**, 139 (1876).

L. BOLTZMANN, Über die Beziehung eines allgemeinen mechanischen Satzes zum zweiten Hauptsatze der Wärmetheorie, *Sitzungsberichte der Akademie der Wissenschaften*, Wien II, **75**, 67 (1877).

H. POINCARÉ, Sur le problème des trois corps et les équations de la dynamique, *Acta Mathematica* **13**, 1 (1890).

E. ZERMELO, Über einen Satz der Dynamik und die mechanische Wärmetheorie, *Wiedemanns Annalen* **57**, 485 (1896).

L. BOLTZMANN, Entgegnung auf die wärmetheoretischen Betrachtungen des Hrn. E. Zermelo, *Wiedemanns Annalen* **57**, 773 (1896).

J. BRICMONT, Science of chaos or chaos in science?, *Physicalia Mag.* **17**, 159 (1995).

J. LEBOWITZ, Microscopic origin of irreversible macroscopic behavior, *Physica A* **263**, 516 (1998).

Chapter 6

Transport coefficients

Solving the kinetic equations is a tricky problem. The main technical difficulties of this resolution arise from the complicated form of the collision integral. This is particularly apparent in the case of the Boltzmann equation for dilute classical gases, whose collision integral is quadratic with respect to the distribution function. Even the kinetic equation of the Lorentz gas, in which the collision integral is linear, cannot in general be solved in a simple way.

We are thus led to devise approximate resolution methods. One of the most commonly used procedures relies on the relaxation time approximation, which allows us to have to solve, instead of an integro-differential equation (if the case arises, non-linear), a linear partial differential equation. The relaxation time approximation is based on the idea that the main effect of the collisions is to produce a relaxation of the distribution towards a local equilibrium distribution, over a time of the order of the collision time. When the amplitudes of the external forces or of the applied gradients are not too important, the solution of the kinetic equation does not depart very much from the local equilibrium distribution towards which it relaxes. It then becomes possible to solve the kinetic equation at first perturbation order.

This resolution method thus relies on two successive linearizations, the relaxation time approximation (which leads to a linearization with respect to the distribution function), and the linearization with respect to the perturbations. This procedure allows us to get an explicit analytic expression for the solution of the kinetic equation. Making use of this solution, we can justify from a microscopic point of view the linear phenomenological laws of transport and compute the transport coefficients. In this chapter, this procedure is applied to the determination of the electrical conductivity and the diffusion coefficient of a Lorentz gas.

1. The relaxation time approximation

The Boltzmann equation and the kinetic equation of the Lorentz gas for the one-particle distribution function $f(\boldsymbol{r},\boldsymbol{p},t)$ are both of the form:

$$\frac{\partial f}{\partial t}+\boldsymbol{v}.\nabla_{\boldsymbol{r}}f+\boldsymbol{F}.\nabla_{\boldsymbol{p}}f=\left(\frac{\partial f}{\partial t}\right)_{\text{coll}}, \tag{6.1.1}$$

where the term $(\partial f/\partial t)_{\text{coll}}$ represents the effect of the collisions on the evolution of the distribution function.[1] In view of the results they allow us to obtain, the kinetic equations of the form (6.1.1) are also called *transport equations.*

The collision integral of the Boltzmann equation is quadratic with respect to the distribution function (which corresponds to the existence of binary collisions within the gas):

$$\left(\frac{\partial f}{\partial t}\right)_{\text{coll}}=\int d\boldsymbol{p}_1\int d\Omega\,\sigma(\Omega)|\boldsymbol{v}-\boldsymbol{v}_1|\big[f(\boldsymbol{r},\boldsymbol{p}',t)f(\boldsymbol{r},\boldsymbol{p}_1',t)-f(\boldsymbol{r},\boldsymbol{p},t)f(\boldsymbol{r},\boldsymbol{p}_1,t)\big]. \tag{6.1.2}$$

In the case of the Lorentz gas, the collision integral is linear (in accordance with the fact that the scatterers are fixed and exterior to the system of particles):

$$\left(\frac{\partial f}{\partial t}\right)_{\text{coll}}=n_i|\boldsymbol{v}|\int d\Omega'\,\sigma(\boldsymbol{p}|\boldsymbol{p}')\big[f(\boldsymbol{r},\boldsymbol{p}',t)-f(\boldsymbol{r},\boldsymbol{p},t)\big]. \tag{6.1.3}$$

Amongst the approximate resolution methods of equations of the type (6.1.1), one of the simplest and most commonly used relies on the *relaxation time approximation.* This approximation is based on the physical idea that the main effect of the collisions as described by the term $(\partial f/\partial t)_{\text{coll}}$ is to induce a relaxation of the distribution function towards a local equilibrium distribution $f^{(0)}(\boldsymbol{r},\boldsymbol{p},t)$ appropriate to the physics of the problem.

1.1. The local equilibrium distributions

The form of the local equilibrium distributions depends on the collisional invariants.

- *Classical gas with binary collisions*

The independent collisional invariants are the mass, the kinetic momentum, and the kinetic energy.

Accordingly, the local equilibrium distributions which make the collision integral vanish are Maxwell–Boltzmann distributions characterized by the local density $n(\boldsymbol{r},t)$, the local mean velocity $\boldsymbol{u}(\boldsymbol{r},t)$, and the local temperature $T(\boldsymbol{r},t)$, all these quantities being slowly varying functions of $\boldsymbol{r}$ and t:

$$f^{(0)}(\boldsymbol{r},\boldsymbol{p},t)=n(\boldsymbol{r},t)\big[2\pi mkT(\boldsymbol{r},t)\big]^{-3/2}\exp\left[-\frac{|\boldsymbol{p}-m\boldsymbol{u}(\boldsymbol{r},t)|^2}{2mkT(\boldsymbol{r},t)}\right]. \tag{6.1.4}$$

In formula (6.1.4), m denotes the mass of one of the gas molecules.

[1] Another evolution equation of the one-particle distribution function, having no collision term and appropriate to the case of plasmas, namely the Vlasov equation, will be studied in Supplement 6A.

• *Lorentz gas*

The kinetic momentum is not a collisional invariant: only its modulus is such an invariant.

Accordingly, the collision integral vanishes for local Maxwell–Boltzmann distributions with no mean velocity parameter[2] ($\boldsymbol{u} = 0$):

$$f^{(0)}(\boldsymbol{r}, \boldsymbol{p}, t) = n(\boldsymbol{r}, t)\big[2\pi mkT(\boldsymbol{r}, t)\big]^{-3/2} \exp\left[-\frac{p^2}{2mkT(\boldsymbol{r}, t)}\right]. \tag{6.1.5}$$

1.2. The transport equation in the relaxation time approximation

In each specific physical situation, we have to determine the parameters of the relevant local equilibrium distribution. Once this particular distribution $f^{(0)}$ is defined, we write the collision term of equation (6.1.1) in the approximate form[3] $-(f - f^{(0)})/\tau(\boldsymbol{v})$, where $\tau(\boldsymbol{v})$ denotes a microscopic relaxation time towards local equilibrium. Equation (6.1.1) then takes the form of a linear partial differential equation:

$$\boxed{\frac{\partial f}{\partial t} + \boldsymbol{v}.\nabla_{\boldsymbol{r}} f + \boldsymbol{F}.\nabla_{\boldsymbol{p}} f = -\frac{f - f^{(0)}}{\tau(\boldsymbol{v})}.} \tag{6.1.6}$$

The transport equation (6.1.6) is designated under the generic name of *Boltzmann equation in the relaxation time approximation*. It can be applied to a gas of molecules undergoing binary collisions,[4] or to a gas of particles carrying out collisions with fixed scattering centers (Lorentz gas).

The remainder of this chapter is devoted to the determination of transport coefficients using equation (6.1.6). Other applications of this equation will be brought into play in the context of the semiclassical theory of electronic transport in solids[5] (the wave vector or the energy dependence of the relaxation time is then related to the details of the electronic collisions mechanisms[6]).

---

[2] The collision integral of the Lorentz model even vanishes, more generally, for any local distribution depending only on the modulus of the kinetic momentum. The thermalization of the gas indeed would imply an exchange of energy between the particles and the scatterers, which the Lorentz model does not account for.

[3] However, the collision integrals (6.1.2) and (6.1.3) vanish for any local equilibrium distribution (from the very definition of a local equilibrium distribution), whereas the approximate form $-(f - f^{(0)})/\tau(\boldsymbol{v})$ of the collision term only vanishes for the particular local equilibrium distribution $f^{(0)}$ under consideration.

[4] See Chapter 7.

[5] The semiclassical theory of electronic transport, known as the *Bloch–Boltzmann theory of transport*, will be presented in Chapter 8.

[6] See Supplement 8A.

2. Linearization with respect to the external perturbations

The local equilibrium distribution $f^{(0)}$ involved on the right-hand side of equation (6.1.6) is not a solution of this equation, since it does not make its left-hand side vanish. However, provided that the amplitudes of the external forces or of the applied gradients are not too large, f remains close to $f^{(0)}$ at any time. Deviations from local equilibrium remaining small, we can look for a distribution function of the form:

$$f \simeq f^{(0)} + f^{(1)}, \qquad f^{(1)} \ll f^{(0)}. \tag{6.2.1}$$

The collision term on the right-hand side of equation (6.1.6) then simply reads $-f^{(1)}/\tau(\boldsymbol{v})$. As for the left-hand side of this equation, we compute it in an approximate way by taking into account the hypothesis $f^{(1)} \ll f^{(0)}$ and retaining only the lowest order terms. We thus solve equation (6.1.6) at first perturbation order.[7]

In the present chapter, we illustrate this general procedure by applying it to a Lorentz gas submitted to a temperature gradient and a chemical potential gradient. We thus obtain microscopic expressions for the kinetic coefficients and the related transport coefficients involved in the phenomenological transport laws. Besides, each one of the transport coefficients may also be computed in a more direct way starting from the solution of equation (6.1.6) in the presence solely of the appropriate external force or applied gradient. As an example, we will show how to recover, in this way, the electrical conductivity and the diffusion coefficient.

3. Kinetic coefficients of a Lorentz gas

Consider a Lorentz gas submitted to a temperature gradient and a chemical potential gradient. There is no applied external force. The local temperature and the local chemical potential, time-independent, are respectively denoted by $T(\boldsymbol{r})$ and $\mu(\boldsymbol{r})$. The local chemical potential is a function of the local density $n(\boldsymbol{r})$ and the local temperature:

$$\exp\left[\frac{\mu(\boldsymbol{r})}{kT(\boldsymbol{r})}\right] = h^3 n(\boldsymbol{r})\big[2\pi mkT(\boldsymbol{r})\big]^{-3/2}. \tag{6.3.1}$$

3.1. The Boltzmann equation

In this context, the relevant local equilibrium distribution is the Maxwell–Boltzmann distribution characterized by $n(\boldsymbol{r})$ and $T(\boldsymbol{r})$:

$$f^{(0)}(\boldsymbol{r},\boldsymbol{p}) = n(\boldsymbol{r})\big[2\pi mkT(\boldsymbol{r})\big]^{-3/2}\exp\left[-\frac{\varepsilon}{kT(\boldsymbol{r})}\right], \qquad \varepsilon = \frac{p^2}{2m}. \tag{6.3.2}$$

Formula (6.3.2) may equivalently be written as:

$$f^{(0)}(\boldsymbol{r},\boldsymbol{p}) = h^{-3}\exp\left[-\frac{\varepsilon-\mu(\boldsymbol{r})}{kT(\boldsymbol{r})}\right]. \tag{6.3.3}$$

[7] The linearization with respect to the perturbations (sometimes called *second linearization* of the Boltzmann equation) must not be confused with the linearization with respect to f resulting from the relaxation time approximation (*first linearization*).

Since there is no applied external force, the Boltzmann equation (6.1.6) reads:

$$\frac{\partial f}{\partial t} + \boldsymbol{v}.\nabla_{\boldsymbol{r}} f = -\frac{f - f^{(0)}}{\tau(\boldsymbol{v})}. \tag{6.3.4}$$

3.2. The first-order distribution function

When the applied gradients are small, we expect the solution f of equation (6.3.4) not to differ too much from $f^{(0)}$. We thus write it in the form (6.2.1), and we look for $f^{(1)}$ by expanding equation (6.3.4) to first order. To begin with, we obtain the lowest-order equation:

$$\frac{\partial f^{(0)}}{\partial t} = 0, \tag{6.3.5}$$

which is actually verified by the chosen distribution $f^{(0)}$ (formulas (6.3.2) or (6.3.3)), since $T(\boldsymbol{r})$ and $\mu(\boldsymbol{r})$ are time-independent. At next order, we get a linear partial differential equation for $f^{(1)}$,

$$\frac{\partial f^{(1)}}{\partial t} + \boldsymbol{v}.\nabla_{\boldsymbol{r}} f^{(0)} = -\frac{f^{(1)}}{\tau(\boldsymbol{v})}, \tag{6.3.6}$$

whose stationary solution is:

$$f^{(1)} = -\tau(\boldsymbol{v})\boldsymbol{v}.\nabla_{\boldsymbol{r}} f^{(0)}. \tag{6.3.7}$$

3.3. Particle and energy fluxes

The particle flux and the energy flux can be obtained from f via the integrals:

$$\begin{cases} \boldsymbol{J}_N = \displaystyle\int f\boldsymbol{v}\, d\boldsymbol{p} \\ \boldsymbol{J}_E = \displaystyle\int f\boldsymbol{v}\varepsilon\, d\boldsymbol{p}. \end{cases} \tag{6.3.8}$$

The distribution $f^{(0)}$ does not contribute to the fluxes (the corresponding integrals vanish for symmetry reasons). To determine the contribution of $f^{(1)}$, whose expression involves the scalar product $\boldsymbol{v}.\nabla_{\boldsymbol{r}} f^{(0)}$ (formula (6.3.7)), we make use of the relation[8] (deduced from the expression (6.3.3) for $f^{(0)}$):

$$\nabla_{\boldsymbol{r}} f^{(0)} = -\frac{1}{k} f^{(0)} \Big[\nabla(-\frac{\mu}{T}) + \varepsilon\nabla(\frac{1}{T})\Big]. \tag{6.3.9}$$

[8] Since there is no ambiguity as far as they are concerned, the spatial gradients of the intensive variables $1/T$ and $-\mu/T$ are simply denoted by $\nabla(1/T)$ and $\nabla(-\mu/T)$ (and not by $\nabla_{\boldsymbol{r}}(1/T)$ and $\nabla_{\boldsymbol{r}}(-\mu/T)$).

This gives the formulas:

$$\begin{cases} \boldsymbol{J}_N = \dfrac{1}{3k}\displaystyle\int \tau(\boldsymbol{v})v^2 f^{(0)}\Big[\nabla(-\frac{\mu}{T}) + \varepsilon\nabla(\frac{1}{T})\Big]\, d\boldsymbol{p} \\ \boldsymbol{J}_E = \dfrac{1}{3k}\displaystyle\int \tau(\boldsymbol{v})v^2 \varepsilon f^{(0)}\Big[\nabla(-\frac{\mu}{T}) + \varepsilon\nabla(\frac{1}{T})\Big]\, d\boldsymbol{p}, \end{cases} \tag{6.3.10}$$

which express the linear response of the particle and energy fluxes to the affinities $\nabla(-\mu/T)$ and $\nabla(1/T)$.

3.4. Kinetic coefficients

The linear response relations (6.3.10) are of the general form:

$$\begin{cases} \boldsymbol{J}_N = L_{NN}\nabla(-\frac{\mu}{T}) + L_{NE}\nabla(\frac{1}{T}) \\ \boldsymbol{J}_E = L_{EN}\nabla(-\frac{\mu}{T}) + L_{EE}\nabla(\frac{1}{T}). \end{cases} \tag{6.3.11}$$

The kinetic coefficients involved in formulas (6.3.11) are expressed as:[9]

$$\begin{cases} L_{NN} = \dfrac{1}{3k}\displaystyle\int \tau(\boldsymbol{v})v^2 f^{(0)}\, d\boldsymbol{p} \\ L_{EN} = L_{NE} = \dfrac{1}{3k}\displaystyle\int \tau(\boldsymbol{v})v^2 \varepsilon f^{(0)}\, d\boldsymbol{p} \\ L_{EE} = \dfrac{1}{3k}\displaystyle\int \tau(\boldsymbol{v})v^2 \varepsilon^2 f^{(0)}\, d\boldsymbol{p}. \end{cases} \tag{6.3.12}$$

The Onsager symmetry relation $L_{EN} = L_{NE}$ is de facto verified.

To compute the integrals involved in formulas (6.3.12), it is necessary to know the law $\tau(\boldsymbol{v})$. We assume for the sake of simplicity that the relaxation time does not depend on the velocity: $\tau(\boldsymbol{v}) = \tau$. Writing $f^{(0)}$ in the form (6.3.2), shows that the kinetic coefficients are expressed in terms of Gaussian integrals.[10] All calculations once

---

[9] Since the medium is isotropic, the kinetic coefficients are scalars.

[10] When the relaxation time does not depend on the velocity, the expressions for L_{NN}, $L_{EN} = L_{NE}$, and L_{EE} involve respectively the integrals $I_n = \int_0^\infty dv\, v^n e^{-mv^2/2kT}$ with $n = 4$, $n = 6$, and $n = 8$. To compute the I_n's, we carry out the change of variable $v = x(2kT/m)^{1/2}$ and we make use of the formula:

$$\int_0^\infty e^{-x^2} x^{2\alpha-1}\, dx = \frac{\Gamma(\alpha)}{2}, \qquad \alpha > 0,$$

where Γ denotes the Euler's Gamma function. We thus get:

$$I_4 = \frac{1.3}{2^3}\pi^{1/2}\Big(\frac{2kT}{m}\Big)^{5/2}, \qquad I_6 = \frac{1.3.5}{2^4}\pi^{1/2}\Big(\frac{2kT}{m}\Big)^{7/2}, \qquad I_8 = \frac{1.3.5.7}{2^5}\pi^{1/2}\Big(\frac{2kT}{m}\Big)^{9/2}.$$

made, we get the following microscopic expressions for the kinetic coefficients:

$$\begin{cases} L_{NN} = \dfrac{n\tau T}{m} \\ L_{EN} = L_{NE} = \dfrac{5}{2}\dfrac{n\tau}{m}kT^2 \\ L_{EE} = \dfrac{35}{4}\dfrac{n\tau}{m}k^2T^3. \end{cases} \tag{6.3.13}$$

3.5. Transport coefficients

We can deduce from formulas (6.3.13) the microscopic expressions for the transport coefficients involved in the linear phenomenological laws.

• *Electrical conductivity*

From the relation:

$$\sigma = \frac{q^2}{T}L_{NN}, \tag{6.3.14}$$

we deduce the electrical conductivity of a Lorentz gas of particles with charge q:

$$\sigma = \frac{nq^2\tau}{m}. \tag{6.3.15}$$

• *Diffusion coefficient*

In the same way, the relation:

$$D = \frac{1}{T}\frac{\partial\mu}{\partial n}\bigg|_T L_{NN} \tag{6.3.16}$$

yields the expression for the diffusion coefficient of the Lorentz gas:

$$D = \frac{kT\tau}{m}. \tag{6.3.17}$$

• *Thermal conductivity*

The number of particles is conserved. The formula relating the thermal conductivity to the kinetic coefficients is thus:

$$\kappa = \frac{1}{T^2}\frac{L_{EE}L_{NN} - L_{NE}L_{EN}}{L_{NN}}. \tag{6.3.18}$$

From formulas (6.3.13), we get the expression for the thermal conductivity of the Lorentz gas:

$$\kappa = \frac{5}{2}\frac{nk^2T\tau}{m}. \tag{6.3.19}$$

Let us now come back in more detail to the electrical conductivity and the diffusion coefficient, which we will directly compute by looking for the solution of the Boltzmann equation (6.1.6) in the presence either of an electric field or of a density gradient.

4. Electrical conductivity

Consider a gas of particles of mass m and charge q in the presence of an applied electric field $\boldsymbol{E}$ spatially uniform and time-independent. The medium is supposed to be macroscopically neutral (the particle density n is thus uniform), and at a uniform temperature T.

The considered particles are classical. These particles may be, for instance, electrons in a non-degenerate semiconductor. The electrons are scattered mainly via their collisions with impurities or other types of defects. The fixed scatterers are exterior to the system of particles. This latter system may thus be treated as a Lorentz gas (insofar as the collisions of the particles between themselves may be neglected with respect to their collisions with the scattering centers).

4.1. The Drude model

When the relaxation time does not depend on the velocity, it is not really necessary to turn to the Boltzmann equation to compute the electrical conductivity. It is actually enough to write down the evolution equation of the mean velocity of the charge carriers in the presence of the electric field, the relaxation being taken into account in the form of a 'fluid friction' term proportional to the mean velocity:

$$m\frac{d\langle\boldsymbol{v}\rangle}{dt}+m\frac{\langle\boldsymbol{v}\rangle}{\tau}=q\boldsymbol{E}. \tag{6.4.1}$$

The stationary mean velocity then directly follows:

$$\langle\boldsymbol{v}\rangle=\frac{q\tau}{m}\boldsymbol{E}. \tag{6.4.2}$$

This model, proposed by P. Drude in 1900, has historically been the basis of the first theory of electronic transport in metals.[11] The drift mobility μ_D of the Drude model is:

$$\mu_D=\frac{q\tau}{m}. \tag{6.4.3}$$

We thus recover very simply for the electrical conductivity $\sigma = nq\mu_D$ the previously obtained expression (6.3.15), known for this reason as the *Drude–Lorentz formula*:

$$\boxed{\sigma=\frac{nq^2\tau}{m}.} \tag{6.4.4}$$

4.2. The Boltzmann equation

The Boltzmann equation (6.1.6) enables us to consider more complex situations than does the Drude model, in which we are simply interested in the evolution of the mean velocity of the charge carriers. Indeed, the Boltzmann equation allows us to take into account the velocity dependence of the relaxation time.

[11] However, to properly describe electronic transport in metals, it is necessary to take into account the quantum character of the electron gas as well as the band structure of the metal (see Chapter 8).

However, to simplify the calculations, we will treat here the Boltzmann equation under the hypothesis $\tau(\boldsymbol{v}) = \tau$. The carrier density and the temperature being uniform, the relevant local equilibrium distribution function $f^{(0)}$ is the Maxwell–Boltzmann distribution characterized by the density n and the temperature T. This distribution is just the thermodynamic equilibrium distribution $f_0(\boldsymbol{p})$:

$$f_0(\boldsymbol{p}) = n(2\pi mkT)^{-3/2} \exp\Big(-\frac{p^2}{2mkT}\Big). \tag{6.4.5}$$

On account of the hypothesis $\tau(\boldsymbol{v}) = \tau$, the Boltzmann equation (6.1.6) reads:

$$\frac{\partial f}{\partial t} + \boldsymbol{v}.\nabla_{\boldsymbol{r}} f + q\boldsymbol{E}.\nabla_{\boldsymbol{p}} f = -\frac{f - f^{(0)}}{\tau}. \tag{6.4.6}$$

4.3. First-order distribution function

When the applied electric field is small, we write the solution f of equation (6.4.6) in the form (6.2.1), and we look for it by means of a perturbation expansion. At lowest order, we obtain the equation:

$$\frac{\partial f^{(0)}}{\partial t} + \boldsymbol{v}.\nabla_{\boldsymbol{r}} f^{(0)} = 0. \tag{6.4.7}$$

Equation (6.4.7) is actually verified by the distribution (6.4.5), since the density and the temperature are uniform and time-independent. At next order, we get a linear partial differential equation for $f^{(1)}$,

$$\frac{\partial f^{(1)}}{\partial t} + \boldsymbol{v}.\nabla_{\boldsymbol{r}} f^{(1)} + q\boldsymbol{E}.\nabla_{\boldsymbol{p}} f_0 = -\frac{f^{(1)}}{\tau}, \tag{6.4.8}$$

whose uniform and stationary solution is:

$$f^{(1)} = -q\tau \boldsymbol{E}.\nabla_{\boldsymbol{p}} f_0. \tag{6.4.9}$$

4.4. Ohm's law: determination of σ

From the expression for the electric current density,

$$\boldsymbol{J} = q \int f \boldsymbol{v} \, d\boldsymbol{p}, \tag{6.4.10}$$

comes:

$$\boldsymbol{J} = q \int f_0 \boldsymbol{v} \, d\boldsymbol{p} - q^2\tau \int \boldsymbol{v}(\boldsymbol{E}.\nabla_{\boldsymbol{p}}) f_0 \, d\boldsymbol{p}. \tag{6.4.11}$$

The first term of the expression (6.4.11) for $\boldsymbol{J}$ vanishes for symmetry reasons (there is no current at equilibrium). The second term may be computed with the help of the formula:

$$\nabla_{\boldsymbol{p}} f_0 = -\frac{1}{kT} \boldsymbol{v} f_0, \tag{6.4.12}$$

giving:

$$\boldsymbol{J} = \frac{q^2\tau}{kT}\int \boldsymbol{v}(\boldsymbol{v}.\boldsymbol{E})f_0\,d\boldsymbol{p}. \tag{6.4.13}$$

Formula (6.4.13) may be identified with Ohm's law $\boldsymbol{J} = \underline{\sigma}.\boldsymbol{E}$. The conductivity tensor $\underline{\sigma}$ here is proportional to the unit matrix: $\sigma_{\alpha\beta} = \sigma\delta_{\alpha\beta}$. The electrical conductivity σ is given by:

$$\sigma = \frac{nq^2\tau}{kT}\langle v_\alpha^2\rangle, \tag{6.4.14}$$

that is, with the help of the equipartition relation $\langle v_\alpha^2\rangle = kT/m$:

$$\boxed{\sigma = \frac{nq^2\tau}{m}.} \tag{6.4.15}$$

Formula (6.4.15), deduced from the Boltzmann equation in the case of a velocity-independent relaxation time, is identical to the Drude–Lorentz formula (6.4.4). However, interestingly enough, the determination of the conductivity via the Boltzmann equation can be generalized to other laws $\tau(\boldsymbol{v})$.

5. Diffusion coefficient

In the same way, we aim here to establish directly a microscopic expression for the diffusion coefficient of a Lorentz gas.[12]

We consider a Lorentz gas at a uniform temperature T, in which a density gradient is maintained, and we assume that there is no applied external force.

5.1. The Boltzmann equation

The temperature being uniform, the relevant local equilibrium distribution $f^{(0)}$ is the Maxwell–Boltzmann distribution characterized by the local density $n(\boldsymbol{r})$ and the temperature T:

$$f^{(0)}(\boldsymbol{r},\boldsymbol{p}) = n(\boldsymbol{r})(2\pi mkT)^{-3/2}\exp\Big(-\frac{p^2}{2mkT}\Big). \tag{6.5.1}$$

The Boltzmann equation (6.1.6) reads, with the hypothesis $\tau(\boldsymbol{v}) = \tau$:

$$\frac{\partial f}{\partial t} + \boldsymbol{v}.\nabla_{\boldsymbol{r}} f = -\frac{f - f^{(0)}}{\tau}. \tag{6.5.2}$$

[12] The following calculation equally applies to the diffusion coefficient of an ideal classical gas of molecules undergoing binary collisions. In this latter case, we are computing the *self-diffusion coefficient*, that is, the diffusion coefficient of given recognizable molecules, which are sometimes designated as *tagged particles*, within a gas consisting of molecules identical, except for the tagging, to those under study. The tagged particles are proportionally small in number. We thus neglect the collisions that they carry out between themselves. The Boltzmann equation (6.1.6) in this case concerns the distribution function of the tagged particles.

5.2. The first-order distribution function

When the density gradient is small, we write the solution f of equation (6.5.2) in the form (6.2.1), and look for it by means of a perturbation expansion. At lowest order, this gives the equation:

$$\frac{\partial f^{(0)}}{\partial t} = 0. \tag{6.5.3}$$

Equation (6.5.3) is actually verified by the distribution (6.5.1), since the local density and the temperature are time-independent. At next order, we obtain a linear partial differential equation for $f^{(1)}$,

$$\frac{\partial f^{(1)}}{\partial t} + \boldsymbol{v}.\nabla_{\boldsymbol{r}} f^{(0)} = -\frac{f^{(1)}}{\tau}, \tag{6.5.4}$$

whose stationary solution is:

$$f^{(1)} = -\tau \boldsymbol{v}.\nabla_{\boldsymbol{r}} f^{(0)}. \tag{6.5.5}$$

Since the temperature is uniform, $f^{(0)}$ depends on $\boldsymbol{r}$ only via the density $n(\boldsymbol{r})$. This gives:[13]

$$f^{(1)} = -\tau \frac{\partial f^{(0)}}{\partial n} \boldsymbol{v}.\nabla n. \tag{6.5.6}$$

5.3. Fick's law: determination of D

The particle current density $\boldsymbol{J}_N(\boldsymbol{r}) = n(\boldsymbol{r})\langle \boldsymbol{v} \rangle$ is expressed in terms of f as:

$$\boldsymbol{J}_N(\boldsymbol{r}) = \int f(\boldsymbol{r}, \mathbf{p}) \boldsymbol{v} \, d\mathbf{p}. \tag{6.5.7}$$

This gives:

$$\boldsymbol{J}_N(\boldsymbol{r}) = \int f^{(0)}(\boldsymbol{r}, \mathbf{p}) \boldsymbol{v} \, d\mathbf{p} - \tau \int \boldsymbol{v} \, (\boldsymbol{v}.\nabla n) \, \frac{\partial f^{(0)}}{\partial n} \, d\mathbf{p}. \tag{6.5.8}$$

The first term of the expression (6.5.8) for $\boldsymbol{J}_N(\boldsymbol{r})$ vanishes for symmetry reasons. The second one can be calculated with the aid of the relation:

$$\frac{\partial f^{(0)}}{\partial n} = \frac{f^{(0)}}{n(\boldsymbol{r})}. \tag{6.5.9}$$

This gives:

$$\boldsymbol{J}_N(\boldsymbol{r}) = -\tau \frac{1}{n(\boldsymbol{r})} \int \boldsymbol{v}(\boldsymbol{v}.\nabla n) f^{(0)} \, d\mathbf{p}. \tag{6.5.10}$$

Equation (6.5.10) may be identified with Fick's law $\boldsymbol{J}_N = -\underline{D}.\nabla n$. The tensor $\underline{D}$ here is proportional to the unit matrix: $D_{\alpha\beta} = D\delta_{\alpha\beta}$. The diffusion coefficient D is given by:

$$D = \tau \langle v_\alpha^2 \rangle, \tag{6.5.11}$$

[13] No confusion being possible, the spatial density gradient is simply denoted here by ∇n (and not by $\nabla_{\boldsymbol{r}} n$).

that is:

$$\boxed{D = \frac{kT\tau}{m},} \tag{6.5.12}$$

an expression which coincides with formula (6.3.17).

Introducing in formula (6.5.11) the root-mean-square velocity $v = \langle v^2 \rangle^{1/2}$, we obtain for the diffusion coefficient in a three-dimensional space an expression in terms of v and of the mean free path defined here as $\ell = v\tau$:

$$\boxed{D = \frac{1}{3} v\ell.} \tag{6.5.13}$$

Formula (6.5.13) displays the role of space dimensionality. In a d-dimensional space, we would have:

$$D = \frac{1}{d} v\ell. \tag{6.5.14}$$

5.4. The Einstein relation

The expressions for the mobility and the diffusion coefficient (formulas (6.4.3) and (6.5.12)) verify the Einstein relation:

$$\frac{D}{\mu_D} = \frac{kT}{q}. \tag{6.5.15}$$

Formulas (6.4.3) and (6.5.12) both involve a microscopic quantity, the collision time τ, which has been taken as an estimation of the relaxation time towards a local equilibrium. This parameter does not appear in the Einstein relation (which, as a matter of fact, can be obtained in the general macroscopic framework of the relations between kinetic coefficients and transport coefficients).

Bibliography

N.W. Ashcroft and N.D. Mermin, *Solid state physics*, W.B. Saunders Company, Philadelphia, 1976.

R. Balian, *From microphysics to macrophysics*, Vol. 2, Springer-Verlag, Berlin, 1992.

K. Huang, *Statistical mechanics*, Wiley, New York, second edition, 1987.

H.J. Kreuzer, *Nonequilibrium thermodynamics and its statistical foundations*, Clarendon Press, Oxford, 1981.

F. Reif, *Fundamentals of statistical and thermal physics*, McGraw-Hill, New York, 1965.

H. Smith and H.H. Jensen, *Transport phenomena*, Oxford Science Publications, Oxford, 1989.

References

P. Drude, Zur Elektronentheorie. I, *Annalen der Physik* **1**, 566 (1900); Zur Elektronentheorie. II, *Annalen der Physik* **3**, 369 (1900).

Supplement 6A

Landau damping

1. Weakly coupled plasma

We are interested here in a *weakly coupled* plasma, which signifies that the Coulomb interaction potential energy between two particles is much smaller than their kinetic energy. To quantify the strength of the Coulomb interactions in a plasma of density n in equilibrium at temperature T, we introduce the *plasma parameter* ρ_p, defined as the ratio of the Coulomb interaction energy at a distance equal to the average interparticle distance $d \sim n^{-1/3}$ to the thermal energy:[1]

$$\rho_p = \frac{e^2 n^{1/3}}{kT}. \tag{6A.1.1}$$

The plasma is considered as weakly coupled when we have $\rho_p \ll 1$.

We will study some properties related to the propagation of electromagnetic waves in a weakly coupled plasma, assumed to be classical and non-relativistic.[2] These properties involve transport coefficients such as the electrical conductivity of the plasma, which we will determine by making use of appropriate evolution equations for the distribution functions of the plasma particles.

2. The Vlasov equations for a collisionless plasma

Consider a *two-component plasma* consisting of electrons with charge e and positive ions of some type, with charge $-Ze$.

To describe this out-of-equilibrium plasma, we have to introduce a distribution function for each of the two types of particles. It is preferable to use distribution functions with gauge-independent arguments. We therefore introduce the distribution functions $f(\boldsymbol{r}, \boldsymbol{v}, t)$ of the electrons and $F(\boldsymbol{r}, \boldsymbol{v}, t)$ of the ions.

[1] For the sake of simplicity, we assume in this estimation that all plasma particles carry a charge of modulus $|e|$.

[2] For the quantum effects to be negligible, the temperature of the plasma must be much higher than the Fermi temperature of the electrons. The plasma is non-relativistic when the condition $mc^2 \gg kT$ is fulfilled (m is the mass of the electron).

2.1. Evolution of the distribution functions

In the presence of an electric field $\boldsymbol{E}$ and a magnetic field $\boldsymbol{H}$, the functions f and F obey evolution equations of the form:

$$\begin{cases} \dfrac{\partial f}{\partial t} + \boldsymbol{v}.\nabla_{\boldsymbol{r}} f + \dfrac{1}{m}\boldsymbol{F}_e.\nabla_{\boldsymbol{v}} f = \left(\dfrac{\partial f}{\partial t}\right)_{\text{coll}} \\ \dfrac{\partial F}{\partial t} + \boldsymbol{v}.\nabla_{\boldsymbol{r}} F + \dfrac{1}{M}\boldsymbol{F}_i.\nabla_{\boldsymbol{v}} F = \left(\dfrac{\partial F}{\partial t}\right)_{\text{coll}}. \end{cases} \tag{6A.2.1}$$

In equations (6A.2.1), $\boldsymbol{F}_e = e(\boldsymbol{E}+\boldsymbol{v}\times\boldsymbol{H}/c)$ and $\boldsymbol{F}_i = -Ze(\boldsymbol{E}+\boldsymbol{v}\times\boldsymbol{H}/c)$ denote the Lorentz forces acting respectively on an electron of mass m and an ion of mass M, whereas the terms $(\partial f/\partial t)_{\text{coll}}$ and $(\partial F/\partial t)_{\text{coll}}$ stand for the influence of the 'collisions' between the particles on the evolution of the distribution functions.

Interestingly enough, an electric field $\boldsymbol{E}$ and a magnetic field $\boldsymbol{H}$ actually appear in equations (6A.2.1) even in the absence of applied external fields. Indeed, the Coulomb interaction between two charged particles is a long-range interaction, whose description uniquely in terms of 'collisions' between the concerned particles is unsuitable. In a plasma, even weakly coupled, there are no well-separated length scales (as the range of the intermolecular forces and the mean intermolecular distance in a classical gas of neutral molecules). The main effect of the Coulomb interactions between the plasma particles is to give rise to an average electric field acting on each of them. This collective effect is taken into account in the driving terms on the left-hand side of the evolution equations (6A.2.1): in the absence of applied external fields, the field $\boldsymbol{E}$ which appears in these equations is the average electric field, created by all the particles and acting on each of them, and the field $\boldsymbol{H}$ is the associated magnetic field.[3] The terms $(\partial f/\partial t)_{\text{coll}}$ and $(\partial F/\partial t)_{\text{coll}}$ describe fluctuations with respect to this average collective effect of the Coulomb interactions.

We will study here phenomena in which these fluctuations do not play any determinant role. Such phenomena may be described in the framework of the model of a *collisionless plasma*, in which both terms $(\partial f/\partial t)_{\text{coll}}$ and $(\partial F/\partial t)_{\text{coll}}$ are discarded. The corresponding evolution equations of the distribution functions read:

$$\begin{cases} \dfrac{\partial f}{\partial t} + \boldsymbol{v}.\nabla_{\boldsymbol{r}} f + \dfrac{e}{m}(\boldsymbol{E} + \dfrac{1}{c}\boldsymbol{v}\times\boldsymbol{H}).\nabla_{\boldsymbol{v}} f = 0 \\ \dfrac{\partial F}{\partial t} + \boldsymbol{v}.\nabla_{\boldsymbol{r}} F - \dfrac{Ze}{M}(\boldsymbol{E} + \dfrac{1}{c}\boldsymbol{v}\times\boldsymbol{H}).\nabla_{\boldsymbol{v}} F = 0. \end{cases} \tag{6A.2.2}$$

Equations (6A.2.2) are invariant under time-reversal, that is, under the change $t \to -t$ (with $\boldsymbol{H} \to -\boldsymbol{H}$). They cannot describe irreversible processes. Accordingly, they cannot, in the strict sense, be qualified as kinetic equations.

---

[3] These two fields are determined in a self-consistent way starting from the distribution functions themselves (see Subsection 2.3).

2.2. The Maxwell equationsb

The evolution equations of the distribution functions must be supplemented by the Maxwell equations. These latter equations may be written in two different ways, depending on whether the plasma is considered as a vacuum in the presence of free charges and currents, or as a dielectric polarizable medium.

- *Vacuum in the presence of free charges and currents*

In the presence of a charge density ρ and a current density $\boldsymbol{J}$, the Maxwell equations read:

$$\begin{cases} \nabla \times \boldsymbol{H} = (1/c)(4\pi \boldsymbol{J} + \partial \boldsymbol{E}/\partial t) & \nabla.\boldsymbol{H} = 0 \\ \nabla.\boldsymbol{E} = 4\pi\rho & \nabla \times \boldsymbol{E} = -(1/c)\partial \boldsymbol{H}/\partial t. \end{cases} \tag{6A.2.3}$$

At low field, the current $\boldsymbol{J}$ and the field $\boldsymbol{E}$ are related through Ohm's law $\boldsymbol{J} = \underline{\sigma}.\boldsymbol{E}$, where $\underline{\sigma}$ is the electrical conductivity tensor of the plasma.

- *Polarizable dielectric medium*

We can also consider the plasma as a polarizable dielectric medium devoid of free charges or currents. The Maxwell equations then read:

$$\begin{cases} \nabla \times \boldsymbol{H} = (1/c)\partial \boldsymbol{D}/\partial t & \nabla.\boldsymbol{H} = 0 \\ \nabla.\boldsymbol{D} = 0 & \nabla \times \boldsymbol{E} = -(1/c)\partial \boldsymbol{H}/\partial t, \end{cases} \tag{6A.2.4}$$

where $\boldsymbol{D}$ is the electrical displacement. At low field, the relation between $\boldsymbol{D}$ and $\boldsymbol{E}$ is linear, so we write:

$$\boldsymbol{D} = \underline{\varepsilon}.\boldsymbol{E}, \tag{6A.2.5}$$

where $\underline{\varepsilon}$ is the electrical permittivity tensor of the plasma.

For a plane electromagnetic wave of wave vector $\boldsymbol{q}$ and angular frequency ω, for which the electric and magnetic fields read:

$$\boldsymbol{E}(\boldsymbol{r},t) = \boldsymbol{E}_0 e^{i(\boldsymbol{q}.\boldsymbol{r}-\omega t)}, \qquad \boldsymbol{H}(\boldsymbol{r},t) = \boldsymbol{H}_0 e^{i(\boldsymbol{q}.\boldsymbol{r}-\omega t)}, \tag{6A.2.6}$$

the equivalence between both descriptions of the plasma is expressed by the following relation[4] between the tensors $\underline{\varepsilon}(\boldsymbol{q},\omega)$ and $\underline{\sigma}(\boldsymbol{q},\omega)$:

$$\boxed{\underline{\varepsilon}(\boldsymbol{q},\omega) = 1 + i\frac{4\pi}{\omega}\underline{\sigma}(\boldsymbol{q},\omega).} \tag{6A.2.7}$$

2.3. Charge and current densities

The average charge and current densities in the out-of-equilibrium plasma are computed with the help of the distribution functions as:

$$\begin{cases} \rho = e\displaystyle\int(-ZF + f)\,d\boldsymbol{v} \\ \boldsymbol{J} = e\displaystyle\int(-ZF + f)\boldsymbol{v}\,d\boldsymbol{v}. \end{cases} \tag{6A.2.8}$$

[4] Formula (6A.2.7) can be derived by comparing the expressions for $\nabla \times \boldsymbol{H}$ in both descriptions (formulas (6A.2.3) and (6A.2.4)).

The evolution equations (6A.2.2), supplemented by the Maxwell equations (formulas (6A.2.3) or (6A.2.4)), and by the definition equations of the average charge and current densities (formulas (6A.2.8)), constitute a closed system of equations. This system allows us in principle to determine, given the initial distribution functions, the distribution functions f and F, as well as the fields $\boldsymbol{E}$ and $\boldsymbol{H}$, at any time $t > 0$. The fields $\boldsymbol{E}$ and $\boldsymbol{H}$ are thus obtained in a self-consistent way. This set of equations, called the *Vlasov equations*, was introduced by A.A. Vlasov in 1938.

3. Conductivity and electrical permittivity of a collisionless plasma

For the sake of simplicity, we assume from now on that the dielectric polarization of the plasma only involves the electrons. As for the ions, they are represented by a positive continuous background which does not play any role in the dynamics and whose effect is simply to maintain the global electric neutrality. The plasma is then a *one-component plasma* described by the electronic distribution function f.

The equations (6A.2.2) reduce in this case to the evolution equation of f:

$$\frac{\partial f}{\partial t} + \boldsymbol{v}.\nabla_{\boldsymbol{r}} f + \frac{e}{m}(\boldsymbol{E} + \frac{1}{c}\boldsymbol{v} \times \boldsymbol{H}).\nabla_{\boldsymbol{v}} f = 0. \tag{6A.3.1}$$

The average charge and current densities are:

$$\begin{cases} \rho = e \displaystyle\int f \, d\boldsymbol{v} \\ \boldsymbol{J} = e \displaystyle\int f \boldsymbol{v} \, d\boldsymbol{v}. \end{cases} \tag{6A.3.2}$$

We aim to compute the electrical conductivity and the dielectric permittivity of this plasma. We do not attempt to solve the problem of the relaxation of a fluctuation of the electronic distribution, which would also involve having to determine the fields $\boldsymbol{E}$ and $\boldsymbol{H}$ in a self-consistent way. Accordingly, $\boldsymbol{E}$ and $\boldsymbol{H}$ are treated here as external fields.

3.1. Linearization of the evolution equation

For low fields, the system does not depart very much from equilibrium. We thus look for a solution of equation (6A.3.1) of the form:

$$f \simeq f_0 + f^{(1)}, \qquad f^{(1)} \ll f_0, \tag{6A.3.3}$$

where $f_0(\boldsymbol{v})$ denotes the equilibrium distribution, assumed to be isotropic, and $f^{(1)}$ is a first-order correction. The equation verified by $f^{(1)}$ reads a priori:

$$\frac{\partial f^{(1)}}{\partial t} + \boldsymbol{v}.\nabla_{\boldsymbol{r}} f^{(1)} + \frac{e}{m}(\boldsymbol{E} + \frac{1}{c}\boldsymbol{v} \times \boldsymbol{H}).\nabla_{\boldsymbol{v}} f_0 = 0. \tag{6A.3.4}$$

The distribution f_0 depending solely on $|\boldsymbol{v}|$, $\nabla_{\boldsymbol{v}} f_0$ is parallel to $\boldsymbol{v}$. The mixed product $(\boldsymbol{v} \times \boldsymbol{H}).\nabla_{\boldsymbol{v}} f_0$ thus vanishes, which leaves the equation:

$$\frac{\partial f^{(1)}}{\partial t} + \boldsymbol{v}.\nabla_{\boldsymbol{r}} f^{(1)} + \frac{e}{m}\boldsymbol{E}.\nabla_{\boldsymbol{v}} f_0 = 0. \tag{6A.3.5}$$

Assuming that the electric field is that of a plane wave of angular frequency ω and wave vector $\boldsymbol{q}$ (formula (6A.2.6)), we look for a solution of equation (6A.3.5) varying in the same manner. For reasons which will appear more clearly later,[5] it is convenient to attribute to the angular frequency ω a finite imaginary part $\epsilon > 0$ and to take the limit $\epsilon \to 0^+$ at the end of the calculations. The equation verified by $f^{(1)}$ is, in this regime:

$$i(\omega + i\epsilon - \boldsymbol{q}.\boldsymbol{v})f^{(1)} = \frac{e}{m}\boldsymbol{E}.\nabla_{\boldsymbol{v}} f_0. \tag{6A.3.6}$$

We get from equation (6A.3.6):

$$f^{(1)} = -i\frac{e}{m}\frac{\boldsymbol{E}.\nabla_{\boldsymbol{v}} f_0}{\omega + i\epsilon - \boldsymbol{q}.\boldsymbol{v}}. \tag{6A.3.7}$$

3.2. Current density and electrical conductivity

The equilibrium distribution being isotropic, the direction of $\boldsymbol{q}$ is a priori the only preferred direction of the system. We will choose the Oz axis parallel to $\boldsymbol{q}$.

In the plasma at equilibrium, the electronic charge density is compensated at each point by the ionic charge density, and the current density vanishes. In the plasma perturbed by the electric field, the charge and current densities at first order are respectively given by the integrals:

$$\rho = e\int f^{(1)}\, d\boldsymbol{v} \tag{6A.3.8}$$

and:

$$\boldsymbol{J} = e\int f^{(1)}\boldsymbol{v}\, d\boldsymbol{v}. \tag{6A.3.9}$$

Making use of the expression (6A.3.7) for $f^{(1)}$, gives:[6]

$$\rho = -i\frac{e^2}{m}\lim_{\epsilon\to 0^+}\int \frac{\boldsymbol{E}.\nabla_{\boldsymbol{v}} f_0}{\omega + i\epsilon - qv_z}\, d\boldsymbol{v} \tag{6A.3.10}$$

and:

$$\boldsymbol{J} = -i\frac{e^2}{m}\lim_{\epsilon\to 0^+}\int \frac{(\boldsymbol{E}.\nabla_{\boldsymbol{v}} f_0)\boldsymbol{v}}{\omega + i\epsilon - qv_z}\, d\boldsymbol{v}. \tag{6A.3.11}$$

Formula (6A.3.11) displays the fact that the conductivity tensor is diagonal. It has two transverse identical components $\sigma_{xx} = \sigma_{yy} = \sigma_T$ and a longitudinal component $\sigma_{zz} = \sigma_L$. In other words, when $\boldsymbol{E}$ is perpendicular (*resp.* parallel) to $\boldsymbol{q}$, the relation between $\boldsymbol{J}$ and $\boldsymbol{E}$ amounts to $\boldsymbol{J} = \sigma_T\boldsymbol{E}$ (*resp.* $\boldsymbol{J} = \sigma_L\boldsymbol{E}$). The dependence with respect to $\boldsymbol{q}$ and ω of σ_T and σ_L reflects the non-local and retarded character of the plasma response.

---

[5] See Chapter 12.

[6] We have set $q = |\boldsymbol{q}|$.

• *Transverse conductivity*

This is given by:

$$\sigma_T(\boldsymbol{q},\omega) = -i\frac{e^2}{m}\lim_{\epsilon\to 0^+}\int \frac{v_x\dfrac{\partial f_0}{\partial v_x}}{\omega + i\epsilon - qv_z}\,d\boldsymbol{v}. \tag{6A.3.12}$$

• *Longitudinal conductivity*

This is given by:

$$\sigma_L(\boldsymbol{q},\omega) = -i\frac{e^2}{m}\lim_{\epsilon\to 0^+}\int \frac{v_z\dfrac{\partial f_0}{\partial v_z}}{\omega + i\epsilon - qv_z}\,d\boldsymbol{v}. \tag{6A.3.13}$$

3.3. Dielectric permittivity

The dielectric permittivity tensor has two transverse components $\varepsilon_{xx} = \varepsilon_{yy} = \varepsilon_T$ and a longitudinal component $\varepsilon_{zz} = \varepsilon_L$, related to the corresponding components of the conductivity tensor through the formulas:

$$\varepsilon_T(\boldsymbol{q},\omega) = 1 + i\frac{4\pi}{\omega}\sigma_T(\boldsymbol{q},\omega) \tag{6A.3.14}$$

and:

$$\varepsilon_L(\boldsymbol{q},\omega) = 1 + i\frac{4\pi}{\omega}\sigma_L(\boldsymbol{q},\omega). \tag{6A.3.15}$$

3.4. Transverse waves and longitudinal wave

Eliminating $\boldsymbol{H}$ between the Maxwell equations (6A.2.3), gives the relation:

$$-\nabla\times(\nabla\times\boldsymbol{E}) = \frac{4\pi}{c^2}\underline{\sigma}\frac{\partial\boldsymbol{E}}{\partial t} + \frac{1}{c^2}\frac{\partial^2\boldsymbol{E}}{\partial t^2}, \tag{6A.3.16}$$

that is:

$$\boldsymbol{q}\times(\boldsymbol{q}\times\boldsymbol{E}) = -\frac{4\pi}{c^2}\underline{\sigma}i\omega\boldsymbol{E} - \frac{\omega^2}{c^2}\boldsymbol{E}, \tag{6A.3.17}$$

or:

$$\boldsymbol{q}\times(\boldsymbol{q}\times\boldsymbol{E}) = -\frac{\omega^2}{c^2}\underline{\varepsilon}(\boldsymbol{q},\omega)\boldsymbol{E} \tag{6A.3.18}$$

(a formula which can also be directly obtained by eliminating $\boldsymbol{H}$ between the Maxwell equations (6A.2.4)).

When projected onto the axes Ox, Oy, and Oz, equation (6A.3.18) leads to the equations:

$$\Big[q^2 - \frac{\omega^2}{c^2}\varepsilon_T(\boldsymbol{q},\omega)\Big]E_x = 0, \qquad \Big[q^2 - \frac{\omega^2}{c^2}\varepsilon_T(\boldsymbol{q},\omega)\Big]E_y = 0, \tag{6A.3.19}$$

and:

$$\frac{\omega^2}{c^2}\varepsilon_L(\boldsymbol{q},\omega)E_z = 0. \tag{6A.3.20}$$

Equations (6A.3.19) and (6A.3.20) admit two families of entirely decoupled solutions. The first one corresponds to transverse waves ($E_x \neq 0$, $E_y \neq 0$, $E_z = 0$), defined by:

$$\boxed{\varepsilon_T(\boldsymbol{q}, \omega) = \frac{q^2 c^2}{\omega^2}.} \tag{6A.3.21}$$

The other family of solutions corresponds to longitudinal waves ($E_x = 0$, $E_y = 0$, $E_z \neq 0$), defined by:

$$\boxed{\varepsilon_L(\boldsymbol{q}, \omega) = 0.} \tag{6A.3.22}$$

These 'new' waves (that is, absent in vacuum), purely longitudinal, are called *electrostatic waves* on the grounds that they have no associated magnetic field ($\boldsymbol{H} = 0$).

4. Longitudinal waves in a Maxwellian plasma

We consider more specifically here the propagation of an electrostatic longitudinal wave, equally designated under the name of *plasma wave* or *Langmuir wave*. The equilibrium distribution f_0 is assumed to be a Maxwellian:

$$f_0(\boldsymbol{v}) = n\Big(\frac{m}{2\pi kT}\Big)^{3/2} \exp\Big(-\frac{mv^2}{2kT}\Big). \tag{6A.4.1}$$

4.1. Longitudinal dielectric permittivity

The longitudinal permittivity of the plasma is a complex quantity. The existence of a non-vanishing imaginary part of $\varepsilon_L(\boldsymbol{q}, \omega)$ corresponds to an energy exchange between the field and the medium. To compute explicitly $\varepsilon_L(\boldsymbol{q}, \omega)$, we start from formulas (6A.3.15) and (6A.3.13), into which we introduce the form (6A.4.1) of f_0, for which:

$$\nabla_{\boldsymbol{v}} f_0 = -\frac{1}{kT} m\boldsymbol{v} f_0. \tag{6A.4.2}$$

The integrations over v_x and v_y in formula (6A.3.13) once carried out, give:

$$\varepsilon_L(\boldsymbol{q}, \omega) = 1 - \frac{4\pi n e^2}{\omega kT}\Big(\frac{m}{2\pi kT}\Big)^{1/2} \lim_{\epsilon \to 0^+} \int \frac{v_z^2 e^{-mv_z^2/2kT}}{\omega + i\epsilon - qv_z} \, dv_z. \tag{6A.4.3}$$

The imaginary part of $\varepsilon_L(\boldsymbol{q}, \omega)$ is:[7]

$$\varepsilon_L''(\boldsymbol{q}, \omega) = \Big(\frac{\pi}{2}\Big)^{1/2} \omega \omega_p^2 \frac{1}{(qv)^3} e^{-\omega^2/2q^2v^2}. \tag{6A.4.4}$$

---

[7] To obtain this imaginary part, make use of the relation:

$$\lim_{\epsilon \to 0^+} \frac{1}{x + i\epsilon} = \mathrm{vp}\frac{1}{x} - i\pi\delta(x),$$

where the symbol vp denotes the Cauchy principal value.

In formula (6A.4.4), the quantity $\omega_p = (4\pi ne^2/m)^{1/2}$ is a characteristic angular frequency called the *plasma angular frequency*. Also, we have set $v = (kT/m)^{1/2}$.

When $q \to 0$, the longitudinal permittivity tends towards a real limit value:

$$\varepsilon(\boldsymbol{q} = 0, \omega) = 1 - \frac{\omega_p^2}{\omega^2}, \tag{6A.4.5}$$

which is also the limit when $q \to 0$ of the transverse permittivity[8] $\varepsilon_T(\boldsymbol{q}, \omega)$.

4.2. Model without spatial dispersion

The result (6A.4.5) may be recovered directly in a model without spatial dispersion in which the $\boldsymbol{q}$-dependence of the wave field is neglected. In a regime of angular frequency ω, we then simply have:

$$\boldsymbol{v} = \frac{ie}{m\omega}\boldsymbol{E} \tag{6A.4.6}$$

and:

$$\boldsymbol{J} = \frac{ie^2}{m\omega}n\boldsymbol{E}. \tag{6A.4.7}$$

We thus get:

$$\sigma(\boldsymbol{q} = 0, \omega) = i\frac{ne^2}{m\omega}. \tag{6A.4.8}$$

The dielectric permittivity $\varepsilon(\boldsymbol{q} = 0, \omega)$ then follows via one or the other of the equations (6A.3.14) or (6A.3.15). This effectively recovers the result (6A.4.5).

4.3. Propagation of the longitudinal wave

The propagation of the longitudinal wave is governed by the condition (6A.3.22), which reads, when formula (6A.4.3) is used:

$$1 - \frac{4\pi ne^2}{\omega kT}\left(\frac{m}{2\pi kT}\right)^{1/2} \lim_{\epsilon \to 0^+} \int \frac{v_z^2 e^{-mv_z^2/2kT}}{\omega + i\epsilon - qv_z}\, dv_z = 0. \tag{6A.4.9}$$

Formula (6A.4.9) yields the dispersion relation of the longitudinal wave, that is, the q-dependence of ω. It has complex roots $\omega = \omega' + i\omega''$, which can be obtained through successive approximations.

If we first neglect the q-depending terms, we get, at lowest order,

$$1 - \frac{4\pi ne^2}{m\omega^2} = 0, \tag{6A.4.10}$$

that is:

$$\omega = \omega_p. \tag{6A.4.11}$$

[8] Indeed, in the limit $q \to 0$, there is no preferred direction in the system.

At this approximation order, the electrostatic wave oscillates without damping at a fixed angular frequency equal to ω_p (it can be excited only if $\omega = \omega_p$). This wave is of a character very different from that of an electromagnetic wave (it does not exist in vacuum).

At next order, we can show that the solution of equation (6A.4.9) actually has a non-vanishing imaginary part, whose existence corresponds to a damping of the plasma wave.

4.4. The Landau damping

In a plasma, especially for an electrostatic wave, energy exchanges between the wave and the particles, leading to a damping of the wave, are likely to occur in the absence of collisions, that is, when the plasma is described by the (reversible) Vlasov equation.

The average power per unit time associated with an electrostatic wave of wave vector $\boldsymbol{q}$ and angular frequency ω is, given the form (6A.2.6) of the field $\boldsymbol{E}$,

$$\overline{\frac{dW}{dt}} = \frac{1}{2} E_0^2 \, \Re e \, \sigma_L(\boldsymbol{q}, \omega), \tag{6A.4.12}$$

that is, according to formula (6A.3.15):

$$\overline{\frac{dW}{dt}} = \frac{\omega}{8\pi} \varepsilon''_L(\boldsymbol{q}, \omega) E_0^2. \tag{6A.4.13}$$

Using expression (6A.4.4) for $\varepsilon''_L(\boldsymbol{q}, \omega)$, we get:

$$\overline{\frac{dW}{dt}} = \frac{1}{8(2\pi)^{1/2}} \omega^2 \omega_p^2 \frac{1}{(qv)^3} e^{-\omega^2/2q^2v^2} E_0^2. \tag{6A.4.14}$$

Formula (6A.4.14) shows that, in the case considered here (Maxwellian plasma), we have $\overline{dW/dt} > 0$. This phenomenon, known as *Landau damping*, was predicted by L.D. Landau in 1946, and experimentally verified later.[9]

Landau damping is not linked to the collisions. It is in fact a resonance phenomenon, entirely due to those electrons whose velocity in the propagation direction of the wave is equal to the phase velocity ω/q. These electrons are resonating particles displacing themselves in phase with the wave. The field of the wave, stationary with respect to these electrons, produces on them a work which does not vanish on average. An electron with a velocity slightly higher than the phase velocity of the wave gives up energy to it, whereas an electron with a velocity slightly smaller receives energy from it. In the case of a Maxwellian plasma, the equilibrium distribution function of the electrons decreases with the modulus of their velocity (formula (6A.4.1)). The balance of these energy exchanges thus corresponds to an energy transfer from the wave towards the individual electrons, that is, to a damping of the wave.

---

[9] This mechanism is used in laboratory plasmas to bring them to the very high temperatures necessary for controlled thermonuclear fusion.

Bibliography

R. BALESCU, *Statistical dynamics. Matter out of equilibrium*, Imperial College Press, London, 1997.

E.M. LIFSHITZ and L.P. PITAEVSKII, *Physical kinetics*, Butterworth-Heinemann, Oxford, 1981.

D. ZUBAREV, V. MOROZOV, and G. RÖPKE, *Statistical mechanics of nonequilibrium processes,* Vol. 1: *Basic concepts, kinetic theory*, Akademie Verlag, Berlin, 1996.

Chapter 7

From the Boltzmann equation to the hydrodynamic equations

In a dilute classical gas of molecules undergoing binary collisions, the Boltzmann equation describes the evolution of the distribution function over a time interval intermediate between the duration of a collision and the relaxation time towards a local equilibrium. This kinetic stage of the evolution is followed by a hydrodynamic one, involving time intervals much longer than the relaxation time. The gas is then close to a local equilibrium state described by a reduced number of relevant variables, namely, the local density, the local mean velocity, and the local temperature. The evolution of these variables is governed by the hydrodynamic equations, which can be deduced from the Boltzmann equation.

For this purpose, we first establish, starting from the Boltzmann equation, local balance equations relative to the collisional invariants. These equations involve averages which are expressed with the aid of the distribution function solution of the Boltzmann equation.

The solutions of the Boltzmann equation which are relevant in this context are functionals of the local density of molecules, of their local mean velocity, and of the local temperature. Such solutions, called the normal solutions, may be obtained by means of the Chapman–Enskog expansion, which allows us to construct them by successive approximations. The zeroth-order approximation of the solution is a local equilibrium distribution. At this order, the dissipative fluxes vanish, and the hydrodynamics is that of a perfect fluid. The first-order correction depends on the local equilibrium distribution and on the affinities. It allows us to obtain the dissipative fluxes and the hydrodynamic equations (namely, the Navier–Stokes equation and the heat equation).

1. The hydrodynamic regime

The evolution of a gas from an initial out-of-equilibrium state comprises first a kinetic stage, which puts into play an evolution time interval Δt much shorter than the relaxation time τ_r towards a local equilibrium (as well as distances Δl much smaller than the mean free path ℓ). The state of the gas is then described by the solution $f(\boldsymbol{r}, \boldsymbol{p}, t)$ of the Boltzmann equation. If the collisions are efficient enough, this kinetic stage is followed by a hydrodynamic one, characterized by an evolution time interval $\Delta t \gg \tau_r$ (and by distances $\Delta l \gg \ell$). During this stage of its evolution, the gas remains close to a local equilibrium state[1] described by the local density of molecules $n(\boldsymbol{r}, t)$, their local mean velocity $\boldsymbol{u}(\boldsymbol{r}, t)$, and the local temperature[2] $T(\boldsymbol{r}, t)$. The evolution equations of these variables are the *hydrodynamic equations*. They involve transport coefficients (in the case of a gas constituted of identical molecules, these coefficients are the thermal conductivity and the viscosity coefficient).

The hydrodynamic equations may be deduced from the Boltzmann equation, which confers on these phenomenological laws a microscopic justification in the context of dilute gases. We have, to begin with, to delineate the validity domain of the hydrodynamic regime.

1.1. Collision time and mean free path

The relaxation time τ_r is at least of the order of the collision time τ. It is common practice to take the estimate $\tau_r \sim \tau$. The collision time τ and the mean free path ℓ are related by:

$$\tau \sim \frac{\ell}{v}. \tag{7.1.1}$$

In formula (7.1.1), v denotes a typical molecular velocity, of the order of the root-mean-square velocity $(3kT/m)^{1/2}$.

We use as an estimate of the mean free path the expression:

$$\ell \sim \frac{1}{n\sigma_{\text{tot}}}, \tag{7.1.2}$$

where n is the gas density and $\sigma_{\text{tot}} \sim r_0^2$ the total collision cross-section (r_0 denotes the range of the intermolecular forces). We deduce from it an estimation of the collision

[1] However, when the mean free path is much larger than the dimensions of the vessel in which the gas is enclosed, the gas, rarefied, cannot reach a local equilibrium state. The corresponding transport regime is called the *ballistic regime* or the *Knudsen regime*. The collisions of the molecules with the walls of the vessel then have an essential role. The ballistic regime plays an important role in electronic transport in very small devices.

[2] In place of the local density of molecules, we rather use in this context the local mass density $\rho(\boldsymbol{r}, t) = mn(\boldsymbol{r}, t)$ (m is the mass of one of the gas molecules). Besides, we often choose as variable, instead of the local temperature $T(\boldsymbol{r}, t)$, the local density of internal energy per unit mass $e_{\text{int}}(\boldsymbol{r}, t)$, related to the local temperature by an equipartition formula:

$$e_{\text{int}}(\boldsymbol{r}, t) = \frac{3}{2m} kT(\boldsymbol{r}, t).$$

time:

$$\tau \sim \frac{1}{n\sigma_{\text{tot}} v}. \tag{7.1.3}$$

The mean free path and the collision time vary inversely proportionally to the gas density.[3]

1.2. Validity domain of the hydrodynamic regime

The hydrodynamic stage of the gas evolution concerns a time interval much longer than τ, and distances much larger than ℓ. In other words, if we denote by ω and $\boldsymbol{q}$ an angular frequency and a wave vector typical of the perturbations imposed to the medium (for instance, by an external force or an applied gradient), the hydrodynamic regime is that of excitations with low angular frequencies and large wavelengths, as characterized by the inequalities:

$$\omega\tau \ll 1, \qquad q\ell \ll 1. \tag{7.1.4}$$

In this regime, it is possible to establish the hydrodynamic equations starting from the Boltzmann equation. For this purpose, we first have to examine the consequences of the Boltzmann equation for the quantities which are conserved in a collision. We thus obtain local balance equations, which involve averages computed with the aid of the distribution function solution of the Boltzmann equation.

1.3. Normal solutions of the Boltzmann equation

The gas being close to a local equilibrium state, we look for those solutions of the Boltzmann equation which depend on $\boldsymbol{r}$ and t only via the local density, the local mean velocity, and the local temperature. These particular solutions are called the *normal solutions*. To obtain them, we can devise a systematic expansion procedure proposed by S. Chapman in 1916 and D. Enskog in 1917. We will successively use two approximation orders of the solution. The zeroth-order approximation is a local equilibrium distribution. The corresponding hydrodynamic equations are those of a 'perfect' fluid, that is, of a fluid in which dissipative processes are neglected. In the first-order approximation, we obtain the hydrodynamic equations of a viscous fluid, which involve dissipative transport coefficients.

2. Local balance equations

From the Boltzmann equation, we can deduce local balance equations for each of the collisional invariants, namely, the mass, the three components of the kinetic momentum, and the kinetic energy in the reference frame moving with the fluid.[4]

---

[3] To evaluate the order of magnitude of ℓ and τ in a gas at room temperature and at atmospheric pressure, we compute the gas density starting from the equation of state. For an ideal gas, we thus obtain, in these conditions, $n \sim 2.5 \times 10^{19}$ molecules.cm^{-3}. The range of the intermolecular forces is $r_0 \sim 10^{-8}$ cm. We get, using the estimation (7.1.2), $\ell \sim 4 \times 10^{-5}$ cm. If the gas is for instance nitrogen, the typical molecular velocity at room temperature is $v \sim 5 \times 10^4$ cm.s^{-1}. The collision time is thus $\tau \sim 8 \times 10^{-10}$ s.

[4] The kinetic energy of a particle in the reference frame linked to the fluid is $\frac{1}{2}m|\boldsymbol{v} - \boldsymbol{u}(\boldsymbol{r},t)|^2$. The choice of this collisional invariant is motivated by the fact that the mean value of $\frac{1}{2}|\boldsymbol{v} - \boldsymbol{u}(\boldsymbol{r},t)|^2$ represents the local density of internal energy per unit mass (see Subsection 2.4).

Let $\chi(\boldsymbol{r},\boldsymbol{p},t)$ be one of these invariants. We can demonstrate the property:[5]

$$\boxed{\int \chi(\boldsymbol{r},\boldsymbol{p},t)\Big(\frac{\partial f}{\partial t}\Big)_{\text{coll}} d\boldsymbol{p} = 0,} \tag{7.2.1}$$

where $(\partial f/\partial t)_{\text{coll}}$ is the collision integral of the Boltzmann equation. The relation (7.2.1) allows us to deduce from the Boltzmann equation a general balance theorem applicable to any one of the collisional invariants.

2.1. General balance theorem

It is obtained by multiplying both sides of the Boltzmann equation:

$$\frac{\partial f}{\partial t} + \boldsymbol{v}.\nabla_{\boldsymbol{r}} f + \boldsymbol{F}.\nabla_{\boldsymbol{p}} f = \Big(\frac{\partial f}{\partial t}\Big)_{\text{coll}} \tag{7.2.2}$$

by a collisional invariant $\chi(\boldsymbol{r},\boldsymbol{p},t)$, and then by integrating over the kinetic momentum. According to formula (7.2.1), the collision term does not contribute. We thus obtain, $\boldsymbol{F}$ denoting the possibly applied external force (assumed independent of the velocity):

$$\int \chi(\boldsymbol{r},\boldsymbol{p},t)\Big(\frac{\partial f}{\partial t} + \boldsymbol{v}.\nabla_{\boldsymbol{r}} f + \boldsymbol{F}.\nabla_{\boldsymbol{p}} f\Big)\, d\boldsymbol{p} = 0. \tag{7.2.3}$$

Equation (7.2.3) may be rewritten in the equivalent form:

$$\frac{\partial}{\partial t}\int \chi f\, d\boldsymbol{p} - \int f\frac{\partial \chi}{\partial t}\, d\boldsymbol{p} + \nabla_{\boldsymbol{r}}.\int \chi\boldsymbol{v} f\, d\boldsymbol{p} - \int \boldsymbol{v} f.\nabla_{\boldsymbol{r}}\chi\, d\boldsymbol{p}$$
$$+ \int \nabla_{\boldsymbol{p}}.(\chi \boldsymbol{F} f)\, d\boldsymbol{p} - \int f\boldsymbol{F}.\nabla_{\boldsymbol{p}}\chi\, d\boldsymbol{p} = 0. \tag{7.2.4}$$

Since the distribution function vanishes when $|\boldsymbol{p}| \to \infty$, the fifth term on the left-hand side of equation (7.2.4) vanishes. That gives the following general balance theorem:[6]

$$\boxed{\frac{\partial}{\partial t}\langle n\chi\rangle - n\langle\frac{\partial \chi}{\partial t}\rangle + \nabla_{\boldsymbol{r}}.\langle n\chi\boldsymbol{v}\rangle - n\langle\boldsymbol{v}.\nabla_{\boldsymbol{r}}\chi\rangle - n\langle\boldsymbol{F}.\nabla_{\boldsymbol{p}}\chi\rangle = 0.} \tag{7.2.5}$$

The averages involved in equation (7.2.5) are defined by the formula:

$$\langle A(\boldsymbol{r},\boldsymbol{p},t)\rangle = \frac{\int f(\boldsymbol{r},\boldsymbol{p},t)A(\boldsymbol{r},\boldsymbol{p},t)\, d\boldsymbol{p}}{\int f(\boldsymbol{r},\boldsymbol{p},t)\, d\boldsymbol{p}} = \frac{1}{n(\boldsymbol{r},t)}\int f(\boldsymbol{r},\boldsymbol{p},t)A(\boldsymbol{r},\boldsymbol{p},t)\, d\boldsymbol{p}. \tag{7.2.6}$$

[5] The derivation of the property (7.2.1) is carried out in an appendix at the end of this chapter.

[6] For short, the local density $n(\boldsymbol{r},t)$ is simply denoted here by n. This quantity, independent of the velocity, may be written either inside or outside the averages (see the definition (7.2.6)). It must however not be confused with the commonly used notation, where n stands for the density at global thermodynamic equilibrium.

They must be computed with the aid of the solution f of the Boltzmann equation.

From the general balance theorem we deduce local balance equations for the mass, the kinetic momentum, and the internal energy.[7]

2.2. The local balance equation for the mass

Choosing $\chi = m$, we obtain from the theorem (7.2.5) the equation:

$$\frac{\partial}{\partial t}(mn) + \nabla.\langle mn\boldsymbol{v}\rangle = 0, \tag{7.2.7}$$

which also reads,[8] in terms of the local mass density $\rho(\boldsymbol{r}, t)$:

$$\boxed{\frac{\partial \rho}{\partial t} + \nabla.(\rho\boldsymbol{u}) = 0.} \tag{7.2.8}$$

The local balance equation for the mass has the form of a continuity equation with no source term. The mass flux $\boldsymbol{J} = \rho\boldsymbol{u}$ is a convective flux.[9]

Equation (7.2.8) may be rewritten in the equivalent form:

$$\boxed{\Big(\frac{\partial}{\partial t} + \boldsymbol{u}.\nabla\Big)\rho + \rho\nabla.\boldsymbol{u} = 0,} \tag{7.2.9}$$

which involves the material or hydrodynamic derivative $d/dt = \partial/\partial t + \boldsymbol{u}.\nabla$.

2.3. The local balance equation for the kinetic momentum

Choosing then $\chi = mv_i$ $(i = x, y, z)$, we get from theorem (7.2.5) the equation:

$$\frac{\partial}{\partial t}\langle \rho v_i\rangle + \nabla.\langle \rho v_i \boldsymbol{v}\rangle - \frac{1}{m}\rho F_i = 0. \tag{7.2.10}$$

Using the equalities:

$$\langle v_i v_j\rangle = \langle (v_i - u_i)(v_j - u_j)\rangle + u_i u_j, \tag{7.2.11}$$

we can rewrite the set of equations (7.2.10) for $i = x, y, z$ in the form of the local balance equation for the kinetic momentum:

$$\boxed{\frac{\partial(\rho\boldsymbol{u})}{\partial t} + \nabla.(\rho\boldsymbol{u}\boldsymbol{u} + \underline{P}) = \frac{\rho}{m}\boldsymbol{F}.} \tag{7.2.12}$$

---

[7] In the rest of this chapter (except when the context could lead to misinterpretation), $\nabla_{\boldsymbol{r}}$ will simply be denoted by ∇.

[8] For short, the local mass density $\rho(\boldsymbol{r}, t)$ is denoted by ρ, and the local mean velocity $\boldsymbol{u}(\boldsymbol{r}, t)$ by $\boldsymbol{u}$. The same will be done, in the following, for some other quantities defined locally.

[9] Since we consider here a gas with only one constituent, there is no diffusive contribution to the mass flux.

In equation (7.2.12), $\rho\boldsymbol{u}\boldsymbol{u}$ denotes the tensor of components $\rho u_i u_j$ and $\underline{\mathcal{P}}$ the pressure tensor, of components:

$$\mathcal{P}_{ij} = \rho\big\langle(v_i - u_i)(v_j - u_j)\big\rangle. \tag{7.2.13}$$

In the presence of an external force $\boldsymbol{F}$, the kinetic momentum is not a conserved quantity. The evolution equation of its local density $\rho\boldsymbol{u}$ involves, besides the flux term $\rho\boldsymbol{u}\boldsymbol{u} + \underline{\mathcal{P}}$, the source term $(\rho/m)\boldsymbol{F}$.

Taking into account equation (7.2.8), we can rewrite equation (7.2.12) in the equivalent form:

$$\boxed{\Big(\frac{\partial}{\partial t} + \boldsymbol{u}.\nabla\Big)\boldsymbol{u} = \frac{1}{m}\boldsymbol{F} - \frac{1}{\rho}\nabla.\underline{\mathcal{P}}.} \tag{7.2.14}$$

2.4. The local balance equation for the internal energy

Finally, for $\chi = \frac{1}{2}m|\boldsymbol{v} - \boldsymbol{u}(\boldsymbol{r},t)|^2$, we have $\langle\partial\chi/\partial t\rangle = 0$ and $\langle\nabla_{\boldsymbol{p}}\chi\rangle = 0$. We thus get from theorem (7.2.5) the equation:

$$\frac{1}{2}\frac{\partial}{\partial t}\big\langle\rho|\boldsymbol{v} - \boldsymbol{u}|^2\big\rangle + \frac{1}{2}\nabla.\big\langle\rho\boldsymbol{v}|\boldsymbol{v} - \boldsymbol{u}|^2\big\rangle - \frac{1}{2}\rho\big\langle\boldsymbol{v}.\nabla|\boldsymbol{v} - \boldsymbol{u}|^2\big\rangle = 0. \tag{7.2.15}$$

We define the local density of internal energy per unit mass,

$$e_{\text{int}}(\boldsymbol{r},t) = \frac{1}{2}\big\langle|\boldsymbol{v} - \boldsymbol{u}(\boldsymbol{r},t)|^2\big\rangle, \tag{7.2.16}$$

and the *heat flux*:

$$\boldsymbol{J}_Q = \frac{1}{2}\rho(\boldsymbol{r},t)\big\langle[\boldsymbol{v} - \boldsymbol{u}(\boldsymbol{r},t)]|\boldsymbol{v} - \boldsymbol{u}(\boldsymbol{r},t)|^2\big\rangle. \tag{7.2.17}$$

The relations:

$$\frac{1}{2}\rho\big\langle\boldsymbol{v}|\boldsymbol{v} - \boldsymbol{u}|^2\big\rangle = \frac{1}{2}\rho\big\langle(\boldsymbol{v} - \boldsymbol{u})|\boldsymbol{v} - \boldsymbol{u}|^2\big\rangle + \frac{1}{2}\rho\boldsymbol{u}\big\langle|\boldsymbol{v} - \boldsymbol{u}|^2\big\rangle = \boldsymbol{J}_Q + \rho e_{\text{int}}\boldsymbol{u} \tag{7.2.18}$$

and:

$$\rho\big\langle v_i(v_j - u_j)\big\rangle = \rho\big\langle(v_i - u_i)(v_j - u_j)\big\rangle = \mathcal{P}_{ij} \tag{7.2.19}$$

allow us to rewrite equation (7.2.15) in the form of the local balance equation for the internal energy:

$$\boxed{\frac{\partial(\rho e_{\text{int}})}{\partial t} + \nabla.(\boldsymbol{J}_Q + \rho e_{\text{int}}\boldsymbol{u}) = -\underline{\mathcal{P}} : \nabla\boldsymbol{u}.} \tag{7.2.20}$$

The internal energy is not a conserved quantity. The evolution equation of its density contains the source term $-\underline{\mathcal{P}} : \nabla\boldsymbol{u}$. The internal energy flux is the sum of the heat flux $\boldsymbol{J}_Q$ (conductive flux) and the convective flux $\rho e_{\text{int}}\boldsymbol{u}$.

Taking into account equation (7.2.8), we can rewrite equation (7.2.20) in the equivalent form:

$$\boxed{\Big(\frac{\partial}{\partial t}+\boldsymbol{u}.\nabla\Big)e_{\text{int}}+\frac{1}{\rho}\nabla.\boldsymbol{J}_Q=-\frac{1}{\rho}\underline{\mathcal{P}}:\nabla\boldsymbol{u}.} \tag{7.2.21}$$

2.5. Passage to the hydrodynamic equations

The local balance equations for the mass, the kinetic momentum, and the internal energy (formulas (7.2.8), (7.2.12) or (7.2.14), and (7.2.20) or (7.2.21)) are formally exact. However, they remain devoid of actual physical content as long as the solution of the Boltzmann equation is not made precise. Indeed, the local balance equations for the kinetic momentum and the internal energy involve the pressure tensor $\underline{\mathcal{P}}$ and the heat flux $\boldsymbol{J}_Q$. These latter quantities are expressed as averages of functions of $\boldsymbol{p}$ which must be computed with the aid of the solution $f(\boldsymbol{r},\boldsymbol{p},t)$ of the Boltzmann equation.

We thus have to determine a convenient form for f, and to deduce from it $\underline{\mathcal{P}}$ and $\boldsymbol{J}_Q$. The local balance equations will then become the hydrodynamic equations. We use for this purpose the Chapman–Enskog expansion.

3. The Chapman–Enskog expansion

The aim of the method of Chapman and Enskog is to build, by means of a systematic expansion procedure, the normal solutions of the Boltzmann equation, that is, the solutions of the form:[10]

$$f(\boldsymbol{r},\boldsymbol{p},t)=\mathcal{F}\big[n(\boldsymbol{r},t),\boldsymbol{u}(\boldsymbol{r},t),T(\boldsymbol{r},t),\boldsymbol{p}\big]. \tag{7.3.1}$$

The averages of any functions of $\boldsymbol{p}$ computed with the aid of a distribution of the type (7.3.1) are determined by the local thermodynamic variables. This is in particular the case of the fluxes. The relations between the fluxes and the gradients of the local thermodynamic variables are the phenomenological laws of transport, also called the *constitutive equations*.

3.1. Principle of the expansion

The normal solutions f of the Boltzmann equation are functionals of the local density, the local mean velocity, and the local temperature. These latter quantities are

[10] The normal solutions constitute a particular class of solutions of the Boltzmann equation. We thus cannot force them to obey arbitrary conditions, either initial or boundary. For instance, near a physical boundary such as the wall of a vessel, the distribution function is not in general of the normal form. Similarly, immediately after a given initial condition, the distribution function is not of the normal form either.

themselves averages over the kinetic momentum computed with the aid of f:

$$\begin{cases} n(\boldsymbol{r},t) = \displaystyle\int f(\boldsymbol{r},\boldsymbol{p},t)\,d\boldsymbol{p} \\ \boldsymbol{u}(\boldsymbol{r},t) = \dfrac{1}{n(\boldsymbol{r},t)}\displaystyle\int f(\boldsymbol{r},\boldsymbol{p},t)\boldsymbol{v}\,d\boldsymbol{p} \\ kT(\boldsymbol{r},t) = \dfrac{1}{n(\boldsymbol{r},t)}\displaystyle\int f(\boldsymbol{r},\boldsymbol{p},t)\dfrac{m}{3}|\boldsymbol{v}-\boldsymbol{u}(\boldsymbol{r},t)|^2\,d\boldsymbol{p}. \end{cases} \tag{7.3.2}$$

In the method of Chapman and Enskog, we look for the normal solutions in the form of an expansion of the type:

$$f = \frac{1}{\xi}(f^{(0)} + \xi f^{(1)} + \xi^2 f^{(2)} + \cdots), \qquad f^{(1)} \ll f^{(0)}, \qquad f^{(2)} \ll f^{(1)} \ldots \tag{7.3.3}$$

In formula (7.3.3), ξ represents an expansion parameter with no particular physical signification (it is only helpful in characterizing the order of terms in the series, and it will be made equal to unity at the end of the calculations). We aim to establish a systematic expansion procedure leading us to decompose the Boltzmann equation (7.2.2) into a series of equations allowing us to successively determine the $f^{(n)}$'s. We will describe here the two first steps of this procedure.

3.2. The two first approximation orders

At lowest order, we assimilate the distribution function to a local equilibrium Maxwell–Boltzmann distribution, denoted $f^{(0)}$. To determine it, we demand that f and $f^{(0)}$ be equivalent as far as the calculation of $n(\boldsymbol{r},t)$, $\boldsymbol{u}(\boldsymbol{r},t)$, and $T(\boldsymbol{r},t)$ (formulas (7.3.2)) is concerned, which amounts to imposing the following requirements on $f^{(1)}$:

$$\begin{cases} \displaystyle\int f^{(1)}\,d\boldsymbol{p} = 0 \\ \displaystyle\int f^{(1)}\boldsymbol{v}\,d\boldsymbol{p} = 0 \\ \displaystyle\int f^{(1)}\boldsymbol{v}^2\,d\boldsymbol{p} = 0. \end{cases} \tag{7.3.4}$$

The conditions (7.3.4) allow us to define unambiguously the local equilibrium distribution associated with f:

$$f^{(0)}(\boldsymbol{r},\boldsymbol{p},t) = n(\boldsymbol{r},t)\big[2\pi mkT(\boldsymbol{r},t)\big]^{-3/2}\exp\left[-\frac{|\boldsymbol{p}-m\boldsymbol{u}(\boldsymbol{r},t)|^2}{2mkT(\boldsymbol{r},t)}\right]. \tag{7.3.5}$$

We then import the expansion (7.3.3) of the normal solution into the Boltzmann equation (7.2.2). For the sake of simplicity, we write the collision term in a condensed form displaying its quadratic character with respect to the distribution function:[11]

$$\left(\frac{\partial f}{\partial t}\right)_{\text{coll}} = \mathcal{I}(f|f). \tag{7.3.6}$$

This gives:

$$\frac{1}{\xi}\left(\frac{\partial^{(0)}}{\partial t}+\cdots\right)(f^{(0)}+\xi f^{(1)}+\cdots)+(\boldsymbol{v}.\nabla_{\boldsymbol{r}}+\boldsymbol{F}.\nabla_{\boldsymbol{p}})\frac{1}{\xi}(f^{(0)}+\xi f^{(1)}+\cdots)$$
$$=\frac{1}{\xi^2}\mathcal{I}(f^{(0)}|f^{(0)})+\frac{1}{\xi}\Big[\mathcal{I}(f^{(0)}|f^{(1)})+\mathcal{I}(f^{(1)}|f^{(0)})\Big]+\cdots \tag{7.3.7}$$

In equation (7.3.7), $\partial^{(0)}/\partial t$ represents the zeroth-order approximation of the derivation operator $\partial/\partial t$. Since f depends on t only via $n(\boldsymbol{r},t)$, $\boldsymbol{u}(\boldsymbol{r},t)$, and $T(\boldsymbol{r},t)$, the operator $\partial/\partial t$ reads:[12]

$$\frac{\partial}{\partial t}=\frac{\partial}{\partial n}\frac{\partial n}{\partial t}+\frac{\partial}{\partial u_i}\frac{\partial u_i}{\partial t}+\frac{\partial}{\partial T}\frac{\partial T}{\partial t}. \tag{7.3.8}$$

We compute the zeroth-order approximation $\partial^{(0)}/\partial t$ of the operator $\partial/\partial t$ by making use of the conservation equations written at zeroth order, which amounts to taking into account the conditions (7.3.4) imposed on $f^{(1)}$. Equation (7.3.7) yields, at lowest order (that is, by identifying the terms in $1/\xi^2$),

$$\mathcal{I}(f^{(0)}|f^{(0)})=0, \tag{7.3.9}$$

then, at next order (by identifying the terms in $1/\xi$):

$$\frac{\partial^{(0)}}{\partial t}f^{(0)}+\boldsymbol{v}.\nabla_{\boldsymbol{r}}f^{(0)}+\boldsymbol{F}.\nabla_{\boldsymbol{p}}f^{(0)}=\mathcal{I}(f^{(0)}|f^{(1)})+\mathcal{I}(f^{(1)}|f^{(0)}). \tag{7.3.10}$$

Equation (7.3.9) simply confirms the fact that the distribution $f^{(0)}$, which makes the collision integral vanish, is a local equilibrium distribution. Equation (7.3.10) allows us to determine $f^{(1)}$. The procedure may be continued at higher orders, but only the two first approximation orders are used in practice.

Even at the two first orders, the Chapman–Enskog method is fairly heavy to use when we retain the exact bilinear structure of the collision integral. It is easier to use when the collision term is written in the relaxation time approximation. We will present here this simpler form of the method.

---

[11] By definition, we set, for any functions $f(\boldsymbol{r},\boldsymbol{p},t)$ and $g(\boldsymbol{r},\boldsymbol{p},t)$:

$$\mathcal{I}(f|g)=\int d\boldsymbol{p}_1\int d\Omega\,\sigma(\Omega)|\boldsymbol{v}-\boldsymbol{v}_1|[f(\boldsymbol{r},\boldsymbol{p}',t)g(\boldsymbol{r},\boldsymbol{p}_1',t)-f(\boldsymbol{r},\boldsymbol{p},t)g(\boldsymbol{r},\boldsymbol{p}_1,t)].$$

[12] We use the standard convention of summation over repeated indices.

4. The zeroth-order approximation

4.1. Pressure tensor and heat flux at zeroth order

At this order, the averages which have to be taken into account for the computation of the pressure tensor, denoted $\underline{\mathcal{P}}^{(0)}$, and of the heat flux, denoted $\boldsymbol{J}_Q^{(0)}$, are computed with the aid of the local equilibrium distribution (7.3.5). We set for further purpose:

$$C = n(\boldsymbol{r},t)\big[2\pi mkT(\boldsymbol{r},t)\big]^{-3/2}, \qquad A = \frac{m}{2kT(\boldsymbol{r},t)}. \tag{7.4.1}$$

• *Pressure tensor*

The components of the pressure tensor defined by the general formula (7.2.13) read, at zeroth order,

$$\mathcal{P}_{ij}^{(0)} = \frac{\rho}{n}C\int (v_i - u_i)(v_j - u_j)e^{-A|\boldsymbol{v}-\boldsymbol{u}|^2}\,d\boldsymbol{p}, \tag{7.4.2}$$

that is, introducing the velocity $\boldsymbol{U}(\boldsymbol{r},t) = \boldsymbol{v} - \boldsymbol{u}(\boldsymbol{r},t)$ in the reference frame moving with the fluid:

$$\mathcal{P}_{ij}^{(0)} = m^4C\int U_iU_je^{-AU^2}\,d\boldsymbol{U}. \tag{7.4.3}$$

The off-diagonal elements of the tensor $\underline{\mathcal{P}}^{(0)}$ vanish. Its diagonal elements are equal to the local hydrostatic pressure $\mathcal{P}(\boldsymbol{r},t) = n(\boldsymbol{r},t)kT(\boldsymbol{r},t)$:

$$\mathcal{P}_{ij}^{(0)}(\boldsymbol{r},t) = \delta_{ij}\mathcal{P}(\boldsymbol{r},t). \tag{7.4.4}$$

• *Heat flux*

The heat flux defined by the general formula (7.2.17) is given, at zeroth order, by:

$$\boldsymbol{J}_Q^{(0)} = \frac{1}{2}m^4C\int \boldsymbol{U}U^2e^{-AU^2}\,d\boldsymbol{U}. \tag{7.4.5}$$

It thus vanishes at this order.

Summing up, in the zeroth-order approximation, the pressure tensor reduces to the hydrostatic pressure term: there is no viscous transfer of kinetic momentum. The heat flux vanishes. Dissipative phenomena are not taken into account at this approximation order: the gas behaves like a perfect fluid.

4.2. Non-dissipative hydrodynamics

The hydrodynamic equations of a perfect fluid are obtained by inserting the expression (7.4.3) for $\underline{\mathcal{P}}^{(0)}$ and by making $\boldsymbol{J}_Q^{(0)} = 0$ in the local balance equations for the kinetic momentum and for the internal energy (equations (7.2.14) and (7.2.21)). We must also take into account the equation of state $\mathcal{P}/\rho = 2e_{\text{int}}/3$.

• *The local balance equation for the kinetic momentum*

In the case of a perfect fluid, this is called the *Euler equation*:

$$\boxed{\Big(\frac{\partial}{\partial t}+\boldsymbol{u}.\nabla\Big)\boldsymbol{u}=\frac{1}{m}\boldsymbol{F}-\frac{1}{\rho}\nabla\mathcal{P}.}\tag{7.4.6}$$

• *The local balance equation for the internal energy*

For a perfect fluid, this equation reads:

$$\boxed{\Big(\frac{\partial}{\partial t}+\boldsymbol{u}.\nabla\Big)e_{\rm int}+\frac{2}{3}e_{\rm int}\nabla.\boldsymbol{u}=0.}\tag{7.4.7}$$

Equation (7.4.7) is equivalent to an equation for the local temperature:

$$\boxed{\Big(\frac{\partial}{\partial t}+\boldsymbol{u}.\nabla\Big)T+\frac{2}{3}T\nabla.\boldsymbol{u}=0.}\tag{7.4.8}$$

The continuity equation (7.2.8), the Euler equation (7.4.6), and the local balance equation for the internal energy (7.4.7) (or the equation for the local temperature (7.4.8)) form the hydrodynamic equations of a perfect fluid. Since dissipative phenomena are not taken into account, the solutions of these equations correspond to indefinitely persisting flows.

Although they have been derived here from the Boltzmann equation (thus for a dilute gas), these equations are valid more generally. They can in fact be established with the aid of phenomenological arguments, also valid in a denser gas or in a liquid. We can deduce from these equations some properties of the fluids (for instance the equation of an adiabatic transformation, the propagation equation of an acoustic wave, the determination of the sound velocity, and so on).

5. The first-order approximation

5.1. The first-order distribution function

We write the Boltzmann equation (7.2.2) in the relaxation time approximation:

$$\frac{\partial f}{\partial t}+\boldsymbol{v}.\nabla_{\boldsymbol{r}}f+\boldsymbol{F}.\nabla_{\boldsymbol{p}}f=-\frac{f-f^{(0)}}{\tau(\boldsymbol{v})}.\tag{7.5.1}$$

To determine $f^{(1)}$, we write, in the spirit of the Chapman-Enskog expansion:[13]

$$f^{(1)}\simeq-\tau(\boldsymbol{v})\Big(\frac{\partial^{(0)}}{\partial t}+\boldsymbol{v}.\nabla_{\boldsymbol{r}}+\boldsymbol{F}.\nabla_{\boldsymbol{p}}\Big)f^{(0)}.\tag{7.5.2}$$

[13] Formula (7.5.2) is the analog of formula (7.3.10) for a collision term written in the relaxation time approximation.

It is convenient to use for $f^{(0)}$ the following notations:[14]

$$f^{(0)} = \frac{\rho}{m}(2\pi mkT)^{-3/2}\exp\Big(-\frac{mU^2}{2kT}\Big), \qquad \boldsymbol{U} = \frac{\boldsymbol{p}}{m} - \boldsymbol{u}(\boldsymbol{r},t). \tag{7.5.3}$$

Since $f^{(0)}$ depends on $\boldsymbol{r}$ and t only via the functions ρ, T, and $\boldsymbol{U}$, it is necessary to compute the partial derivatives of $f^{(0)}$ with respect to these latter quantities:

$$\begin{cases} \dfrac{\partial f^{(0)}}{\partial\rho} = \dfrac{f^{(0)}}{\rho} \\ \dfrac{\partial f^{(0)}}{\partial T} = \dfrac{1}{T}\Big(\dfrac{m}{2kT}U^2 - \dfrac{3}{2}\Big)f^{(0)} \\ \dfrac{\partial f^{(0)}}{\partial U_i} = -\dfrac{m}{kT}U_i f^{(0)}. \end{cases} \tag{7.5.4}$$

We also have:

$$\frac{\partial f^{(0)}}{\partial v_i} = -\frac{m}{kT}U_i f^{(0)}. \tag{7.5.5}$$

We deduce from these results:

$$f^{(1)} = -\tau(\boldsymbol{v})f^{(0)}\Big[\frac{1}{\rho}D^{(0)}(\rho)+\frac{1}{T}\Big(\frac{m}{2kT}U^2-\frac{3}{2}\Big)D^{(0)}(T)+\frac{m}{kT}U_jD^{(0)}(u_j)-\frac{1}{kT}\boldsymbol{F}.\boldsymbol{U}\Big], \tag{7.5.6}$$

with $D^{(0)}(X) = (\partial^{(0)}/\partial t + \boldsymbol{v}.\nabla)X$. The quantities $D^{(0)}(\rho)$, $D^{(0)}(u_j)$, and $D^{(0)}(T)$ are evaluated with the aid of the zeroth-order hydrodynamic equations (formulas (7.2.8), (7.4.6), and (7.4.8)):

$$\begin{cases} D^{(0)}(\rho) = -\rho\nabla.\boldsymbol{u} + \boldsymbol{U}.\nabla\rho \\ D^{(0)}(u_j) = -\dfrac{1}{\rho}\dfrac{\partial\mathcal{P}}{\partial x_j} + \dfrac{1}{m}F_j + \boldsymbol{U}.\nabla u_j \\ D^{(0)}(T) = -\dfrac{2}{3}T\nabla.\boldsymbol{u} + \boldsymbol{U}.\nabla T. \end{cases} \tag{7.5.7}$$

These expressions once imported into formula (7.5.6), give:

$$f^{(1)} = -\tau(\boldsymbol{v})f^{(0)}\Big[\frac{1}{T}(\boldsymbol{U}.\nabla T)\Big(\frac{m}{2kT}U^2 - \frac{5}{2}\Big) + \frac{m}{kT}U_iU_j\frac{\partial u_j}{\partial x_i} - \frac{1}{3}\frac{m}{kT}U^2(\nabla.\boldsymbol{u})\Big]. \tag{7.5.8}$$

Introducing the symmetric tensor $\underline{\Lambda}$, of components:

$$\Lambda_{ij} = \frac{m}{2}\Big(\frac{\partial u_j}{\partial x_i} + \frac{\partial u_i}{\partial x_j}\Big), \tag{7.5.9}$$

[14] For the sake of simplicity, we use the same notation $f^{(0)}$ for the distribution (7.3.5) and for the function of ρ, T, and $\boldsymbol{U}$ defined by formula (7.5.3).

we rewrite formula (7.5.8) in the form:

$$f^{(1)} = -\tau(\boldsymbol{v})f^{(0)}\Big[\frac{1}{T}(\boldsymbol{U}.\nabla T)\Big(\frac{m}{2kT}U^2 - \frac{5}{2}\Big) + \frac{1}{kT}\Lambda_{ij}\Big(U_iU_j - \frac{1}{3}\delta_{ij}U^2\Big)\Big]. \quad (7.5.10)$$

Neither the applied force nor the density gradient are involved in formula (7.5.10). This is consistent with the conditions (7.3.4) which impose the absence of a dissipative particle flux at this order, as well as with the fact that no diffusion takes place in a fluid with only one constituent.

5.2. Pressure tensor and heat flux at first order

For the sake of simplicity, the relaxation time is assumed independent of the velocity.

- *Pressure tensor*

Computed with the aid of the approximation $f \simeq f^{(0)} + f^{(1)}$ of the distribution function, its components read:

$$\mathcal{P}_{ij} = m\int (v_i - u_i)(v_j - u_j)(f^{(0)} + f^{(1)})\,d\boldsymbol{p} = nkT\delta_{ij} + \mathcal{P}_{ij}^{(1)}. \quad (7.5.11)$$

Only the second term of the expression (7.5.10) for $f^{(1)}$ contributes to $\mathcal{P}_{ij}^{(1)}$:

$$\mathcal{P}_{ij}^{(1)} = -\frac{\tau m^4}{kT}\Lambda_{kl}\int U_iU_j\Big(U_kU_l - \frac{1}{3}\delta_{kl}U^2\Big)f^{(0)}\,d\boldsymbol{U}. \quad (7.5.12)$$

The tensor of components $\mathcal{P}_{ij}^{(1)}$ is a symmetric tensor of trace $\sum_{i=1}^{3}\mathcal{P}_{ii}^{(1)} = 0$, which depends linearly on the symmetric tensor $\underline{\Lambda}$. It thus must be of the form:

$$\mathcal{P}_{ij}^{(1)} = -\frac{2\eta}{m}\Big(\Lambda_{ij} - \frac{m}{3}\delta_{ij}\nabla.\boldsymbol{u}\Big). \quad (7.5.13)$$

In the constitutive equation (7.5.13), $m\nabla.\boldsymbol{u}$ is the trace of the tensor $\underline{\Lambda}$, and η is the viscosity coefficient.[15] Identifying formulas (7.5.12) and (7.5.13), we show that the viscosity coefficient is expressed in terms of Gaussian integrals.[16] We finally obtain:

$$\boxed{\eta = nkT\tau.} \quad (7.5.14)$$

Thus, in the first-order approximation, there is a dissipative contribution to the pressure tensor $\underline{\mathcal{P}}$, whose components read:

$$\mathcal{P}_{ij} = \delta_{ij}\mathcal{P} - \frac{2\eta}{m}\Big(\Lambda_{ij} - \frac{m}{3}\delta_{ij}\nabla.\boldsymbol{u}\Big), \quad (7.5.15)$$

---

[15] The direct calculation of the viscosity coefficient of a dilute classical gas from its experimental definition (*Newton's law*) is carried out in an appendix at the end of this chapter. This calculation allows us in particular to identify the quantity η involved in the expression for $\mathcal{P}_{ij}^{(1)}$ with the viscosity coefficient of Newton's law.

[16] See the details of the calculations in the appendix.

that is:

$$\mathcal{P}_{ij} = \delta_{ij}\mathcal{P} - \eta\Big(\frac{\partial u_j}{\partial x_i} + \frac{\partial u_i}{\partial x_j} - \frac{2}{3}\delta_{ij}\frac{\partial u_l}{\partial x_l}\Big). \qquad (7.5.16)$$

• *Heat flux*

The first-order heat flux, denoted $\boldsymbol{J}_Q^{(1)}$, only involves the first term of the expression (7.5.10) for $f^{(1)}$:

$$\boldsymbol{J}_Q^{(1)} = -\frac{\tau m^4}{2}\int \boldsymbol{U}U^2\Big(\frac{m}{2kT}U^2 - \frac{5}{2}\Big)\frac{1}{T}U_i\frac{\partial T}{\partial x_i}f^{(0)}\,d\boldsymbol{U}. \qquad (7.5.17)$$

The constitutive equation (7.5.17) identifies with Fourier's law:

$$\boldsymbol{J}_Q^{(1)} = -\underline{\kappa}.\nabla T. \qquad (7.5.18)$$

The thermal conductivity tensor $\underline{\kappa}$ is here proportional to the unit matrix: $\kappa_{\alpha\beta} = \kappa\delta_{\alpha\beta}$. The thermal conductivity κ deduced from formula (7.5.17) is expressed in terms of Gaussian integrals.[17] After the calculations, we obtain:

$$\boxed{\kappa = \frac{5}{2}\frac{nk^2T\tau}{m}.} \qquad (7.5.19)$$

5.3. First-order hydrodynamic equations

To obtain the hydrodynamic equations in the first-order approximation, we must introduce the expressions (7.5.16) and (7.5.18) for the pressure tensor and for the heat flux in the local balance equations for the kinetic momentum and for the internal energy (equations (7.2.14) and (7.2.21)).

• *The local balance equation for the kinetic momentum*

In the first-order approximation, this equation reads:

$$\boxed{\Big(\frac{\partial}{\partial t} + \boldsymbol{u}.\nabla\Big)\boldsymbol{u} = \frac{1}{m}\boldsymbol{F} - \frac{1}{\rho}\nabla\mathcal{P} + \frac{\eta}{\rho}\nabla^2\boldsymbol{u} + \frac{\eta}{3\rho}\nabla(\nabla.\boldsymbol{u}).} \qquad (7.5.20)$$

---

[17] The expression for κ involves the integrals $I_n = \int_0^\infty dv\, v^n e^{-mv^2/2kT}$ with $n = 6$ and $n = 8$. We have:

$$I_6 = \frac{1.3.5}{2^4}\pi^{1/2}\Big(\frac{2kT}{m}\Big)^{1/2}, \qquad I_8 = \frac{1.3.5.7}{2^5}\pi^{1/2}\Big(\frac{2kT}{m}\Big)^{3/2}.$$

• *The local balance equation for the internal energy*

This may be written in the form of an equation for the local temperature:

$$\rho c_v\left[\left(\frac{\partial}{\partial t}+\boldsymbol{u}.\nabla\right)T+\frac{2}{3}T(\nabla.\boldsymbol{u})\right]=\nabla.(\kappa\nabla T)+\frac{\eta}{2}\left(\frac{2\Lambda_{ij}}{m}-\frac{2}{3}\delta_{ij}\nabla.\boldsymbol{u}\right)^2, \tag{7.5.21}$$

having set[18] $c_v = 3k/2m$. Equation (7.5.21) is the *heat equation.*

The first-order approximation allows us to take into account the dissipative phenomena within the gas, and to obtain microscopic expressions for the viscosity coefficient and the thermal conductivity. The validity of this approximation is linked to the smallness of $f^{(1)}$ as compared to $f^{(0)}$, and thus in particular to the smallness of the mean free path as compared to the typical evolution distances of the local density, the mean local velocity, and the local temperature.

The hydrodynamic equations (7.5.20) and (7.5.21) may thus, in the case of a dilute gas, be deduced from the Boltzmann equation. Their validity domain is actually much larger than that of the Boltzmann equation. Accordingly, they may be established phenomenologically in denser gases and in liquids.[19]

5.4. The thermal diffusion equation

The heat equation (7.5.21) takes, in some cases, a much simpler form. For instance, in a liquid in which the local mean velocity is much smaller than the sound velocity, the heat equation reads:

$$\rho c_p\frac{\partial T}{\partial t}-\nabla.(\kappa\nabla T)=0. \tag{7.5.22}$$

In equation (7.5.22), c_p denotes the specific heat at constant pressure per unit mass.[20] Considering κ as independent of the point of space, equation (7.5.22) takes the form of a diffusion equation:

$$\rho c_p\frac{\partial T}{\partial t}-\kappa\nabla^2 T=0. \tag{7.5.23}$$

[18] The notation c_v does not stand here for the specific heat at constant volume per particle, but for the specific heat at constant volume per unit mass.

[19] In an incompressible fluid ($\nabla.\boldsymbol{u}=0$), the local balance equation for the kinetic momentum is called the *Navier–Stokes equation*:

$$\left(\frac{\partial}{\partial t}+\boldsymbol{u}.\nabla\right)\boldsymbol{u}=\frac{1}{m}\boldsymbol{F}-\frac{1}{\rho}\nabla\mathcal{P}+\frac{\eta}{\rho}\nabla^2\boldsymbol{u}.$$

[20] The fact that it is c_p, and not c_v, which is involved in equation (7.5.22), comes from the fact that the equations for the local density and the local temperature are not decoupled (See Supplement 16B).

The *thermal diffusion coefficient*, also called the *thermal diffusivity*, is:

$$\boxed{D_{\rm th} = \frac{\kappa}{\rho c_p}.} \tag{7.5.24}$$

Equation (7.5.23) is experimentally verified in liquids (if the local mean velocity is small compared to the sound velocity), as well as in solids.

Appendices

7A. A property of the collision integral

We aim here to demonstrate the property:

$$\boxed{\int \chi(\boldsymbol{r},\boldsymbol{p},t)\left(\frac{\partial f}{\partial t}\right)_{\text{coll}} d\boldsymbol{p} = 0,} \tag{7A.1}$$

where $(\partial f/\partial t)_{\text{coll}}$ is the collision integral of the Boltzmann equation and $\chi(\boldsymbol{r},\boldsymbol{p},t)$ a collisional invariant.

Coming back to the definition of $(\partial f/\partial t)_{\text{coll}}$ for molecules undergoing binary collisions, namely:

$$\left(\frac{\partial f}{\partial t}\right)_{\text{coll}} = \int d\boldsymbol{p}_1 \int d\Omega\, \sigma(\Omega)|\boldsymbol{v}-\boldsymbol{v}_1|(f'f_1'-ff_1), \tag{7A.2}$$

we write the left-hand side of equation (7A.1) in the form:

$$\int \chi(\boldsymbol{r},\boldsymbol{p},t)\left(\frac{\partial f}{\partial t}\right)_{\text{coll}} d\boldsymbol{p} = \int d\boldsymbol{p}\int d\boldsymbol{p}_1 \int d\Omega\, \sigma(\Omega)|\boldsymbol{v}-\boldsymbol{v}_1|\chi(\boldsymbol{r},\boldsymbol{p},t)(f'f_1'-ff_1). \tag{7A.3}$$

To demonstrate formula (7A.1), we first make use of the fact that the integral on the right-hand side of formula (7A.3) is invariant under the permutation of the kinetic momenta $\boldsymbol{p}=m\boldsymbol{v}$ and $\boldsymbol{p}_1=m\boldsymbol{v}_1$, which allows us to write:

$$\int \chi(\boldsymbol{r},\boldsymbol{v},t)\left(\frac{\partial f}{\partial t}\right)_{\text{coll}} d\boldsymbol{p} = \frac{1}{2}\int d\boldsymbol{p}\int d\boldsymbol{p}_1 \int d\Omega\, \sigma(\Omega)|\boldsymbol{v}-\boldsymbol{v}_1|(\chi+\chi_1)(f'f_1'-ff_1). \tag{7A.4}$$

We then carry out the change of variables $(\boldsymbol{p},\boldsymbol{p}_1)\to(\boldsymbol{p}',\boldsymbol{p}_1')$, which amounts to considering the collision $\{\boldsymbol{p}',\boldsymbol{p}_1'\}\to\{\boldsymbol{p},\boldsymbol{p}_1\}$, inverse of the collision $\{\boldsymbol{p},\boldsymbol{p}_1\}\to\{\boldsymbol{p}',\boldsymbol{p}_1'\}$. The cross-sections relative to both collisions are identical. Besides, we have the equalities:

$$|\boldsymbol{v}-\boldsymbol{v}_1| = |\boldsymbol{v}'-\boldsymbol{v}_1'| \tag{7A.5}$$

as well as:

$$d\boldsymbol{p}d\boldsymbol{p}_1 = d\boldsymbol{p}'d\boldsymbol{p}_1'. \tag{7A.6}$$

This gives:

$$\int \chi(\boldsymbol{r},\boldsymbol{v},t)\left(\frac{\partial f}{\partial t}\right)_{\text{coll}} d\boldsymbol{p} = \frac{1}{2}\int d\boldsymbol{p}\int d\boldsymbol{p}_1 \int d\Omega\, \sigma(\Omega)|\boldsymbol{v}-\boldsymbol{v}_1|(\chi'+\chi_1')(ff_1-f'f_1'). \tag{7A.7}$$

Taking the half-sum of the expressions (7A.4) and (7A.7) for $\int \chi(\boldsymbol{r},\boldsymbol{v},t)(\partial f \partial t)_{\text{coll}}\, d\boldsymbol{p}$, finally gives:

$$\int \chi(\boldsymbol{r},\boldsymbol{v},t)\left(\frac{\partial f}{\partial t}\right)_{\text{coll}} d\boldsymbol{p} = \frac{1}{4}\int d\boldsymbol{p}\int d\boldsymbol{p}_1\int d\Omega\,\sigma(\Omega)|\boldsymbol{v}-\boldsymbol{v}_1|(\chi+\chi_1-\chi'-\chi_1')(f'f_1'-ff_1). \tag{7A.8}$$

As χ is a collisional invariant, the right-hand side of equation (7A.8) vanishes, which demonstrates the property (7A.1).

This property of the collision integral is valid for any distribution function f, whether or not it is a solution of the Boltzmann equation. It is a consequence of the fact that the collisions, which modify the kinetic momenta of the molecules, are considered as local and instantaneous.

7B. Newton's law and viscosity coefficient

Consider a gas of spatially uniform and time-independent density and temperature, in which there is a time-independent local mean velocity $\boldsymbol{u}(\boldsymbol{r})$, of components:

$$u_x = A + By, \qquad u_y = 0, \qquad u_z = 0, \tag{7B.1}$$

where A and B are constants. We can think of this gas as made up of different layers sliding one on top of another, as schematically depicted in Fig. 7.1.

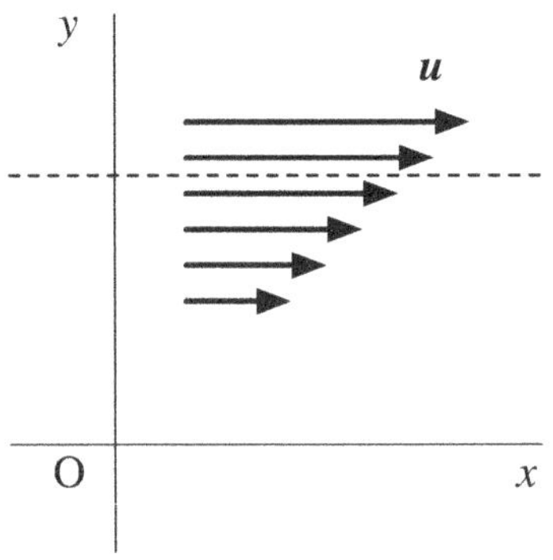

Fig. 7.1 Flow of gas layers sliding one on top of another.

We are trying to find the friction force Φ per unit surface sustained by the gas situated above a given plane (represented by the dashed line in Fig. 7.1). When the local mean velocity gradient is small, the friction force depends linearly on it. This is *Newton's law*:

$$\boxed{\Phi = -\eta\frac{\partial u_x}{\partial y},} \tag{7B.2}$$

where η is the viscosity coefficient of the gas. The gas above the plane loses some x-component of its kinetic momentum in favor of the gas below the plane. The resulting tangential friction force per unit surface area is given by:

$$\Phi = mn\langle (v_x - u_x)(v_y - u_y)\rangle. \tag{7B.3}$$

This force identifies with the off-diagonal element $\mathcal{P}_{xy}$ of the pressure tensor. To compute the average value involved in formula (7B.3), it is convenient to make use of the distribution function $F(\boldsymbol{r}, \boldsymbol{v}, t)$. This gives:

$$\Phi = m \int F(\boldsymbol{r}, \boldsymbol{v}, t)(v_x - u_x)(v_y - u_y)\, d\boldsymbol{v}. \tag{7B.4}$$

Let us now compute the viscosity coefficient with the aid of the Boltzmann equation in the relaxation time approximation.

7B.1. The Boltzmann equation

The relevant local equilibrium distribution $F^{(0)}$ is the Maxwell–Boltzmann function characterized by the local mean velocity $\boldsymbol{u}(\boldsymbol{r})$:

$$F^{(0)}(\boldsymbol{r}, \boldsymbol{v}) = n\Big(\frac{m}{2\pi kT}\Big)^{3/2} \exp\Big[-\frac{m|\boldsymbol{v} - \boldsymbol{u}(\boldsymbol{r})|^2}{2kT}\Big]. \tag{7B.5}$$

We denote by $\boldsymbol{U}(\boldsymbol{r})$ the velocity $\boldsymbol{v} - \boldsymbol{u}(\boldsymbol{r})$ of the particles in the reference frame moving with the fluid: $U_x = v_x - u_x$, $U_y = v_y$, $U_z = v_z$. The distribution (7B.5) is actually a function[21] of $\boldsymbol{U}(\boldsymbol{r})$:

$$F^{(0)}(\boldsymbol{U}) = n\Big(\frac{m}{2\pi kT}\Big)^{3/2} \exp\Big(-\frac{mU^2}{2kT}\Big). \tag{7B.6}$$

The distribution $F^{(0)}$ depends on y through U_x. The Boltzmann equation in the relaxation time approximation reads:[22]

$$\frac{\partial F}{\partial t} + \boldsymbol{v}.\nabla_{\boldsymbol{r}} F = -\frac{F - F^{(0)}}{\tau}. \tag{7B.7}$$

7B.2. First-order distribution function

When the gradient $\partial u_x/\partial y$ is small, we write the solution of equation (7B.7) in the form:

$$F \simeq F^{(0)} + F^{(1)}, \qquad F^{(1)} \ll F^{(0)}, \tag{7B.8}$$

---

[21] For the sake of simplicity, we use the same notation $F^{(0)}$ for the distribution (7B.5) and for the function of $\boldsymbol{U}$ defined by formula (7B.6).

[22] It is assumed that the relaxation time does not depend on the velocity.

and we look for $F^{(1)}$ by expanding equation (7B.7) at first order. To begin with, we obtain the lowest-order equation:

$$\frac{\partial F^{(0)}}{\partial t} = 0. \tag{7B.9}$$

Equation (7B.9) is actually verified by the chosen distribution $F^{(0)}$ (formulas (7B.5) or (7B.6)), since the local mean velocity $\boldsymbol{u}(\boldsymbol{r})$ is time-independent. At next order, we get a partial-differential equation for $F^{(1)}$,

$$\frac{\partial F^{(1)}}{\partial t} + \boldsymbol{v}.\nabla_{\boldsymbol{r}} F^{(0)} = -\frac{F^{(1)}}{\tau}, \tag{7B.10}$$

whose stationary solution is:

$$F^{(1)} = -\tau v_y \frac{\partial F^{(0)}}{\partial y}. \tag{7B.11}$$

7B.3. Newton's law: determination of η

The friction force is expressed with the aid of F (formula (7B.4)). Using the approximation $F \simeq F^{(0)} + F^{(1)}$ for the distribution function, we get:

$$\Phi = m \int (F^{(0)} + F^{(1)}) U_x U_y \, d\boldsymbol{U}. \tag{7B.12}$$

The first term of the expression (7B.12) for Φ vanishes for symmetry reasons. To compute the second term of this expression, we make use of the equality:

$$\frac{\partial F^{(0)}}{\partial y} = \frac{m U_x}{kT} \frac{\partial u_x}{\partial y} F^{(0)}, \tag{7B.13}$$

which allows us to rewrite the expression (7B.11) for $F^{(1)}$ in the form:

$$F^{(1)} = -\frac{m\tau}{kT} U_x U_y \frac{\partial u_x}{\partial y} F^{(0)}. \tag{7B.14}$$

The friction force at first order is thus given by:

$$\Phi^{(1)} = -\frac{m^2 \tau}{kT} \frac{\partial u_x}{\partial y} \int F^{(0)} U_x^2 U_y^2 \, d\boldsymbol{U}. \tag{7B.15}$$

Equation (7B.15) identifies with Newton's law:

$$\Phi^{(1)} = -\eta \frac{\partial u_x}{\partial y}. \tag{7B.16}$$

The viscosity coefficient η is expressed in terms of Gaussian integrals.[23] After the calculations, we get:

$$\boxed{\eta = nkT\tau.} \tag{7B.17}$$

At a given temperature, the viscosity coefficient is proportional to the product $n\tau$. The relaxation time, of the same order of magnitude as the collision time, varies, as the latter, inversely proportionally to the gas density. The viscosity coefficient computed with the aid of the Boltzmann equation thus does not depend, at a given temperature, on the gas density. This result was established and experimentally verified by J.C. Maxwell in 1860.

The above calculation is valid provided that the Boltzmann equation itself is applicable, which implies that the gas must be dilute, but not rarefied. The mean free path must, in particular, remain much smaller than the characteristic dimensions of the vessel containing the gas. Otherwise the collisions with the walls would prevail over the collisions between the molecules, and the very notion of viscosity would then lose any signification.

[23] The expression for η involves the integrals $I_n = \int_0^\infty dv\, v^n e^{-mv^2/2kT}$ with $n = 0$ and $n = 2$. We have:

$$I_0 = \frac{1}{2}\pi^{1/2}\Big(\frac{2kT}{m}\Big)^{1/2}, \qquad I_2 = \frac{1}{2^2}\pi^{1/2}\Big(\frac{2kT}{m}\Big)^{3/2}.$$

Bibliography

R. BALESCU, *Statistical dynamics. Matter out of equilibrium*, Imperial College Press, London, 1997.

R. BALIAN, *From microphysics to macrophysics*, Vol. 2, Springer-Verlag, Berlin, 1992.

S. CHAPMAN and T.G. COWLING, *The mathematical theory of non-uniform gases*, Cambridge University Press, Cambridge, third edition, 1970.

K. HUANG, *Statistical mechanics*, Wiley, New York, second edition, 1987.

H.J. KREUZER, *Nonequilibrium thermodynamics and its statistical foundations*, Clarendon Press, Oxford, 1981.

L.D. LANDAU and E.M. LIFSHITZ, *Fluid mechanics*, Butterworth-Heinemann, Oxford, second edition, 1987.

J.A. MCLENNAN, *Introduction to nonequilibrium statistical mechanics*, Prentice-Hall, Englewood Cliffs, 1989.

D.A. MCQUARRIE, *Statistical mechanics*, University Science Books, Sausalito, second edition, 2000.

F. REIF, *Fundamentals of statistical and thermal physics*, McGraw-Hill, New York, 1965.

D. ZUBAREV, V. MOROZOV, and G. RÖPKE, *Statistical mechanics of nonequilibrium processes,* Vol. 1: *Basic concepts, kinetic theory*, Akademie Verlag, Berlin, 1996.

R. ZWANZIG, *Nonequilibrium statistical mechanics*, Oxford University Press, Oxford, 2001.

Chapter 8

The Bloch–Boltzmann theory of electronic transport

The Boltzmann equation is at the heart of the semiclassical Bloch–Boltzmann theory of electronic transport in solids. The kinetic theory of gases based on the Boltzmann equation, originally developed for dilute classical gases, has since been applied successfully to the electron gas in metals and in semiconductors, despite of the fact that this latter gas is neither classical (except in non-degenerate semiconductors) nor dilute.

The possibility of writing a Boltzmann equation for electrons in solids relies on the semiclassical model of electron dynamics, in which each electron is described by its position $\boldsymbol{r}$, its wave vector $\boldsymbol{k}$, and a band index n, whereas the applied fields are treated classically. The interband transitions are neglected, and we look for the distribution function $f(\boldsymbol{r}, \boldsymbol{k}, t)$ of the electrons belonging to a given band. In such conditions, we can write for $f(\boldsymbol{r}, \boldsymbol{k}, t)$ an evolution equation analogous to the Boltzmann equation, the collision term being formulated in a way appropriate to the description of the different scattering mechanisms involving the electrons.

The most important of them are the electron–impurity collisions and the electron–phonon collisions (that is, the interactions of the electrons with the lattice vibrations). In some cases, it is possible to describe these collisions by means of the relaxation time approximation. The microscopic expression for the latter depends on the details of the scattering process.

Making use of the transport equation thus obtained, it is possible to compute the transport coefficients of the electron gas in metals and in semiconductors, and to discuss some of their characteristics, in particular their temperature dependence. We can also treat, within this framework, the specific effects which arise in the presence of an applied magnetic field (Hall effect and magnetoresistance), provided however that this latter field is not too intense.

1. The Boltzmann equation for the electron gas

1.1. Drawbacks of the Drude model

The Drude model, established in 1900, has historically been the first model describing electronic transport in metals. The electrons are considered as free particles of mass m and charge e undergoing collisions with fixed ions. These collisions make the mean velocity of the electrons relax with a characteristic time τ:

$$m\frac{d\langle \boldsymbol{v}\rangle}{dt} + m\frac{\langle \boldsymbol{v}\rangle}{\tau} = e\boldsymbol{E} \tag{8.1.1}$$

($\boldsymbol{E}$ is an applied electric field). The sole parameter of the model is the relaxation time, which, in Drude's original idea, corresponded to the collision time of the electron with the fixed ions.

Despite notable successes such as the Drude–Lorentz formula for the electrical conductivity, the Drude model suffers from important drawbacks. In particular, it does not allow us to explain the temperature dependence of the conductivity of a normal metal[1] (that is, non-superconducting). The shortcomings of the Drude model are essentially related to the fact that the electrons are treated like a classical gas. Between two successive collisions, the electrons are considered as free, except for their confinement within the sample. The periodicity of the ionic array, which gives rise to the band structure of the metal, is thus disregarded. Moreover, the physical nature of the collision mechanisms accounted for by the relaxation time is not correctly elucidated.

To remedy these deficiencies, it is necessary to take into account both the quantum character of the electron gas and the metal band structure, and to properly describe the collision processes. This is the aim of the semiclassical theory of transport relying on the Bloch theorem and the Boltzmann equation.

1.2. The semiclassical model of electron dynamics

The dynamics of *Bloch electrons*, evolving in the periodic potential of the lattice ions in the presence of an electric field and, possibly, of a magnetic field, may be studied in the more elaborate framework of the *semiclassical model*. Then, the use of the Boltzmann equation for the electronic distribution function enables us to relate the transport properties to the band structure. The whole procedure constitutes the *Bloch–Boltzmann theory* of electronic transport.

The electrons are treated in the independent electron approximation. In the absence of applied fields, the electronic states $|n\boldsymbol{k}\rangle$ are labeled by the band index n and the wave vector $\boldsymbol{k}$. These electronic states are the eigenstates of the one-electron Schrödinger equation in the presence of the periodic potential of the ions. In contrast to Drude's idea, the electrical conductivity is not limited by the 'collisions' of the electrons on the lattice ions.

[1] The resistivity (inverse of the conductivity) of a normal metal varies proportionally to T at room temperature, but follows a law in T^5, called the *Bloch–Grüneisen law*, at low temperatures (see Supplement 8A).

In the semiclassical model, an electron[2] is labeled by its position $\boldsymbol{r}$, its wave vector $\boldsymbol{k}$, and a band index n. The index n is considered as a constant of motion (otherwise stated, the interband transitions are not taken into account). Between two collisions, the motion of an electron of a given band of index n submitted to an electric field $\boldsymbol{E}(\boldsymbol{r}, t)$ and a magnetic field $\boldsymbol{H}(\boldsymbol{r}, t)$ is governed by the semiclassical equations:

$$\begin{cases} \dot{\boldsymbol{r}} = \boldsymbol{v}_n(\boldsymbol{k}) = \dfrac{1}{\hbar}\nabla_{\boldsymbol{k}}\varepsilon_n(\boldsymbol{k}) \\ \hbar\dot{\boldsymbol{k}} = e\Big[\boldsymbol{E}(\boldsymbol{r}, t) + \dfrac{1}{c}\boldsymbol{v}_n(\boldsymbol{k}) \times \boldsymbol{H}(\boldsymbol{r}, t)\Big], \end{cases} \tag{8.1.2}$$

in which $\boldsymbol{v}_n(\boldsymbol{k})$ and $\varepsilon_n(\boldsymbol{k})$ respectively denote the mean velocity and the energy of the electron in the state $|n\boldsymbol{k}\rangle$. The semiclassical model of electron dynamics thus relies on the knowledge of the band structure of the solid. This model is applicable provided that the external fields are not too intense and vary sufficiently slowly in space and in time so as to allow us to consider electronic wave packets and to neglect interband transitions.

1.3. The distribution function of the electrons pertaining to a given band

We then introduce the distribution function of the electrons pertaining to a given band (in practice, the conduction band of a metal or a semiconductor). Given the form of the equations of motion (8.1.2), it is convenient to make use of the distribution function[3] $f(\boldsymbol{r}, \boldsymbol{k}, t)$, defined so that $2(2\pi)^{-3} f(\boldsymbol{r}, \boldsymbol{k}, t)\, d\boldsymbol{r} d\boldsymbol{k}$ represents the mean number of electrons which, at time t, are in the phase space volume element $d\boldsymbol{r} d\boldsymbol{k}$ about the point $(\boldsymbol{r}, \boldsymbol{k})$, the factor 2 accounting for the two possible orientations of the electron spin.[4] The local density of the electrons of the considered band is given by:

$$n(\boldsymbol{r}, t) = \frac{2}{(2\pi)^3} \int f(\boldsymbol{r}, \boldsymbol{k}, t)\, d\boldsymbol{k}. \tag{8.1.3}$$

1.4. The Boltzmann equation

The Boltzmann equation for the distribution function $f(\boldsymbol{r}, \boldsymbol{k}, t)$ is of the form:[5]

$$\frac{\partial f}{\partial t} + \boldsymbol{v}_{\boldsymbol{k}}.\nabla_{\boldsymbol{r}} f + \frac{\boldsymbol{F}}{\hbar}.\nabla_{\boldsymbol{k}} f = \left(\frac{\partial f}{\partial t}\right)_{\text{coll}}, \tag{8.1.4}$$

where $\boldsymbol{F}(\boldsymbol{r}, t) = e\big[\boldsymbol{E}(\boldsymbol{r}, t) + \boldsymbol{v}_{\boldsymbol{k}} \times \boldsymbol{H}(\boldsymbol{r}, t)/c\big]$ is the Lorentz force acting on the electrons.

---

[2] We are actually considering an electronic wave packet localized around a mean position $\boldsymbol{r}$ and a mean wave vector $\boldsymbol{k}$.

[3] The band index being considered as a constant of motion, it will not be explicitly displayed in the expression for the distribution function.

[4] The effect of the magnetic field on the electronic energies is considered as negligible.

[5] The band index n being fixed, we will simply denote from now on by $\boldsymbol{v}_{\boldsymbol{k}}$ and $\varepsilon_{\boldsymbol{k}}$ the mean velocity and the energy of an electron in the state $|\boldsymbol{k}\rangle$.

Since the electrons are fermions, the global equilibrium distribution at temperature T is the Fermi–Dirac distribution,

$$f_0(\varepsilon_{\boldsymbol{k}}) = \frac{1}{e^{(\varepsilon_{\boldsymbol{k}}-\mu)/kT}+1}, \tag{8.1.5}$$

where μ denotes the chemical potential of the electrons.

In order to solve equation (8.1.4) and to compute the transport coefficients of the electron gas, it is necessary to make precise the expression for the collision integral $(\partial f/\partial t)_{\text{coll}}$ for the different types of electronic collisions. We have to take into account the fact that the electrons, being fermions, obey Pauli's exclusion principle.

2. The Boltzmann equation's collision integral

2.1. The collision processes

In the independent electron approximation, the Coulomb interaction between electrons is taken into account on average in the one-electron Hamiltonian. Departures from this average correspond to electron–electron interactions. These interactions play a minor role in the conduction.[6] Indeed, because of the screening due to the presence of the other electrons, and because of the limitations in the phase space due to Pauli's principle, the electrons may be considered as only weakly interacting. At high temperatures, the electron–electron interactions play a less important role than the interactions of the electrons with the thermal vibrations of the ions. At low temperatures, except in extremely pure crystals, the conductivity is limited by the collisions of the electrons with the impurities and the lattice defects. However, in metallic or semiconducting devices with reduced dimensions and high carrier densities, the role of the interactions between electrons is more important, without however becoming dominant.

The most important electronic collision processes are, on the one hand, the collisions with the impurities and the crystalline defects, and, on the other hand, the interactions with the lattice vibrations (the electron–phonon collisions). As a general rule, the collision processes lead to a change of the electron wave vector (together with, in some cases, a change of the electron energy). They give rise to a scattering of the electron.

2.2. The collision integral

The collision integral appearing in equation (8.1.4) involves the probability per unit time for a Bloch electron, initially in a state $|\boldsymbol{k}\rangle$, to be scattered in a state $|\boldsymbol{k}'\rangle$ (and vice versa). These transitions are induced by a new, non-periodic, potential term in the one-electron Hamiltonian.

In the Born approximation, the transition probability between a state $|\boldsymbol{k}\rangle$ and a state $|\boldsymbol{k}'\rangle$ is proportional to the square of the modulus of the matrix element of the

[6] This differentiates the electron gas in a solid from an ordinary gas, in which the collisions between particles are the only ones to take into account (except for the collisions with the walls).

perturbation potential between states $|\boldsymbol{k}\rangle$ and $|\boldsymbol{k}'\rangle$. In such a transition, the kinetic momentum and the energy are globally conserved.

We shall now provide some indications of the structure of the collision integral, first for impurity scattering, then for lattice vibration scattering. Note that, if two or more independent scattering processes come into play, the collision integral is the sum of the integrals associated with each process.

2.3. Impurity scattering

An impurity atom acts on the electrons as a localized scattering center. In the course of a collision with an impurity of mass M, the electron undergoes a transition from a state $|\boldsymbol{k}\rangle$ to a state $|\boldsymbol{k}'\rangle$, whereas the impurity energy changes by $(\hbar|\boldsymbol{k}-\boldsymbol{k}'|)^2/2M$. Since $M \gg m$, this energy change is much smaller than the initial electron energy. This latter energy is thus practically unmodified by this type of scattering, which for this reason is qualified as elastic.

We assume that the impurities are dilute enough to allow us to consider that an electron interacts with only one impurity at a time. Such a collision is considered as local and instantaneous. The only parameter necessary to its description is the electron wave vector[7] $\boldsymbol{k}$. The distribution function $f(\boldsymbol{r},\boldsymbol{k},t)$, denoted for short by $f_{\boldsymbol{k}}$, represents the probability for the state $|\boldsymbol{k}\rangle$ to be occupied by an electron (the probability for the state $|\boldsymbol{k}\rangle$ to be empty is thus $1-f_{\boldsymbol{k}}$). To write down the electron–impurity collision integral, we have to introduce the conditional probability $W_{\boldsymbol{k}',\boldsymbol{k}}dt$ for a transition from state $|\boldsymbol{k}\rangle$ to state $|\boldsymbol{k}'\rangle$ to take place during the time interval dt. If the interaction potential is weak enough, we can treat the scattering in the Born approximation (or, which amounts to the same thing, we can make use of the Fermi golden rule of time-dependent perturbation theory). We thus obtain for the transition rate from state $|\boldsymbol{k}\rangle$ to state $|\boldsymbol{k}'\rangle$ the expression:

$$W_{\boldsymbol{k}',\boldsymbol{k}} = \frac{2\pi}{\hbar}|\langle\boldsymbol{k}'|V_i|\boldsymbol{k}\rangle|^2\delta(\varepsilon_{\boldsymbol{k}}-\varepsilon_{\boldsymbol{k}'}). \tag{8.2.1}$$

In formula (8.2.1), V_i denotes the electron–impurity interaction potential, and the function $\delta(\varepsilon_{\boldsymbol{k}}-\varepsilon_{\boldsymbol{k}'})$ expresses the fact that the electron energy is not modified by the collision. The transition rates verify the microreversibility property:[8]

$$W_{\boldsymbol{k},\boldsymbol{k}'} = W_{\boldsymbol{k}',\boldsymbol{k}}. \tag{8.2.2}$$

To obtain the probability of a transition from state $|\boldsymbol{k}\rangle$ to state $|\boldsymbol{k}'\rangle$ during the time interval dt, given that initially the electron is in the state $|\boldsymbol{k}\rangle$ and that the

---

[7] We limit ourselves here to the case of a collision with a non-magnetic impurity. Accordingly, the electron spin is not modified by the collision.

[8] This symmetry property can directly be checked in the expression (8.2.1) for the transition rates in the Born approximation, since the perturbation potential V_i is Hermitean. However, the equality of the rates $W_{\boldsymbol{k}',\boldsymbol{k}}$ and $W_{\boldsymbol{k},\boldsymbol{k}'}$, which stems from the time-reversal invariance of the microscopic equations of motion, is valid independently of the calculation method (thus, possibly, beyond the Born approximation).

state $|\boldsymbol{k}'\rangle$ is empty, we have to multiply $W_{\boldsymbol{k}',\boldsymbol{k}}$ by $f_{\boldsymbol{k}}(1-f_{\boldsymbol{k}'})$. The collision integral for the electron–impurity collisions thus reads:

$$\left(\frac{\partial f}{\partial t}\right)_{\text{coll}} = \sum_{\boldsymbol{k}'}\Big[W_{\boldsymbol{k},\boldsymbol{k}'} f_{\boldsymbol{k}'}(1-f_{\boldsymbol{k}}) - W_{\boldsymbol{k}',\boldsymbol{k}} f_{\boldsymbol{k}}(1-f_{\boldsymbol{k}'})\Big], \tag{8.2.3}$$

or, using a continuous description for the wave vectors:

$$\left(\frac{\partial f}{\partial t}\right)_{\text{coll}} = \frac{V}{(2\pi)^3}\int\Big[W_{\boldsymbol{k},\boldsymbol{k}'} f_{\boldsymbol{k}'}(1-f_{\boldsymbol{k}}) - W_{\boldsymbol{k}',\boldsymbol{k}} f_{\boldsymbol{k}}(1-f_{\boldsymbol{k}'})\Big]\,d\boldsymbol{k}'. \tag{8.2.4}$$

In equation (8.2.4), V denotes the volume of the sample.

The microreversibility property leads to a noticeable simplification of the collision integral, which becomes a linear functional of the electronic distribution function:

$$\boxed{\left(\frac{\partial f}{\partial t}\right)_{\text{coll}} = \frac{V}{(2\pi)^3}\int W_{\boldsymbol{k}',\boldsymbol{k}}(f_{\boldsymbol{k}'} - f_{\boldsymbol{k}})\,d\boldsymbol{k}'.} \tag{8.2.5}$$

The exclusion principle therefore has no effect on the structure of the collision integral corresponding to elastic collisions (the expression for $(\partial f/\partial t)_{\text{coll}}$ would have been the same in the absence of restrictions on the occupation of the final state[9]).

2.4. Lattice vibration scattering

The lattice ions are not rigorously fixed in a periodic array, but they undergo oscillations (corresponding to lattice vibrations) with respect to their equilibrium positions. The amplitude of these oscillations increases with the temperature. They give rise to a scattering of the electrons, which we usually describe in terms of electron–phonon interaction.[10]

The collision of an electron with a phonon is an inelastic scattering process, in which the electron energy is not conserved. We do not intend here to treat in detail the electron–phonon interaction, but we simply aim to show how, using a schematic model for the lattice vibrations, it is possible to get an idea about the structure of $(\partial f/\partial t)_{\text{coll}}$ in the case of inelastic collisions.[11] We consider the *Einstein model*, in which every ion is assumed to vibrate like a harmonic oscillator, independently of the other ions. The states $|\nu\rangle$ of such an oscillator have energies E_ν. The occupation probability of each of these states is denoted by p_ν. At equilibrium at temperature T, we have $p_\nu \sim e^{-E_\nu/kT}$. The transition probability per unit time of the scattering

---

[9] The form (8.2.5) of the collision integral is identical to that of a classical Lorentz gas.

[10] The electron–phonon interaction constitutes the main mechanism at the origin of the temperature dependence of the d.c. resistivity. When the temperature is lowered, the amplitude of the lattice vibrations decreases. The scattering of electrons by impurities and defects then becomes the dominant collision process.

[11] See also Supplement 8A.

process $\{|\boldsymbol{k}\rangle, \nu\} \to \{|\boldsymbol{k}'\rangle, \nu'\}$ is of the form $W_{\boldsymbol{k}',\boldsymbol{k};\nu',\nu}$ (with $\nu' = \nu \pm 1$ in the most simple case in which we neglect the processes involving more than one quantum). It is proportional to $\delta(\varepsilon_{\boldsymbol{k}'} - \varepsilon_{\boldsymbol{k}} + E_{\nu'} - E_\nu)$, which expresses the conservation of the total energy. The collision integral thus reads:

$$\left(\frac{\partial f}{\partial t}\right)_{\text{coll}} = \sum_{\boldsymbol{k}',\nu,\nu'} \left[W_{\boldsymbol{k},\boldsymbol{k}';\nu,\nu'} p_{\nu'} f_{\boldsymbol{k}'}(1 - f_{\boldsymbol{k}}) - W_{\boldsymbol{k}',\boldsymbol{k};\nu',\nu} p_\nu f_{\boldsymbol{k}}(1 - f_{\boldsymbol{k}'})\right], \tag{8.2.6}$$

or, using a continuous description for the wave vectors:

$$\left(\frac{\partial f}{\partial t}\right)_{\text{coll}} = \frac{V}{(2\pi)^3} \int \left[\Lambda_{\boldsymbol{k},\boldsymbol{k}'} f_{\boldsymbol{k}'}(1 - f_{\boldsymbol{k}}) - \Lambda_{\boldsymbol{k}',\boldsymbol{k}} f_{\boldsymbol{k}}(1 - f_{\boldsymbol{k}'})\right] d\boldsymbol{k}'. \tag{8.2.7}$$

In formula (8.2.7), we have, since $\nu' = \nu \pm 1$:

$$\Lambda_{\boldsymbol{k}',\boldsymbol{k}} = \sum_\nu \left(W_{\boldsymbol{k}',\boldsymbol{k};\nu+1,\nu}\, p_\nu + W_{\boldsymbol{k}',\boldsymbol{k};\nu,\nu+1}\, p_{\nu+1}\right). \tag{8.2.8}$$

The collision integral (8.2.7) is formally analogous to the collision integral (8.2.4) for the electron–impurity collisions. However, whereas in formula (8.2.4), $W_{\boldsymbol{k}',\boldsymbol{k}}$ is proportional to $\delta(\varepsilon_{\boldsymbol{k}} - \varepsilon_{\boldsymbol{k}'})$, this is not the case for the quantity $\Lambda_{\boldsymbol{k}',\boldsymbol{k}}$ involved in formula (8.2.7). This is due to the inelasticity of the electron–phonon collisions. The symmetry of the transition rates, that is, the relation:

$$W_{\boldsymbol{k},\boldsymbol{k}';\nu,\nu'} = W_{\boldsymbol{k}',\boldsymbol{k};\nu',\nu}, \tag{8.2.9}$$

actually does not imply the symmetry of the quantities $\Lambda_{\boldsymbol{k}',\boldsymbol{k}}$ and $\Lambda_{\boldsymbol{k},\boldsymbol{k}'}$. Taking into account the relation $p_{\nu'}/p_\nu = e^{(E_\nu - E_{\nu'})/kT}$ and the conservation of the total energy $\varepsilon_{\boldsymbol{k}'} - \varepsilon_{\boldsymbol{k}} + E'_\nu - E_\nu = 0$, gives the relation between $\Lambda_{\boldsymbol{k}',\boldsymbol{k}}$ and $\Lambda_{\boldsymbol{k},\boldsymbol{k}'}$:

$$\Lambda_{\boldsymbol{k},\boldsymbol{k}'} = \Lambda_{\boldsymbol{k}',\boldsymbol{k}} e^{(\varepsilon_{\boldsymbol{k}'} - \varepsilon_{\boldsymbol{k}})/kT}. \tag{8.2.10}$$

This formula does not allow further simplification of the collision integral, which remains a non-linear functional of the electronic distribution function.

3. Detailed balance

The global equilibrium distribution $f_0(\varepsilon_{\boldsymbol{k}})$ is the time-independent solution of the Boltzmann equation (8.1.4). Besides, the distribution f_0 is also a solution of the equation $(\partial f/\partial t)_{\text{coll}} = 0$. Let us examine the consequences of this property, first in the case of impurity scattering, then in the case of lattice vibration scattering.

3.1. Impurity scattering

The fact that the equilibrium distribution f_0 makes the collision integral (8.2.5) vanish is expressed through the following global balance relation:

$$\sum_{\boldsymbol{k}'} W_{\boldsymbol{k},\boldsymbol{k}'} f_0(\varepsilon_{\boldsymbol{k}'}) = \sum_{\boldsymbol{k}'} W_{\boldsymbol{k}',\boldsymbol{k}} f_0(\varepsilon_{\boldsymbol{k}}). \tag{8.3.1}$$

The equality (8.3.1) is not only a global property of the considered sums, but it is realized term by term. Indeed, we have the stronger property, called the *detailed balance relation*:[12]

$$W_{\boldsymbol{k},\boldsymbol{k}'} f_0(\varepsilon_{\boldsymbol{k}'}) = W_{\boldsymbol{k}',\boldsymbol{k}} f_0(\varepsilon_{\boldsymbol{k}}). \tag{8.3.2}$$

Formula (8.3.2) expresses the fact that the transitions equilibrate separately for each pair of states $\{|\boldsymbol{k}\rangle, |\boldsymbol{k}'\rangle\}$. This property is a consequence, on the one hand, of the microreversibility property (8.2.2), and, on the other hand, of the fact that the equilibrium distribution f_0 depends solely on the electron energy, which is conserved in the collision with an impurity.

3.2. Lattice vibration scattering

For collision processes in which the electronic energy is not conserved, such as the interaction of electrons with lattice vibrations schematically accounted for by the collision integral (8.2.7), the global balance relation reads:

$$\sum_{\boldsymbol{k}'} \Lambda_{\boldsymbol{k},\boldsymbol{k}'} f_0(\varepsilon_{\boldsymbol{k}'})[1 - f_0(\varepsilon_{\boldsymbol{k}})] = \sum_{\boldsymbol{k}'} \Lambda_{\boldsymbol{k}',\boldsymbol{k}} f_0(\varepsilon_{\boldsymbol{k}})[1 - f_0(\varepsilon_{\boldsymbol{k}'})]. \tag{8.3.3}$$

We can check that the equality (8.3.3) is actually realized term by term:

$$\Lambda_{\boldsymbol{k},\boldsymbol{k}'} f_0(\varepsilon_{\boldsymbol{k}'})[1 - f_0(\varepsilon_{\boldsymbol{k}})] = \Lambda_{\boldsymbol{k}',\boldsymbol{k}} f_0(\varepsilon_{\boldsymbol{k}})[1 - f_0(\varepsilon_{\boldsymbol{k}'})]. \tag{8.3.4}$$

The detailed balance relation (8.3.4) is a consequence of the relation (8.2.10) between $\Lambda_{\boldsymbol{k},\boldsymbol{k}'}$ and $\Lambda_{\boldsymbol{k}',\boldsymbol{k}}$, and of the form (8.1.5) of the equilibrium Fermi–Dirac distribution.

4. The linearized Boltzmann equation

To solve the Boltzmann equation (8.1.4), we use when possible, as in the case of dilute classical gases, the relaxation time approximation, according to which the main effect of the term $(\partial f/\partial t)_{\text{coll}}$ is to make the distribution function relax towards a local equilibrium distribution adapted to the physics of the problem. In the case of the electron gas, the local equilibrium distribution is a Fermi–Dirac function characterized by a local temperature $T(\boldsymbol{r},t)$ and a local chemical potential $\mu(\boldsymbol{r},t)$, and denoted for short by $f^{(0)}(\varepsilon_{\boldsymbol{k}})$:

$$f^{(0)}(\varepsilon_{\boldsymbol{k}}) = \frac{1}{e^{[\varepsilon_{\boldsymbol{k}} - \mu(\boldsymbol{r},t)]/kT(\boldsymbol{r},t)} + 1}. \tag{8.4.1}$$

In the evolution equation of $f_{\boldsymbol{k}}$, the collision integral is thus written approximately as:

$$\left(\frac{\partial f}{\partial t}\right)_{\text{coll}} \simeq -\frac{f_{\boldsymbol{k}} - f^{(0)}(\varepsilon_{\boldsymbol{k}})}{\tau(\boldsymbol{k})}, \tag{8.4.2}$$

where $\tau(\boldsymbol{k})$ is a relaxation time towards the local equilibrium described by $f^{(0)}(\varepsilon_{\boldsymbol{k}})$. This time generally depends on the wave vector $\boldsymbol{k}$ (or merely on the energy $\varepsilon_{\boldsymbol{k}}$ in an

[12] The detailed balance relation is a very general property of systems at thermodynamic equilibrium (see Chapter 12).

isotropic system). The $\boldsymbol{k}$ (or $\varepsilon_{\boldsymbol{k}}$) dependence of the relaxation time is related to the details of the collision mechanisms.[13]

The electron gas may be submitted to electric and magnetic fields, as well as to a chemical potential gradient and a temperature gradient. Note that the electric field and the magnetic field act very differently on the electrons; indeed, the electric field releases energy to the electrons, while the magnetic field does not. If the departures from local equilibrium remain small, we look for a solution of the Boltzmann equation of the form:

$$f_{\boldsymbol{k}} \simeq f^{(0)}(\varepsilon_{\boldsymbol{k}}) + f_{\boldsymbol{k}}^{(1)}, \qquad f_{\boldsymbol{k}}^{(1)} \ll f^{(0)}(\varepsilon_{\boldsymbol{k}}). \tag{8.4.3}$$

The term $f_{\boldsymbol{k}}^{(1)}$ is of first order with respect to the perturbation created by the electric field, the chemical potential gradient, and the temperature gradient, but it contains terms of any order with respect to the magnetic field. In the term of equation (8.1.4) explicitly involving the magnetic field, the distribution function $f^{(0)}$ does not play any role. Indeed, $f^{(0)}$ depending on $\boldsymbol{k}$ via the energy $\varepsilon_{\boldsymbol{k}}$, gives:

$$\nabla_{\boldsymbol{k}} f^{(0)} = \hbar \boldsymbol{v}_{\boldsymbol{k}} \frac{\partial f^{(0)}}{\partial \varepsilon_{\boldsymbol{k}}}. \tag{8.4.4}$$

The mixed product $(\boldsymbol{v}_{\boldsymbol{k}} \times \boldsymbol{H}).\nabla_{\boldsymbol{k}} f^{(0)}$ thus vanishes.

When identifying the first-order terms, we get from equation (8.1.4) for $f_{\boldsymbol{k}}$ the linear partial differential equation obeyed by $f_{\boldsymbol{k}}^{(1)}$:

$$\frac{\partial f_{\boldsymbol{k}}^{(1)}}{\partial t} + \boldsymbol{v}_{\boldsymbol{k}}.\Big(\frac{\varepsilon_{\boldsymbol{k}} - \mu}{T} \nabla_{\boldsymbol{r}} T + \nabla_{\boldsymbol{r}} \mu - e\boldsymbol{E}\Big)\left(-\frac{\partial f^{(0)}}{\partial \varepsilon_{\boldsymbol{k}}}\right) + \frac{e}{\hbar c}(\boldsymbol{v}_{\boldsymbol{k}} \times \boldsymbol{H}).\nabla_{\boldsymbol{k}} f_{\boldsymbol{k}}^{(1)} = -\frac{f_{\boldsymbol{k}}^{(1)}}{\tau(\boldsymbol{k})}. \tag{8.4.5}$$

The linearized Boltzmann equation (8.4.5) is the basis of the study of linear transport phenomena in the semiclassical regime. In particular, in the presence of an electric field and a temperature gradient, it enables us to compute the electrical conductivity, the thermal conductivity, and the thermoelectric coefficients.[14] It also allows us to analyze the transport phenomena in the presence of a magnetic field (Hall effect, longitudinal or transverse magnetoresistance) in the semiclassical regime.

5. Electrical conductivity

The Bloch–Boltzmann theory relies on the knowledge of the band structure. We will consider here the simple case of electrons belonging to one band not completely filled, the conduction band of a metal or a semiconductor. This band is described in the effective mass approximation, the effective mass tensor being assumed to be proportional to the unit matrix. The dispersion law of the electronic energy reads $\varepsilon_{\boldsymbol{k}} = \hbar^2 k^2/2m^*$, the bottom of the conduction band being taken as the origin of energies. The mean velocity of an electron in state $|\boldsymbol{k}\rangle$ is $\boldsymbol{v}_{\boldsymbol{k}} = \hbar\boldsymbol{k}/m^*$.

---

13 The microscopic calculation of the relaxation times will be carried out in Supplement 8A.

14 The calculation for the thermal conductivity and the thermoelectric coefficients is carried out in Supplement 8B.

We assume that there is an applied electric field $\boldsymbol{E}$, spatially uniform and time-independent, and that there is neither temperature nor chemical potential gradient, nor applied magnetic field. In stationary regime, the linearized Boltzmann equation (8.4.5) reads in this case:

$$(\boldsymbol{v_k}.e\boldsymbol{E})\frac{\partial f^{(0)}}{\partial \varepsilon_{\boldsymbol{k}}} = -\frac{f_{\boldsymbol{k}}^{(1)}}{\tau(\varepsilon_{\boldsymbol{k}})} \tag{8.5.1}$$

(we assume here that the relaxation time is a function of the electronic energy).

The relevant local equilibrium distribution is the Fermi–Dirac distribution $f_0(\varepsilon_{\boldsymbol{k}})$. We thus obtain[15] from equation (8.5.1):

$$f_{\boldsymbol{k}}^{(1)} = e\tau(\varepsilon)(\boldsymbol{v}.\boldsymbol{E})\Big(-\frac{\partial f_0}{\partial \varepsilon}\Big). \tag{8.5.2}$$

The electric current density $\boldsymbol{J}$ can be obtained from $f_{\boldsymbol{k}}$ via the integral:

$$\boldsymbol{J} = \frac{2e}{(2\pi)^3}\int f_{\boldsymbol{k}}\boldsymbol{v}\,d\boldsymbol{k}. \tag{8.5.3}$$

The equilibrium distribution does not contribute to the current, which reads:

$$\boldsymbol{J} = \frac{e^2}{4\pi^3}\int \tau(\varepsilon)\boldsymbol{v}(\boldsymbol{v}.\boldsymbol{E})\Big(-\frac{\partial f_0}{\partial \varepsilon}\Big)\,d\boldsymbol{k}. \tag{8.5.4}$$

5.1. General expression for the conductivity

Formula (8.5.4) allows us to compute the conductivity tensor $\underline{\sigma}$ involved in Ohm's law $\boldsymbol{J} = \underline{\sigma}.\boldsymbol{E}$. With the chosen dispersion relation, this tensor is proportional to the unit matrix: $\sigma_{\alpha\beta} = \sigma\delta_{\alpha\beta}$.

To compute σ, it is convenient to introduce the electronic density of states, denoted by $n(\varepsilon)$, and to write any component of $\boldsymbol{J}$, such as for instance J_x, in the form of an integral over the energy:

$$J_x = e^2\int_0^\infty v_x^2 E_x\tau(\varepsilon)\Big(-\frac{\partial f_0}{\partial \varepsilon}\Big)n(\varepsilon)d\varepsilon. \tag{8.5.5}$$

The dispersion relation $\varepsilon = m^* v^2/2$ being taken into account, the electrical conductivity reads:

$$\sigma = \frac{2e^2}{3m^*}\int_0^\infty \varepsilon n(\varepsilon)\tau(\varepsilon)\Big(-\frac{\partial f_0}{\partial \varepsilon}\Big)\,d\varepsilon. \tag{8.5.6}$$

At this stage, it is convenient to bring into the expression for σ the electronic density $n = \int_0^\infty n(\varepsilon)f_0(\varepsilon)d\varepsilon$, which can also be expressed, in terms of the integrated density of states $N(\varepsilon) = \int_0^\varepsilon n(\varepsilon')d\varepsilon'$, as $n = \int_0^\infty N(\varepsilon)(-\partial f_0/\partial\varepsilon)\,d\varepsilon$. We thus obtain:

$$\sigma = \frac{2}{3}\frac{ne^2}{m^*}\frac{\int_0^\infty \varepsilon n(\varepsilon)\tau(\varepsilon)(-\frac{\partial f_0}{\partial \varepsilon})\,d\varepsilon}{\int_0^\infty N(\varepsilon)(-\frac{\partial f_0}{\partial \varepsilon})\,d\varepsilon}. \tag{8.5.7}$$

[15] From now on, the energy and the mean velocity of the electron in state $|\boldsymbol{k}\rangle$ are simply denoted by ε and $\boldsymbol{v}$ (instead of $\varepsilon_{\boldsymbol{k}}$ and $\boldsymbol{v_k}$).

In the present model, the density of states and the integrated density of states are respectively of the form $n(\varepsilon) = C\varepsilon^{1/2}$ and $N(\varepsilon) = (2C/3)\varepsilon^{3/2}$. We usually rewrite the expression (8.5.7) for the conductivity in a form analogous to the Drude–Lorentz formula:

$$\boxed{\sigma = \frac{ne^2\langle\tau\rangle}{m^*}.} \tag{8.5.8}$$

The drift mobility of the electrons then reads:

$$\mu_D = \frac{e\langle\tau\rangle}{m^*}. \tag{8.5.9}$$

In formulas (8.5.8) and (8.5.9), the notation $\langle\tau\rangle$ represents an average relaxation time defined by:

$$\boxed{\langle\tau\rangle = \frac{\int_0^\infty \tau(\varepsilon)\varepsilon^{3/2}\left(-\frac{\partial f_0}{\partial\varepsilon}\right)d\varepsilon}{\int_0^\infty \varepsilon^{3/2}\left(-\frac{\partial f_0}{\partial\varepsilon}\right)d\varepsilon}.} \tag{8.5.10}$$

Let us now make precise the expression for $\langle\tau\rangle$ in both cases of a highly degenerate electron gas (metal) and of a non-degenerate electron gas (non-degenerate semiconductor).

5.2. Metal

On account of the analytic form of the Fermi–Dirac function $f_0(\varepsilon)$, we can, for a function of energy $\Phi(\varepsilon)$ slowly varying over an energy interval $|\varepsilon - \mu| \simeq kT$ and not increasing more rapidly than a power of ε when $\varepsilon \to \infty$, write the *Sommerfeld expansion*:

$$\int_0^\infty \Phi(\varepsilon)\left(-\frac{\partial f_0}{\partial\varepsilon}\right)d\varepsilon = \Phi(\mu) + \frac{\pi^2}{6}(kT)^2\frac{\partial^2\Phi}{\partial\varepsilon^2}\bigg|_{\varepsilon=\mu} + \cdots, \qquad kT \ll \mu. \tag{8.5.11}$$

At low temperature, the opposite of the derivative of the Fermi–Dirac function approximately acts like a delta function centered at the chemical potential at zero temperature, that is, at the Fermi level ε_F:

$$-\frac{\partial f_0}{\partial\varepsilon} \simeq \delta(\varepsilon - \varepsilon_F), \qquad kT \ll \varepsilon_F. \tag{8.5.12}$$

In the case of a metal, the integrals involved in formula (8.5.10) may thus be evaluated at $T = 0$ up to terms of order $(kT/\varepsilon_F)^2$. We get approximately:

$$\langle\tau\rangle \simeq \tau(\varepsilon_F). \tag{8.5.13}$$

It follows that the expression for the electrical conductivity of a metal only involves the value of the relaxation time at the Fermi level:

$$\boxed{\sigma = \frac{ne^2\tau(\varepsilon_F)}{m^*}.} \tag{8.5.14}$$

5.3. Non-degenerate semiconductor

The distribution function $f_0(\varepsilon)$ is then the Maxwell–Boltzmann function, of the form:

$$f_0(\varepsilon) = Ae^{-\beta\varepsilon}, \qquad \beta = (kT)^{-1}. \tag{8.5.15}$$

We have:

$$-\frac{\partial f_0}{\partial \varepsilon} = A\beta e^{-\beta\varepsilon}. \tag{8.5.16}$$

The relaxation time generally depends on the energy according to a power law of the type $\tau(\varepsilon) \propto \varepsilon^s$, the value of the exponent s being related to the nature of the collision process.[16] Formula (8.5.10) reads accordingly:

$$\langle\tau\rangle \propto \frac{\int_0^\infty \varepsilon^s \varepsilon^{3/2} e^{-\beta\varepsilon}\, d\varepsilon}{\int_0^\infty \varepsilon^{3/2} e^{-\beta\varepsilon}\, d\varepsilon}. \tag{8.5.17}$$

We thus get:

$$\boxed{\langle\tau\rangle \propto (kT)^s \frac{\Gamma(\frac{5}{2}+s)}{\Gamma(\frac{5}{2})},} \tag{8.5.18}$$

where Γ denotes Euler's Gamma function. The temperature dependence of $\langle\tau\rangle$ and μ_D reflects the energy dependence of the relaxation time.

6. Semiclassical transport in the presence of a magnetic field

The transport properties of a metal or a semiconductor are deeply modified by the application of a magnetic field. We shall be concerned here by the transverse configuration, in which an electric field $\boldsymbol{E} = (E_x, E_y, 0)$ and a magnetic field $\boldsymbol{H} = (0, 0, H)$, perpendicular to $\boldsymbol{E}$, are applied to the sample. The hypotheses about the charge carriers and the band structure remain the same as previously. The temperature and the chemical potential are assumed to be uniform.

In the presence of a magnetic field $\boldsymbol{H}$, a free electron moves along an helix, called the *cyclotron helix*, with an angular frequency $\omega_c = |e|H/mc$, called the *cyclotron pulsation*. In the case of a band structure described in the effective mass approximation,

16 See Supplement 8A.

m must be replaced by m^* in the expression for ω_c. In the presence of a magnetic field, the semiclassical theory of transport is applicable provided that we have $\hbar\omega_c \ll kT$. If this condition is not fulfilled, the Landau quantification of the electronic levels must be taken into account. The Boltzmann equation, in which the magnetic field only appears in the Lorentz force, is then no longer valid, and we must have recourse to a purely quantum theory of electronic transport.[17]

We will limit ourselves here to the semiclassical transport as described by the linearized Boltzmann equation (8.4.5). The field $\boldsymbol{E}$ being perpendicular to $\boldsymbol{H}$, the same is true of the current $\boldsymbol{J}$. The linear relation between $\boldsymbol{J}$ and $\boldsymbol{E}$ may be written either in the form of Ohm's law $\boldsymbol{J} = \underline{\sigma}.\boldsymbol{E}$ or, inversely, in the form:

$$\boldsymbol{E} = \underline{\rho}.\boldsymbol{J}, \tag{8.6.1}$$

where $\underline{\rho}$ is the resistivity tensor. The components of the tensors $\underline{\sigma}$ and $\underline{\rho}$ depend on the magnetic field. Introducing the collision time[18] τ, we can, while remaining within the semiclassical regime defined by the condition $\hbar\omega_c \ll kT$, distinguish two different magnetic field regimes. If $\omega_c\tau \gg 1$, the electron can move along several pitches of the cyclotron helix between two successive collisions. The magnetic field is then considered as being strong.[19] In the opposite case $\omega_c\tau \ll 1$, the magnetic field is said to be weak.

6.1. Drude model in transverse magnetic field

When the relaxation time does not depend on the electronic energy, the tensor $\underline{\sigma}$ may be computed in the Drude model's framework. We write the evolution equation of the mean velocity of the electrons:

$$m^*\frac{d\langle\boldsymbol{v}\rangle}{dt} + m^*\frac{\langle\boldsymbol{v}\rangle}{\tau} = e\Big[\boldsymbol{E} + \frac{1}{c}\langle\boldsymbol{v}\rangle \times \boldsymbol{H}\Big], \tag{8.6.2}$$

from which we deduce the current $\boldsymbol{J} = ne\langle\boldsymbol{v}\rangle$. In stationary regime, we thus obtain the expressions for the conductivity tensor,

$$\underline{\sigma} = \frac{ne^2\tau}{m^*}\begin{pmatrix} \dfrac{1}{1+\omega_c^2\tau^2} & -\dfrac{\omega_c\tau}{1+\omega_c^2\tau^2} \\ \dfrac{\omega_c\tau}{1+\omega_c^2\tau^2} & \dfrac{1}{1+\omega_c^2\tau^2} \end{pmatrix}, \tag{8.6.3}$$

and for the resistivity tensor:

$$\underline{\rho} = \frac{m^*}{ne^2\tau}\begin{pmatrix} 1 & \omega_c\tau \\ -\omega_c\tau & 1 \end{pmatrix}. \tag{8.6.4}$$

---

[17] The foundations of the quantum theory of electronic transport will be expounded in Chapter 15.

[18] The notation τ stands here for a typical value of the relaxation time $\tau(\varepsilon)$.

[19] Even though this regime of magnetic field is said to be strong, it allows us to remain within the semiclassical transport domain. The two conditions $\hbar\omega_c \ll kT$ and $\omega_c\tau \gg 1$ must be simultaneously fulfilled, which implies that we must have $\hbar/\tau \ll kT$ (see the discussion in Section 7).

The *longitudinal conductivity* σ_{xx} is an even function of H, whereas the *transverse conductivity* σ_{xy} is an odd function of H. The Onsager relation:

$$\sigma_{xy}(\boldsymbol{H}) = \sigma_{yx}(-\boldsymbol{H}) \tag{8.6.5}$$

is verified.

In the absence of collisions, which, in the Drude model, corresponds to τ infinite, the conductivity tensor simply reads:[20]

$$\underline{\sigma} = \frac{ne^2}{m^*}\begin{pmatrix} 0 & -\omega_c^{-1} \\ \omega_c^{-1} & 0 \end{pmatrix}. \tag{8.6.6}$$

6.2. The first-order distribution function

To take into account the energy dependence of the relaxation time, it is necessary to use the Boltzmann equation. In stationary regime, the linearized Boltzmann equation (8.4.5) reads, for this problem:

$$\boldsymbol{v}.e\boldsymbol{E}\left(\frac{\partial f_0}{\partial \varepsilon}\right) + \frac{e}{\hbar c}(\boldsymbol{v}\times\boldsymbol{H}).\nabla_{\boldsymbol{k}} f_{\boldsymbol{k}}^{(1)} = -\frac{f_{\boldsymbol{k}}^{(1)}}{\tau(\varepsilon)}. \tag{8.6.7}$$

By analogy with the solution (8.5.2) in the presence of the electric field alone, we are looking for an expression for $f_{\boldsymbol{k}}^{(1)}$ of the form:

$$f_{\boldsymbol{k}}^{(1)} = \boldsymbol{v}.\boldsymbol{C}(\varepsilon)\left(-\frac{\partial f_0}{\partial \varepsilon}\right), \tag{8.6.8}$$

where $\boldsymbol{C}(\varepsilon)$ is a vector to be determined. Introducing the expression (8.6.8) for $f_{\boldsymbol{k}}^{(1)}$ in equation (8.6.7), we obtain for the vector $\boldsymbol{C}$ the equation:

$$\frac{e}{m^*c}\boldsymbol{C}.(\boldsymbol{v}\times\boldsymbol{H}) + \frac{\boldsymbol{C}.\boldsymbol{v}}{\tau(\varepsilon)} = e\boldsymbol{E}.\boldsymbol{v}. \tag{8.6.9}$$

Equation (8.6.9) must be verified whatever the velocity $\boldsymbol{v}$. Consequently, $\boldsymbol{C}$ must be a solution of the equation:

$$\boldsymbol{\omega_c}\times\boldsymbol{C} = \frac{\boldsymbol{C}}{\tau(\varepsilon)} - e\boldsymbol{E}, \qquad \boldsymbol{\omega_c} = -e\boldsymbol{H}/m^*c. \tag{8.6.10}$$

In the transverse configuration, the solution of equation (8.6.10) is:

$$\boldsymbol{C} = \frac{e\tau^2(\varepsilon)}{1+\omega_c^2\tau^2(\varepsilon)}\left[\frac{\boldsymbol{E}}{\tau(\varepsilon)} + \boldsymbol{\omega_c}\times\boldsymbol{E}\right]. \tag{8.6.11}$$

20 The situation described here is very different from the situation without magnetic field, in which case the conductivity $\sigma = ne^2\tau/m^*$ becomes infinite in the absence of collisions.

This gives:

$$f_{\boldsymbol{k}}^{(1)} = \left(-\frac{\partial f_0}{\partial \varepsilon}\right)\frac{e\tau^2(\varepsilon)}{1+\omega_c^2\tau^2(\varepsilon)}\left[\frac{\boldsymbol{E}.\boldsymbol{v}}{\tau(\varepsilon)} + (\boldsymbol{\omega_c}\times\boldsymbol{E}).\boldsymbol{v}\right]. \tag{8.6.12}$$

6.3. The conductivity tensor

Using formula (8.5.3) for the current, in which only $f_{\boldsymbol{k}}^{(1)}$ contributes to the integral, as well as the expression (8.6.12) for $f_{\boldsymbol{k}}^{(1)}$, we obtain the expression for the conductivity tensor in transverse magnetic field:

$$\underline{\sigma} = \frac{ne^2}{m^*}\begin{pmatrix} \left\langle \dfrac{\tau}{1+\omega_c^2\tau^2}\right\rangle & -\left\langle \dfrac{\omega_c\tau^2}{1+\omega_c^2\tau^2}\right\rangle \\ \left\langle \dfrac{\omega_c\tau^2}{1+\omega_c^2\tau^2}\right\rangle & \left\langle \dfrac{\tau}{1+\omega_c^2\tau^2}\right\rangle \end{pmatrix}. \tag{8.6.13}$$

In formula (8.6.13), the symbol $\langle\ldots\rangle$ has the same meaning as in formula (8.5.10). The result (8.6.13) has the same structure as the Drude model's result (formula (8.6.3)). However, in contrast to the latter, it accounts for the energy dependence of the relaxation time.

Let us now analyze, in the semiclassical regime, the specific effects due to the presence of the magnetic field, namely, the Hall effect and the transverse magnetoresistance.

6.4. The Hall effect

Consider a bar of parallelepipedic geometry, directed along the Ox axis. The electric field is applied perpendicularly to Ox, and the magnetic field is parallel to Oz (Fig. 8.1).

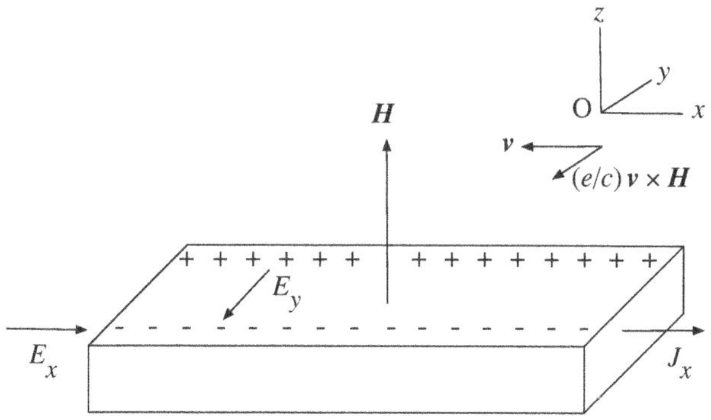

Fig. 8.1 Schematic drawing of the Hall experiment.

The geometry imposes the condition $J_y = 0$. As a consequence, an electric field E_y, called the *Hall field*, is induced by the presence of the magnetic field. The existence of this electric field constitutes the *Hall effect*, discovered by E.H. Hall in 1879. The charge carriers (electrons or holes[21]), deflected by the magnetic field in the direction of Oy, tend to accumulate on the sides of the sample. This results in the setting-up of an electric field along Oy which opposes the motion of the charge carriers and their further accumulation on the sides of the sample.

We assume here that the magnetic field is weak ($\omega_c\tau \ll 1$). The tensor (8.6.13) then reads:

$$\underline{\sigma} = \frac{ne^2}{m^*}\begin{pmatrix} \langle\tau\rangle & -\omega_c\langle\tau^2\rangle \\ \omega_c\langle\tau^2\rangle & \langle\tau\rangle \end{pmatrix}. \tag{8.6.14}$$

The value taken in these conditions by the Hall field E_y can be obtained by writing that the current $J_y = \sigma_{yx}E_x + \sigma_{yy}E_y$ vanishes. We usually set:

$$E_y = -E_x \tan\theta_H, \tag{8.6.15}$$

where θ_H is the *Hall angle*. For $\omega_c\tau \ll 1$, this gives:

$$\boxed{\theta_H = \omega_c\frac{\langle\tau^2\rangle}{\langle\tau\rangle}.} \tag{8.6.16}$$

The measure of the Hall effect is actually a measure of the component ρ_{yx} of the resistivity tensor. Indeed, as $J_y = 0$ we have:

$$\rho_{yx} = \frac{E_y}{J_x}. \tag{8.6.17}$$

We usually introduce the *Hall coefficient* R_H defined by:

$$\rho_{yx} = R_H H. \tag{8.6.18}$$

For $\omega_c\tau \ll 1$, this gives:

$$\boxed{R_H = \frac{1}{nqc}\frac{\langle\tau^2\rangle}{\langle\tau\rangle^2}.} \tag{8.6.19}$$

Let us now provide the explicit expressions for R_H in a metal and in a non-degenerate semiconductor.

- *Metal*

For a metal, we have:

$$\langle\tau\rangle = \tau(\varepsilon_F), \qquad \langle\tau^2\rangle = \tau^2(\varepsilon_F). \tag{8.6.20}$$

[21] We assume here that there is only one type of charge carriers, which may be either electrons with charge $q = e$ or holes with charge $q = -e$.

The charge carriers being electrons, we get:

$$R_H = \frac{1}{nec}. \tag{8.6.21}$$

The Hall coefficient of a metal thus does not depend on any parameter other than the electron density.

• *Non-degenerate semiconductor*

The sign of the Hall coefficient allows us to determine the sign of the charge carriers. Taking for the energy dependence of the relaxation time a power law of the type $\tau(\varepsilon) \propto \varepsilon^s$, gives:

$$R_H = \frac{1}{nqc} \frac{\Gamma(\frac{5}{2} + 2s)\Gamma(\frac{5}{2})}{\left[\Gamma(\frac{5}{2} + s)\right]^2}. \tag{8.6.22}$$

The value of R_H depends on the exponent characterizing the law $\tau(\varepsilon)$.

6.5. Transverse magnetoresistance

The force due to the Hall field compensates the Lorentz force: the current J_y thus vanishes. This is only an average compensation. Actually, when a magnetic field is present, the distribution of the velocity lines opens in space. The resistance parallel to Ox increases. This is the *magnetoresistance* phenomenon. In the considered geometry, the magnetoresistance is said to be *transverse*.

The magnetoresistance is an effect even in H. To compute it, we have to come back to formula (8.6.13) for the conductivity tensor, from which we deduce the expressions for J_x and J_y:

$$\begin{cases} J_x = \dfrac{ne^2}{m^*}\left[\langle\dfrac{\tau}{1+\omega_c^2\tau^2}\rangle E_x - \langle\dfrac{\omega_c\tau^2}{1+\omega_c^2\tau^2}\rangle E_y\right] \\ J_y = \dfrac{ne^2}{m^*}\left[\langle\dfrac{\omega_c\tau^2}{1+\omega_c^2\tau^2}\rangle E_x + \langle\dfrac{\tau}{1+\omega_c^2\tau^2}\rangle E_y\right]. \end{cases} \tag{8.6.23}$$

In the case $\omega_c\tau \ll 1$, the magnetoresistance is, at lowest order in magnetic field, an effect of order H^2. Developing formulas (8.6.23) at second order in H and taking into account the fact that, as previously, the current J_y vanishes, gives:

$$J_x = \frac{ne^2}{m^*}\left[\langle\tau\rangle - \omega_c^2\langle\tau^3\rangle + \omega_c^2\frac{\langle\tau^2\rangle^2}{\langle\tau\rangle}\right] E_x. \tag{8.6.24}$$

The relative modification, due to the presence of the magnetic field, of the longitudinal conductivity is of order H^2:

$$\frac{\Delta\sigma}{\sigma} = -\omega_c^2\frac{\langle\tau\rangle\langle\tau^3\rangle - \langle\tau^2\rangle^2}{\langle\tau\rangle^2}, \qquad \omega_c\tau \ll 1. \tag{8.6.25}$$

In the case $\omega_c\tau \gg 1$, the magnetoresistance saturates. The relative modification of the longitudinal conductivity becomes independent of the magnetic field:

$$\frac{\Delta\sigma}{\sigma} = \frac{\langle\tau^{-1}\rangle^{-1} - \langle\tau\rangle}{\langle\tau\rangle}, \qquad \omega_c\tau \gg 1. \tag{8.6.26}$$

The Drude model, in which the relaxation time is considered as a constant, does not allow us to describe this effect (formulas (8.6.25) and (8.6.26) with constant τ both yield $\Delta\sigma = 0$).

7. Validity limits of the Bloch–Boltzmann theory

7.1. Separation of time scales

In order for the Boltzmann equation for Bloch electrons to be valid, it is necessary to be able to consider, as in the case of dilute classical gases, the collisions as local and instantaneous. In particular, the duration τ_0 of a collision must be much smaller than the collision time τ: $\tau_0 \ll \tau$. In other words, these two time scales must be well separated.

In practice, τ_0 becomes comparable to τ only in very impure metals or in liquids. These are therefore systems for which the transport theory is more tricky.

7.2. Quantum limitations

In addition, it is also necessary to take into account the existence of purely quantum limitations.

The transition rate from a state $|\boldsymbol{k}\rangle$ to a state $|\boldsymbol{k}'\rangle$ is calculated with the aid of the Born approximation or of the Fermi golden rule. In order to be allowed to consider a unique collision, we have to assume that the time interval Δt over which we study the evolution of the distribution function is much smaller than the collision time τ, otherwise stated, that $\hbar/\tau$ is a very small quantity as compared to a typical electronic energy. In a non-degenerate semiconductor, this condition takes the form of the *Peierls criterion*:

$$\frac{\hbar}{\tau} \ll kT. \tag{8.7.1}$$

In a metal, inequality (8.7.1) is violated. However, according to a remark of L. Landau, instead of kT, it is the Fermi energy ε_F which has to be taken in consideration in this case when the sample is of macroscopic dimensions. The validity condition of the Bloch–Boltzmann theory thus reads in a metal:

$$\frac{\hbar}{\tau(\varepsilon_F)} \ll \varepsilon_F, \tag{8.7.2}$$

a much less restrictive condition than would be inequality (8.7.1). Condition (8.7.2) may be rewritten equivalently in the form of the *Ioffe–Regel criterion*:

$$k_F\ell \gg 1. \tag{8.7.3}$$

In formula (8.7.3), k_F is the Fermi wave vector, and $\ell \sim \hbar k_F \tau(\varepsilon_F)/m$ the *elastic mean free path*[22] of the electrons. When criterion (8.7.3) is verified, the Bloch–Boltzmann theory is applicable. When disorder increases in such a way that this condition is no longer fulfilled, quantum interference effects are likely to appear and to profoundly modify the transport properties.[23] These effects have been observed in systems of small dimensions, called *mesoscopic systems*, in which the spectrum of electronic states is discrete and the electronic motion coherent (which signifies that, if an electron can propagate in the system without undergoing inelastic collisions, its wave function keeps a definite phase).

7.3. The Pauli master equation and the Boltzmann equation

The quantum theory of transport, developed in particular by R. Kubo, relies on the electron density matrix, whose evolution in the presence of external fields and of the various interactions has to be determined.[24] The diagonal elements of the density matrix may be interpreted in terms of average occupation probabilities of the electronic states. Their semiclassical analog is the electronic distribution function. The discussion of the validity of the Boltzmann equation thus amounts to the discussion of the hypotheses allowing us to discard the off-diagonal elements of the density matrix.

The first derivation of the Boltzmann equation from quantum mechanics, proposed by W. Pauli in 1928, relied on the *random phase hypothesis*,[25] according to which the phases of the quantum amplitudes are distributed at random at any time. W. Pauli obtained in this way, for the diagonal elements of the density matrix, an irreversible evolution equation, the *Pauli master equation*. This equation only involves the diagonal elements of the density matrix. Its semiclassical analog is the Boltzmann equation for the distribution function.

---

[22] The elastic mean free path concerns the collisions which do not modify the electron energy, such as the electron–impurity collisions.

[23] See Supplement 15A.

[24] See Chapter 15.

[25] The random phase hypothesis is the quantum analog of the molecular chaos hypothesis of the kinetic theory of dilute classical gases (see Chapter 9).

Bibliography

N.W. Ashcroft and N.D. Mermin, *Solid state physics*, W.B. Saunders Company, Philadelphia, 1976.

C. Cohen-Tannoudji, B. Diu, and F. Laloë, *Quantum mechanics*, Vol. 2, Hermann and Wiley, Paris, second edition, 1977.

D.K. Ferry, *Semiconductor transport*, Taylor & Francis, London, 2000.

Y. Imry, *Introduction to mesoscopic physics*, Oxford University Press, Oxford, second edition, 2002.

W. Jones and N.H. March, *Theoretical solid-state physics: non-equilibrium and disorder*, Vol. 2, Wiley, New York, 1973. Reprinted, Dover Publications, New York, 1985.

P.S. Kireev, *Semiconductor physics*, Mir Publishers, second edition, Moscow, 1978.

R. Peierls, *Some simple remarks on the basis of transport theory*, Lecture Notes in Physics 31 (G. Kirczenow and J. Marro editors), Springer-Verlag, Berlin, 1974.

H. Smith and H.H. Jensen, *Transport phenomena*, Oxford Science Publications, Oxford, 1989.

P.Y. Yu and M. Cardona, *Fundamentals of semiconductors*, Springer-Verlag, Berlin, third edition, 2003.

J.M. Ziman, *Electrons and phonons: the theory of transport phenomena in solids*, Clarendon Press, Oxford, 1960. Reissued, Oxford Classic Texts in the Physical Sciences, Oxford, 2001.

J.M. Ziman, *Principles of the theory of solids*, Cambridge University Press, Cambridge, second edition, 1979.

References

P. Drude, Zur Elektronentheorie. I, *Annalen der Physik* **1**, 566 (1900); Zur Elektronentheorie. II, *Annalen der Physik* **3**, 369 (1900).

D.A. Greenwood, The Boltzmann equation in the theory of electrical conduction in metals, *Proc. Phys. Soc. London* **71**, 585 (1958).

W. Pauli, in *Probleme der Modernen Physik*, S. Hirzel, Leipzig, 1928. Reprinted in *Collected scientific papers by W. Pauli* (R. Kronig and V.E. Weisskopf editors), Interscience, New York, 1964.

Supplement 8A

Collision processes

1. Introduction

One of the aims of the study of collision processes is to elucidate the microscopic nature of the mechanisms limiting the electronic transport and to derive, when possible, an expression for the relaxation time associated with the collision integral.

For elastic collisions in an isotropic system, such as electron–impurity or electron–long-wavelength-acoustic-phonon collisions, this time can be defined and microscopically computed, provided that the transition rate $W_{\boldsymbol{k}',\boldsymbol{k}}$ from state $|\boldsymbol{k}\rangle$ to state $|\boldsymbol{k}'\rangle$ only depends on the modulus of $\boldsymbol{k}$ and on the angle between the vectors $\boldsymbol{k}$ and $\boldsymbol{k}'$. The relaxation time is then a function of the electron energy $\varepsilon_{\boldsymbol{k}}$, a quantity conserved in the collision. Once the law $\tau(\varepsilon_{\boldsymbol{k}})$ is determined, the transport coefficients can be explicitly computed.

2. Electron–impurity scattering

As previously, we consider here electrons belonging to one band not completely filled, described in the effective mass approximation. The effective mass tensor is assumed to be proportional to the unit matrix.

2.1. The relaxation time for elastic collisions

In the case of elastic collisions such as the electron–impurity collisions, the collision integral of the Boltzmann equation is of the form:

$$\left(\frac{\partial f}{\partial t}\right)_{\text{coll}} = \frac{V}{(2\pi)^3}\int W_{\boldsymbol{k}',\boldsymbol{k}}(f_{\boldsymbol{k}'} - f_{\boldsymbol{k}})\,d\boldsymbol{k}'. \tag{8A.2.1}$$

If the deviations from local equilibrium remain small, we look for a solution of the Boltzmann equation of the form $f_{\boldsymbol{k}} \simeq f^{(0)}(\varepsilon_{\boldsymbol{k}}) + f_{\boldsymbol{k}}^{(1)}$, where $f_{\boldsymbol{k}}^{(1)}$ is a correction of first order with respect to the perturbations which make the system depart from the local equilibrium described by $f^{(0)}(\varepsilon_{\boldsymbol{k}})$. The elasticity of the collisions implying the equality $f^{(0)}(\varepsilon_{\boldsymbol{k}}) = f^{(0)}(\varepsilon_{\boldsymbol{k}'})$, leaves us with:

$$\left(\frac{\partial f}{\partial t}\right)_{\text{coll}} = \frac{V}{(2\pi)^3}\int W_{\boldsymbol{k}',\boldsymbol{k}}(f_{\boldsymbol{k}'}^{(1)} - f_{\boldsymbol{k}}^{(1)})\,d\boldsymbol{k}'. \tag{8A.2.2}$$

In the relaxation time approximation, we write:

$$\left(\frac{\partial f}{\partial t}\right)_{\text{coll}} \simeq -\frac{f_{\boldsymbol{k}}^{(1)}}{\tau(\boldsymbol{k})}. \tag{8A.2.3}$$

Comparing formulas (8A.2.2) and (8A.2.3), gives a formal expression for the inverse relaxation time:

$$\frac{1}{\tau(\boldsymbol{k})} = \frac{V}{(2\pi)^3}\int W_{\boldsymbol{k}',\boldsymbol{k}}\left(1-\frac{f_{\boldsymbol{k}'}^{(1)}}{f_{\boldsymbol{k}}^{(1)}}\right)d\boldsymbol{k}'. \tag{8A.2.4}$$

However, the expression (8A.2.4) for $1/\tau(\boldsymbol{k})$ involves the solution, which remains to be determined, of the linearized Boltzmann equation.

We can get rid of this drawback by studying the situation in which the only perturbation present is due to a uniform and time-independent applied electric field $\boldsymbol{E}$. The linearized Boltzmann equation reads, in this case:

$$(\boldsymbol{v}_{\boldsymbol{k}}.e\boldsymbol{E})\frac{\partial f^{(0)}}{\partial \varepsilon_{\boldsymbol{k}}} = -\frac{f_{\boldsymbol{k}}^{(1)}}{\tau(\boldsymbol{k})}. \tag{8A.2.5}$$

The relevant local equilibrium distribution $f^{(0)}$ being the Fermi–Dirac function $f_0(\varepsilon_{\boldsymbol{k}})$, we have:

$$f_{\boldsymbol{k}}^{(1)} = e\tau(\boldsymbol{k})(\boldsymbol{v}_{\boldsymbol{k}}.\boldsymbol{E})\left(-\frac{\partial f_0}{\partial \varepsilon_{\boldsymbol{k}}}\right). \tag{8A.2.6}$$

To determine $\tau(\boldsymbol{k})$, we make the assumption that, the system being isotropic, the relaxation time is a function of $\varepsilon_{\boldsymbol{k}}$ (or, which amounts to the same thing, a function of $k = |\boldsymbol{k}|$). This assumption will be verified at the end of the calculation. The collisions being elastic, gives us $\tau(k) = \tau(k')$. Since $\boldsymbol{v}_{\boldsymbol{k}} = \hbar\boldsymbol{k}/m^*$, we deduce from formula (8A.2.6) for $f_{\boldsymbol{k}}^{(1)}$, together with its analog for $f_{\boldsymbol{k}'}^{(1)}$, the identity:

$$\frac{f_{\boldsymbol{k}'}^{(1)}}{f_{\boldsymbol{k}}^{(1)}} = \frac{\boldsymbol{k}'.\boldsymbol{E}}{\boldsymbol{k}.\boldsymbol{E}}. \tag{8A.2.7}$$

The inverse relaxation time is thus given by the formula:

$$\frac{1}{\tau(k)} = \frac{V}{(2\pi)^3}\int W_{\boldsymbol{k}',\boldsymbol{k}}\left(1-\frac{\boldsymbol{k}'.\,\boldsymbol{E}}{\boldsymbol{k}.\boldsymbol{E}}\right)d\boldsymbol{k}', \tag{8A.2.8}$$

in which the electronic distribution function does not appear.

To make explicit the integral on the right-hand side of equation (8A.2.8), we take the direction of $\boldsymbol{k}$ as the polar axis Oz and choose the Ox axis so as to have the field $\boldsymbol{E}$ contained in the plane xOz (Fig. 8A.1). This gives:[1]

$$\begin{cases} \boldsymbol{k}.\boldsymbol{E} = kE\cos\alpha \\ \boldsymbol{k}'.\boldsymbol{E} = k'E(\cos\alpha\cos\theta + \sin\alpha\sin\theta\cos\phi), \end{cases} \tag{8A.2.9}$$

[1] The notations are those of Fig. 8A.1, setting $E = |\boldsymbol{E}|$.

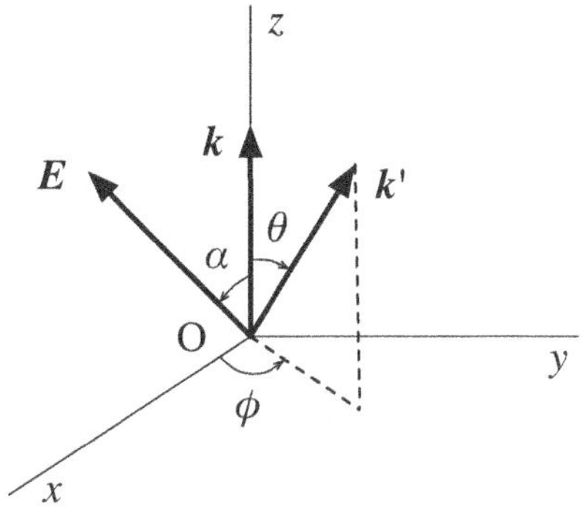

Fig. 8A.1 Electron–impurity interaction ($k = k'$) in the presence of an electric field.

and, therefore:

$$\frac{\boldsymbol{k}'.\boldsymbol{E}}{\boldsymbol{k}.\boldsymbol{E}} = \cos\theta + \tan\alpha \sin\theta \cos\phi. \tag{8A.2.10}$$

At the Born approximation of scattering, the transition rate $W_{\boldsymbol{k}',\boldsymbol{k}}$ reads:

$$W_{\boldsymbol{k}',\boldsymbol{k}} = \frac{2\pi}{\hbar} |\langle \boldsymbol{k}'|V_i|\boldsymbol{k}\rangle|^2 \delta(\varepsilon_{\boldsymbol{k}} - \varepsilon_{\boldsymbol{k}'}), \tag{8A.2.11}$$

where V_i denotes the electron–impurity interaction potential. The corresponding differential scattering cross-section is:[2]

$$\sigma(\theta) = \frac{m^{*2}V^2}{4\pi^2\hbar^4} |\langle \boldsymbol{k}'|V_i|\boldsymbol{k}\rangle|^2. \tag{8A.2.12}$$

This gives:

$$W_{\boldsymbol{k}',\boldsymbol{k}} = \frac{(2\pi)^3}{V^2} \frac{\hbar}{m^* k} \delta(k - k')\sigma(\theta). \tag{8A.2.13}$$

Since the rate $W_{\boldsymbol{k}',\boldsymbol{k}}$ depends only, besides k, on the scattering angle θ (and not on the absolute orientations of the vectors $\boldsymbol{k}$ and $\boldsymbol{k}'$), formula (8A.2.8) involves an integration over the surface of constant energy $\varepsilon_{\boldsymbol{k}}$. This integration once carried out, leaves us with:

$$\frac{1}{\tau(k)} = \frac{v}{V} \int \sigma(\theta)(1 - \cos\theta - \tan\alpha \sin\theta \cos\phi)\, d\Omega, \qquad v = \frac{\hbar k}{m^*}. \tag{8A.2.14}$$

In the factor $(1 - \cos\theta - \tan\alpha \sin\theta \cos\phi)$, only the term $(1 - \cos\theta)$ provides a non-vanishing contribution to the above integral. We thus have:

$$\frac{1}{\tau(k)} = \frac{v}{V} \int \sigma(\theta)(1 - \cos\theta)\, d\Omega. \tag{8A.2.15}$$

[2] With the hypotheses made for $W_{\boldsymbol{k}',\boldsymbol{k}}$, the differential cross-section depends only on θ (and not on both angles θ and ϕ). It is thus denoted by $\sigma(\theta)$.

In fact, the sample does not contain a unique impurity, but a concentration n_i of impurities. If these impurities are randomly distributed in space,[3] the inverse relaxation time is obtained by adding the contributions of the various impurities:

$$\boxed{\frac{1}{\tau(k)} = n_i v \int \sigma(\theta)(1 - \cos\theta)\, d\Omega.} \tag{8A.2.16}$$

Formula (8A.2.16) can be compared to the estimation $\tau^{-1} \sim n\sigma_{\text{tot}} v$ of the inverse relaxation time in a dilute classical gas (n is the gas density, $\sigma_{\text{tot}} = \int \sigma(\Omega)\, d\Omega$ the total scattering cross-section, and v a typical molecular velocity). In the formula (8A.2.16) concerning electronic collisions, the factor $(1 - \cos\theta)$ appearing in the integrand accounts for the greater relative importance of collisions with a large scattering angle. The quantity:

$$\ell_{\text{tr}} = v\tau(k) = \frac{1}{n_i \int \sigma(\theta)(1 - \cos\theta)\, d\Omega} \tag{8A.2.17}$$

is called the *transport mean free path*.[4]

Let us now provide the expressions for $\sigma(\theta)$ and $\tau(k)$ for different types of electron–impurity collisions.

2.2. Electron–impurity relaxation time in semiconductors

In a semiconductor, the impurities may be either neutral or ionized according to the temperature. Let us take the example of an n-type semiconductor. At low enough temperatures, the donor levels are occupied and the impurities are neutral. When kT becomes comparable to the energy difference between the impurity levels and the bottom of the conduction band, the donors ionize, losing their electrons for the benefit of the conduction band.

We will successively treat the scattering by these two types of impurities.

• *Ionized impurities*

If there are few charge carriers, the screening effect is negligible. The potential interaction energy of an electron of charge e with an impurity of charge $\pm Ze$ is thus Coulombian. For an impurity assumed to be situated at the origin, we have:[5]

$$V_i(r) = \pm \frac{Ze^2}{\varepsilon_d r}, \tag{8A.2.18}$$

where ε_d is the relative dielectric permittivity of the medium.

---

[3] This hypothesis amounts to considering the scattering as incoherent, that is, to discarding interferences between the waves scattered by different impurities.

[4] The transport mean free path plays an important role in the diffusive transport of light waves in a disordered medium (see Supplement 16A).

[5] We are using Gauss units, in which the vacuum permittivity equals unity.

To compute the differential scattering cross-section, we start from the quantum formula (8A.2.12), in which the Coulombian potential (8A.2.18) may be considered as the limit of a screened Coulombian potential[6] of infinite range. The calculations having been made, we obtain for $\sigma(\theta)$ a result identical to the classical Rutherford scattering formula:

$$\sigma(\theta) = \frac{R^2}{4\sin^4(\theta/2)}. \tag{8A.2.19}$$

Formula (8A.2.19), in which the quantity $R = Ze^2/\varepsilon_d m^* v^2$ represents a length linked to the impact parameter b through the equality $b = R\cot(\theta/2)$, shows that the scattering by ionized impurities is very anisotropic. The angular integral on the right-hand side of formula (8A.2.16), which reads:

$$\int_0^{\pi} \frac{1-\cos\theta}{\sin^4(\theta/2)} \sin\theta \, d\theta, \tag{8A.2.20}$$

diverges in the limit $\theta \to 0$ which corresponds to a very large impact parameter. Now, in a solid, the impact parameter has a natural upper limit $b_m = n_i^{-1/3}/2$, equal to half the typical distance between two impurities. A lower limit $\theta_{\min}$ of the deviation angle, given by the equality $\tan(\theta_{\min}/2) = R/b_m$, corresponds to the upper limit b_m of the impact parameter. Taking into account this lower limit of θ, we rewrite formula (8A.2.16) as:

$$\frac{1}{\tau(k)} = n_i v 2\pi \frac{R^2}{4} \int_{\theta_{\min}}^{\pi} \frac{1-\cos\theta}{\sin^4(\theta/2)} \sin\theta \, d\theta. \tag{8A.2.21}$$

This gives:

$$\int_{\theta_{\min}}^{\pi} \frac{1-\cos\theta}{\sin^4(\theta/2)} \sin\theta \, d\theta = -8\log\sin\frac{\theta_{\min}}{2} = 4\log\Big(1+\cot^2\frac{\theta_{\min}}{2}\Big). \tag{8A.2.22}$$

In view of the definition of $\theta_{\min}$, we finally get:

$$\frac{1}{\tau(k)} = 2\pi n_i R^2 v \log\left[1+\left(\frac{b_m}{R}\right)^2\right]. \tag{8A.2.23}$$

Formula (8A.2.23) was established by E. Conwell and V.F. Weisskopf in 1950.

Coming back to the expression for R, we deduce from formula (8A.2.23) that the relaxation time for scattering by ionized impurities varies as the third power of the velocity. Considering the relaxation time as a function of the energy ε, we have:

$$\tau(\varepsilon) \propto \varepsilon^{3/2}. \tag{8A.2.24}$$

In a non-degenerate semiconductor, the temperature dependence of the average relaxation time and of the mobility reflects the energy dependence of $\tau(\varepsilon)$. From formula (8A.2.24) it follows that the mobility limited by ionized impurity scattering depends on the temperature according to a law $\mu_D \propto T^{3/2}$.

[6] See formula (8A.2.25).

• *Neutral impurities*

For the sake of simplicity, we consider that, as far as the scattering of an electron is concerned, a neutral impurity may be treated like a very strongly screened ionized impurity. We therefore write:

$$V_i(r) = \pm \frac{Ze^2 e^{-k_0 r}}{\varepsilon_d r}, \tag{8A.2.25}$$

where k_0^{-1} denotes the screening length.[7]

The corresponding differential scattering cross-section is:

$$\sigma(\theta) = 4\left(\frac{Ze^2 m^*}{\varepsilon_d \hbar^2}\right)^2 \frac{1}{(k_0^2 + K^2)^2}, \tag{8A.2.26}$$

where $\boldsymbol{K} = \boldsymbol{k}' - \boldsymbol{k}$ represents the wave vector change in the course of the scattering process. The collision being elastic, we have:

$$k = k', \qquad K = 2k \sin\frac{\theta}{2}. \tag{8A.2.27}$$

Formula (8A.2.26) can thus be rewritten in the form:

$$\sigma(\theta) = 4\left(\frac{Ze^2 m^*}{\varepsilon_d \hbar^2}\right)^2 \left(k_0^2 + 4k^2 \sin^2\frac{\theta}{2}\right)^{-2}. \tag{8A.2.28}$$

The divergence of the integral (8A.2.16) giving $1/\tau(k)$ is automatically avoided by the introduction of the screening length.

If the velocity is not too large and the screening strong enough, we can neglect k^2 as compared to k_0^2 in formula (8A.2.28). In this case, the differential cross-section does not depend on θ (the scattering is isotropic) or on the velocity. We have:

$$\frac{1}{\tau(k)} = n_i v \sigma_{\text{tot}}. \tag{8A.2.29}$$

The relaxation time thus varies inversely proportionally to the velocity, that is, in terms of energy:

$$\tau(\varepsilon) \propto \varepsilon^{-1/2}. \tag{8A.2.30}$$

The above calculation relies on the hypothesis that the scattering of an electron by a neutral impurity may be described using the screened potential (8A.2.25) in the limit of very strong screening. However, the actual problem of scattering by a neutral impurity is notably more complicated. A more refined calculation shows that actually the scattering cross-section is inversely proportional to the velocity. The corresponding relaxation time is thus energy-independent. As a result, the mobility limited by the scattering on neutral impurities is temperature-independent.

---

[7] The screening of a charged impurity in a semiconductor is due to the gas of charge carriers. For a non-degenerate electron gas of density n, the screening length is given by the Debye expression: $k_0^{-1} = (kT/4\pi ne^2)^{1/2}$.

2.3. Electron–impurity relaxation time in metals

In a metal, an impurity is screened by the conduction electrons. The electron–impurity interaction potential energy is thus given by formula[8] (8A.2.25).

The impurity ion, 'dressed' with the mean number of electrons per atom of the metal, carries an effective charge $-Ze$ (Z is the valence difference between the impurity ion and a standard lattice ion). The differential scattering cross-section is given by formula (8A.2.28). This gives:

$$\frac{1}{\tau(k)} = n_i v 2\pi \times 4\left(\frac{Ze^2m^*}{\hbar^2}\right)^2 \int_0^\pi (1-\cos\theta)\left(k_0^2 + 4k^2\sin^2\frac{\theta}{2}\right)^{-2} \sin\theta\, d\theta. \quad (8A.2.31)$$

To determine the resistivity of the metal, it is enough to compute the relaxation time at the Fermi level. Formula (8A.2.31) shows that the contribution of the impurities to the resistivity of the metal, called the *residual resistivity*,[9] is proportional to Z^2. In view of the possible contributions of other types of impurities, the residual resistivity reads (*Linde's rule*):

$$\rho \sim a + bZ^2. \quad (8A.2.32)$$

The foregoing calculation actually overestimates the scattering cross-sections and, consequently, the resistivity. The electron–impurity interaction potential in a metal is indeed not weak enough for the Born approximation to be valid. It is necessary to proceed to a more careful analysis, in particular for explaining the resistivity of metallic alloys.[10]

3. Electron–phonon scattering

The inelastic scattering of electrons by lattice vibrations induces a temperature dependence of the electrical conductivity. To give a correct account of this effect, it is necessary to deal with the correlation of ionic motions, which can only be described in terms of phonons.

3.1. The electron–phonon collision integral

The electron–phonon interaction Hamiltonian is of the generic form:[11]

$$H_{\text{el-ph}} = \sum_{\boldsymbol{k},\boldsymbol{k}'} g(\boldsymbol{k},\boldsymbol{k}') a^\dagger_{\boldsymbol{k}'} a_{\boldsymbol{k}} (b^\dagger_{-\boldsymbol{q}} + b_{\boldsymbol{q}}), \qquad \boldsymbol{q} = \boldsymbol{k}' - \boldsymbol{k}. \quad (8A.3.1)$$

---

[8] The screening length is here the Thomas–Fermi screening length: $k_0^{-1} = (kT_F/4\pi ne^2)^{1/2}$ (T_F denotes the Fermi temperature).

[9] In the presence of non-magnetic scattering impurities, the resistivity of a metal decreases monotonically with the temperature towards a temperature-independent term, which constitutes, by definition, the residual resistivity. It is the only remaining contribution to the resistivity at very low temperatures, when scattering by lattice vibrations has become negligible.

[10] This can be achieved with the aid of a *phase shift method* due to J. Friedel (1956).

[11] We neglect here the *Umklapp processes* for which $\boldsymbol{k}' - \boldsymbol{k} = \boldsymbol{q} + \boldsymbol{K}$, where $\boldsymbol{K}$ is a reciprocal lattice vector. These processes, whose effect is much more difficult to calculate, play an important role in some transport properties, in particular in thermal conduction (however, they do not qualitatively modify the temperature dependence of the resistivity of normal metals). It can be shown that, when these processes are discarded, the only phonon modes to consider are the longitudinal modes for which the polarization vector is parallel to the wave vector.

In formula (8A.3.1), $a_{\boldsymbol{k}}$ and $a^\dagger_{\boldsymbol{k}'}$ are respectively the annihilation operator of an electron in state $|\boldsymbol{k}\rangle$ and the creation operator of an electron in state $|\boldsymbol{k}'\rangle$, whereas $b_{\boldsymbol{q}}$ and $b^\dagger_{-\boldsymbol{q}}$ denote respectively the annihilation operator of a phonon of wave vector $\boldsymbol{q}$ and the creation operator of a phonon of wave vector $-\boldsymbol{q}$.

The Hamiltonian (8A.3.1) induces electronic transitions between the different electronic states $|\boldsymbol{k}\rangle$. The corresponding collision integral is of the form:

$$\left(\frac{\partial f}{\partial t}\right)_{\text{coll}} = \frac{2\pi}{\hbar}\sum_{\boldsymbol{k}'}|g(\boldsymbol{k},\boldsymbol{k}')|^2 \times \Big[f_{\boldsymbol{k}'}(1-f_{\boldsymbol{k}})\Big\{n_{-\boldsymbol{q}}\delta(\varepsilon_{\boldsymbol{k}}-\varepsilon_{\boldsymbol{k}'}-\hbar\omega_{-\boldsymbol{q}})+(1+n_{\boldsymbol{q}})\delta(\varepsilon_{\boldsymbol{k}}-\varepsilon_{\boldsymbol{k}'}+\hbar\omega_{\boldsymbol{q}})\Big\} - f_{\boldsymbol{k}}(1-f_{\boldsymbol{k}'})\Big\{n_{\boldsymbol{q}}\delta(\varepsilon_{\boldsymbol{k}'}-\varepsilon_{\boldsymbol{k}}-\hbar\omega_{\boldsymbol{q}})+(1+n_{-\boldsymbol{q}})\delta(\varepsilon_{\boldsymbol{k}}-\varepsilon_{\boldsymbol{k}'}+\hbar\omega_{-\boldsymbol{q}})\Big\}\Big], \tag{8A.3.2}$$

where $n_{\boldsymbol{q}}$ and $\omega_{\boldsymbol{q}}$ denote the mean occupation number and the angular frequency of mode $\boldsymbol{q}$. In formula (8A.3.2), the two contributions to the entering term in $f_{\boldsymbol{k}'}(1-f_{\boldsymbol{k}})$ correspond respectively to the absorption of a phonon of wave vector $-\boldsymbol{q}$ and the emission of a phonon of wave vector $\boldsymbol{q}$, whereas the two contributions to the leaving term in $f_{\boldsymbol{k}}(1-f_{\boldsymbol{k}'})$ correspond respectively to the absorption of a phonon of wave vector $\boldsymbol{q}$ and the emission of a phonon of wave vector $-\boldsymbol{q}$.

3.2. Electron–acoustic-phonon relaxation time

We assume that the phonons are at thermodynamic equilibrium. We thus have $n_{\boldsymbol{q}} = n^0_{\boldsymbol{q}}$, where $n^0_{\boldsymbol{q}} = (e^{\beta\hbar\omega_{\boldsymbol{q}}}-1)^{-1}$ denotes the Bose–Einstein distribution function at temperature[12] $T=(k_B\beta)^{-1}$. For crystals with a symmetry center, we have $n^0_{\boldsymbol{q}} = n^0_{-\boldsymbol{q}}$.

The collisions of electrons with long wavelength acoustic phonons are quasielastic, as pictured by the inequality $\hbar|\omega_{\boldsymbol{q}}| \ll \varepsilon_{\boldsymbol{k}}$. The corresponding collision integral has the same structure as the electron–impurity collision integral (8A.2.1). It reads:

$$\left(\frac{\partial f}{\partial t}\right)_{\text{coll}} = \frac{2\pi}{\hbar}\sum_{\boldsymbol{k}'}|g(\boldsymbol{k},\boldsymbol{k}')|^2(f_{\boldsymbol{k}'}-f_{\boldsymbol{k}})(2n^0_{\boldsymbol{q}}+1)\delta(\varepsilon_{\boldsymbol{k}'}-\varepsilon_{\boldsymbol{k}}), \qquad \boldsymbol{k}'=\boldsymbol{k}+\boldsymbol{q}, \tag{8A.3.3}$$

with:

$$g(\boldsymbol{k},\boldsymbol{k}') = -i\gamma_{\boldsymbol{q}}, \qquad \gamma_{\boldsymbol{q}} = \frac{1}{V}\left(\frac{N\hbar}{2M\omega_{\boldsymbol{q}}}\right)^{1/2} qU(\boldsymbol{q}). \tag{8A.3.4}$$

In formula (8A.3.4), N is the number of atoms in the sample, M the mass of one of these atoms, and $U(\boldsymbol{q})$ the Fourier transform of the electron–ion interaction potential. We obtain for the inverse relaxation time:

$$\frac{1}{\tau(k)} = \frac{2\pi}{\hbar}\sum_{\boldsymbol{q}}(1-\cos\theta)|\gamma_{\boldsymbol{q}}|^2(2n^0_{\boldsymbol{q}}+1)\delta(\varepsilon_{\boldsymbol{k}+\boldsymbol{q}}-\varepsilon_{\boldsymbol{k}}), \tag{8A.3.5}$$

[12] In order to avoid any confusion with the electron wave vector modulus k, the Boltzmann constant is denoted here by k_B.

that is, since $n_{\boldsymbol{q}}^0 \gg 1$:

$$\frac{1}{\tau(k)} = \frac{4\pi}{\hbar}\sum_{\boldsymbol{q}}(1-\cos\theta)|\gamma_{\boldsymbol{q}}|^2 n_{\boldsymbol{q}}^0 \delta(\varepsilon_{\boldsymbol{k}+\boldsymbol{q}} - \varepsilon_{\boldsymbol{k}}). \qquad (8A.3.6)$$

The vectors $\boldsymbol{k}$ and $\boldsymbol{k}+\boldsymbol{q}$ having identical modulus, we have $1-\cos\theta = 2\cos^2\alpha$ (Fig. 8A.2). Using a continuous description for the wave vectors, we get:[13]

$$\frac{1}{\tau(k)} = \frac{4\pi}{\hbar}\frac{V}{(2\pi)^3}\int_0^\infty q^2\,dq \int_0^\pi 2\pi\sin\alpha\,d\alpha\; 2\cos^2\alpha|\gamma_{\boldsymbol{q}}|^2 n_{\boldsymbol{q}}^0\delta(\varepsilon_{\boldsymbol{k}+\boldsymbol{q}} - \varepsilon_{\boldsymbol{k}}). \qquad (8A.3.7)$$

Making use of the relation:

$$\delta(\varepsilon_{\boldsymbol{k}+\boldsymbol{q}} - \varepsilon_{\boldsymbol{k}}) = \frac{m^*}{kq\hbar^2}\delta\left(\cos\alpha + \frac{q}{2k}\right), \qquad (8A.3.8)$$

we obtain:

$$\frac{1}{\tau(k)} = \frac{4\pi}{\hbar}\frac{V}{8\pi^3}\int_0^\infty \frac{2\pi q^2 m^*}{kq\hbar^2}\,dq \int_0^\pi 2\sin\alpha\cos^2\alpha|\gamma_{\boldsymbol{q}}|^2 n_{\boldsymbol{q}}^0\delta\left(\cos\alpha + \frac{q}{2k}\right)d\alpha. \qquad (8A.3.9)$$

The angular integral yields a non-vanishing contribution only if $q < 2k$. We thus have:

$$\frac{1}{\tau(k)} = \frac{m^* V}{2\pi\hbar^3 k^3}\int_0^{2k} q^3|\gamma_{\boldsymbol{q}}|^2 n_{\boldsymbol{q}}^0\,dq. \qquad (8A.3.10)$$

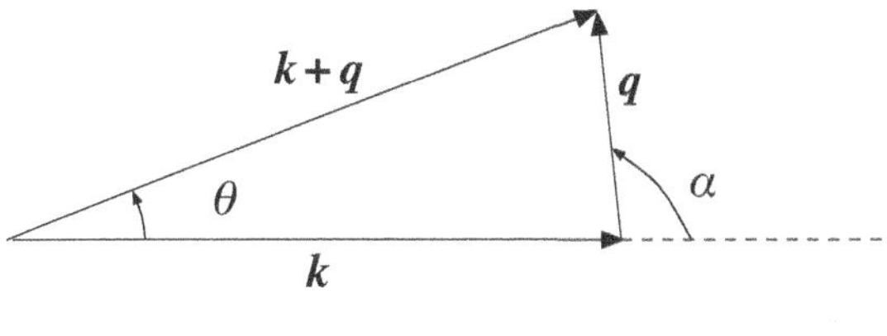

Fig. 8A.2 Electron–acoustic-phonon interaction.

The electron–ion interaction is generally described by a screened Coulombian potential,

$$U(r) \propto \frac{e^{-k_0 r}}{r}, \qquad (8A.3.11)$$

of Fourier transform:[14]

$$U(q) \propto \frac{4\pi}{k_0^2 + q^2}. \qquad (8A.3.12)$$

---

[13] The notations are those of Fig. 8A.2.

[14] For the sake of simplicity, we use the same notation $U(.)$ for the function $U(r)$ and its Fourier transform $U(q)$.

The function $U(q)$ tends towards a constant as $q \to 0$. Therefore, for the acoustic phonon modes for which $\omega_{\boldsymbol{q}} \sim c_s q$, where c_s denotes the sound velocity, the parameter $\gamma_{\boldsymbol{q}}$ introduced in formula (8A.3.4) behaves proportionally to $q^{1/2}$ as $q \to 0$. Taking as integration variable $x = \hbar\omega_{\boldsymbol{q}}/k_BT = \hbar c_s q/k_BT$, and setting, as $q \to 0$, $|\gamma_{\boldsymbol{q}}|^2 = A^2 q$, we finally get the following approximate expression,

$$\frac{1}{\tau(k)} \simeq \frac{m^* V}{2\pi\hbar^3 k^3}\left(\frac{k_BT}{\hbar c_s}\right)^5 A^2 \int_0^{2\hbar ck/k_BT} x^4 n(x)\, dx, \qquad (8A.3.13)$$

where $n(x) = (e^x - 1)^{-1}$.

3.3. Temperature dependence of the resistivity of metals

The resistivity of normal metals being governed by the relaxation time at the Fermi level, its temperature dependence reflects that of $\tau(\varepsilon_F)$, which can be deduced from formula (8A.3.13).

For $k = k_F$ (k_F denoting the modulus of the Fermi wave vector), the lower bound $2\hbar c_s k/k_BT$ of the integral on the right-hand side of equation (8A.3.13) is equal to $2\Theta/T$, where the temperature Θ is defined by[15] $\Theta = \hbar c_s k_F/k_B$. There are two main regimes of temperature dependence of $\tau(\varepsilon_F)$, as follows.

- *Low temperature regime*

At low temperatures ($T \ll \Theta$), the upper bound of the integral (8A.3.13) is much larger than 1, and we can extend the integration interval to infinity. The whole temperature dependence of $1/\tau(\varepsilon_F)$ comes from the prefactor $\propto T^5$.

- *High temperature regime*

At high temperatures ($T \gg \Theta$), the values of x which come into play in the integral (8A.3.13) being much smaller than 1, the function $n(x)$ behaves like x^{-1}, and the integrand like x^3. Its integration up to a bound inversely proportional to the temperature introduces a factor $\propto T^{-4}$. The inverse relaxation time at the Fermi level then varies as T.

Two distinct regimes are actually observed in the temperature dependence of the resistivity of normal metals: the resistivity first increases according to a law in T^5, known as the *Bloch–Grüneisen law*, at low temperatures, then it increases according to a law in T at high temperatures.

---

[15] The order of magnitude of Θ is the same as that of the Debye temperature $\Theta_D = \hbar c_s q_D/k_B$ (q_D is the Debye wave vector). Indeed, we have $k_F^3 = 3\pi^2 n$, where n is the density of conduction electrons, and $q_D^3 = 6\pi^2 N/V$, where N/V is the ionic density. For a metal of valence Z, we thus have $q_D = (2/Z)^{1/3} k_F$.

Bibliography

N.W. ASHCROFT and N.D. MERMIN, *Solid state physics*, W.B. Saunders Company, Philadelphia, 1976.

C. COHEN-TANNOUDJI, B. DIU, and F. LALOË, *Quantum mechanics*, Vol. 2, Hermann and Wiley, Paris, second edition, 1977.

W. JONES and N.H. MARCH, *Theoretical solid-state physics: non-equilibrium and disorder*, Vol. 2, Wiley, New York, 1973. Reprinted, Dover Publications, New York, 1985.

P.S. KIREEV, *Semiconductor physics*, Mir Publishers, second edition, Moscow, 1978.

L.D. LANDAU and E.M. LIFSHITZ, *Mechanics*, Butterworth-Heinemann, Oxford, third edition, 1976.

H. SMITH and H.H. JENSEN, *Transport phenomena*, Oxford Science Publications, Oxford, 1989.

P.Y. YU and M. CARDONA, *Fundamentals of semiconductors*, Springer-Verlag, Berlin, third edition, 2003.

J.M. ZIMAN, *Electrons and phonons: the theory of transport phenomena in solids*, Clarendon Press, Oxford, 1960. Reissued, Oxford Classic Texts in the Physical Sciences, Oxford, 2001.

J.M. ZIMAN, *Principles of the theory of solids*, Cambridge University Press, Cambridge, second edition, 1979.

References

E. CONWELL and V.F. WEISSKOPF, Theory of impurity scattering in semiconductors, *Phys. Rev.* **77**, 388 (1950).

Supplement 8B

Thermoelectric coefficients

1. Particle and heat fluxes

Let us consider a metal or an n-type semiconductor in which there is a uniform density n of conduction electrons with charge e. The conduction band is described in the effective mass approximation, with an effective mass tensor proportional to the unit matrix.

We assume that the temperature $T(\boldsymbol{r})$, time-independent, can vary from one point to another inside the sample. The latter is also submitted to a uniform and time-independent electric field $\boldsymbol{E} = -\nabla\phi$. There is no applied magnetic field. In stationary regime, the linearized Boltzmann equation reads, in this case:

$$\boldsymbol{v_k}.\Big(\frac{\varepsilon_{\boldsymbol{k}} - \mu}{T}\nabla_{\boldsymbol{r}}T + \nabla_{\boldsymbol{r}}\mu - e\boldsymbol{E}\Big)\Big(-\frac{\partial f^{(0)}}{\partial \varepsilon_{\boldsymbol{k}}}\Big) = -\frac{f_{\boldsymbol{k}}^{(1)}}{\tau(\varepsilon_{\boldsymbol{k}})}. \tag{8B.1.1}$$

The solution of equation (8B.1.1) is:[1]

$$f_{\boldsymbol{k}}^{(1)} = \tau(\varepsilon)\boldsymbol{v}.\Big[(e\boldsymbol{E} - \nabla\mu) - \frac{\varepsilon - \mu}{T}\nabla T\Big]\Big(-\frac{\partial f^{(0)}}{\partial \varepsilon}\Big). \tag{8B.1.2}$$

A particle flux $\boldsymbol{J}_N$ and a heat flux $\boldsymbol{J}_Q^* = \boldsymbol{J}_E - \overline{\mu}\boldsymbol{J}_N$, where $\boldsymbol{J}_E$ is the energy flux and $\overline{\mu}(\boldsymbol{r}) = \mu[T(\boldsymbol{r})] + e\phi(\boldsymbol{r})$ the local electrochemical potential, appear inside the system. In linear regime, the particle flux can be calculated using the distribution function $f_{\boldsymbol{k}}^{(1)}$ via the integral:

$$\boldsymbol{J}_N = \frac{2}{(2\pi)^3}\int f_{\boldsymbol{k}}^{(1)}\boldsymbol{v}\,d\boldsymbol{k}. \tag{8B.1.3}$$

The heat flux $\boldsymbol{J}_Q^*$ can in principle be calculated with the aid of the formula:

$$\boldsymbol{J}_Q^* = \frac{2}{(2\pi)^3}\int f_{\boldsymbol{k}}^{(1)}(\varepsilon - \overline{\mu})\boldsymbol{v}\,d\boldsymbol{k}, \tag{8B.1.4}$$

which, in the linear regime, reduces to:

$$\boldsymbol{J}_Q^* = \frac{2}{(2\pi)^3}\int f_{\boldsymbol{k}}^{(1)}(\varepsilon - \mu)\boldsymbol{v}\,d\boldsymbol{k}. \tag{8B.1.5}$$

---

[1] Since there is no ambiguity, $\nabla_{\boldsymbol{r}}$ will from now on be simply denoted by ∇. The energy and the mean velocity of the electron in state $|\boldsymbol{k}\rangle$ are simply denoted by ε and $\boldsymbol{v}$.

2. General expression for the kinetic coefficients

The expressions for $\boldsymbol{J}_N$ and $\boldsymbol{J}_Q^*$ involve the following integrals,

$$K_p = \frac{1}{3}\int_0^\infty v^2(\varepsilon-\mu)^p n(\varepsilon)\tau(\varepsilon)\Big(-\frac{\partial f^{(0)}}{\partial\varepsilon}\Big)\,d\varepsilon, \qquad p=0,1,2, \tag{8B.2.1}$$

where $n(\varepsilon)$ is the density of states in energy of the electrons. More precisely, the fluxes $\boldsymbol{J}_N$ and $\boldsymbol{J}_Q^*$ are given by the formulas:

$$\begin{cases} \boldsymbol{J}_N = K_0(e\boldsymbol{E}-\nabla\mu) - K_1\dfrac{1}{T}\nabla T \\[2ex] \boldsymbol{J}_Q^* = K_1(e\boldsymbol{E}-\nabla\mu) - K_2\dfrac{1}{T}\nabla T. \end{cases} \tag{8B.2.2}$$

The linear response relations (8B.2.2) are of the general form:

$$\begin{cases} \boldsymbol{J}_N = -L_{11}\dfrac{1}{T}\nabla\overline{\mu} + L_{12}\nabla(\dfrac{1}{T}) \\[2ex] \boldsymbol{J}_Q^* = -L_{21}\dfrac{1}{T}\nabla\overline{\mu} + L_{22}\nabla(\dfrac{1}{T}), \end{cases} \tag{8B.2.3}$$

with:

$$\begin{cases} L_{11} = K_0T \\[1ex] L_{12} = L_{21} = K_1T \\[1ex] L_{22} = K_2T. \end{cases} \tag{8B.2.4}$$

The Onsager symmetry relation $L_{12}=L_{21}$ is therefore verified, as expected.

3. Thermal conductivity

The isothermal electrical conductivity and the open-circuit thermal conductivity are expressed respectively as $\sigma = e^2K_0$ and $\kappa = (K_0K_2-K_1^2)/K_0T$. According to formula (8B.2.1) for $p=0$, we have:

$$K_0 = \frac{n}{m^*}\langle\tau\rangle, \tag{8B.3.1}$$

where:

$$\langle\tau\rangle = \frac{\int_0^\infty \tau(\varepsilon)\varepsilon^{3/2}\big(-\frac{\partial f^{(0)}}{\partial\varepsilon}\big)\,d\varepsilon}{\int_0^\infty \varepsilon^{3/2}\big(-\frac{\partial f^{(0)}}{\partial\varepsilon}\big)\,d\varepsilon} \tag{8B.3.2}$$

(we have taken as usual $n(\varepsilon) = C\varepsilon^{1/2}$). Similarly we have, according to formula (8B.2.1) for $p=1$ and $p=2$,

$$K_1 = \frac{n}{m^*}\big\langle(\varepsilon-\mu)\tau\big\rangle, \tag{8B.3.3}$$

and:

$$K_2 = \frac{n}{m^*}\langle(\varepsilon - \mu)^2\tau\rangle. \tag{8B.3.4}$$

In formulas (8B.3.3) and (8B.3.4), the symbol $\langle\ldots\rangle$ has the same meaning as in formula (8B.3.2). We deduce from these results the expression for the open-circuit thermal conductivity:

$$\kappa = \frac{n}{m^*T}\left[\langle\varepsilon^2\tau\rangle - \frac{\langle\varepsilon\tau\rangle^2}{\langle\tau\rangle}\right]. \tag{8B.3.5}$$

Let us now make precise the form of the integrals K_0, K_1, K_2, and deduce from them the expression for κ in a metal and in a non-degenerate n-type semiconductor.

3.1. Metal

The Sommerfeld expansion at lowest order of the integrals K_0, K_1, and K_2 yields the approximate expressions:

$$K_0 \simeq k_0(\varepsilon_F), \qquad K_1 \simeq \frac{\pi^2}{3}(kT)^2 \frac{dk_0(\varepsilon)}{d\varepsilon}\bigg|_{\varepsilon=\varepsilon_F}, \qquad K_2 \simeq \frac{\pi^2}{3}(kT)^2 k_0(\varepsilon_F), \tag{8B.3.6}$$

where:

$$k_0(\varepsilon) = \frac{2}{3m^*}\varepsilon n(\varepsilon)\tau(\varepsilon). \tag{8B.3.7}$$

We have $K_1^2 \ll K_0K_2$ and thus $\kappa \simeq K_2/T$. Since $\sigma = e^2K_0$, there is a proportionality relation between κ and σT, known as the *Wiedemann–Franz law*:

$$\boxed{\frac{\kappa}{\sigma T} \simeq \frac{\pi^2}{3}\frac{k^2}{e^2}.} \tag{8B.3.8}$$

The ratio $\mathcal{L} = \kappa/\sigma T$, called the *Lorenz number*, is thus in this model a temperature-independent constant:[2]

$$\mathcal{L} = \frac{\pi^2}{3}\frac{k^2}{e^2}. \tag{8B.3.9}$$

Relation (8B.3.8) was established empirically in 1853 for numerous metals. The value of the Lorenz number, obtained here in an extremely simple band structure model, does not in fact depend on it. In the general case in which the electrical conductivity and the thermal conductivity are tensors, we can show that the components of the same indices of these tensors are in the ratio $\pi^2k^2T/3e^2$.

3.2. Non-degenerate semiconductor

In this case too, there is a proportionality relation between κ and σT. To get it, it is necessary to compute the integrals K_p with $f^{(0)}(\varepsilon)$ the local equilibrium Maxwell–Boltzmann distribution, the energy dependence of the relaxation time being taken into account.

---

[2] We have computed here the contribution of the electrons to the thermal conductivity of the metal. The electronic contribution to the thermal conductivity being one or two orders of magnitude higher than that of the lattice, we can consider that it actually represents the thermal conductivity of the metal.

For a relaxation time depending on the energy according to a power law of the type $\tau(\varepsilon) \propto \varepsilon^s$, we get for the Lorenz number the value:[3]

$$\mathcal{L} = \left(\frac{5}{2} + s\right)\frac{k^2}{e^2}. \tag{8B.3.10}$$

The ratio $\mathcal{L} = \kappa/\sigma T$ is not universal as it is the case in metals, but it reflects the energy dependence of the relaxation time.

The foregoing calculation relies on the hypothesis of the existence of a unique relaxation time governing both electrical conduction and thermal conduction. Such a relaxation time does exist when the collision processes are elastic (which means in practice that the energy change of an electron in the course of a collision is small compared to kT). The scattering of electrons by lattice vibrations obeys this condition at high temperatures. At low temperatures, the mechanism limiting the conductivity is the elastic scattering of electrons by impurities. The Wiedemann–Franz law is indeed well verified at high as well as at low temperatures.

4. The Seebeck and Peltier coefficients

The linear response relations (8B.2.2) also allow us to explicitly evaluate the thermoelectric coefficients.

4.1. The Seebeck coefficient

In open circuit, a temperature gradient is accompanied by an electrochemical potential gradient given, according to the first of equations (8B.2.3) and to formulas (8B.2.4), by:

$$\nabla\overline{\mu} = -\frac{K_1}{K_0}\frac{1}{T}\nabla T. \tag{8B.4.1}$$

The thermoelectric power η of the considered material, defined by the relation $\nabla\overline{\mu} = -e\eta\nabla T$, is thus:

$$\eta = \frac{1}{eT}\frac{K_1}{K_0}. \tag{8B.4.2}$$

In a metal, in view of the expressions (8B.3.6) for K_0 and K_1, we obtain:

$$\eta = \frac{\pi^2}{3}\frac{k}{e}kT\left.\frac{d\log k_0(\varepsilon)}{d\varepsilon}\right|_{\varepsilon=\varepsilon_F}. \tag{8B.4.3}$$

With the chosen band structure, the thermoelectric power is negative. As a general rule, its sign depends on the band structure of the material.

---

3 In a semiconductor, the contribution of the electrons to the thermal conductivity is small compared to the contribution of the phonons. Moreover, the above calculation is oversimplified, since it takes only the electrons into account, and not the holes. If we take into account the contribution of the holes, a supplementary term adds to the expression (8B.3.10) for the Lorenz number in a semiconductor.

4.2. The Peltier coefficient

The Peltier effect consists in the appearance, at uniform temperature, of a heat flux accompanying an electric current. In these conditions, equations (8B.2.3) and formulas (8B.2.4) lead to:

$$\begin{cases} \boldsymbol{J}_N = -K_0 \nabla \overline{\mu} \\ \boldsymbol{J}_Q^* = -K_1 \nabla \overline{\mu}. \end{cases} \tag{8B.4.4}$$

The Peltier coefficient π of the material, defined by the relation $\boldsymbol{J}_Q^* = e\pi \boldsymbol{J}_N$, is thus:

$$\pi = \frac{1}{e}\frac{K_1}{K_0}. \tag{8B.4.5}$$

The microscopic expressions (8B.4.3) and (8B.4.5) for η and π verify as expected the second Kelvin relation:

$$\pi = \eta T. \tag{8B.4.6}$$

Bibliography

N.W. Ashcroft and N.D. Mermin, *Solid state physics*, W.B. Saunders Company, Philadelphia, 1976.

W. Jones and N.H. March, *Theoretical solid-state physics: non-equilibrium and disorder*, Vol. 2, Wiley, New York, 1973. Reprinted, Dover Publications, New York, 1985.

P.S. Kireev, *Semiconductor physics*, Mir Publishers, second edition, Moscow, 1978.

H. Smith and H.H. Jensen, *Transport phenomena*, Oxford Science Publications, Oxford, 1989.

J.M. Ziman, *Electrons and phonons: the theory of transport phenomena in solids*, Clarendon Press, Oxford, 1960. Reissued, Oxford Classic Texts in the Physical Sciences, Oxford, 2001.

J.M. Ziman, *Principles of the theory of solids*, Cambridge University Press, Cambridge, second edition, 1979.

Chapter 9

Master equations

This chapter is devoted to the master equations, which play a very important role in the description of the evolution of out-of-equilibrium physical systems.

To begin with, we consider those physical phenomena likely to be modelized by random processes. The processes which are most commonly used for this type of modelization are the Markov processes, often called picturesquely 'processes without memory'. A Markov process is governed by its transition probabilities, which allow us, step by step, to determine its evolution from its initial distribution. The transition probabilities of a Markov process obey a non-linear functional equation, the Chapman–Kolmogorov equation. If we have independent information about the behavior of the transition probabilities over short time intervals, we can deduce from the Chapman–Kolmogorov equation a linear equation, called the master equation, describing the evolution of the transition probabilities over much longer time intervals.

Then, adopting a more physical point of view, we investigate how such an evolution equation can be established starting from the Liouville–von Neumann equation for the density matrix. In particular, we study the evolution in the presence of a perturbation of a system with a very large number of degrees of freedom. If the density matrix of the system is diagonal at the initial time, its diagonal elements obey afterward –provided that some hypotheses are verified– an irreversible evolution equation, similar to the master equation for a Markov process, called the Pauli master equation.

The Pauli master equation, which is valid in the weak-coupling long-time limit, allows us to describe the irreversible behavior of a physical system, in particular the approach of equilibrium starting from an initial out-of-equilibrium distribution.

1. Markov processes: the Chapman–Kolmogorov equation

1.1. Conditional probabilities

To study the time development of a random process, and in particular to determine in what measure its present is influenced or conditioned by its past, it is convenient to have recourse to the notion of conditional probability introduced for a set of random variables.

The *elementary conditional probability*[1] $p_{1|1}(x_2, t_2|x_1, t_1)$ is defined as the probability density of the process $X(t)$ at time t_2, given that the value taken by $X(t)$ at time t_1 was x_1. This definition implies that the times t_1 and t_2 are ordered: $t_1 < t_2$. More generally, we can fix the values of $X(t)$ at k different times $t_1, \ldots, t_k$ and consider the joint probability at n other times $t_{k+1}, \ldots, t_{k+n}$, the times $t_1, \ldots, t_{k+n}$ being ordered: $t_1 < \ldots < t_k < t_{k+1} < \ldots < t_{k+n}$. We thus define the conditional probability $p_{n|k}(x_{k+1}, t_{k+1}; \ldots; x_{k+n}, t_{k+n}|x_1, t_1; \ldots; x_k, t_k)$. According to Bayes' rule, it can be expressed as the ratio of the two joint probabilities p_{k+n} and p_k:

$$p_{n|k}(x_{k+1}, t_{k+1}; \ldots; x_{k+n}, t_{k+n}|x_1, t_1; \ldots; x_k, t_k) = \frac{p_{k+n}(x_1, t_1; \ldots; x_k, t_k; x_{k+1}, t_{k+1}; \ldots; x_{k+n}, t_{k+n})}{p_k(x_1, t_1; \ldots; x_k, t_k)}. \tag{9.1.1}$$

Inversely, the joint probability p_n reads, in terms of p_1 and of the conditional probabilities $p_{1|1}(x_2, t_2|x_1, t_1), \ldots, p_{1|n-1}(x_n, t_n|x_1, t_1; \ldots; x_{n-1}, t_{n-1})$:

$$p_n(x_1, t_1; x_2, t_2; \ldots; x_n, t_n) = p_1(x_1, t_1) \ldots p_{1|n-1}(x_n, t_n|x_1, t_1; \ldots; x_{n-1}, t_{n-1}). \tag{9.1.2}$$

Let us now come back to the elementary conditional probability $p_{1|1}(x_2, t_2|x_1, t_1)$. The consistency condition:

$$\int p_2(x_1, t_1; x_2, t_2)\, dx_1 = p_1(x_2, t_2) \tag{9.1.3}$$

implies the relation:

$$p_1(x_2, t_2) = \int p_{1|1}(x_2, t_2|x_1, t_1) p_1(x_1, t_1)\, dx_1, \qquad t_1 < t_2. \tag{9.1.4}$$

Equation (9.1.4) allows us to obtain the distribution of $X(t)$ at a time $t_2 > t_1$ given the distribution of $X(t)$ at time t_1, provided that the elementary conditional probability $p_{1|1}(x_2, t_2|x_1, t_1)$ is known. However, the elementary conditional probability may depend on the past history of the process. In such a case, equation (9.1.4) is of no practical use.

---

[1] For convenience, the probability densities (be they conditional or joint densities) are simply here called probabilities.

1.2. Characterization of random processes

To fully characterize a stochastic process $X(t)$, it is generally necessary to know the whole set of joint probabilities p_n, that is, in view of the relation (9.1.2), the distribution p_1 and the whole set of conditional probabilities $p_{1|n}$. However, some stochastic processes, whose present is either not at all or only weakly influenced by their past history, may be characterized in a more simple way.

A first example is that of the *completely random processes*, characterized by p_1. Another example is that of the *Markov processes*, characterized by p_1 and $p_{1|1}$. The Markov processes are by far the most frequently used in the modelization of physical phenomena by random processes.[2]

1.3. Completely random processes

In the case of a completely random process, the value taken by the variable at a given time is independent of the values it took before. The conditional probabilities then reduce to non-conditional probabilities:

$$p_{n|k}(x_{k+1},t_{k+1};\ldots;x_{k+n},t_{k+n}|x_1,t_1;\ldots;x_k,t_k) = p_n(x_{k+1},t_{k+1};\ldots;x_{k+n},t_{k+n}). \tag{9.1.5}$$

In particular, the general formula (9.1.2) can be simplified into:

$$p_n(x_1,t_1;\ldots;x_n,t_n) = p_1(x_1,t_1)\ldots p_1(x_n,t_n). \tag{9.1.6}$$

Formula (9.1.6) expresses the fact that the random variables corresponding to the values taken by the process $X(t)$ at times $t_1,\ldots,t_n$ are statistically independent. The present of a completely random process is not at all influenced by its past.

1.4. Markov processes

In a Markov process, the value taken by the variable at a given time is influenced only by its most recent past values. More precisely, a stochastic process $X(t)$ is a Markov process if, for any times $t_1 < t_2 < \ldots < t_n$, and for any n, we have:

$$p_{1|n-1}(x_n,t_n|x_1,t_1;x_2,t_2;\ldots;x_{n-1},t_{n-1}) = p_{1|1}(x_n,t_n|x_{n-1},t_{n-1}). \tag{9.1.7}$$

Once 'arrived' at the value x_{n-1} at time t_{n-1}, after being 'passed' by x_1 at time t_1, x_2 at time t_2 ..., the process then evolves in a way which only depends on x_{n-1}. The evolution of the process after a given time depends only on the value it took at this time, and not on its previous history.[3] The central quantity for the description

---

[2] In particular, they play a role in the description of Brownian motion (see Chapter 11).

[3] Equation (9.1.7) defines a *first-order Markov process*. More generally, we can define *higher-order Markov processes*. For instance, if we have, for any times $t_1 < t_2 < \ldots < t_n$, and for any n,

$$p_{1|n-1}(x_n,t_n|x_1,t_1;x_2,t_2;\ldots;x_{n-1},t_{n-1}) = p_{1|2}(x_n,t_n|x_{n-1},t_{n-1};x_{n-2},t_{n-2}),$$

otherwise stated, if the conditional probability $p_{1|n-1}$ only depends on the two most recent past values of the variable, the process $X(t)$ is a *second-order Markov process*. Higher-order Marlov processes may be defined in an analogous way. The essential characteristic of a Markov process (whatever its order) is that the memory of its past history does not persist indefinitely, but eventually disappears. The first-order Markov processes being the only ones considered here, they will be simply called Markov processes (with no explicit mention of their order).

of a Markov process is the elementary conditional probability $p_{1|1}$. On account of equation (9.1.7), the general formula (9.1.2) for the joint probabilities indeed reads, in the Markovian case:

$$p_n(x_1, t_1; x_2, t_2; \dots; x_n, t_n) = p_1(x_1, t_1) \dots p_{1|1}(x_n, t_n | x_{n-1}, t_{n-1}). \tag{9.1.8}$$

All the joint probabilities are thus determined if we know the probability p_1 and the elementary conditional probability $p_{1|1}$, called *transition probability*. The transition probability of a Markov process is independent from the past history of the process. Equation (9.1.4) actually allows us, in this case, to determine $p_1(x_2, t_2)$ given $p_1(x_1, t_1)$ $(t_2 > t_1)$, and thus to establish the evolution equation of $p_1(x, t)$.

Amongst the Markov processes, the stationary processes play a particularly important role. For a Markov process to be stationary, it is necessary and sufficient that p_1 be time-independent and that $p_{1|1}$ only depend on the time interval involved. Most often, for a stationary process, the probability p_1 represents the distribution which is attained after a sufficiently long time τ, whatever the initial state x_0. In this case, we have:

$$p_1(x) = \lim_{\tau \to \infty} p_{1|1}(x, \tau | x_0). \tag{9.1.9}$$

The process is then entirely defined by the knowledge of the transition probability.

1.5. The Chapman–Kolmogorov equation

We have, in a general way, the identity:

$$p_{1|1}(x_3, t_3 | x_1, t_1) = \int p_{1|1}(x_2, t_2 | x_1, t_1) p_{1|2}(x_3, t_3 | x_1, t_1; x_2, t_2)\, dx_2, \quad t_1 < t_2 < t_3. \tag{9.1.10}$$

In the case of a Markov process, the definition (9.1.7) being taken into account, identity (9.1.10) involves only $p_{1|1}$. It reads:

$$\boxed{p_{1|1}(x_3, t_3 | x_1, t_1) = \int p_{1|1}(x_2, t_2 | x_1, t_1) p_{1|1}(x_3, t_3 | x_2, t_2)\, dx_2,} \tag{9.1.11}$$

with $t_1 < t_2 < t_3$. Equation (9.1.11) expresses a necessary condition for a random process to be Markovian. This functional equation, which expresses a constraint that the transition probability of a Markov process must obey, is known under the name of *Chapman–Kolmogorov equation*, or of *Smoluchowski equation*.[4] It was established by M. Smoluchowski in 1906, then by S. Chapman in 1916 and A. Kolmogorov in 1931.

A specific Markov process is fully determined by p_1 and $p_{1|1}$. These functions must obey the Chapman–Kolmogorov equation (9.1.11), as well as the consistency condition (9.1.4). Two non-negative functions p_1 and $p_{1|1}$ which verify these two conditions uniquely define a Markov process.

---

[4] For instance it is under this latter name that it is referred to in the context of Brownian motion (see Chapter 11).

2. Master equation for a Markovian random process

Consider a Markovian process $X(t)$. The Chapman–Kolmogorov equation (9.1.11) being non-linear, its solution is not unique. Thus this equation does not allow us, per se, to specify the transition probability. However, if we have information about the behavior of $p_{1|1}$ over short time intervals, it is possible to deduce from the Chapman–Kolmogorov equation a linear evolution equation for $p_{1|1}$, valid over much longer time intervals, called the *master equation*.

2.1. The master equation

Consider a Markov process whose transition probability only depends on the difference of the times involved:[5]

$$p_{1|1}(x_2, t_2|x_1, t_1) = p_{1|1}(x_2, \Delta t|x_1), \qquad \Delta t = t_2 - t_1. \tag{9.2.1}$$

We assume that, for $\Delta t \ll \tau$ (the time scale τ remaining to be defined from a physical point of view), the behavior of $p_{1|1}(x_2, \Delta t|x_1)$ is of the form:[6]

$$p_{1|1}(x_2, \Delta t|x_1) = (1 - \lambda\Delta t)\delta(x_2 - x_1) + \Delta t W(x_2|x_1) + o(\Delta t). \tag{9.2.2}$$

In formula (9.2.2), the quantity $W(x_2|x_1) \geq 0$ is the transition probability per unit time from x_1 to $x_2 \neq x_1$. The parameter λ is determined by the normalization condition $\int p_{1|1}(x_2, \Delta t|x_1)\, dx_2 = 1$:

$$\lambda = \int W(x_2|x_1)\, dx_2. \tag{9.2.3}$$

The important point here is the existence of a transition rate $W(x_2|x_1)$. As displayed by equation (9.2.2), $W(x_2|x_1)$ is linked to $p_{1|1}(x_2, \Delta t|x_1)$ (more precisely, to the behavior of $p_{1|1}(x_2, \Delta t|x_1)$ for $\Delta t \ll \tau$).

On account of the stationarity property of $p_{1|1}$, we can rewrite equation (9.1.11) in the form:

$$p_{1|1}(x_3, t + \Delta t|x_1) = \int p_{1|1}(x_2, t|x_1) p_{1|1}(x_3, \Delta t|x_2)\, dx_2, \tag{9.2.4}$$

where we have set $t = t_2 - t_1$ and $\Delta t = t_3 - t_2$. On the right-hand side of equation (9.2.4), we can make use of formula (9.2.2) to rewrite $p_{1|1}(x_3, \Delta t|x_2)$. This gives:[7]

$$p_{1|1}(x_3, t+\Delta t|x_1) = (1-\lambda\Delta t) p_{1|1}(x_3, t|x_1) + \Delta t \int p_{1|1}(x_2, t|x_1) W(x_3|x_2)\, dx_2. \tag{9.2.5}$$

---

[5] No hypothesis being made about the distribution p_1 (which may be time-dependent), such a process is not necessarily stationary (only its transition probability is stationary).

[6] The symbol $o(\Delta t)$ denotes an unspecified term such that $o(\Delta t)/\Delta t \to 0$ as $\Delta t \to 0$.

[7] This implies $\Delta t \ll \tau$. However, no upper bound is imposed to the time t, which may be arbitrarily large.

We now assume that $\Delta t \ll t$ and we formally take the limit $\Delta t \to 0$ in equation (9.2.5). This gives the integro-differential equation:[8]

$$\frac{\partial p_{1|1}(x_3, t|x_1)}{\partial t} = -\lambda\, p_{1|1}(x_3, t|x_1) + \int p_{1|1}(x_2, t|x_1) W(x_3|x_2)\, dx_2, \qquad (9.2.6)$$

which can also be rewritten, using the expression (9.2.3) for λ (with modified notations):

$$\frac{\partial p_{1|1}(x, t|x_1)}{\partial t} = \int \big[W(x|x')p_{1|1}(x', t|x_1) - W(x'|x)p_{1|1}(x, t|x_1)\big]\, dx'. \qquad (9.2.7)$$

In the transition probabilities involved in equation (9.2.7), the initial value x_1 is given. We can write equivalently, simplifying the notations,[9]

$$\boxed{\frac{\partial p_{1|1}(x, t)}{\partial t} = \int \big[W(x|x')p_{1|1}(x', t) - W(x'|x)p_{1|1}(x, t)\big]\, dx',} \qquad (9.2.8)$$

with the initial condition:

$$p_{1|1}(x, 0) = \delta(x - x_1). \qquad (9.2.9)$$

Equation (9.2.8) is called the master equation. Its solution determined for $t \geq 0$ by the initial condition (9.2.9) is the transition probability $p_{1|1}(x, t|x_1)$.

The master equation has the structure of a balance equation[10] for $p_{1|1}$. The entering term in $p_{1|1}(x', t)$ involves the transition rate $W(x|x')$, whereas the leaving term in $p_{1|1}(x, t)$ involves the rate $W(x'|x)$. The master equation is an integro-differential equation of first order with respect to time. It enables us to determine the transition probability at any arbitrarily large time $t > 0$, and thus, in principle, to study the approach of equilibrium.

2.2. Time scales

From a physical point of view, a random process inducing transitions between the states of a system can be considered as a Markov process only over a time interval of

---

[8] The left-hand side of equation (9.2.6) involves a mathematical derivative. Actually, from a physical point of view, Δt cannot become arbitrarily small (see Subsection 2.2). Otherwise stated, the derivative $\partial p_{1|1}(x_3, t|x_1)/\partial t$ stands in fact for $[p_{1|1}(x_3, t + \Delta t|x_1) - p_{1|1}(x_3, t|x_1)]/\Delta t$.

[9] The quantity $p_{1|1}(x, t|x_1)$ is denoted simply by $p_{1|1}(x, t)$, the information about the initial condition being provided separately (formula (9.2.9)).

[10] Since the rates are themselves related to the transition probabilities (formula (9.2.2)), we are in principle in the presence –with the master equation as well as with the Chapman–Kolmogorov equation from which it is deduced– of an equation involving the sole transition probabilities. However, in practice, the master equation is used in a somewhat different way. Within a known physical context, we consider that the transition rates characterizing the behavior of $p_{1|1}$ at short times are data about which we can have independent information. The master equation then takes a meaning different from that of the Chapman–Kolmogorov equation, since this latter equation does not contain any specific information about the considered Markov process.

evolution much longer than the duration τ_c of a transition: $\Delta t \gg \tau_c$. These transitions being in general due to microscopic interactions, the modelization of a physical phenomenon by a Markov process can only be done over a time interval Δt much longer than the typical duration of such an interaction.

The existence of another characteristic time scale, much longer, must also be taken into account. Indeed, when we write the formula (9.2.2) relative to the short-time behavior of $p_{1|1}$, we are in fact interested in an evolution time interval much shorter than the typical time τ separating two microscopic interactions: $\Delta t \ll \tau$. The probability for two microscopic interactions to take place during the time interval Δt is then actually negligible.

The description of the evolution of $p_{1|1}$ in terms of a master equation thus only makes sense if the two time scales τ_c and τ are clearly separated, as pictured by the inequality:

$$\tau_c \ll \tau. \tag{9.2.10}$$

2.3. The evolution equation of p_1

We can deduce from the master equation an evolution equation for the one-time distribution. Taking the derivative with respect to t of the relation (9.1.4), rewritten in the form:

$$p_1(x,t) = \int p_{1|1}(x,t|x_1,0)p_1(x_1,0)\,dx_1, \qquad t > 0, \tag{9.2.11}$$

and re-expressing $\partial p_{1|1}(x,t|x_1,0)/\partial t$ by making use of the master equation (9.2.7), we deduce from equation (9.2.8) the equation:

$$\frac{\partial p_1(x,t)}{\partial t} = \iint \Big[W(x|x')p_{1|1}(x',t|x_1) - W(x'|x)p_{1|1}(x,t|x_1)\Big]p_1(x_1,t_1)\,dx'dx_1. \tag{9.2.12}$$

Carrying out the integration over x_1, we deduce from equation (9.2.12) the evolution equation of p_1:

$$\boxed{\frac{\partial p_1(x,t)}{\partial t} = \int \big[W(x|x')p_1(x',t) - W(x'|x)p_1(x,t)\big]\,dx'.} \tag{9.2.13}$$

Equation (9.2.13) for p_1 is formally identical to the master equation (9.2.8) for $p_{1|1}$. It is a linear first-order differential equation, from which we can deduce $p_1(x,t)$ from the initial distribution $p_1(x,0)$ if the transition rates[11] are known.

[11] The equation (9.2.13) for p_1 is often also called, by extension, a master equation. This denomination may however be somewhat confusing. The fact that the distribution $p_1(x,t)$ obeys equation (9.2.13) does not guarantee that the same is true for the transition probability $p_{1|1}(x,t)$ (and thus, that the process under consideration is Markovian).

3. The Pauli master equation

Consider a physical system which continuously passes from one microscopic state to another, due to the effect of microscopic interactions. In some cases, the resulting macroscopic evolution may be described by an equation for the average occupation probabilities of the states formally analogous to the balance equation (9.2.13). Such an equation may be deduced from the evolution equation of the distribution function or of the density operator of the system by means of convenient approximations.

The evolution equations of this type are generically called master equations. Historically, the first 'physical' master equation was obtained by W. Pauli in 1928.

3.1. Evolution equation of the average occupation probabilities

Consider a physical system likely to be found in a set of states $\{n\}$ with average probabilities $p_n(t)$. If the transition rates $W_{n',n}$ between states n and n' only depend on n and n', we can consider that the transitions between states are induced by a Markov process and write down the corresponding master equation, called the *Pauli master equation*:

$$\frac{dp_n(t)}{dt} = \sum_{n'} \big[W_{n,n'} p_{n'}(t) - W_{n',n} p_n(t) \big]. \tag{9.3.1}$$

Equation (9.3.1) is a balance equation for the average probabilities. It allows us in principle –if the transition rates are known– to deduce the probabilities $p_n(t)$ for $t > 0$ from their initial values.

The Pauli master equation seems to present a universal character. However, in the case of a quantum system, this equation is actually only valid under very peculiar circumstances. Consider for instance a system of Hamiltonian H_0 with eigenstates $\{|\phi_n\rangle\}$ of energies ε_n, between which transitions are induced by a perturbation Hamiltonian[12] λH_1 (the real coupling parameter λ measuring the strength of the perturbation). We denote by $\rho^I(t) = e^{iH_0 t/\hbar} \rho(t) e^{-iH_0 t/\hbar}$ the density operator of the system in the interaction picture (that is, in the Heisenberg picture with respect to the unperturbed Hamiltonian H_0). Under the effect of the pertubation λH_1, $\rho^I(t)$ evolves between times t and $t + \Delta t$ according to the equation:

$$\rho^I(t + \Delta t) = U^I(t + \Delta t, t) \rho^I(t) U^{I\dagger}(t + \Delta t, t), \tag{9.3.2}$$

where $U^I(t + \Delta t, t)$ is the elementary evolution operator $e^{iH_0 t/\hbar} e^{-i\lambda H_1 \Delta t/\hbar} e^{-iH_0 t/\hbar}$. From formula (9.3.2), coming back to the Schrödinger picture, and taking the limit $\Delta t \to 0$, we get the evolution equation of the diagonal element $\rho_{nn}(t)$ of the density matrix:

$$\frac{d\rho_{nn}(t)}{dt} = \lim_{\Delta t \to 0} \frac{1}{\Delta t} \Big\{ \sum_m \sum_p [U_{nm} \rho_{mp}(t) U^*_{pn}] - \rho_{nn}(t) \Big\}, \tag{9.3.3}$$

where $U(t)$ denotes the elementary evolution operator $e^{-i\lambda H_1 \Delta t/\hbar}$. Equation (9.3.3) does not reduce to a balance equation for the diagonal elements of the density matrix.

---

12 The Hamiltonian H_1 is assumed to have no diagonal elements on the eigenbase of H_0, a hypothesis which can always be made possibly through an appropriate redefinition of H_0.

The terms with $p \neq m$ give oscillating contributions which cannot be discarded without further assumptions. If it is possible to make, at any time, the *random phase assumption*, according to which the phases of the quantum coherences are distributed at random (which amounts to assuming that the density matrix is diagonal at any time), equation (9.3.3) takes the form of a balance equation involving solely the diagonal elements of $\rho(t)$:

$$\frac{d\rho_{nn}(t)}{dt} = \lim_{\Delta t \to 0} \frac{1}{\Delta t} \sum_m \left[|U_{nm}|^2 \rho_{mm}(t) - |U_{mn}|^2 \rho_{nn}(t) \right]. \tag{9.3.4}$$

If the energy spectrum is continuous or densely packed, we can introduce the transition rates $W_{n',n}$ as defined by:

$$W_{n',n} = \lim_{\Delta t \to 0} \frac{|U_{n'n}|^2}{\Delta t}, \tag{9.3.5}$$

and which may be expressed with the aid of the Fermi golden rule:

$$W_{n',n} = \frac{2\pi}{\hbar} \lambda^2 |\langle \phi_{n'} | H_1 | \phi_n \rangle|^2 \delta(\varepsilon_n - \varepsilon_{n'}). \tag{9.3.6}$$

We thus get for the diagonal elements of the density matrix an evolution equation of the form of the Pauli master equation:[13]

$$\boxed{\frac{d\rho_{nn}(t)}{dt} = \sum_{n'} \left[W_{n,n'} \rho_{n'n'}(t) - W_{n',n} \rho_{nn}(t) \right].} \tag{9.3.7}$$

3.2. Irreversibility

In contrast to the microscopic equations of motion, the Pauli master equation is not time-reversal invariant.[14] Thus it may conveniently describe the irreversible behavior of a macroscopic system. However, as shown by the general evolution equation (9.3.3), a quantum system obeys the Pauli master equation only if its density matrix may be considered as diagonal on the base $\{|\phi_n\rangle\}$. However, even if the density matrix is diagonal at a given time, it is no longer diagonal later on, due to the interaction λH_1.

The main difficulty we encounter when we try to establish the Pauli master equation is precisely the way of passing from the microscopic description, reversible, to a macroscopic one, irreversible. In the following, we will present a method in which the approximations allowing us to arrive to an irreversible description appear explicitly. In a first step, we write, via a second-order perturbation calculation, an evolution equation for the diagonal elements of the density matrix. This equation, reversible, called the *generalized master equation*, involves only the diagonal elements of $\rho(t)$. We then show how, owing to proper hypotheses (thermodynamic limit, short memory approximation), we can deduce from the generalized master equation (reversible), the Pauli master equation (irreversible).

---

[13] The diagonal elements $\rho_{nn}(t)$ of the density matrix have to be identified with the average occupation probabilities $p_n(t)$.

[14] Indeed, all its terms are real, and time is linearly involved in the first-order derivative.

4. The generalized master equation

We consider here a quantum system with a very large number of degrees of freedom, described by a Hamiltonian $H = H_0 + \lambda H_1$. For instance, in the case of a gas of interacting molecules, H_0 may be the ideal gas Hamiltonian, the term λH_1 representing the interaction between the molecules. Or, in the case of a system in contact with a thermostat, H_0 may be the Hamiltonian describing the global set of the uncoupled system and the thermostat, the term λH_1 representing the coupling. We aim at studying the statistical evolution of the system resulting from the transitions between the eigenstates of H_0 induced by the interaction λH_1.

4.1. Evolution equation of the diagonal elements of $\rho(t)$

We are interested in the evolution of the diagonal elements of the density matrix $\rho(t)$ on the eigenbase $\{|\phi_n\rangle\}$ of H_0. We assume that the density matrix is diagonal at time $t = 0$ (initial random phase hypothesis): at the initial time, the system can be found in the state $|\phi_n\rangle$ with the average probability $\rho_{nn}(0)$.

Carrying out a second-order perturbation expansion of the Liouville–von Neumann equation $i\hbar d\rho(t)/dt = [H, \rho(t)]$, we can show[15] that $\rho_{nn}(t)$ obeys an evolution equation of the form:

$$\frac{d\rho_{nn}(t)}{dt} = \lambda^2 \sum_{n'\neq n} \int_0^t dt'\, \Omega_{n,n'}(t-t')\big[\rho_{n'n'}(t') - \rho_{nn}(t')\big] + O(\lambda^3\rho), \tag{9.4.1}$$

where we have introduced the functions:

$$\Omega_{n,n'}(t) = \Omega_{n',n}(t) = \frac{2}{\hbar^2}|\langle\phi_{n'}|H_1|\phi_n\rangle|^2 \cos\left(\frac{\varepsilon_n - \varepsilon_{n'}}{\hbar}t\right). \tag{9.4.2}$$

Equation (9.4.1), which involves only the diagonal elements of the density matrix, is a *retarded* or *generalized*[16] master equation. It contains a *memory kernel* defined by the functions $\Omega_{n,n'}(t)$.

4.2. Time-reversal invariance

The generalized master equation does not allow us to describe an irreversible evolution. It is indeed time-reversal invariant. To display this invariance, let us change t into $-t$ in equation (9.4.1) (retaining on the right-hand side only the term in λ^2). We get:

$$-\frac{d\rho_{nn}(-t)}{dt} = \lambda^2 \sum_{n'\neq n} \int_0^{-t} dt'\, \Omega_{n,n'}(-t-t')\big[\rho_{n'n'}(t') - \rho_{nn}(t')\big]. \tag{9.4.3}$$

---

[15] The derivation of equation (9.4.1) is purely mathematical and does not rely on physical arguments. It will not be detailed here.

[16] The generalized master equation (9.4.1) is sometimes qualified as 'non-Markovian', owing to the fact that it involves the diagonal elements of the density matrix at times t' prior to t. Such a terminology is however ambiguous, since the generalized master equation does not contain enough information to allow us to determine whether the process it describes is, or is not, Markovian.

As displayed by formula (9.4.2), $\Omega_{nn'}(t)$ is an even function. We can therefore write:

$$-\frac{d\rho_{nn}(-t)}{dt} = \lambda^2 \sum_{n' \neq n} \int_0^{-t} dt' \, \Omega_{n,n'}(t+t') \big[\rho_{n'n'}(t') - \rho_{nn}(t')\big], \tag{9.4.4}$$

that is, setting $t' = -t''$:

$$\frac{d\rho_{nn}(-t)}{dt} = \lambda^2 \sum_{n' \neq n} \int_0^{t} dt'' \, \Omega_{n,n'}(t-t'') \big[\rho_{n'n'}(-t'') - \rho_{nn}(-t'')\big]. \tag{9.4.5}$$

Consequently, if a set of diagonal elements $\{\rho_{nn}(t)\}$ is a solution of equation (9.4.1), the set of elements $\{\rho_{nn}(-t)\}$ is also a solution of this equation. The generalized master equation is thus time-reversal invariant. It does not allow us to account for an irreversible evolution.

5. From the generalized master equation to the Pauli master equation

5.1. Passage to an instantaneous master equation

In the weak-coupling limit, provided that t is sufficiently small as compared to the characteristic evolution time τ of the diagonal elements of the density matrix (a time which, according to equation (9.4.1), is of order λ^{-2}), we can discard the evolution of ρ between times t' and t, and replace the diagonal elements of the density matrix on the right-hand side of equation (9.4.1) by their values at time t. Otherwise stated, for $t \ll \tau$, we can rewrite equation (9.4.1) in a non-retarded form:

$$\frac{d\rho_{nn}(t)}{dt} = \lambda^2 \sum_{n' \neq n} \left[\int_0^{t} dt' \, \Omega_{nn'}(t')\right] \big[\rho_{n'n'}(t) - \rho_{nn}(t)\big]. \tag{9.5.1}$$

We have:

$$\int_0^t dt' \, \Omega_{nn'}(t') = \frac{2}{\hbar^2} |\langle \phi_{n'}|H_1|\phi_n\rangle|^2 \frac{\hbar}{\varepsilon_n - \varepsilon_{n'}} \sin\left(\frac{\varepsilon_n - \varepsilon_{n'}}{\hbar} t\right). \tag{9.5.2}$$

In equation (9.5.1), the coefficients of the diagonal elements of the density matrix are time-dependent. Besides, this equation remains invariant under time-reversal. Both features differentiate it from the Pauli master equation (9.3.6). However, in the case of a system with a very large number of degrees of freedom, we can, by means of an approximation about the form of the memory kernel, obtain, starting from (9.5.1), the Pauli master equation.

5.2. The thermodynamic limit

In a system with a large number of degrees of freedom, the unperturbed eigenstates depend in general on several quantum numbers. We will label every state of energy ε by the pair (ε, α). The matrix elements of the perturbation and the diagonal elements of the density matrix will be denoted respectively by $H_1(\varepsilon, \alpha; \varepsilon', \alpha')$ and $\rho(\varepsilon, \alpha; t)$. With these notations, and on account of formula (9.5.2), equation (9.5.1) may be rewritten in the form:

$$\frac{d\rho(\varepsilon,\alpha;t)}{dt} = \lambda^2 \sum_{\alpha'} \sum_{\varepsilon'} \frac{2}{\hbar^2} |H(\varepsilon,\alpha;\varepsilon',\alpha')|^2 \, \frac{\hbar}{\varepsilon-\varepsilon'} \sin\left(\frac{\varepsilon-\varepsilon'}{\hbar}t\right) \left[\rho(\varepsilon',\alpha';t) - \rho(\varepsilon,\alpha;t)\right]. \tag{9.5.3}$$

In the thermodynamic limit in which the size of the system tends towards infinity, we can introduce the density of states in energy $n(\varepsilon)$ and replace the discrete sum over ε' involved on the right-hand side of equation (9.5.3) by an integral over the energy. We can then write for any time t (bounded however by τ):

$$\frac{\partial\rho(\varepsilon,\alpha;t)}{\partial t} = \lambda^2 \sum_{\alpha'} \int d\varepsilon' \, n(\varepsilon') \frac{2}{\hbar^2} |H(\varepsilon,\alpha;\varepsilon',\alpha')|^2 \times \frac{\hbar}{\varepsilon-\varepsilon'} \sin\left(\frac{\varepsilon-\varepsilon'}{\hbar}t\right) \left[\rho(\varepsilon',\alpha';t) - \rho(\varepsilon,\alpha;t)\right]. \tag{9.5.4}$$

Note that, at this stage, the evolution equation of the diagonal elements of the density matrix is still reversible.

5.3. Short memory approximation

Let us introduce the microscopic time scale $\tau_c = \hbar/\Delta$, where Δ characterizes the energy width of the function $f(\varepsilon') = \sum_{\alpha'} n(\varepsilon')|H(\varepsilon,\alpha;\varepsilon',\alpha')|^2$. For $t \gg \tau_c$, the function $f(\varepsilon')$ varies much more slowly than the function $[\hbar/(\varepsilon-\varepsilon')]\sin[(\varepsilon-\varepsilon')t/\hbar]$. This latter function may thus be considered in this limit as a delta function of weight $\pi\hbar$ centered[17] at ε. Equation (9.5.4) then reads:

$$\frac{d\rho(\varepsilon,\alpha;t)}{dt} = \lambda^2 \sum_{\alpha'} \int d\varepsilon' \, n(\varepsilon') \frac{2\pi}{\hbar} \delta(\varepsilon-\varepsilon') |H(\varepsilon,\alpha;\varepsilon',\alpha')|^2 \left[\rho(\varepsilon',\alpha';t) - \rho(\varepsilon,\alpha;t)\right]. \tag{9.5.5}$$

The integration over ε' gives:

$$\frac{\partial\rho(\varepsilon,\alpha;t)}{\partial t} = \sum_{\alpha'} \left[W(\alpha,\alpha')\rho(\varepsilon,\alpha';t) - W(\alpha',\alpha)\rho(\varepsilon,\alpha;t)\right]. \tag{9.5.6}$$

In equation (9.5.6), we have set:

$$W(\alpha',\alpha) = W(\alpha,\alpha') = \frac{2\pi}{\hbar}\lambda^2 |H(\varepsilon,\alpha';\varepsilon,\alpha)|^2 n(\varepsilon). \tag{9.5.7}$$

[17] We make use of the formula:

$$\lim_{t\to\infty} \frac{\hbar}{\varepsilon-\varepsilon'} \sin\left(\frac{\varepsilon-\varepsilon'}{\hbar}t\right) = \pi\hbar\delta(\varepsilon-\varepsilon').$$

Equation (9.5.6), with the transition rates (9.5.7), is the Pauli master equation.

In contrast to the generalized master equation, the Pauli master equation is not time-reversal invariant. It describes an irreversible evolution of the system. An irreversible evolution equation is thus obtained as soon as we make the hypothesis $t \gg \tau_c$, called the *short memory approximation*. It is only then that the evolution of the diagonal elements of the density matrix becomes analogous to that of the probabilities $p_n(t)$ of a Markov process.

6. Discussion

6.1. Validity domain

Let us come back on the hypotheses needed to establish the Pauli master equation. Amongst these, the hypotheses made to obtain the generalized master equation (9.4.1) (perturbation calculation, initial random phase hypothesis) are made first. Then, we have the passage to the thermodynamic limit, followed by the passage to the limit $t \gg \tau_c$ (short memory approximation), the order in which these two limits are taken being crucial. When the coupling parameter λ is finite, the Pauli master equation is only valid for times $t \ll \tau$, where $\tau = O(\lambda^{-2})$. For the Pauli master equation to be valid at any time t, we have to take the limit $(\lambda \to 0,\ t \to \infty)$, the product $\lambda^2 t$ remaining finite.

As in the case of the master equations governing the transition probabilities of Markovian processes, we again encounter, for the derivation of the Pauli master equation, the need for the existence of two well-separated time scales. The short time scale τ_c characterizes the duration of a microscopic interaction inducing a transition from one state to another. The large time scale τ corresponds to the typical time separating two microscopic interactions, and depends on the coupling strength. The description of the evolution in terms of a master equation is only possible if $\tau_c \ll \tau$. It is then valid for times t such that $\tau_c \ll t \ll \tau$.

6.2. Coarse-grained description

It remains to determine in what conditions and under what hypotheses the Pauli master equation, established for times between τ_c and τ, may be used at arbitrarily large times, and in particular to describe the approach of equilibrium (which typically takes place over durations of the order of several τ).

The Pauli master equation is a linear first-order differential equation with constant coefficients. Thus, the density matrix being assumed to be diagonal at time t, the Pauli equation allows us, through a perturbative calculation, to determine the evolution of its diagonal elements up to times of order $t + \Delta t$, with $\tau_c \ll \Delta t \ll \tau$. If, again, we assume that the density matrix is diagonal at time $t + \Delta t$, we can, in the same way, use the master equation to obtain its diagonal elements at time $t + 2\Delta t$, and so on. The Pauli master equation allows us in this way to obtain the density matrix at any time,[18] provided that we work using a *coarse-grained* time scale, that is, a

[18] The derivative $\partial \rho_{nn}(t)/\partial t$ then represents in fact $(1/\Delta t)[\rho_{nn}(t + \Delta t) - \rho_{nn}(t)]$.

time resolution limited by Δt. We can, in this way, make use of the Pauli master equation for times t much larger than τ, and thus in particular to study the approach of equilibrium of a macroscopic system.

6.3. The van Hove limit

The conditions necessary to the derivation of the Pauli master equation were re-examined, in particular by L. van Hove in 1955 and I. Prigogine in 1962, with the aim of getting rid of the repeated initial random phase assumption.

Van Hove considers the evolution operator $\exp(-iHt/\hbar)$, where $H = H_0 + \lambda H_1$. A perturbation expansion truncated after some orders in λ is only valid at very short times. To describe the approach of equilibrium, we have to study times at least of the order of the typical time interval separating two microscopic interactions. This latter time interval is proportional to λ^{-2} if the individual interactions are described in the Born approximation. This suggests that λ should be considered as small and t as large, the product $\lambda^2 t$ remaining finite. We thus keep the terms in powers of $\lambda^2 t$ and neglect those of the type $\lambda^m t^n$ with $m \neq 2n$. Using these concepts, van Hove has been able to show that the sum of the relevant terms in the evolution operator leads to a quantity whose temporal dependence is governed by the Pauli master equation.

Bibliography

C. COHEN-TANNOUDJI, J. DUPONT-ROC, and G. GRYNBERG, *Atom–photon interactions*, Wiley, New York, 1992.

W. FELLER, *An introduction to probability theory and its applications*, Vol. 1, Wiley, New York, third edition, 1968; Vol. 2, Wiley, New York, second edition, 1971.

C.W. GARDINER, *Handbook of stochastic methods*, Springer-Verlag, Berlin, third edition, 2004.

N.G. VAN KAMPEN, *Stochastic processes in physics and chemistry*, North-Holland, Amsterdam, third edition, 2007.

H.J. KREUZER, *Nonequilibrium thermodynamics and its statistical foundations*, Clarendon Press, Oxford, 1981.

R. KUBO, M. TODA, and N. HASHITSUME, *Statistical physics II: nonequilibrium statistical mechanics*, Springer-Verlag, Berlin, second edition, 1991.

L. MANDEL and E. WOLF, *Optical coherence and quantum optics*, Cambridge University Press, Cambridge, 1995.

R.M. MAZO, *Brownian motion: fluctuations, dynamics, and applications*, Oxford University Press, Oxford, 2002.

A. PAPOULIS, *Probability, random variables, and stochastic processes*, McGraw-Hill, New York, 1984.

I. PRIGOGINE, *Non-equilibrium statistical mechanics*, Interscience Publishers, New York, 1962.

F. REIF, *Fundamentals of statistical and thermal physics*, McGraw-Hill, New York, 1965.

D. ZUBAREV, V. MOROZOV, and G. RÖPKE, *Statistical mechanics of nonequilibrium processes,* Vol. 2: *Relaxation and hydrodynamic processes*, Akademie Verlag, Berlin, 1996.

R. ZWANZIG, *Nonequilibrium statistical mechanics*, Oxford University Press, Oxford, 2001.

References

W. PAULI, in *Probleme der Modernen Physik*, S. Hirzel, Leipzig, 1928. Reprinted in *Collected scientific papers by W. Pauli* (R. KRONIG and V.F. WEISSKOPF editors), Interscience, New York, 1964.

L. VAN HOVE, Quantum-mechanical perturbations giving rise to a statistical transport equation, *Physica* **21**, 517 (1955).

L. VAN HOVE, The approach to equilibrium in quantum statistics. A perturbation treatment to general order, *Physica* **23**, 441 (1957).

R. ZWANZIG, Statistical mechanics of irreversibility, in *Lectures in theoretical physics*, Vol. 3 (W.E. BRITTIN, B.W. DOWNS, and J. DOWNS editors), Interscience, New York, 1961.

R. ZWANZIG, On the identity of three generalized master equations, *Physica* **30**, 1109 (1964).

I. OPPENHEIM and K.E. SHULER, Master equations and Markov processes, *Phys. Rev. B* **138**, 1007 (1965).

O. PENROSE, Foundations of statistical mechanics, *Rep. Prog. Phys.* **42**, 129 (1979).

Chapter 10

Brownian motion: the Langevin model

In 1827, the botanist R. Brown discovered under the microscope the incessant and irregular motion of small pollen particles suspended in water. He also remarked that small mineral particles behave exactly in the same way (such an observation is important, since it precludes to attributing this phenomenon to some 'vital force' specific to biological objects). In a general way, a particle in suspension in a fluid executes a Brownian motion when its mass is much larger than the mass of one of the fluid's molecules.

The idea according to which the motion of a Brownian particle is a consequence of the motion of the lighter molecules of the surrounding fluid became widespread during the second half of the nineteenth century. The first theoretical explanation of this phenomenon was given by A. Einstein in 1905. The direct experimental checking of the Einstein's theory led to the foundation of the atomic theory of matter (in particular the measurement of the Avogadro's number by J. Perrin in 1908). A more achieved theory of Brownian motion was proposed by P. Langevin in 1908.

However, slightly before A. Einstein, and in a completely different context, L. Bachelier had already obtained the law of Brownian motion in his thesis entitled "La théorie de la spéculation" (1900). Models having recourse to Brownian motion or to its generalizations are widely used nowadays in financial mathematics. In a more general setting, Brownian motion played an important role in mathematics: historically, it was to represent the displacement of a Brownian particle that a stochastic process was constructed for the first time (N. Wiener, 1923).

The outstanding importance of Brownian motion in out-of-equilibrium statistical physics stems from the fact that the concepts and methods used in its study are not restricted to the description of the motion of a particle immersed in a fluid of lighter molecules, but are general and may be applied to a wide class of physical phenomena.

1. The Langevin model

Brownian motion is the complicated motion, of an erratic type, carried out by a 'heavy'[1] particle immersed in a fluid under the effect of the collisions it undergoes with the molecules of this fluid.

The first theoretical explanations of Brownian motion were given, independently, by A. Einstein in 1905 and M. Smoluchowski in 1906. In these first models, the inertia of the Brownian particle was not taken into account. A more elaborate description of Brownian motion, accounting for the effects of inertia, was proposed by P. Langevin in 1908. This latter theory will be presented here first.

1.1. The Langevin equation

The Langevin model is a classical phenomenological model. Reasoning for the sake of simplicity in one dimension, we associate with the Brownian particle's position a coordinate x. Two forces, both characterizing the effect of the fluid, act on the particle of mass m: a viscous friction force $-m\gamma(dx/dt)$, characterized by the friction coefficient $\gamma > 0$, and a fluctuating force $F(t)$, representing the unceasing impacts of the fluid's molecules on the particle. The fluctuating force, assumed to be independent of the particle's velocity, is considered as an external force, called the *Langevin force*.

In the absence of a potential, the Brownian particle is said to be 'free'. Its equation of motion, the *Langevin equation*, reads:

$$m\frac{d^2x}{dt^2} = -m\gamma\frac{dx}{dt} + F(t), \tag{10.1.1}$$

or:

$$\boxed{m\frac{dv}{dt} = -m\gamma v + F(t), \qquad v = \frac{dx}{dt}\cdot} \tag{10.1.2}$$

The Langevin equation is historically the first example of a stochastic differential equation, that is, a differential equation involving a random term $F(t)$ with specified statistical properties. The solution $v(t)$ of equation (10.1.2) for a given initial condition is itself a stochastic process.

In the Langevin model, the friction force $-m\gamma v$ and the fluctuating force $F(t)$ represent two consequences of the same physical phenomenon (namely, the collisions of the Brownian particle with the fluid's molecules). To fully define the model, we have to characterize the statistical properties of the random force.

1.2. Hypotheses concerning the Langevin force

The fluid, also called the *bath*, is supposed to be in a stationary state.[2] As regards the bath, no instant plays a privilegiate role. Accordingly, the fluctuating force acting on

[1] We understand here by 'heavy' a particle with a mass much larger than that of one of the fluid's molecules.

[2] Most often, it will be considered that the bath is in thermodynamic equilibrium.

the Brownian particle is conveniently modelized by a stationary random process. As a result, the one-time average[3] $\langle F(t) \rangle$ does not depend on t and the two-time average $\langle F(t)F(t') \rangle$ depends only on the time difference $t - t'$.

Besides these minimal characteristics, the Langevin model requires some supplementary hypotheses about the random force.

• *Average value*

We assume that the average value of the Langevin force vanishes:

$$\langle F(t) \rangle = 0. \tag{10.1.3}$$

This hypothesis is necessary to have the average value of the Brownian particle's velocity vanishing at equilibrium (as it should, since there is no applied external force).

• *Autocorrelation function*

The autocorrelation function of the random force,

$$g(\tau) = \langle F(t)F(t+\tau) \rangle, \tag{10.1.4}$$

is an even function of τ, decreasing over a characteristic time τ_c (correlation time). We set:

$$\int_{-\infty}^{\infty} g(\tau)\, d\tau = 2\mathcal{D}m^2 \tag{10.1.5}$$

(the signification of the parameter $\mathcal{D}$ will be made precise later). The correlation time is of the order of the mean time interval separating two successive collisions of the fluid's molecules on the Brownian particle. If this time is much shorter than the other characteristic times, such as for instance the relaxation time of the average velocity from a well-defined initial value,[4] we can assimilate $g(\tau)$ to a delta function of weight $2\mathcal{D}m^2$:

$$g(\tau) = 2\mathcal{D}m^2\delta(\tau). \tag{10.1.6}$$

• *Gaussian character of the Langevin force*

Most often, we also assume for convenience that $F(t)$ is a Gaussian process. All the statistical properties of the Langevin force are then calculable given only its average and its autocorrelation function.[5]

---

[3] The averages taking place here are defined as ensemble averages computed with the aid of the distribution function of the bath (see Supplement 10B).

[4] See Subsection 2.2.

[5] This hypothesis may be justified on account of the central limit theorem: indeed, due to the numerous collisions undergone by the Brownian particle, the force $F(t)$ may be considered as resulting from the superposition of a very large number of identically distributed random functions.

2. Response and relaxation

The Langevin equation is a stochastic linear differential equation. This linearity enables us to compute exactly the average response and relaxation properties of the Brownian particle.

2.1. Response to an external perturbation: mobility

Assume that an external time-dependent applied force, independent of the coordinate, is exerted on the particle. This force $F_{\text{ext}}(t)$ adds to the random force $F(t)$. The equation of motion of the Brownian particle then reads:

$$m\frac{dv}{dt} = -m\gamma v + F(t) + F_{\text{ext}}(t), \qquad v = \frac{dx}{dt}. \tag{10.2.1}$$

On average, we have:

$$m\frac{d\langle v\rangle}{dt} = -m\gamma\langle v\rangle + F_{\text{ext}}(t), \qquad \langle v\rangle = \frac{d\langle x\rangle}{dt}. \tag{10.2.2}$$

For a harmonic applied force $F_{\text{ext}}(t) = \Re e(Fe^{-i\omega t})$, the solution of equation (10.2.2) is, in stationary regime, of the form:

$$\langle v(t)\rangle = \Re e\big(\langle v\rangle e^{-i\omega t}\big). \tag{10.2.3}$$

We have:

$$\langle v\rangle = \mathcal{A}(\omega)F, \tag{10.2.4}$$

where the quantity:

$$\boxed{\mathcal{A}(\omega) = \frac{1}{m}\frac{1}{\gamma - i\omega}} \tag{10.2.5}$$

is the complex admittance of the Langevin model.

More generally, for an external force $F_{\text{ext}}(t)$ of Fourier transform[6] $F_{\text{ext}}(\omega)$, the stationary solution $\langle v(t)\rangle$ of equation (10.2.2) has the Fourier transform:

$$\langle v(\omega)\rangle = \mathcal{A}(\omega)F_{\text{ext}}(\omega). \tag{10.2.6}$$

The average velocity of the Brownian particle responds linearly to the external applied force. We can associate with this response a transport coefficient. The Brownian particle, if it carries a charge q, acquires under the effect of a static electric field E the limit velocity $\langle v\rangle = qE/m\gamma$. Its mobility $\mu = \langle v\rangle/E$ is thus:[7]

$$\mu = \frac{q}{m\gamma} = q\mathcal{A}(\omega = 0). \tag{10.2.7}$$

---

[6] For the sake of simplicity, we use the same notation $F_{\text{ext}}(.)$ for the force $F_{\text{ext}}(t)$ and its Fourier transform $F_{\text{ext}}(\omega)$, as well as the same notation $\langle v(.)\rangle$ for the average velocity $\langle v(t)\rangle$ and its Fourier transform $\langle v(\omega)\rangle$.

[7] No confusion being possible here with a chemical potential, the drift mobility of the Brownian particle is simply denoted by μ (and not by μ_D).

2.2. Evolution of the velocity from a well-defined initial value

Assume now that there is no applied external force, and that at time $t = 0$ the Brownian particle's velocity has a well-defined value, non-random, denoted by v_0:

$$v(0) = v_0. \tag{10.2.8}$$

The solution of equation (10.1.2) corresponding to the initial condition (10.2.8) reads:

$$v(t) = v_0 e^{-\gamma t} + \frac{1}{m} \int_0^t F(t') e^{-\gamma(t-t')}\, dt', \qquad t > 0. \tag{10.2.9}$$

The velocity $v(t)$ of the Brownian particle is a random process. In the above defined conditions, this process is not stationary. We will compute the average value and the variance of $v(t)$ at any time $t > 0$.

- *Average velocity*

Since the fluctuating force vanishes on average, we obtain, from formula (10.2.9):

$$\langle v(t) \rangle = v_0 e^{-\gamma t}, \qquad t > 0. \tag{10.2.10}$$

The average velocity relaxes exponentially towards zero with a relaxation time $\tau_r = \gamma^{-1}$.

- *Velocity variance*

The variance of the velocity is defined for instance by the formula:

$$\sigma_v^2(t) = \langle [v(t) - \langle v(t) \rangle]^2 \rangle. \tag{10.2.11}$$

We get, from formulas (10.2.9) and (10.2.10):

$$\sigma_v^2(t) = \frac{1}{m^2} \int_0^t dt' \int_0^t dt'' \langle F(t') F(t'') \rangle e^{-\gamma(t-t')} e^{-\gamma(t-t'')}. \tag{10.2.12}$$

When the autocorrelation function of the Langevin force is given by the simplified formula (10.1.6), we obtain:

$$\sigma_v^2(t) = 2\mathcal{D} \int_0^t e^{-2\gamma(t-t')}\, dt', \tag{10.2.13}$$

that is:

$$\sigma_v^2(t) = \frac{\mathcal{D}}{\gamma}(1 - e^{-2\gamma t}), \qquad t > 0. \tag{10.2.14}$$

At time $t = 0$, the variance of the velocity vanishes (the initial velocity is a non-random variable). Under the effect of the Langevin force, velocity fluctuations arise, and the variance $\sigma_v^2(t)$ increases with time. At first, this increase is linear:

$$\sigma_v^2(t) \simeq 2\mathcal{D}t, \qquad t \ll \tau_r. \tag{10.2.15}$$

We can interpret formula (10.2.15) as describing a phenomenon of diffusion in the velocity space. The parameter $\mathcal{D}$, which has been introduced in the definition of $g(\tau)$ (formula (10.1.6)), takes the meaning of a *diffusion coefficient in the velocity space*. The variance of the velocity does not however increase indefinitely, but ends up saturating at the value $\mathcal{D}/\gamma$:

$$\sigma_v^2(t) \simeq \frac{\mathcal{D}}{\gamma}, \qquad t \gg \tau_r. \tag{10.2.16}$$

2.3. Second fluctuation-dissipation theorem

We can also write the variance of the velocity in the form:

$$\sigma_v^2(t) = \langle v^2(t) \rangle - \langle v(t) \rangle^2. \tag{10.2.17}$$

For $t \gg \tau_r$, the average velocity tends towards zero (formula (10.2.10)). Equations (10.2.16) and (10.2.17) show that $\langle v^2(t) \rangle$ then tends towards a limit value $\mathcal{D}/\gamma$ independent of v_0. The average energy $\langle E(t) \rangle = m\langle v^2(t) \rangle/2$ tends towards the corresponding limit $\langle E \rangle = m\mathcal{D}/2\gamma$. Then, the Brownian particle is in equilibrium with the bath.

If the bath is itself in thermodynamic equilibrium at temperature T, the average energy of the particle in equilibrium with it takes its equipartition value $\langle E \rangle = kT/2$. Comparing both expressions for $\langle E \rangle$, we get a relation between the diffusion coefficient $\mathcal{D}$ in the velocity space, associated with the velocity fluctuations, and the friction coefficient γ, which characterizes the dissipation:

$$\boxed{\gamma = \frac{m}{kT}\mathcal{D}.} \tag{10.2.18}$$

Using formula (10.1.5), we can rewrite equation (10.2.18) in the form:[8]

$$\boxed{\gamma = \frac{1}{2mkT} \int_{-\infty}^{\infty} \langle F(t)F(t+\tau) \rangle \, d\tau.} \tag{10.2.19}$$

Equation (10.2.19) relates the friction coefficient to the autocorrelation function of the Langevin force. It is known as the *second fluctuation-dissipation theorem*.[9] This theorem expresses here the fact that the friction force and the fluctuating force represent

---

[8] It will be shown in Subsection 4.3 that this relation can be extended to the case in which the autocorrelation function of the Langevin force is not a delta function but a function of finite width characterized by the correlation time τ_c, provided that we have $\tau_c \ll \tau_r$. See also on this question Supplement 10A.

[9] Generally speaking, the fluctuation-dissipation theorem, which can be formulated in different ways, constitutes the heart of the linear response theory (see Chapter 14). In the case of Brownian motion as described by the Langevin model, the terminology of 'second' fluctuation-dissipation theorem, associated with formula (10.2.19) for the integral of the random force autocorrelation function, is due to R. Kubo.

two aspects of the same physical phenomenon, namely, the collisions of the Brownian particle with the molecules of the fluid which surrounds it.

2.4. Evolution of the displacement from a well-defined initial position: diffusion of the Brownian particle

Assume that at time $t = 0$ the particle's coordinate has a well-defined value:

$$x(0) = x_0. \tag{10.2.20}$$

Integrating the expression (10.2.9) for the velocity between times 0 and t, we get, given the initial condition (10.2.20):

$$x(t) = x_0 + \frac{v_0}{\gamma}(1 - e^{-\gamma t}) + \frac{1}{m}\int_0^t \frac{1 - e^{-\gamma(t-t')}}{\gamma} F(t')\, dt', \qquad t > 0. \tag{10.2.21}$$

The displacement $x(t) - x_0$ of the Brownian particle is also a random process. This process is not stationary. We will calculate the average and the variance of the displacement as functions of time, as well as the second moment $\langle [x(t) - x_0]^2 \rangle$.

• *Average displacement*

We have:

$$\langle x(t) \rangle = x_0 + \frac{v_0}{\gamma}(1 - e^{-\gamma t}), \qquad t > 0. \tag{10.2.22}$$

For $t \gg \tau_r$, the average displacement $\langle x(t) \rangle - x_0$ tends towards the finite limit v_0/γ.

• *Variance of the displacement*

The variance of the displacement $x(t) - x_0$ is also the variance of $x(t)$, defined for instance by the formula:

$$\sigma_x^2(t) = \langle [x(t) - \langle x(t) \rangle]^2 \rangle. \tag{10.2.23}$$

From formulas (10.2.21) and (10.2.22), we get:

$$\sigma_x^2(t) = \frac{1}{m^2\gamma^2}\int_0^t dt' \int_0^t dt'' \, \langle F(t')F(t'') \rangle [1 - e^{-\gamma(t-t')}][1 - e^{-\gamma(t-t'')}], \tag{10.2.24}$$

that is, taking for the autocorrelation function of the Langevin force the simplified expression (10.1.6):

$$\sigma_x^2(t) = \frac{2\mathcal{D}}{\gamma^2}\int_0^t (1 - e^{-\gamma t'})^2 \, dt'. \tag{10.2.25}$$

When the integration is carried out, we get:

$$\sigma_x^2(t) = \frac{2\mathcal{D}}{\gamma^2}\left(t - 2\frac{1 - e^{-\gamma t}}{\gamma} + \frac{1 - e^{-2\gamma t}}{2\gamma}\right), \qquad t > 0. \tag{10.2.26}$$

Starting from its vanishing initial value, the variance of the displacement increases, first as $2\mathcal{D}t^3/3$ for $t \ll \tau_r$, then as $2\mathcal{D}t/\gamma^2$ for $t \gg \tau_r$.

On the other hand, since $x(t) - x_0 = x(t) - \langle x(t)\rangle + \langle x(t)\rangle - x_0$, we have:

$$\langle [x(t) - x_0]^2 \rangle = \sigma_x^2(t) + \frac{v_0^2}{\gamma^2}(1 - e^{-\gamma t})^2, \qquad t > 0. \tag{10.2.27}$$

For $t \gg \tau_r$, we therefore have:

$$\langle [x(t) - x_0]^2 \rangle \simeq 2\frac{\mathcal{D}}{\gamma^2}t. \tag{10.2.28}$$

Formulas (10.2.26) and (10.2.28) show that the Brownian particle diffuses at large times. The diffusion coefficient D is related to the diffusion coefficient in the velocity space $\mathcal{D}$ by the formula:

$$\boxed{D = \frac{\mathcal{D}}{\gamma^2}.} \tag{10.2.29}$$

2.5. Viscous limit

In the first theories of Brownian motion, proposed by A. Einstein in 1905 and M. Smoluchowski in 1906, the diffusive behavior of the Brownian particle was obtained in a simpler way. In this approach, we consider a unique dynamical variable, the particle's displacement. We do not take into account the inertia term in the equation of motion, which we write in the following approximate form:

$$\boxed{\eta\frac{dx}{dt} = F(t).} \tag{10.2.30}$$

The autocorrelation function of the random force is written in the form:

$$\langle F(t)F(t')\rangle = 2D\eta^2\delta(t - t'). \tag{10.2.31}$$

Equation (10.2.30), supplemented by equation (10.2.31), describes Brownian motion in the *viscous limit*, in which the friction is strong enough so that the inertia term may be neglected.[10] The Brownian motion is then said to be *overdamped*. This description, valid for sufficiently large evolution time intervals, corresponds well to the experimental observations of J. Perrin in 1908.

In the viscous limit, the displacement of the Brownian particle can be directly obtained by integrating equation (10.2.30). With the initial condition (10.2.20), it reads:[11]

$$x(t) - x_0 = \frac{1}{\eta}\int_0^t F(t')\,dt'. \tag{10.2.32}$$

---

[10] More precisely, equation (10.2.30) can be deduced from the Langevin equation (10.1.2) in the limit $m \to 0$, $\gamma \to \infty$, the viscosity coefficient $\eta = m\gamma$ remaining finite. Equation (10.2.31) corresponds to equation (10.1.6), written in terms of the relevant parameters D and η.

[11] When the force $F(t)$ is modelized by a random stationary Gaussian process of autocorrelation function $g(\tau) = 2D\eta^2\delta(\tau)$, the process $x(t) - x_0$ defined by formula (10.2.32) is called the *Wiener process* (see Chapter 11).

Using the expression (10.2.31) for the random force autocorrelation function, we get, for any t:

$$\langle [x(t) - x_0]^2 \rangle = 2Dt. \tag{10.2.33}$$

In this description, the motion of the Brownian particle is diffusive at any time.

2.6. The Einstein relation

From formulas (10.2.7) and (10.2.29), we obtain a relation between the mobility and the diffusion coefficient of the Brownian particle,

$$\frac{D}{\mu} = \frac{m\mathcal{D}}{q\gamma}, \tag{10.2.34}$$

which also reads, on account of the second fluctuation-dissipation theorem (10.2.18):

$$\boxed{\frac{D}{\mu} = \frac{kT}{q}.} \tag{10.2.35}$$

Formula (10.2.35) is the Einstein relation between the diffusion coefficient D, associated with the displacement fluctuations, and the mobility μ, related to the dissipation. The Einstein relation is a formulation of the *first fluctuation-dissipation theorem.*[12] It may also be written in the form of a relation between D and η:

$$\boxed{D = \frac{kT}{\eta}.} \tag{10.2.36}$$

3. Equilibrium velocity fluctuations

We are interested here in the dynamics of the velocity fluctuations of a Brownian particle in equilibrium with the bath. We assume, as previously, that the latter is in thermodynamic equilibrium at temperature T.

To obtain the expression for the velocity of the Brownian particle at equilibrium, we first write the solution $v(t)$ of the Langevin equation for the initial condition[13] $v(t_0) = v_0$:

$$v(t) = v_0 e^{-\gamma(t-t_0)} + \frac{1}{m}\int_{t_0}^{t} F(t') e^{-\gamma(t-t')}\, dt'. \tag{10.3.1}$$

We then take the limit $t_0 \to -\infty$. As shown by formula (10.3.1), the initial value of the velocity is 'forgotten' and $v(t)$ reads:

$$v(t) = \frac{1}{m}\int_{-\infty}^{t} F(t') e^{-\gamma(t-t')}\, dt'. \tag{10.3.2}$$

[12] This theorem will be established in a more general way in Subsection 3.4.

[13] Equation (10.3.1) is thus the generalization of equation (10.2.9) to any initial time t_0.

In these conditions, at any finite time t, the particle is in equilibrium with the bath. Its velocity $v(t)$ is a stationary random process.[14] Since the average value of the velocity vanishes at equilibrium, the autocorrelation function of $v(t)$, which we will now compute, represents the dynamics of the equilibrium velocity fluctuations.

3.1. Correlation function between the Langevin force and the velocity

To begin with, starting from formula (10.3.2), it is possible to compute the correlation function $\langle v(t)F(t')\rangle$:

$$\langle v(t)F(t')\rangle = \frac{1}{m}\int_{-\infty}^{t} \langle F(t'')F(t')\rangle e^{-\gamma(t-t'')}\,dt''. \tag{10.3.3}$$

When the autocorrelation function of the Langevin force is of the form (10.1.6), equation (10.3.3) reads:

$$\langle v(t)F(t')\rangle = 2\mathcal{D}m\int_{-\infty}^{t} \delta(t'-t'')e^{-\gamma(t-t'')}\,dt''. \tag{10.3.4}$$

From formula (10.3.4), we get:

$$\langle v(t)F(t')\rangle = \begin{cases} 2\mathcal{D}me^{-\gamma(t-t')}, & t' < t \\ 0, & t' > t. \end{cases} \tag{10.3.5}$$

Formula (10.3.5) displays the fact that the Brownian particle velocity at time t is not correlated with the Langevin force at a subsequent time $t' > t$.

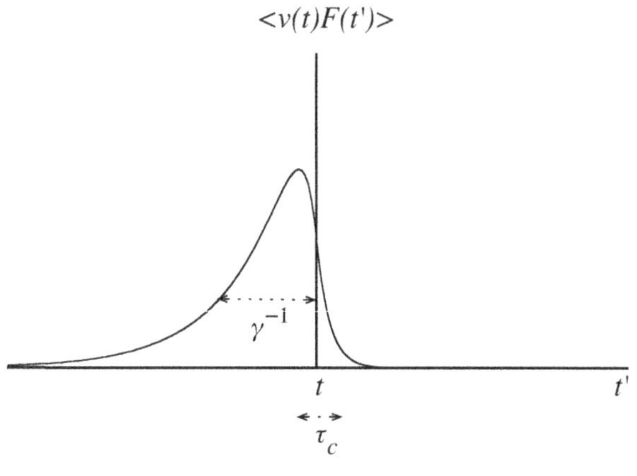

Fig. 10.1 Correlation function between the Langevin force and the velocity at finite τ_c.

---

[14] When the force $F(t)$ is modelized by a stationary Gaussian random process of autocorrelation function $g(\tau) = 2\mathcal{D}m^2\delta(\tau)$, the stationary process $v(t)$ defined by formula (10.3.2) is called the *Ornstein–Uhlenbeck process* (see Chapter 11 and Supplement 11B).

Actually, the correlation time τ_c of the Langevin force does not vanish, and the result (10.3.5) is correct only for $|t - t'| \gg \tau_c$. Taking the finite correlation time into account results in the smoothing out of the discontinuity exhibited by formula (10.3.5), the correlation function $\langle v(t)F(t')\rangle$ passing in fact continuously from its maximum value to zero over a time interval of order τ_c. The shape at finite τ_c of the curve representing $\langle v(t)F(t')\rangle$ as a function of t', the time t being fixed, is shown in Fig. 10.1.

3.2. Equilibrium velocity autocorrelation function

When the velocity $v(t)$ is replaced by expression (10.3.2), the autocorrelation function $\langle v(t)v(t')\rangle$ reads:

$$\langle v(t)v(t')\rangle = \frac{1}{m}\int_{-\infty}^{t} \langle F(t'')v(t')\rangle e^{-\gamma(t-t'')}\, dt''. \tag{10.3.6}$$

If we neglect τ_c, taking formula (10.3.5) into account, we get:

$$\langle v(t)v(t')\rangle = \frac{\mathcal{D}}{\gamma} e^{-\gamma|t-t'|}. \tag{10.3.7}$$

or, setting for convenience $t' = 0$ in formula (10.3.7):

$$\boxed{\langle v(t)v\rangle = \frac{\mathcal{D}}{\gamma} e^{-\gamma|t|}.} \tag{10.3.8}$$

The decrease of the velocity autocorrelation function is described by an exponential of time constant $\tau_r = \gamma^{-1}$.

Neglecting τ_c thus yields an autocorrelation function of the velocity at equilibrium in a 'tent' shape. Such a function is not differentiable at the cusp. This singularity disappears when we consider the fact that τ_c is actually finite. The small $|t|$-behavior of the velocity autocorrelation function $\langle v(t)v\rangle$ is then parabolic[15] (Fig. 10.2).

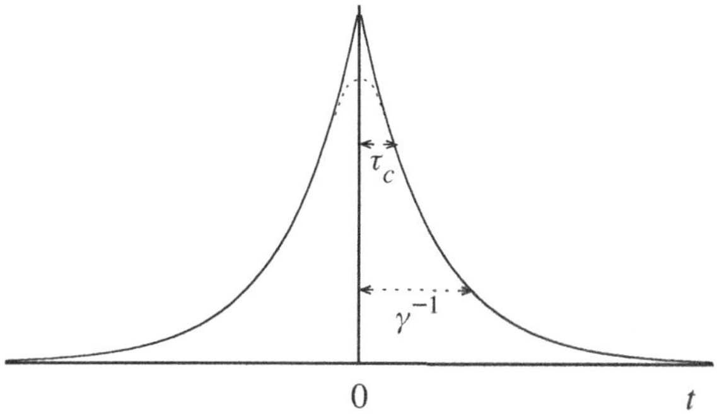

Fig. 10.2 Equilibrium velocity correlation function, at vanishing τ_c (full line), and at finite τ_c (dotted line).

[15] This property will be established later (see formula (10.4.13)).

3.3. The regression theorem

When τ_c is neglected, the evolution for $t \geq t'$ of the autocorrelation function $\langle v(t)v(t')\rangle$ is thus described by the following differential equation:

$$\frac{d}{dt}\langle v(t)v(t')\rangle = -\gamma\langle v(t)v(t')\rangle, \qquad t \geq t'. \tag{10.3.9}$$

Equation (10.3.9) is of the same form as the differential equation $d\langle v(t)\rangle/dt = -\gamma\langle v(t)\rangle$ $(t \geq 0)$ describing the relaxation of $\langle v(t)\rangle$ from a well-defined initial value $v(t = 0)$. This property, according to which the velocity fluctuations regress (that is, disappear) according to the same law as the average velocity, is called the *regression theorem.*

3.4. The first fluctuation-dissipation theorem

The bath being at thermodynamic equilibrium at temperature T, we can make use of the relation (10.2.18) between $\mathcal{D}$ and γ, and rewrite the equilibrium velocity autocorrelation function given by formula (10.3.8) in the form:

$$\langle v(t)v\rangle = \frac{kT}{m}e^{-\gamma|t|}. \tag{10.3.10}$$

Using the *Fourier–Laplace transformation,*[16] we deduce from formula (10.3.10) the equality:

$$\int_0^\infty \langle v(t)v\rangle e^{i\omega t}\,dt = \frac{kT}{m}\frac{1}{\gamma - i\omega}. \tag{10.3.11}$$

Coming back to the definition (10.2.5) of the complex admittance, we get the identity:

$$\boxed{\mathcal{A}(\omega) = \frac{1}{kT}\int_0^\infty \langle v(t)v\rangle e^{i\omega t}\,dt.} \tag{10.3.12}$$

Formula (10.3.12) is the expression for the *first fluctuation-dissipation theorem.*[17] It relates the complex admittance describing the response to an external harmonic perturbation to the equilibrium velocity autocorrelation function. The Einstein relation (10.2.36) corresponds to the particular case of a static external perturbation.

[16] The Fourier–Laplace transformation, also called the *unilateral Fourier transformation*, is defined over the integration interval $(0, \infty)$ (in contrast to the ordinary Fourier transformation, defined over the interval $(-\infty, \infty)$). The usual Laplace transformation, also defined over the interval $(0, \infty)$, uses, in place of $-i\omega$, a complex parameter z.

[17] The terminology of 'first' fluctuation-dissipation theorem, associated with formula (10.3.12) for the Fourier–Laplace transform of the equilibrium velocity autocorrelation function, is due to R. Kubo. We can also get this result by applying the Kubo formulas of the general theory of linear response to the isolated system made up of the Brownian particle coupled with the bath (see Chapter 14).

4. Harmonic analysis of the Langevin model

The Langevin equation is a linear stochastic differential equation. A standard method of resolution of this type of equation is the harmonic analysis, which applies to stationary random processes. The Langevin force $F(t)$ is, by hypothesis, such a process. The same is true for the velocity $v(t)$ of the Brownian particle, provided that the particle has been in contact with the bath for a sufficiently long time to be itself in equilibrium at any finite time t.

Using this method, we will study anew the equilibrium velocity fluctuations and discuss in details the case of finite τ_c.

4.1. Relation between the spectral densities of the random force and of the velocity

The Fourier transforms $F(\omega)$ of the random force, on the one hand, and $v(\omega)$ of the Brownian particle's velocity, on the other hand, are defined by the formulas:[18]

$$F(\omega) = \int_{-\infty}^{\infty} F(t)e^{i\omega t}\, dt \tag{10.4.1}$$

and:

$$v(\omega) = \int_{-\infty}^{\infty} v(t)e^{i\omega t}\, dt. \tag{10.4.2}$$

Given that $F(t)$ and $v(t)$ are stationary random processes, $F(\omega)$ and $v(\omega)$ are in fact obtained by integrating over a large interval of finite width T of the time axis starting from any origin, the limit $T \to \infty$ being taken at the end of the calculations. In the framework of the Langevin model, $F(\omega)$ and $v(\omega)$ are related by the formula:

$$v(\omega) = \frac{1}{m}\frac{1}{\gamma - i\omega}F(\omega). \tag{10.4.3}$$

The spectral densities $S_F(\omega)$ and $S_v(\omega)$ are defined by the formulas:

$$S_F(\omega) = \lim_{T\to\infty} \frac{1}{T}\langle |F(\omega)|^2 \rangle, \qquad S_v(\omega) = \lim_{T\to\infty} \frac{1}{T}\langle |v(\omega)|^2 \rangle. \tag{10.4.4}$$

According to equation (10.4.3), we have:

$$S_v(\omega) = \frac{1}{m^2}\frac{1}{\gamma^2 + \omega^2}S_F(\omega). \tag{10.4.5}$$

The spectral density of the Brownian particle's velocity is thus the product of the spectral density of the random force by a Lorentzian of width $\sim \gamma$.

[18] For the sake of simplicity, we use the same notation $F(.)$ for the random force $F(t)$ and its Fourier transform $F(\omega)$, as well as the same notation $v(.)$ for the velocity $v(t)$ of the Brownian particle and its Fourier transform $v(\omega)$.

According to the Wiener–Khintchine theorem, the spectral density and the autocorrelation function of a stationary random process form a Fourier transform pair. The autocorrelation function $g(\tau)$ of the random force being a very 'peaked' function (of width $\sim \tau_c$) around $\tau = 0$, the spectral density $S_F(\omega)$ is a very 'broad' function (of width $\sim \tau_c^{-1}$). The bath being at thermodynamic equilibrium, $S_F(\omega)$ is referred to as the thermal noise.[19]

4.2. White noise case

Let us assume that the spectral density $S_F(\omega)$ is independent of the angular frequency (white noise):

$$S_F(\omega) = S_F, \qquad S_F = 2\mathcal{D}m^2. \tag{10.4.6}$$

According to the Wiener–Khintchine theorem, $g(\tau)$ is in this case a delta function of weight $2\mathcal{D}m^2$ (formula (10.1.6)).

Using once again the Wiener–Khintchine theorem, we then deduce from equation (10.4.5) the velocity autocorrelation function of the particle in equilibrium with the bath:

$$\langle v(t)v \rangle = \frac{1}{2\pi} \int_{-\infty}^{\infty} \frac{1}{m^2} \frac{1}{\gamma^2 + \omega^2} 2\mathcal{D}m^2 e^{-i\omega t}\, d\omega. \tag{10.4.7}$$

After doing the integration, we recover formula (10.3.10).

4.3. Generalization to a colored noise

The correlation time τ_c being finite, the spectral density of the random force is in fact, not a constant, but a function of the angular frequency, decreasing at large angular frequencies and of width $\sim \tau_c^{-1}$. Such a noise is said to be *colored.*

Let us take for instance for $S_F(\omega)$ a Lorentzian of width $\sim \omega_c$ (with $\omega_c = \tau_c^{-1}$):

$$S_F(\omega) = S_F \frac{\omega_c^2}{\omega_c^2 + \omega^2}, \qquad S_F = 2\mathcal{D}m^2. \tag{10.4.8}$$

The Langevin force autocorrelation function has then an exponential form:[20]

$$g(\tau) = \mathcal{D}m^2 \omega_c e^{-\omega_c|\tau|}. \tag{10.4.9}$$

We have, as in the white noise case:

$$\int_{-\infty}^{\infty} g(\tau)\, d\tau = 2\mathcal{D}m^2. \tag{10.4.10}$$

---

[19] We will come back in more detail to the study of the thermal noise (in an electric conductor at equilibrium) in Supplement 10C.

[20] Such expressions for $S_F(\omega)$ and $g(\tau)$ may be justified starting from certain microscopic models for the interaction of the Brownian particle with the bath (see Supplement 10B).

The autocorrelation function $\langle v(t)v \rangle$ is given by:

$$\langle v(t)v \rangle = \frac{1}{2\pi} \int_{-\infty}^{\infty} \frac{1}{m^2} \frac{1}{\gamma^2 + \omega^2} 2\mathcal{D}m^2 \frac{\omega_c^2}{\omega_c^2 + \omega^2} e^{-i\omega t} \, d\omega. \tag{10.4.11}$$

After integration, we get:

$$\langle v(t)v \rangle = \frac{\mathcal{D}}{\gamma} \frac{\omega_c^2}{\omega_c^2 - \gamma^2} \left(e^{-\gamma|t|} - \frac{\gamma}{\omega_c} e^{-\omega_c|t|} \right). \tag{10.4.12}$$

The equilibrium velocity autocorrelation function as given by equation (10.4.12) behaves in a parabolic way for $|t| \ll \tau_c$. The cusped singularity at the origin which exists when τ_c is neglected is no longer present (Fig. 10.2).

Formula (10.4.12) allows us to show that, even in the case of a colored noise, the second fluctuation-dissipation theorem (10.2.19) still holds, provided that we have $\tau_c \ll \tau_r \; (= \gamma^{-1})$. Setting $t = 0$ in formula (10.4.12), we have indeed:

$$\langle v^2 \rangle = \frac{\mathcal{D}}{\gamma} \frac{\omega_c}{\omega_c + \gamma}, \tag{10.4.13}$$

that is, on account of formula (10.4.10):

$$\langle v^2 \rangle = \frac{1}{2\gamma m^2} \frac{\omega_c}{\omega_c + \gamma} \int_{-\infty}^{\infty} g(\tau) \, d\tau. \tag{10.4.14}$$

The bath being in thermodynamic equilibrium at temperature T, we deduce from formula (10.4.14), in the limit $\gamma \ll \omega_c$, the second fluctuation-dissipation theorem (equation (10.2.19)).[21]

5. Time scales

Thus, as shown for instance by formula (10.4.12), two time scales come into play in the dynamics of the equilibrium velocity fluctuations of a Brownian particle. The first one, very short, is the correlation time τ_c of the random force, whereas the other one, much longer, is the relaxation time $\tau_r = \gamma^{-1}$ of the average velocity.[22]

For the velocity fluctuations to regress (that is, to decrease noticeably), a time at least of the order of the longest time scale τ_r is needed. The Brownian particle's velocity is thus essentially a slow variable. The random force, whose autocorrelation

---

[21] The arguments presented here are approximate. Indeed, it is not fully consistent to take into account the finite correlation time of the random force while retaining the instantaneous character of the friction term as it appears in the Langevin equation (10.1.2). In the case of colored noise, we must in fact write down a generalized Langevin equation with a retarded friction term (see Supplement 10A).

[22] Despite the fact that formula (10.4.12) has been established in the particular case of an exponential autocorrelation function $g(\tau)$, the result concerning the characteristic times of the dynamics of the equilibrium velocity fluctuations has a more general scope.

function decreases over a much shorter time of order τ_c, is a rapid variable. This separation of time scales, as pictured by the inequality:

$$\tau_c \ll \tau_r, \tag{10.5.1}$$

is crucial in the Langevin model. It can be shown, using microscopic models, that inequality (10.5.1) is actually verified when the particle under study is much heavier than the molecules of the fluid which surrounds it. It is only in this case that a particle moving within a fluid may be qualified as 'Brownian' and its evolution properly described by the Langevin equation (10.1.2).

Moreover, in the term in dv/dt involved in the Langevin equation (10.1.2), dt does not stand for an infinitesimal time interval but for a finite time interval Δt during which a finite change Δv of the particle velocity takes place. The interval Δt is necessarily much longer than the collision time τ_c, since the evolution of the Brownian particle's velocity results from the many collisions that the particle undergoes with the fluid's molecules. Besides, equation (10.1.2) (averaged) describes the relaxation of the average velocity fluctuations, and accounts for a significant evolution only when Δt remains small as compared to the relaxation time τ_r. These considerations show that the Langevin equation describes the evolution of the velocity of a Brownian particle over a time interval Δt between τ_c and τ_r:

$$\tau_c \ll \Delta t \ll \tau_r. \tag{10.5.2}$$

In the viscous limit, in which interest is focused on the evolution of the Brownian particle's displacement (Einstein–Smoluchowski description), the evolution time interval of interest verifies the inequality:

$$\Delta t \gg \tau_r. \tag{10.5.3}$$

Bibliography

P.M. CHAIKIN and T.C. LUBENSKY, *Principles of condensed matter physics*, Cambridge University Press, Cambridge, 1995.

S. CHANDRASEKHAR, Stochastic problems in physics and astronomy, Rev. Mod. Phys. **15**, 1 (1943). Reprinted in *Selected papers on noise and stochastic processes* (N. WAX editor), Dover Publications, New York, 2003.

C.W. GARDINER, *Handbook of stochastic methods*, Springer-Verlag, Berlin, third edition, 2004.

N.G. VAN KAMPEN, *Stochastic processes in physics and chemistry*, North-Holland, Amsterdam, third edition, 2007.

R. KUBO, M. TODA, and N. HASHITSUME, *Statistical physics II: nonequilibrium statistical mechanics*, Springer-Verlag, Berlin, second edition, 1991.

R.M. MAZO, *Brownian motion: fluctuations, dynamics, and applications*, Oxford University Press, Oxford, 2002.

F. REIF, *Fundamentals of statistical and thermal physics*, McGraw-Hill, New York, 1965.

R. ZWANZIG, *Nonequilibrium statistical mechanics*, Oxford University Press, Oxford, 2001.

References

L. BACHELIER, *Théorie de la spéculation*, Thesis published in *Annales Scientifiques de l'École Normale Supérieure* **17**, 21 (1900). Reprinted, Éditions J. Gabay, Paris, 1995.

A. EINSTEIN, Über die von der molekularkinetischen Theorie der Wärme geforderte Bewegung von in ruhenden Flüssigkeiten suspendierten Teilchen, *Annalen der Physik* **17**, 549 (1905).

M. SMOLUCHOWSKI, Zur kinetischen Theorie der Brownschen Molekularbewegung und der Suspensionen, *Annalen der Physik* **21**, 756 (1906).

P. LANGEVIN, Sur la théorie du mouvement brownien, *Comptes rendus de l'Académie des Sciences* (Paris), **146**, 530 (1908).

J. PERRIN, *Les atomes*, Félix Alcan, Paris, 1913. Reprinted, Flammarion, Paris, 1991.

G.E. UHLENBECK and L.S. ORNSTEIN, On the theory of the Brownian motion, Phys. Rev. **36**, 823 (1930). Reprinted in *Selected papers on noise and stochastic processes* (N. WAX editor), Dover Publications, New York, 2003.

R. KUBO, The fluctuation-dissipation theorem and Brownian motion, 1965, *Tokyo Summer Lectures in Theoretical Physics* (R. KUBO editor), Syokabo, Tokyo and Benjamin, New York, 1966.

R. KUBO, The fluctuation-dissipation theorem, *Rep. Prog. Phys.* **29**, 255 (1966).

Supplement 10A

The generalized Langevin model

1. The generalized Langevin equation

1.1. The non-retarded Langevin model

The Langevin model relies on the equation of motion of the Brownian particle, written in the form:

$$m\frac{dv}{dt} = -m\gamma v + F(t), \qquad v = \frac{dx}{dt}, \tag{10A.1.1}$$

where γ denotes the friction coefficient and $F(t)$ the Langevin force. In equation (10A.1.1), the friction force $-m\gamma v$ is fully determined by the instantaneous value of the particle's velocity. Also, the correlation time τ_c of the random force $F(t)$ is considered as much shorter than the other characteristic times,[1] in particular the relaxation time $\tau_r = \gamma^{-1}$ of the average velocity. This model, known as the *simple* or *non-retarded* Langevin model,[2] is well adapted to the description of the motion of a particle much heavier than the molecules of the fluid which surrounds it. The inequality $\gamma\tau_c \ll 1$ is then indeed verified.

1.2. Response function of the velocity

According to equation (10A.1.1), the velocity of the particle in equilibrium with the bath reads:

$$v(t) = \frac{1}{m}\int_{-\infty}^{t} F(t')e^{-\gamma(t-t')}\,dt'. \tag{10A.1.2}$$

The *response function* $\mathcal{A}(t)$ of the velocity,[3] defined by the relation:

$$v(t) = \int_{-\infty}^{\infty} \mathcal{A}(t-t')F(t')\,dt', \tag{10A.1.3}$$

---

[1] If we neglect τ_c, and if we assume that $F(t)$ is a Gaussian process, the evolution of the velocity from time t, as described by equation (10A.1.1), depends only on its value at this time, and not on the values it took at prior times. The velocity $v(t)$ is thus in this case a Markov process. This property remains approximately true if we take into account the finite character of τ_c, provided that the inequality $\tau_c \ll \tau_r$ is verified (see Chapter 11).

[2] The reason for this designation will be made clear later.

[3] For the sake of simplicity, we use the same notation $\mathcal{A}(.)$ for the response function $\mathcal{A}(t)$ and its Fourier transform $\mathcal{A}(\omega)$.

is thus:

$$\mathcal{A}(t) = \Theta(t)\frac{1}{m}e^{-\gamma t}. \tag{10A.1.4}$$

In formula (10A.1.4), $\Theta(t)$ denotes the Heaviside function:

$$\Theta(t) = \begin{cases} 0, & t < 0 \\ 1, & t > 0. \end{cases} \tag{10A.1.5}$$

The Fourier transform of $\mathcal{A}(t)$ is the complex admittance of the Langevin model:

$$\mathcal{A}(\omega) = \frac{1}{m}\frac{1}{\gamma - i\omega}. \tag{10A.1.6}$$

In some respects, this description of the motion of a particle immersed within a fluid is too schematic. In particular, the friction cannot be established instantaneously. Its setting up requires a time at least equal to the collision time τ_c of the particle with the fluid's molecules. Retardation effects are thus necessarily present in the friction term. Their consideration leads to a modification of the form of the equation of motion and, accordingly, of the velocity response function, as well as to a modification of the random force autocorrelation function.

1.3. The generalized Langevin model

To take into account the retardation effects, we replace the differential equation (10A.1.1) by the following integro-differential equation,

$$\boxed{m\frac{dv}{dt} = -m\int_{-\infty}^{t}\gamma(t-t')v(t')\,dt' + F(t), \qquad v = \frac{dx}{dt},} \tag{10A.1.7}$$

in which the friction force at time t is determined by the values of the velocity at times $t' < t$. Equation (10A.1.7) is called the *generalized* or *retarded* Langevin equation.

The friction force $-m\int_{-\infty}^{t}\gamma(t-t')v(t')\,dt'$ involves a memory kernel defined by the function $\gamma(t)$ for $t > 0$. It is in fact convenient to consider that the memory kernel $\gamma(t)$ is defined for any t as a decreasing function of $|t|$, of width $\sim \tau_c$, and such that $\int_{-\infty}^{\infty}\gamma(t)\,dt = 2\gamma$. Introducing the *causal* or *retarded* memory kernel $\tilde{\gamma}(t) = \Theta(t)\gamma(t)$, we can rewrite equation (10A.1.7) in the following equivalent form:

$$\boxed{m\frac{dv}{dt} = -m\int_{-\infty}^{\infty}\tilde{\gamma}(t-t')v(t')\,dt' + F(t), \qquad v = \frac{dx}{dt}.} \tag{10A.1.8}$$

The Langevin force $F(t)$ is modelized here, as in the non-retarded Langevin model, by a random stationary process of zero mean. This process is, most often, assumed to be Gaussian. Since we now take into account the retarded character of the friction, it is necessary, for the consistency of the model, to take equally into account the non-vanishing correlation time of the random force. Accordingly, we assume that the autocorrelation function $g(\tau) = \langle F(t)F(t+\tau)\rangle$ decreases over a finite time $\sim \tau_c$. In contrast to the non-retarded case, we do not make the hypothesis $\gamma\tau_c \ll 1$.

2. Complex admittance

In the presence of an applied external force $F_{\rm ext}(t)$, the retarded equation of motion of the particle reads:

$$m\frac{dv}{dt} = -m\int_{-\infty}^{\infty}\tilde{\gamma}(t-t')v(t')\,dt' + F(t) + F_{\rm ext}(t), \qquad v = \frac{dx}{dt}. \tag{10A.2.1}$$

On average, we have:

$$m\frac{d\langle v\rangle}{dt} = -m\int_{-\infty}^{\infty}\tilde{\gamma}(t-t')\langle v(t')\rangle\,dt' + F_{\rm ext}(t), \qquad \langle v\rangle = \frac{d\langle x\rangle}{dt}. \tag{10A.2.2}$$

For a harmonic applied force $F_{\rm ext}(t) = \Re e(Fe^{-i\omega t})$, the solution of equation (10A.2.2) is, in stationary regime, of the form:

$$\langle v(t)\rangle = \Re e(\langle v\rangle e^{-i\omega t}), \tag{10A.2.3}$$

with:

$$\langle v\rangle = \mathcal{A}(\omega)F. \tag{10A.2.4}$$

The quantity:

$$\mathcal{A}(\omega) = \frac{1}{m}\frac{1}{\gamma(\omega) - i\omega} \tag{10A.2.5}$$

is the complex admittance of the generalized Langevin model. In formula (10A.2.5), the generalized friction coefficient $\gamma(\omega)$, defined by the formula:

$$\gamma(\omega) = \int_0^{\infty}\gamma(t)e^{i\omega t}\,dt, \tag{10A.2.6}$$

denotes the Fourier–Laplace transform of the memory kernel $\gamma(t)$ (or the Fourier transform of the retarded memory kernel $\tilde{\gamma}(t)$). Note that $\gamma(\omega = 0) = \gamma$.

More generally, for an external force $F_{\rm ext}(t)$ of Fourier transform $F_{\rm ext}(\omega)$, the Fourier transform $\langle v(\omega)\rangle$ of the stationary solution $\langle v(t)\rangle$ of equation (10A.2.2) is:

$$\langle v(\omega)\rangle = \mathcal{A}(\omega)F_{\rm ext}(\omega). \tag{10A.2.7}$$

3. Harmonic analysis of the generalized Langevin model

The generalized Langevin equation (10A.1.8) is a linear stochastic integro-differential equation, to which we can apply harmonic analysis. Indeed, the initial time in the retarded friction term having been taken equal to $-\infty$, the particle finds itself, at any finite time t, in equilibrium with the bath. Its velocity is thus a stationary random process. Both the random force $F(t)$ and the velocity $v(t)$ can be expanded in Fourier series. The spectral densities $S_v(\omega)$ and $S_F(\omega)$ are related by the formula:

$$S_v(\omega) = \frac{1}{m^2}\frac{1}{|\gamma(\omega) - i\omega|^2}S_F(\omega). \tag{10A.3.1}$$

3.1. First fluctuation-dissipation theorem: spectral densities

According to the first fluctuation-dissipation theorem,[4] the complex admittance is the Fourier–Laplace transform of the equilibrium velocity autocorrelation function:

$$\boxed{\mathcal{A}(\omega) = \frac{1}{kT}\int_0^\infty \langle v(t)v\rangle e^{i\omega t}\,dt.} \qquad (10A.3.2)$$

The Fourier transform of the velocity autocorrelation function is thus:

$$\int_{-\infty}^\infty \langle v(t)v\rangle e^{i\omega t}\,dt = 2kT\,\Re e\,\mathcal{A}(\omega), \qquad (10A.3.3)$$

that is, on account of formula (10A.2.5):

$$\int_{-\infty}^\infty \langle v(t)v\rangle e^{i\omega t}\,dt = \frac{2kT}{m}\frac{\Re e\,\gamma(\omega)}{|\gamma(\omega) - i\omega|^2}. \qquad (10A.3.4)$$

Using the Wiener–Khintchine theorem, we deduce from equation (10A.3.4) the spectral density of the velocity,

$$S_v(\omega) = \frac{2kT}{m}\frac{\Re e\,\gamma(\omega)}{|\gamma(\omega) - i\omega|^2}, \qquad (10A.3.5)$$

then, using formula (10A.3.1), the spectral density of the random force:

$$S_F(\omega) = 2mkT\,\Re e\,\gamma(\omega). \qquad (10A.3.6)$$

3.2. Second fluctuation-dissipation theorem

At this stage, using once more the Wiener–Khintchine theorem, we can write, from formula (10A.3.6), the real part of the generalized friction coefficient in the form of a Fourier integral:

$$\Re e\,\gamma(\omega) = \frac{1}{2mkT}\int_{-\infty}^\infty \langle F(t)F(t+\tau)\rangle e^{i\omega\tau}\,d\tau. \qquad (10A.3.7)$$

By inverse Fourier transformation, we deduce from equation (10A.3.7) a proportionality relation between the memory kernel and the random force autocorrelation function:

$$\boxed{\gamma(\tau) = \frac{1}{mkT}g(\tau).} \qquad (10A.3.8)$$

[4] This result can be demonstrated by applying the Kubo formulas of the linear response theory to the isolated system made up of the particle coupled with the bath (see also Chapter 14).

The decrease of the memory kernel is thus characterized by the correlation time[5] τ_c.

The generalized friction coefficient is proportional to the Fourier–Laplace transform of the random force autocorrelation function:

$$\boxed{\gamma(\omega) = \frac{1}{mkT}\int_0^\infty \langle F(t)F(t+\tau)\rangle e^{i\omega\tau}\, d\tau.} \tag{10A.3.9}$$

Formula (10A.3.9) constitutes the expression for the second fluctuation-dissipation theorem in the generalized Langevin model.

4. An analytical model

If we have explicit analytical expressions for the memory kernel and the random force autocorrelation function, we can derive analytical expressions for the complex admittance, and, possibly, for the velocity response function in the generalized Langevin model.

4.1. Complex admittance

If the Langevin force autocorrelation function has an exponential form,

$$g(\tau) = \mathcal{D}m^2\omega_c e^{-\omega_c|\tau|}, \tag{10A.4.1}$$

with $\mathcal{D}m^2 = \gamma mkT$, the parameter ω_c denoting an angular frequency characteristic of the bath, the memory kernel, too, has an exponential form (see formula (10A.3.8)). Coming back to the variable t, we can write:[6]

$$\gamma(t) = \gamma\omega_c e^{-\omega_c|t|}. \tag{10A.4.2}$$

In the limit $\omega_c \to \infty$, we recover the expressions corresponding to the non-retarded Langevin equation: $g(\tau) = 2\mathcal{D}m^2\delta(\tau)$, $\gamma(t) = 2\gamma\delta(t)$.

According to formula[7] (10A.4.2), we have:

$$\gamma(\omega) = \gamma\frac{\omega_c}{\omega_c - i\omega}. \tag{10A.4.3}$$

[5] By contrast, formula (10A.3.8) shows that it is inconsistent to take into account the finite correlation time of the random force while retaining the instantaneous character of the friction term as it is displayed in the non-retarded Langevin equation (10A.1.1).

[6] Expressions of this type can be obtained in the framework of certain microscopic models of the interaction of the particle with the bath. This in particular the case in the Caldeira–Leggett model (see Supplement 10B). It is then seen that ω_c can be given the significance of an angular frequency characteristic of the bath.

[7] The generalized friction coefficient $\gamma(\omega)$ is the Fourier transform of the causal function $\tilde{\gamma}(t) = \Theta(t)\gamma(t)$.

The corresponding complex admittance reads:

$$\mathcal{A}(\omega) = \frac{1}{m} \frac{1}{\gamma \dfrac{\omega_c}{\omega_c - i\omega} - i\omega}. \tag{10A.4.4}$$

The poles of $\mathcal{A}(\omega)$ give access to the characteristic relaxation times of the average velocity from a well-defined initial value. In the weak-coupling case $\omega_c/\gamma > 4$, these poles are of the form $-i\omega_\pm$, with:

$$\omega_\pm = \frac{\omega_c}{2}\left[1 \pm \left(1 - 4\gamma\omega_c^{-1}\right)^{1/2}\right]. \tag{10A.4.5}$$

4.2. The velocity response function

With the model chosen for $\mathcal{A}(\omega)$ (formula (10A.4.4)), in the case $\omega_c/\gamma > 4$, we have:

$$\mathcal{A}(t) = \Theta(t)\frac{1}{m}\left(1 - 4\gamma\omega_c^{-1}\right)^{-1/2}\left(\frac{\omega_+}{\omega_c}e^{-\omega_- t} - \frac{\omega_-}{\omega_c}e^{-\omega_+ t}\right). \tag{10A.4.6}$$

The expression (10A.4.6) for $\mathcal{A}(t)$ has to be compared with the corresponding expression in the non-retarded Langevin model (formula (10A.1.4)). In the generalized Langevin model, we do not make the assumption $\gamma\tau_c \ll 1$. Accordingly, there is no clear-cut separation of time scales between the random force and the velocity of the particle.

Bibliography

R. KUBO, M. TODA, and N. HASHITSUME, *Statistical physics II: nonequilibrium statistical mechanics*, Springer-Verlag, Berlin, second edition, 1991.

R. ZWANZIG, *Nonequilibrium statistical mechanics*, Oxford University Press, Oxford, 2001.

References

R. KUBO, The fluctuation-dissipation theorem and Brownian motion, 1965, *Tokyo Summer Lectures in Theoretical Physics* (R. KUBO editor), Syokabo, Tokyo and Benjamin, New York, 1966.

R. KUBO, The fluctuation-dissipation theorem, *Rep. Prog. Phys.* **29**, 255 (1966).

Supplement 10B

Brownian motion in a bath of oscillators

1. The Caldeira–Leggett model

In order to have at our disposal a microscopic basis for the generalized Langevin equation, we can study the dynamics of a free particle interacting with an environment made up of an infinite number of independent harmonic oscillators in thermal equilibrium. In the case of a linear coupling with each environment mode, the effect of the environment can be eliminated and the particle's equation of motion can be established exactly. After an appropriate modelization, it takes the form of a generalized Langevin equation in which determined microscopic expressions are assigned to both the memory kernel and the random force autocorrelation function.

This model of dissipation is known as the *Caldeira–Leggett model.*[1] It is widely used to describe the dissipative dynamics of classical or quantum systems.

1.1. The Caldeira–Leggett Hamiltonian

Consider a particle of mass m, described by its coordinate x and the conjugate momentum p, evolving in a potential $\phi(x)$. The particle is coupled with a bath of N independent harmonic oscillators of masses m_n, described by the coordinates x_n and the conjugate momenta p_n ($n = 1, \ldots, N$). The coupling between the particle and each oscillator of the bath is assumed bilinear. The Hamiltonian of the global system made up of the particle and the set of the oscillators to which it is coupled reads:

$$H_{\mathrm{C-L}} = \frac{p^2}{2m} + \phi(x) + \frac{1}{2}\sum_{n=1}^{N}\left[\frac{p_n^2}{m_n} + m_n\omega_n^2\Big(x_n - \frac{c_n}{m_n\omega_n^2}x\Big)^2\right]. \qquad (10B.1.1)$$

The constants c_n measure the strength of the coupling.

In the case of a free particle ($\phi(x) = 0$), the model described by the Caldeira–Leggett Hamiltonian is exactly solvable.[2]

---

[1] Although it had previously been proposed by other authors, it is after the work of A.O. Caldeira and A.J. Leggett on decoherence that this model became famous.

[2] The same is true for a particle evolving in a harmonic potential (the dissipative dynamics of a harmonic oscillator is studied in Supplement 14A).

1.2. The dissipative free particle

Using the Hamiltonian (10B.1.1) with $\phi(x) = 0$, we can write the Hamilton's equations for all the degrees of freedom of the global system, that is, for the particle,

$$\frac{dx}{dt} = \frac{p}{m}, \qquad \frac{dp}{dt} = \sum_n c_n \left(x_n - \frac{c_n}{m_n \omega_n^2} x \right), \tag{10B.1.2}$$

and, for the bath's oscillators:

$$\frac{dx_n}{dt} = \frac{p_n}{m_n}, \qquad \frac{dp_n}{dt} = -m_n \omega_n^2 x_n + c_n x. \tag{10B.1.3}$$

Equations (10B.1.3) can formally be solved by considering the particle's position $x(t)$ as known. This gives:

$$x_n(t) = x_n(t_0) \cos \omega_n (t - t_0) + \frac{p_n(t_0)}{m_n \omega_n} \sin \omega_n (t - t_0) + c_n \int_{t_0}^{t} \frac{\sin \omega_n (t - t')}{m_n \omega_n} x(t')\, dt', \tag{10B.1.4}$$

where t_0 denotes the initial time at which the coupling is established. Integrating by parts the integral on the right-hand side of equation (10B.1.4), we can write the equality:

$$x_n(t) - \frac{c_n}{m_n \omega_n^2} x(t) = \left[x_n(t_0) - \frac{c_n}{m_n \omega_n^2} x(t_0) \right] \cos \omega_n (t - t_0) + \frac{p_n(t_0)}{m_n \omega_n} \sin \omega_n (t - t_0) - c_n \int_{t_0}^{t} \frac{\cos \omega_n (t - t')}{m_n \omega_n^2} \frac{p(t')}{m}\, dt'. \tag{10B.1.5}$$

The equation of motion of the particle coupled with the bath, which, according to equations (10B.1.2), is of the form:

$$m\ddot{x}(t) = \sum_n c_n \left[x_n(t) - \frac{c_n}{m_n \omega_n^2} x(t) \right], \tag{10B.1.6}$$

may be reformulated, on account of the equality (10B.1.5), in the form of a closed integro-differential equation for $x(t)$:

$$\boxed{m\ddot{x}(t) + m \int_{t_0}^{t} \gamma(t - t') \dot{x}(t')\, dt' = -m x(t_0) \gamma(t - t_0) + F(t).} \tag{10B.1.7}$$

The functions $\gamma(t)$ and $F(t)$ involved in equation (10B.1.7) are defined in terms of the microscopic parameters of the model by the formulas:

$$\gamma(t) = \frac{1}{m} \sum_n \frac{c_n^2}{m_n \omega_n^2} \cos \omega_n t \tag{10B.1.8}$$

and:

$$F(t)=\sum_n c_n\left[x_n(t_0)\cos\omega_n(t-t_0)+\frac{p_n(t_0)}{m_n\omega_n}\sin\omega_n(t-t_0)\right]. \tag{10B.1.9}$$

In equation (10B.1.7), $\gamma(t)$ acts as a memory kernel and $F(t)$ as a random force. Using, instead of $\gamma(t)$, the retarded memory kernel $\tilde{\gamma}(t)=\Theta(t)\gamma(t)$, we can rewrite equation (10B.1.7) in the following equivalent form:

$$\boxed{m\ddot{x}(t)+m\int_{t_0}^{\infty}\tilde{\gamma}(t-t')\dot{x}(t')\,dt'=-mx(t_0)\tilde{\gamma}(t-t_0)+F(t).} \tag{10B.1.10}$$

Equations (10B.1.7) and (10B.1.10) have been deduced without approximation from the Hamilton's equations (10B.1.2) and (10B.1.3). Both the memory kernel and the random force are expressed in terms of the parameters of the microscopic Caldeira–Leggett Hamiltonian. In particular, $F(t)$ is a linear combination of the variables $x_n(t_0)$ and $p_n(t_0)$ associated with the initial state of the oscillators' bath (formula (10B.1.9)). If we assume that at time t_0 the bath is in thermal equilibrium at temperature T, its distribution function is:[3]

$$\rho_B=Z^{-1}\exp\left[-\beta\sum_n\left(\frac{p_n^2}{2m_n}+\frac{m_n\omega_n^2}{2}x_n^2\right)\right],\qquad \beta=(kT)^{-1}. \tag{10B.1.11}$$

The Gaussian character of the distribution (10B.1.11) leads us to consider $F(t)$ as a random stationary Gaussian process, characterized by its average and its autocorrelation function:[4]

$$\begin{cases}\langle F(t)\rangle=0\\ \langle F(t)F(t+\tau)\rangle=mkT\gamma(\tau).\end{cases} \tag{10B.1.12}$$

1.3. The spectral density of the coupling

The equations of motion (10B.1.7) or (10B.1.10) do not allow us, per se, to describe an irreversible dynamics. This is only possible if the number N of the bath's oscillators tends towards infinity, their angular frequencies forming a continuum in this limit.

A central ingredient in the model is the Fourier transform of the retarded memory kernel $\tilde{\gamma}(t)$. We calculate it by attributing to ω a small imaginary part $\epsilon>0$ and by letting ϵ tend towards zero at the end of the calculation. We thus first define:

$$\gamma(\omega+i\epsilon)=\int_0^{\infty}\tilde{\gamma}(t)e^{i\omega t}e^{-\epsilon t}\,dt,\qquad \epsilon>0, \tag{10B.1.13}$$

[3] This is the classical distribution function of the bath in the phase space. The quantum generalization will be outlined in Section 3.

[4] The averages involved here are computed with the aid of the bath's distribution function (10B.1.11).

then the Fourier transform[5] $\gamma(\omega)$:

$$\gamma(\omega) = \lim_{\epsilon \to 0^+} \gamma(\omega + i\epsilon). \tag{10B.1.14}$$

Proceeding in this way, from the expression for $\tilde{\gamma}(t)$,

$$\tilde{\gamma}(t) = \Theta(t) \frac{1}{m} \sum_n \frac{c_n^2}{m_n \omega_n^2} \cos \omega_n t, \tag{10B.1.15}$$

we obtain that for $\gamma(\omega)$:

$$\gamma(\omega) = \frac{i}{2m} \sum_n \frac{c_n^2}{m_n \omega_n^2} \lim_{\epsilon \to 0^+} \left(\frac{1}{\omega - \omega_n + i\epsilon} + \frac{1}{\omega + \omega_n + i\epsilon} \right). \tag{10B.1.16}$$

We deduce[6] in particular from equation (10B.1.16):

$$\Re e\, \gamma(\omega) = \frac{\pi}{2m} \sum_n \frac{c_n^2}{m_n \omega_n^2} \big[\delta(\omega - \omega_n) + \delta(\omega + \omega_n) \big]. \tag{10B.1.17}$$

At this stage, we generally introduce the *spectral density of the coupling with the environment*, defined for $\omega > 0$ by the formula:

$$\boxed{J(\omega) = \frac{\pi}{2} \sum_n \frac{c_n^2}{m_n \omega_n} \delta(\omega - \omega_n), \qquad \omega > 0.} \tag{10B.1.18}$$

For $\omega > 0$, we have the relation:

$$\Re e\, \gamma(\omega) = \frac{J(\omega)}{m\omega}. \tag{10B.1.19}$$

In the continuum limit, the quantities $\Re e\, \gamma(\omega)$ and $J(\omega)$ may be considered as continuous functions of ω.

---

[5] We thus define the Fourier transform of $\tilde{\gamma}(t)$ in the distribution sense. This procedure is commonly used when computing the generalized susceptibilities as Fourier transforms of the response functions (see Chapter 12).

[6] We make use of the relation:

$$\lim_{\epsilon \to 0^+} \frac{1}{x + i\epsilon} = \mathrm{vp}\, \frac{1}{x} - i\pi\delta(x),$$

where the symbol vp denotes the Cauchy principal value.

1.4. Ohmic dissipation

The dynamics of the particle coupled with the bath is determined by the above defined spectral density. In particular, the large-time dynamics is controlled by the behavior of $J(\omega)$ at low angular frequencies. In many cases, this behavior is described by a power law of the type $J(\omega) \propto \omega^\delta$. The exponent $\delta > 0$ is most often an integer, whose value depends on the dimensionality of the space corresponding to the considered environment.[7]

The value $\delta = 1$ is especially important. Indeed, in this case, the equation of motion of the particle coupled with the bath contains a friction term proportional to the velocity (in a certain limit).[8] The corresponding dissipation model is known as the *Ohmic model*.[9] It is defined by the relations:

$$\boxed{J(\omega) = m\gamma\omega \quad (\omega > 0), \qquad \Re e\,\gamma(\omega) = \gamma.} \tag{10B.1.20}$$

The expressions (10B.1.20) for $J(\omega)$ and $\Re e\,\gamma(\omega)$ are only valid at low angular frequencies. Indeed, the spectral density does not in fact increase without bounds, but it decreases towards zero as $\omega \to \infty$. To account for this behavior, we write, in place of the relations (10B.1.20), the formulas:

$$J(\omega) = m\gamma\omega f_c\Big(\frac{\omega}{\omega_c}\Big) \quad (\omega > 0), \qquad \Re e\,\gamma(\omega) = \gamma f_c\Big(\frac{\omega}{\omega_c}\Big), \tag{10B.1.21}$$

in which $f_c(\omega/\omega_c)$ is a cut-off function tending towards zero more rapidly than ω^{-1} as[10] $\omega \to \infty$. We often choose for convenience a Lorentzian cut-off function:

$$f_c\Big(\frac{\omega}{\omega_c}\Big) = \frac{\omega_c^2}{\omega_c^2 + \omega^2}. \tag{10B.1.22}$$

We then have:

$$J(\omega) = m\gamma\omega\frac{\omega_c^2}{\omega_c^2 + \omega^2} \quad (\omega > 0), \qquad \Re e\,\gamma(\omega) = \gamma\frac{\omega_c^2}{\omega_c^2 + \omega^2}. \tag{10B.1.23}$$

In this case, the memory kernel is modelized by an exponential function of time constant ω_c^{-1}:

$$\gamma(t) = \gamma\omega_c e^{-\omega_c|t|}. \tag{10B.1.24}$$

The corresponding modelization has to be done on the Langevin force autocorrelation function, which yields:

$$\big\langle F(t)F(t+\tau)\big\rangle = mkT\gamma\omega_c\, e^{-\omega_c|\tau|}. \tag{10B.1.25}$$

[7] The exponent δ may possibly take non-integer values, for instance in the case of an interaction with a disordered or fractal environment.

[8] This limit is the infinitely short memory limit (see Subsection 1.6)

[9] The reason for this designation will be made clear in Subsection 1.6.

[10] The angular frequency ω_c characterizes the width of the angular frequency band of the bath oscillators effectively coupled with the particle.

1.5. The generalized Langevin equation

Let us now come back to the equation of motion (10B.1.10). Its right-hand side involves, besides the Langevin force $F(t)$, a term $-mx(t_0)\tilde{\gamma}(t-t_0)$ depending on the particle's initial coordinate. In the Ohmic model, the function $\tilde{\gamma}(t)$ decreases over a characteristic time $\sim \omega_c^{-1}$. The quantity $\tilde{\gamma}(t-t_0)$ is thus negligible if $\omega_c(t-t_0) \gg 1$. Mathematically, this condition can be fulfilled by taking the limit $t_0 \to -\infty$. Then the particle's initial coordinate does not play any role in the equation of motion, which takes the form of the generalized Langevin equation:

$$\boxed{m\frac{dv}{dt} + m\int_{-\infty}^{+\infty} \tilde{\gamma}(t-t')v(t')\,dt' = F(t), \qquad v = \frac{dx}{dt}.} \tag{10B.1.26}$$

1.6. Infinitely short memory limit

The memory kernel of the Ohmic model admits an infinitely short memory limit $\gamma(t) = 2\gamma\delta(t)$, which may for instance be obtained by taking the limit $\omega_c \to \infty$ in equation (10B.1.24). The corresponding limit must be taken in the Langevin force autocorrelation function (formula (10B.1.25)), which then reads $\langle F(t)F(t+\tau)\rangle = 2mkT\gamma\,\delta(\tau)$.

In this limit, equation (10B.1.26) takes a non-retarded form:

$$m\frac{dv}{dt} + m\gamma v(t) = F(t), \qquad v = \frac{dx}{dt}. \tag{10B.1.27}$$

The formal similarity between the non-retarded Langevin equation (10B.1.27) and Ohm's law in an electrical circuit[11] justifies the designation of Ohmic model given to the dissipation model defined by the spectral density (10B.1.20).

2. Dynamics of the Ohmic free particle

2.1. The velocity autocorrelation function

Since the Langevin force $F(t)$ may be considered as a stationary random process, the same is true of the solution $v(t)$ of equation (10B.1.26). Therefore we can use harmonic analysis and the Wiener–Khintchine theorem to determine the equilibrium velocity autocorrelation function:

$$\langle v(t)v\rangle = \frac{2kT}{m}\frac{1}{2\pi}\int_{-\infty}^{\infty} \frac{\Re e\,\gamma(\omega)}{|\gamma(\omega)-i\omega|^2} e^{-i\omega t}\,d\omega. \tag{10B.2.1}$$

In the Ohmic model with a Lorenzian cut-off function, we have:

$$\gamma(\omega) = \gamma\frac{\omega_c}{\omega_c - i\omega}, \tag{10B.2.2}$$

[11] The analogy with an electrical problem will be studied in more details in the Supplement 10C devoted to the Nyquist theorem.

so that equation (10B.2.1) takes the form:

$$\langle v(t)v\rangle = \frac{kT\gamma}{m\pi}\int_{-\infty}^{\infty}\frac{\omega_c^2}{(\gamma\omega_c-\omega^2)^2+\omega^2\omega_c^2}e^{-i\omega t}\,d\omega. \tag{10B.2.3}$$

Consider the case $\omega_c/\gamma > 4$, which corresponds to a weak coupling between the particle and the bath. We can then write:

$$\langle v(t)v\rangle = \frac{kT\gamma}{m\pi}\left(1-4\gamma\omega_c^{-1}\right)^{-1/2}\int_{-\infty}^{\infty}\left(\frac{1}{\omega^2+\omega_-^2}-\frac{1}{\omega^2+\omega_+^2}\right)e^{-i\omega t}\,d\omega, \tag{10B.2.4}$$

with:

$$\omega_\pm = \frac{\omega_c}{2}\left[1\pm\left(1-4\gamma\omega_c^{-1}\right)^{1/2}\right]. \tag{10B.2.5}$$

After the integration, we have:

$$\langle v(t)v\rangle = \frac{kT}{m}\left(1-4\gamma\omega_c^{-1}\right)^{-1/2}\left(\frac{\omega_+}{\omega_c}e^{-\omega_-|t|}-\frac{\omega_-}{\omega_c}e^{-\omega_+|t|}\right). \tag{10B.2.6}$$

In the infinitely short memory limit $\omega_c \to \infty$, we retrieve the result of the non-retarded Langevin model:

$$\langle v(t)v\rangle = \frac{kT}{m}e^{-\gamma|t|}. \tag{10B.2.7}$$

2.2. The time-dependent diffusion coefficient

The *time-dependent diffusion coefficient* $D(t)$ is defined by:

$$D(t) = \frac{1}{2}\frac{d}{dt}\langle[x(t)-x]^2\rangle, \qquad t>0. \tag{10B.2.8}$$

It is obtained by integration of the velocity autocorrelation function:

$$D(t) = \int_0^t \langle v(t')v\rangle\,dt'. \tag{10B.2.9}$$

The limit value at large times of $D(t)$ is, at any non-vanishing temperature, the Einstein value $D = kT/\eta$ of the diffusion coefficient. This result holds regardless of the value, finite or not, of ω_c.

In the non-retarded model, we have:

$$D(t) = \frac{kT}{\eta}(1-e^{-\gamma t}), \qquad t>0. \tag{10B.2.10}$$

The time-dependent diffusion coefficient increases monotonically towards its limit.

3. The quantum Langevin equation

The Caldeira–Leggett model may be extended to the quantum case. To this end, we have to take into account the quantum character of the noise and to modify accordingly the spectral density $S_F(\omega)$ of the random force. We write:[12]

$$S_F(\omega) = \hbar\omega \coth\frac{\beta\hbar\omega}{2} m\,\Re e\,\gamma(\omega). \tag{10B.3.1}$$

The memory kernel being independent of the temperature, its modelization by an exponential (formula (10B.1.24)) may be maintained in the quantum case, which amounts to keep the expression (10B.1.23) for $\Re e\,\gamma(\omega)$. However, the relation (10B.1.12) between the memory kernel and the random force autocorrelation function has to be modified. Equation (10B.1.26), which is still formally valid, is then referred to as a *quantum Langevin equation*. It allows us to describe at any temperature, including $T = 0$, the dynamics of the Ohmic free particle.

The main characteristics of the dynamics are the following:[13] below some *crossover temperature* linked to the bath, the description of the dynamics in terms of Brownian motion, that is, with well-separated time scales for the random force, on the one hand, and for the particle's velocity, on the other hand, becomes inadequate. Indeed, at large times, both the random force autocorrelation function and the velocity autocorrelation function exhibit a long negative time tail $\propto -t^{-2}$.

3.1. Velocity autocorrelation function

The detailed study of the velocity autocorrelation function shows that there is a crossover temperature $T_c = \hbar\omega_-/\pi k$ (or $T_c = \hbar\gamma/\pi k$ in the infinitely short memory limit), above and below which $\langle v(t)v\rangle$ displays qualitatively different behaviors. For $T > T_c$, $\langle v(t)v\rangle$ is positive at any time. Despite the existence of quantum corrections, this regime may be qualified as classical. For $T < T_c$, $\langle v(t)v\rangle$ is first positive, then vanishes, and eventually becomes negative at large times. This regime is qualified as quantal.

3.2. Time-dependent diffusion coefficient

For $T > T_c$, $D(t)$ increases monotonically towards its limit kT/η. However, for $T < T_c$, $D(t)$ first increases, then passes through a maximum, and eventually slowly decreases towards kT/η. Thus, in the quantum regime, the time-dependent diffusion coefficient may exceed its stationary value. The diffusive regime is only attained very slowly, that is, after a time $t \gg t_{\text{th}}$ (the 'thermal time' $t_{\text{th}} = \hbar/2\pi kT$ is a time linked to the temperature, all the longer as the temperature is lower).

---

[12] Equation (10B.3.1) constitutes the quantum formulation of the second fluctuation-dissipation theorem (see Chapter 14). It is the quantum generalization of the classical expression for the noise spectral density:

$$S_F(\omega) = 2mkT\,\Re e\,\gamma(\omega),$$

which is obtained by Fourier transformation of formula (10B.1.12) for $\langle F(t)F(t+\tau)\rangle$.

[13] The detailed calculations, fairly intricate, will not be reproduced here.

At $T = 0$, $D(t)$ passes through a maximum at a time $t_m \sim \gamma^{-1}$. For $\gamma t \gg 1$, we have:

$$D_{T=0}(t) \sim \frac{\hbar}{\pi m}\frac{1}{\gamma t}, \qquad \gamma t \gg 1. \tag{10B.3.2}$$

The diffusion is then logarithmic:

$$\langle [x(t) - x]^2 \rangle \sim 2\frac{\hbar}{\pi m}\log \gamma t, \qquad \gamma t \gg 1. \tag{10B.3.3}$$

The curves representing $D(t)$ (and the corresponding classical diffusion coefficient) as a function of γt for different temperatures are plotted in Fig. 10B.1.

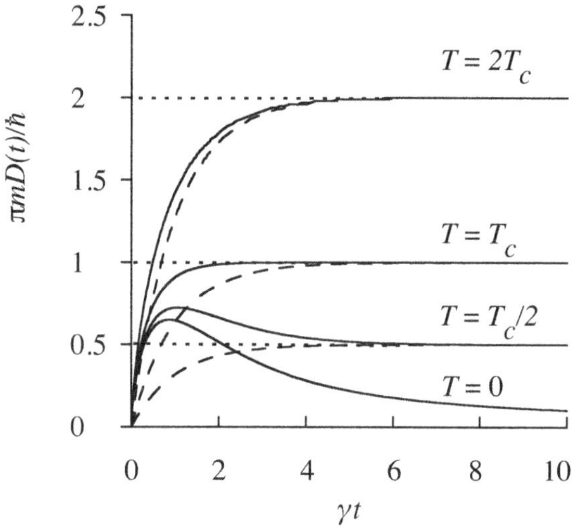

Fig. 10B.1 The coefficient $D(t)$ (in dashed lines, its classical counterpart).

Bibliography

U. WEISS, *Quantum dissipative systems*, World Scientific, Singapore, third edition, 2008.

R. ZWANZIG, *Nonequilibrium statistical mechanics*, Oxford University Press, Oxford, 2001.

References

I.R. SENITZKY, Dissipation in quantum mechanics. The harmonic oscillator, *Phys. Rev.* **119**, 670 (1960).

G.W. FORD, M. KAC, and P. MAZUR, Statistical mechanics of assemblies of coupled oscillators, *J. Math. Phys.* **6**, 504 (1965).

P. ULLERSMA, An exactly solvable model for Brownian motion, *Physica* **32**, 27, 56, 74, 90 (1966).

A.O. CALDEIRA and A.J. LEGGETT, Quantum tunnelling in a dissipative system, *Ann. Phys.* **149**, 374 (1983).

V. HAKIM and V. AMBEGAOKAR, Quantum theory of a free particle interacting with a linearly dissipative environment, *Phys. Rev. A* **32**, 423 (1985).

C. ASLANGUL, N. POTTIER, and D. SAINT-JAMES, Time behavior of the correlation functions in a simple dissipative quantum model, *J. Stat. Phys.* **40**, 167 (1985).

A.J. LEGGETT, S. CHAKRAVARTY, A.T. DORSEY, M.P.A. FISHER, A. GARG, and W. ZWERGER, Dynamics of the dissipative two-state system, *Rev. Mod. Phys.* **59**, 1 (1987).

G.W. FORD and M. KAC, On the quantum Langevin equation, *J. Stat. Phys.* **46**, 803 (1987).

Supplement 10C

The Nyquist theorem

1. Thermal noise in an electrical circuit

The charge carriers in a conductor in thermodynamic equilibrium are in a state of permanent thermal agitation. This thermal noise manifests itself in particular through fluctuations of the potential difference existing between the extremities of the conductor.

The thermal noise in a linear electrical system was experimentally studied by J.B. Johnson in 1928. These measurements allowed him to establish that the variance of the fluctuating potential difference is proportional to the electric resistance and to the temperature of the conductor (it depends neither on the shape of the latter nor on the material which it is made of). From the theoretical point of view, the relation between the variance of the fluctuating potential difference, the resistance of the conductor, and the temperature was established by H. Nyquist in 1928 (*Nyquist formula*). The experimental checking of the Nyquist formula allowed for the determination of the Boltzmann constant. The thermal noise in a conductor is also designated as the *Johnson noise* or as the *Nyquist noise*.

The Nyquist theorem can be extended to a general class of linear dissipative systems other than electrical ones. Historically, it constitutes one of the first statements of the fluctuation-dissipation theorem.[1]

2. The Nyquist theorem

Consider an electrical circuit made up of a resistance R and an inductance L in series. Under the effect of thermal agitation, the electrons of the circuit give rise to a fluctuating current $I(t)$. We represent the interactions responsible for this current by a fluctuating potential difference $V(t)$. We assume that the circuit is linear, in other words, that the relation between $I(t)$ and $V(t)$ is given by Ohm's law.

---

[1] In a general way, the fluctuation-dissipation theorem expresses a relation between the admittance of a linear dissipative system and the equilibrium fluctuations of relevant generalized forces (see Chapter 14).

2.1. Ohm's law

In the absence of an external potential difference, Ohm's law reads:

$$L\frac{dI}{dt} + RI = V(t). \tag{10C.2.1}$$

Interestingly, equation (10C.2.1) is formally analogous to the Langevin equation of Brownian motion:

$$m\frac{dv}{dt} + m\gamma v = F(t). \tag{10C.2.2}$$

The correspondence between equations (10C.2.1) and (10C.2.2) relies on the usual analogies between electrical quantities and mechanical ones:

$$\begin{cases} L \longleftrightarrow m \\ \dfrac{R}{L} \longleftrightarrow \gamma \\ I(t) \longleftrightarrow v(t) \\ V(t) \longleftrightarrow F(t). \end{cases} \tag{10C.2.3}$$

By analogy with the hypotheses made about the Langevin force, we assume that the fluctuating potential difference $V(t)$ is a centered stationary random process, with fluctuations characterized by a correlation time τ_c. We denote by $g(\tau) = \langle V(t)V(t+\tau)\rangle$ the autocorrelation function of $V(t)$ and set:

$$\int_{-\infty}^{\infty} g(\tau)\, d\tau = 2\mathcal{D}L^2. \tag{10C.2.4}$$

If τ_c is much shorter than the other characteristic times, such as for instance the relaxation time $\tau_r = L/R$, we assimilate $g(\tau)$ to a delta function of weight $2\mathcal{D}L^2$:

$$g(\tau) = 2\mathcal{D}L^2\delta(\tau). \tag{10C.2.5}$$

2.2. Evolution of the current from a well-defined initial value

If, at time $t = 0$, the current is perfectly determined and equals I_0, its expression at a time $t > 0$ is:

$$I(t) = I_0 e^{-t/\tau_r} + \frac{1}{L}\int_0^t V(t')e^{-(t-t')/\tau_r}\, dt', \qquad t > 0. \tag{10C.2.6}$$

The average current is given by:

$$\langle I(t)\rangle = I_0 e^{-t/\tau_r}, \qquad t > 0. \tag{10C.2.7}$$

The variance of the current evolves with time as:

$$\sigma_I^2(t) = \mathcal{D}\tau_r(1 - e^{-2t/\tau_r}), \qquad t > 0. \tag{10C.2.8}$$

It saturates at the value $\mathcal{D}\tau_r$ for times $t \gg \tau_r$. In this limit $\langle I^2 \rangle = \mathcal{D}\tau_r$.

It can be shown[2] that at equilibrium the average value $L\langle I^2 \rangle/2$ of the energy stored in the inductance is equal to $kT/2$. Once this result is established, we can follow step by step the approach adopted in the Brownian motion context to derive the second fluctuation-dissipation theorem. At equilibrium, we have $\langle I^2 \rangle = kT/L$ and $\langle I^2 \rangle = \mathcal{D}\tau_r$ as well. We deduce from the identity of both expressions for $\langle I^2 \rangle$ the relation:

$$\frac{1}{\tau_r} = \frac{L}{kT}\mathcal{D}, \tag{10C.2.9}$$

which also reads, on account of the expression for τ_r and of formula (10C.2.4):

$$\boxed{R = \frac{1}{kT}\int_0^\infty \big\langle V(t)V(t+\tau)\big\rangle\, d\tau.} \tag{10C.2.10}$$

Formula (10C.2.10) relates the resistance R of the circuit to the autocorrelation function of the fluctuating potential difference. It constitutes the expression in the present problem for the second fluctuation-dissipation theorem.

2.3. Spectral density of the thermal noise: the Nyquist theorem

The problem can also be solved by harmonic analysis. The spectral density $S_V(\omega)$ of the fluctuating potential difference is the Fourier transform of the autocorrrelation function of $V(t)$.

In the context of the Nyquist theorem, we rather use, instead of $S_V(\omega)$, the spectral density $J_V(\omega)$, defined for positive angular frequencies and related to $S_V(\omega)$ by:

$$J_V(\omega) = \begin{cases} 2S_V(\omega), & \omega \geq 0 \\ 0, & \omega < 0. \end{cases} \tag{10C.2.11}$$

[2] This property results from the aforementioned analogy between electrical quantities and mechanical ones. It can also be directly demonstrated by considering the current as a macroscopic variable of the system. The probability $P(I)dI$ for the current to have a value between I and $I + dI$ is:

$$P(I)\,dI \sim \exp\Big(-\frac{\Delta F}{kT}\Big)\,dI,$$

where ΔF is the variation of the free energy with respect to its value at vanishing current. A global motion of the charges giving rise to a current I, but leaving their relative states of motion unchanged, has a negligible effect on their entropy. We thus have $\Delta F = \Delta E = LI^2/2$, so that:

$$P(I) \sim \exp\Big(-\frac{1}{2}\frac{LI^2}{kT}\Big).$$

The average energy stored in the inductance is thus:

$$\frac{1}{2}L\langle I^2 \rangle = \frac{1}{2}kT.$$

We can rewrite formula (10C.2.10) in the equivalent form:

$$J_V(\omega = 0) = 4RkT. \qquad (10C.2.12)$$

If we assume that the autocorrelation function of the fluctuating potential difference is a delta function, the associated spectral density is independent of the angular frequency: thermal noise is white. If this autocorrelation function has a width $\sim \tau_c$, the associated spectral density is constant up to angular frequencies $\sim {\tau_c}^{-1}$:

$$\boxed{J_V(\omega) = 4RkT, \qquad \omega \ll {\tau_c}^{-1}.} \qquad (10C.2.13)$$

Formula (10C.2.13) constitutes the *Nyquist theorem.*[3]

2.4. Generalization to any linear circuit

The complex admittance of the considered circuit is:

$$\mathcal{A}(\omega) = \frac{1}{R - iL\omega}. \qquad (10C.2.14)$$

The complex impedance $\mathcal{Z}(\omega) = 1/\mathcal{A}(\omega)$ has an ω-independent real part R. For a more general linear circuit, the real part of $\mathcal{Z}(\omega)$ may depend on the angular frequency. It is thus denoted $R(\omega)$. The Nyquist theorem then reads:

$$\boxed{J_V(\omega) = 4R(\omega)kT, \qquad \omega \ll {\tau_c}^{-1}.} \qquad (10C.2.15)$$

The Nyquist theorem is very important in experimental physics and in electronics. It provides a quantitative expression for the noise due to thermal fluctuations in a linear circuit. Therefore it plays a role in any evaluation of the signal-to-noise ratio which limits the performances of a device. The spectral density of the thermal noise is proportional to the temperature. We thus reduce the noise of thermal origin by lowering the temperature.

2.5. Measurement of the Boltzmann constant

The variance of the fluctuating current is given by the integral:

$$\langle I^2(t) \rangle = \frac{1}{2\pi} \int_0^\infty |\mathcal{A}(\omega)|^2 J_V(\omega)\, d\omega, \qquad (10C.2.16)$$

that is, on account of the Nyquist theorem (10C2.13):

$$\langle I^2(t) \rangle = \frac{2kT}{\pi} \int_0^\infty |\mathcal{A}(\omega)|^2 R(\omega)\, d\omega. \qquad (10C.2.17)$$

[3] In practice the correlation time τ_c of the fluctuating potential difference is $\sim 10^{-14}$ s. The spectral density of the thermal noise is thus constant up to angular frequencies $\sim 10^{14}$ s^{-1}.

The values of $R(\omega)$ and of $|\mathcal{A}(\omega)|^2$ are determined experimentally. The measurement of $\langle I^2(t)\rangle$ thus allows us in principle to obtain the value of the Boltzmann constant.

In practice, we are more interested in the variance of $V(t)$. We measure the contribution $\langle V^2(t)\rangle_\omega$ to the variance of the angular frequencies belonging to a given range $(\omega, \omega + \Delta\omega)$ of positive angular frequencies:

$$\langle V^2(t)\rangle_\omega = \frac{1}{2\pi} J_V(\omega)\Delta\omega. \tag{10C.2.18}$$

Using the Nyquist theorem (10C.2.13), we get:

$$\langle V^2(t)\rangle_\omega = \frac{2}{\pi} R(\omega) kT\Delta\omega. \tag{10C.2.19}$$

The measurement of the ratio $\langle V^2(t)\rangle_\omega / R(\omega)$ allows us experimental access to the value of the Boltzmann constant.

Bibliography

R. KUBO, M. TODA, and N. HASHITSUME, *Statistical physics II: nonequilibrium statistical mechanics*, Springer-Verlag, Berlin, second edition, 1991.

F. REIF, *Fundamentals of statistical and thermal physics*, McGraw-Hill, New York, 1965.

References

J.B. JOHNSON, Thermal agitation of electricity in conductors, *Phys. Rev.* **32**, 97 (1928).

H. NYQUIST, Thermal agitation of electric charge in conductors, *Phys. Rev.* **32**, 110 (1928).

R. KUBO, The fluctuation-dissipation theorem and Brownian motion, 1965, *Tokyo Summer Lectures in Theoretical Physics* (R. KUBO editor), Syokabo, Tokyo and Benjamin, New York, 1966.

R. KUBO, The fluctuation-dissipation theorem, *Rep. Prog. Phys.* **29**, 255 (1966).

Chapter 11

Brownian motion: the Fokker–Planck equation

Like the preceding one, this chapter deals with the Brownian motion of a free particle, but from a different point of view. We aim here to determine the temporal evolution of the distribution function of the Brownian particle's velocity. This 'à la Schrödinger' approach is complementary to the previously adopted 'à la Heisenberg' one, which consists in determining the solution of the Langevin equation.

The 'à la Schrödinger' approach relies crucially on the notion of Markov process. If the Langevin force is assumed to be Gaussian and delta-correlated (in other words, if the noise is a Gaussian white noise), the velocity of the Brownian particle can indeed be considered as a Markov process. This property remains approximately true even if the random force is not Gaussian and/or has a finite correlation time (colored noise). The Markovian character of the velocity allows us to express its distribution function at time $t + \Delta t$ in terms of its distribution function at time t, provided that Δt is sufficiently large to be compatible with the Markovian hypothesis ($\Delta t \gg \tau_c$, where τ_c is the correlation time of the Langevin force). Under certain conditions, we can deduce from this relation a second-order partial differential equation for the distribution function, known as the Fokker–Planck equation. Its characteristics are obtained by studying the evolution of the velocity over the time interval Δt via the Langevin equation.

Over much longer evolution time intervals ($\Delta t \gg \tau_r$, where τ_r is the relaxation time of the average velocity), the displacement of the Brownian particle may in turn be represented by a Markov process. The distribution function of the displacement then obeys an evolution equation similar to the Fokker–Planck equation, known as the Einstein–Smoluchowski equation, which identifies with the usual diffusion equation.

1. Evolution of the velocity distribution function

We consider a free Brownian particle, and aim to determine the temporal evolution of the velocity distribution function $f(v,t)$, defined as the probability density that at time t the particle's velocity is between v and $v+dv$. The function $f(v,t)$ gives access, at any time, to quantities such as the average value or the variance of the velocity.

1.1. Markovian character of $v(t)$

The velocity of the Brownian particle obeys the Langevin equation:

$$m\frac{dv}{dt} = -m\gamma v + F(t), \tag{11.1.1}$$

in which $F(t)$ is the Langevin force, whose average value vanishes, and whose autocorrelation function $g(\tau)$ decreases over a characteristic time τ_c. Equation (11.1.1) is only valid if τ_c is much shorter than the relaxation time $\tau_r = \gamma^{-1}$ of the average velocity. It must be interpreted as describing the evolution of the velocity over a time interval Δt between τ_c and τ_r. We assume moreover that $F(t)$ is a Gaussian process. Since equation (11.1.1) is linear, $v(t)$ is also a Gaussian process.[1]

When $g(\tau)$ can be assimilated to a delta function (white noise), we have the following decorrelation property:

$$\langle v(t)F(t')\rangle = 0, \qquad t' > t. \tag{11.1.2}$$

Since $F(t)$ and $v(t)$ are Gaussian processes, it results from formula (11.1.2) that $v(t)$ and $F(t')$ are statistically independent quantities for $t' > t$. Then, since equation (11.1.1) is a first-order differential equation, the evolution of the velocity from a given time depends only on its value at this time (and not on its values at prior times). The Brownian particle's velocity is thus, in this case, a Markov process.[2] This process is called the *Ornstein–Uhlenbeck process*.

The transition probability $p_{1|1}$ of the Markov process associated with $v(t)$ satisfies the Chapman–Kolmogorov equation (or Smoluchowski equation), which reads:

$$p_{1|1}(v_3,t_3|v_1,t_1) = \int p_{1|1}(v_2,t_2|v_1,t_1)p_{1|1}(v_3,t_3|v_2,t_2)\,dv_2, \qquad t_1 < t_2 < t_3. \tag{11.1.3}$$

1.2. Evolution equation of $f(v,t)$

We are interested in the evolution of the distribution function $f(v,t)$ over a time interval $\Delta t \gg \tau_c$. Given that the distribution function at initial time t_0 is $f(v_0,t_0)$, we have, at time t,

$$f(v,t) = \int p_{1|1}(v,t|v_0,t_0)f(v_0,t_0)\,dv_0, \tag{11.1.4}$$

[1] See Supplement 11B.

[2] If the Gaussian hypothesis is not verified, the Markovian character of $v(t)$ is only approximate. The same is true when we take into account the finite correlation time τ_c of the Langevin force (that is, when the noise is colored). In this case, the Brownian particle's velocity may be considered as a Markov process provided that we study its evolution over a time interval $\Delta t \gg \tau_c$.

and, similarly, at time $t + \Delta t$:

$$f(v, t + \Delta t) = \int p_{1|1}(v, t + \Delta t | v_0, t_0) f(v_0, t_0) \, dv_0. \tag{11.1.5}$$

Our aim is to directly relate $f(v, t + \Delta t)$ and $f(v, t)$, without passing through the initial distribution. Since $v(t)$ is a Markov process, the transition probability $p_{1|1}$ verifies equation (11.1.3), which we can rewrite, changing the notations, in the form:

$$p_{1|1}(v, t + \Delta t | v_0, t_0) = \int p_{1|1}(v', t | v_0, t_0) p_{1|1}(v, t + \Delta t | v', t) \, dv'. \tag{11.1.6}$$

Importing the expression (11.1.6) for $p_{1|1}(v, t + \Delta t | v_0, t_0)$ into equation (11.1.5), we get:

$$f(v, t + \Delta t) = \iint p_{1|1}(v', t | v_0, t_0) p_{1|1}(v, t + \Delta t | v', t) \, f(v_0, t_0) \, dv' dv_0. \tag{11.1.7}$$

Making use of the integral expression (11.1.4) for the distribution function (written for $f(v', t)$), we deduce from equation (11.1.7) a direct relation between $f(v, t + \Delta t)$ and $f(v, t)$:

$$\boxed{f(v, t + \Delta t) = \int f(v', t) p_{1|1}(v, t + \Delta t | v', t) \, dv', \qquad \Delta t \gg \tau_c.} \tag{11.1.8}$$

In equation (11.1.8), the transition probability $p_{1|1}(v, t + \Delta t | v', t)$ does not depend separately on times t and $t + \Delta t$, but only on their difference[3] Δt.

It is possible, under certain conditions, to deduce from the integral equation (11.1.8) a partial differential equation for $f(v, t)$. For this purpose, we carry out a systematic expansion of equation (11.1.8), called the *Kramers–Moyal expansion*. This procedure relies on the physical idea that the Brownian particle's velocity variations produced by the collisions with the lighter fluid's molecules are weak in relative value. We thus get from equation (11.1.8) a partial differential equation involving, besides the time-derivative $\partial f(v, t)/\partial t$, partial derivatives of $f(v, t)$ of any order with respect to v. Owing to certain hypotheses, it is possible to keep in this equation only the two first derivatives of $f(v, t)$ with respect to v. The evolution equation thus obtained is the *Fokker–Planck equation*. It represents a particular form of master equation.

2. The Kramers–Moyal expansion

The random motion of the Brownian particle results from the incessant agitation of the molecules of the fluid which surrounds it. The collisions with these molecules

[3] This property, which will be demonstrated in Subsection 2.1, is not enough to ensure the stationarity of the process $v(t)$. As a supplementary requirement, the Brownian particle has to be in equilibrium with the bath.

modify the velocity of the particle. However, the Brownian particle is much heavier than the fluid's molecules, so that the relative variations of its velocity produced by the collisions are small, at least as we consider the evolution of the velocity over a time interval $\Delta t \ll \tau_r$.

Thus, considering equation (11.1.8) over such a time interval, we have to take into account the fact that the velocity variation $w = v - v'$ is much smaller than the velocity v.

2.1. Transition probability

We write the transition probability $p_{1|1}(v, t + \Delta t|v', t)$ in the form of a function $p_{1|1}(w, t + \Delta t|v', t)$ of the velocity variation w and the initial velocity v' (this function represents the conditional distribution of w at fixed v'). It can be deduced from the moments of the velocity variation, this latter quantity being expressed using the Langevin equation (11.1.1) integrated between times t and $t + \Delta t$.

For the sake of simplicity, the argument of $p_{1|1}$ corresponding to the initial velocity, supposed to be fixed, will be denoted by v (instead of v'). The velocity variation w reads:

$$w = -\gamma \int_t^{t+\Delta t} v(t')\,dt' + \frac{1}{m} \int_t^{t+\Delta t} F(t')\,dt'. \tag{11.2.1}$$

From equation (11.2.1), we deduce the average of w at fixed v,

$$\langle w \rangle = -\gamma v \Delta t + O\big[(\gamma \Delta t)^2\big], \tag{11.2.2}$$

as well as the average of w^2 at fixed v, which reads:

$$\langle w^2 \rangle = \gamma^2 \bigg\langle \bigg[\int_t^{t+\Delta t} v(t')\,dt' \bigg]^2 \bigg\rangle - \frac{2\gamma}{m} \int_t^{t+\Delta t} dt' \int_t^{t+\Delta t} dt'' \big\langle v(t') F(t'') \big\rangle + \frac{1}{m^2} \int_t^{t+\Delta t} dt' \int_t^{t+\Delta t} dt'' \big\langle F(t') F(t'') \big\rangle. \tag{11.2.3}$$

For $\Delta t \gg \tau_c$, we can use the simplified expression $g(\tau) = 2\mathcal{D} m^2 \delta(\tau)$ for the Langevin force autocorrelation function. We then have:

$$\big\langle v(t') F(t'') \big\rangle = \begin{cases} 2\mathcal{D} m e^{-\gamma(t'-t'')}, & t'' < t' \\ 0, & t'' > t'. \end{cases} \tag{11.2.4}$$

From equations (11.2.2) and (11.2.3), the variance of w is:

$$\langle w^2 \rangle - \langle w \rangle^2 = 2\mathcal{D} \Delta t + O\big[(\gamma \Delta t)^2\big]. \tag{11.2.5}$$

The higher-order moments of w can be calculated in an analogous way. They all contain, a priori, a contribution of first-order in Δt:

$$\langle w^n \rangle \simeq M_n \Delta t, \qquad \Delta t \ll \tau_r. \tag{11.2.6}$$

In formula (11.2.6), the coefficients M_n may depend on v. In particular, from formulas (11.2.2) and (11.2.5), we have:

$$M_1 = -\gamma v, \qquad M_2 = 2\mathcal{D}. \tag{11.2.7}$$

The Langevin force being Gaussian, the velocity variation w computed from the integrated Langevin equation (11.2.1) is itself a Gaussian random variable.[4] The transition probability $p_{1|1}$, considered as a function of w at fixed v, is thus a Gaussian law. It is characterized by the mean and the variance of w, whose expressions for $\Delta t \ll \tau_r$ are given by formulas (11.2.2) and (11.2.5). These quantities do not depend separately on t and $t + \Delta t$, but only on the time interval Δt. The same is true of the transition probability, which can thus be written as a function[5] $p_{1|1}(w, v, \Delta t)$:

$$\boxed{p_{1|1}(w, v, \Delta t) = \big(4\pi\mathcal{D}\Delta t\big)^{-1/2} \exp\left[-\frac{(w + \gamma v \Delta t)^2}{4\mathcal{D}\Delta t}\right], \qquad \tau_c \ll \Delta t \ll \tau_r.} \tag{11.2.8}$$

When $p_{1|1}$ has the Gaussian form (11.2.8), only the coefficients M_1 and M_2 differ from zero, whereas the coefficients M_n with $n > 2$ vanish.

2.2. Expansion of the evolution equation of $f(v, t)$

To deduce from equation (11.1.8) a partial differential equation for $f(v, t)$, we first rewrite this evolution equation using the previously introduced notations:

$$f(v, t + \Delta t) = \int f(v - w, t) p_{1|1}(w, v - w, \Delta t)\, dw. \tag{11.2.9}$$

By virtue of the inequality $|w| \ll |v|$, the product $f(v - w, t)p_{1|1}(w, v - w, \Delta t)$ can be expanded in a Taylor series[6] of w:

$$f(v - w, t) p_{1|1}(w, v - w, \Delta t) = \sum_{n=0}^{\infty} \frac{(-1)^n}{n!} w^n \frac{\partial^n}{\partial v^n}\Big[f(v, t) p_{1|1}(w, v, \Delta t)\Big]. \tag{11.2.10}$$

Importing the series expansion (11.2.10) into the right-hand side of equation (11.2.9), we obtain the relation:

$$f(v, t + \Delta t) = \sum_{n=0}^{\infty} \frac{(-1)^n}{n!} \int w^n \frac{\partial^n}{\partial v^n}\Big[f(v, t) p_{1|1}(w, v, \Delta t)\Big]\, dw, \tag{11.2.11}$$

---

[4] See Supplement 11B.

[5] Even if we do not assume that $F(t)$ is a Gaussian process, the central limit theorem indicates that $p_{1|1}(w, v, \Delta t)$, considered as a function of w at fixed v, is a Gaussian law provided that we have $\Delta t \gg \tau_c$. The velocity variation during the time interval Δt results in fact from a very large number of statistically independent velocity variations due to the many collisions that the Brownian particle undergoes during the time interval Δt.

[6] Note that the dependence of $p_{1|1}$ with respect to its first argument w must be kept as it is. Actually, it is not possible to make an expansion with respect to this argument since $p_{1|1}(w, v, \Delta t)$ varies rapidly with w.

which also reads, with the aid of the moments $\langle w^n \rangle$ of $p_{1|1}(w, v, \Delta t)$:

$$f(v, t+\Delta t) = \sum_{n=0}^{\infty} \frac{(-1)^n}{n!} \frac{\partial^n}{\partial v^n} \Big[\langle w^n \rangle f(v,t) \Big], \qquad \Delta t \gg \tau_c. \tag{11.2.12}$$

Formula (11.2.12) is the Kramers–Moyal expansion of $f(v, t+\Delta t)$. This expansion was established by H.A. Kramers in 1940 and J.E. Moyal in 1949.

For $\Delta t \ll \tau_r$, the moments $\langle w^n \rangle$ are proportional to Δt (formula (11.2.6)). Keeping only the terms of order Δt, we get from equation (11.2.12):

$$f(v, t+\Delta t) - f(v,t) = \sum_{n=1}^{\infty} \frac{(-1)^n}{n!} \frac{\partial^n}{\partial v^n} (M_n f) \Delta t. \tag{11.2.13}$$

Taking formally the limit $\Delta t \to 0$, we deduce from equation (11.2.13) the partial diffferential equation obeyed by $f(v,t)$:

$$\boxed{\frac{\partial f}{\partial t} = \sum_{n=1}^{\infty} \frac{(-1)^n}{n!} \frac{\partial^n}{\partial v^n} (M_n f).} \tag{11.2.14}$$

3. The Fokker–Planck equation

The structure of equation (11.2.14) suggests that we should study in which cases it is possible to use only a limited numbers of terms in the series appearing on its right-hand side. The Fokker–Planck approximation consists in assuming that the terms of order $n \geq 3$ of this series may be neglected. We then obtain the simpler partial differential equation:

$$\boxed{\frac{\partial f}{\partial t} = -\frac{\partial}{\partial v}(M_1 f) + \frac{1}{2}\frac{\partial^2}{\partial v^2}(M_2 f),} \tag{11.3.1}$$

proposed by A.D. Fokker in 1913 and M. Planck in 1917.

As for the velocity distribution function of a Brownian particle, the Fokker–Planck equation is exact if the Langevin force is Gaussian. Indeed, in this case, the moments of order higher than two of the velocity transfer are of order higher than or equal to $(\Delta t)^2$, so that there are only two non-vanishing terms in the Kramers–Moyal expansion.

3.1. Conservation equation in the velocity space

In the case of Brownian motion as described by the Langevin equation, M_1 and M_2 are given by formula (11.2.7). The Fokker–Planck equation thus reads:

$$\frac{\partial f(v,t)}{\partial t} = \frac{\partial \big[\gamma v f(v,t)\big]}{\partial v} + \frac{\partial^2 \big[\mathcal{D} f(v,t)\big]}{\partial v^2}, \tag{11.3.2}$$

which has the form of a generalized[7] diffusion equation in the velocity space.

It may also be written in the form of a continuity equation,

$$\frac{\partial f}{\partial t} + \frac{\partial J}{\partial v} = 0, \tag{11.3.3}$$

in which the probability current J is defined by:

$$J = -\gamma v f - \mathcal{D}\frac{\partial f}{\partial v}. \tag{11.3.4}$$

The evolution of the solution of the Fokker–Planck equation is thus described by the hydrodynamic picture of a continuous flow in the velocity space. As displayed by formula (11.3.4), the corresponding probability current is the sum of a 'convective' current $-\gamma v f$ and a 'diffusive' current $-\mathcal{D}\partial f/\partial v$.

3.2. Stationary solution

In the stationary regime, f does not depend on t. Consequently, according to equation (11.3.3), J does not depend on v. We can then integrate the differential equation (11.3.4) for f using for instance the variation of parameters method. The only normalizable solution is obtained by making $J = 0$ (in one dimension, a stationary state is thus a state with vanishing current[8]). This gives:

$$f(v) \sim \exp\Big(-\frac{\gamma v^2}{2\mathcal{D}}\Big). \tag{11.3.5}$$

If the bath is in thermodynamic equilibrium at temperature T, the stationary solution (11.3.5) corresponds to the Maxwell–Boltzmann distribution at this temperature, as shown by the second fluctuation-dissipation theorem $\gamma = m\mathcal{D}/kT$: the Brownian particle is then thermalized.

3.3. Fundamental solution

We assume that the initial velocity has a well-defined, non-random, initial value, denoted by v_0. We want the fundamental solution of the Fokker–Planck equation (11.3.2), that is, the solution $f(v,t)$ corresponding to the initial condition:

$$f(v,0) = \delta(v - v_0). \tag{11.3.6}$$

We introduce the Fourier transform[9] with respect to v of $f(v,t)$, defined by:

$$f(\xi,t) = \int_{-\infty}^{\infty} f(v,t)e^{i\xi v}\,dv. \tag{11.3.7}$$

[7] Equation (11.3.2) indeed contains, as well as the diffusion term $\partial^2[\mathcal{D}f(v,t)]/\partial v^2$, the drift term $\partial[\gamma v f(v,t)]/\partial v$.

[8] This property disappears in higher dimensions, where there are stationary states with non-vanishing current.

[9] For the sake of simplicity, we use the same notation $f(.,t)$ for the distribution function $f(v,t)$ and its Fourier transform $f(\xi,t)$.

The function $f(\xi, t)$ obeys the initial condition:

$$f(\xi, 0) = e^{i\xi v_0}. \tag{11.3.8}$$

The Fokker–Planck equation (11.3.2) becomes, after Fourier transformation with respect to v, a first-order partial differential equation for $f(\xi, t)$:

$$\frac{\partial f(\xi, t)}{\partial t} + \gamma\xi \frac{\partial f(\xi, t)}{\partial \xi} = -\mathcal{D}\xi^2 f(\xi, t). \tag{11.3.9}$$

Equation (11.3.9) may be identified with the total differential:

$$\frac{\partial f}{\partial t} dt + \frac{\partial f}{\partial \xi} d\xi = df, \tag{11.3.10}$$

setting:

$$\frac{dt}{1} = \frac{d\xi}{\gamma\xi} = -\frac{df}{\mathcal{D}\xi^2 f}. \tag{11.3.11}$$

From equations (11.3.11), we get:

$$\xi = \xi_1 e^{\gamma t}, \qquad f = f_1 \exp\Big(-\frac{\mathcal{D}}{\gamma}\frac{\xi^2}{2}\Big). \tag{11.3.12}$$

The quantity $\xi e^{-\gamma t}$ is thus a first integral (that is, a quantity conserved in the course of the evolution). The general solution of equation (11.3.9) is of the form:

$$f(\xi, t) = \psi(\xi e^{-\gamma t}) \exp\Big(-\frac{\mathcal{D}}{\gamma}\frac{\xi^2}{2}\Big). \tag{11.3.13}$$

In equation (11.3.13), ψ stands for a function which must be chosen so that $f(\xi, t)$ obeys the initial condition (11.3.8):

$$\psi(\xi) = e^{i\xi v_0} \exp\Big(\frac{\mathcal{D}}{\gamma}\frac{\xi^2}{2}\Big). \tag{11.3.14}$$

This gives:

$$f(\xi, t) = \exp(i\xi e^{-\gamma t} v_0) \exp\Big[-\frac{\mathcal{D}}{\gamma}\frac{\xi^2}{2}(1 - e^{-2\gamma t})\Big]. \tag{11.3.15}$$

The fundamental solution $f(v, t)$ is obtained from expression (11.3.15) for $f(\xi, t)$ by inverse Fourier transformation:

$$\boxed{f(v, t) = \Big(\frac{\gamma}{2\pi\mathcal{D}}\Big)^{1/2} (1 - e^{-2\gamma t})^{-1/2} \exp\Big[-\frac{\gamma}{2\mathcal{D}}\frac{(v - v_0 e^{-\gamma t})^2}{1 - e^{-2\gamma t}}\Big], \qquad t > 0.} \tag{11.3.16}$$

At any time $t > 0$, the distribution (11.3.16) is a Gaussian law of mean:

$$\langle v(t) \rangle = v_0 e^{-\gamma t}, \tag{11.3.17}$$

and of variance:

$$\sigma_v^2(t) = \frac{\mathcal{D}}{\gamma}(1 - e^{-2\gamma t}). \tag{11.3.18}$$

These results for the mean value and the variance of the velocity coincide with the formulas deduced directly from the Langevin equation. At large times ($\gamma t \gg 1$), the fundamental solution approaches the stationary distribution as given by:

$$f(v) = (\gamma/2\pi\mathcal{D})^{1/2} \exp(-\gamma v^2/2\mathcal{D}). \tag{11.3.19}$$

As shown by formula (11.1.4), the fundamental solution $f(v,t)$ identifies with the transition probability $p_{1|1}(v,t|v_0,t_0 = 0)$. In the case $\gamma t \ll 1$, we actually recover[10] from formula (11.3.16) the expression (11.2.8) for $p_{1|1}$.

4. Brownian motion and Markov processes

4.1. Free particle diffusion: the Einstein–Smoluchowski equation

Markov processes plays a central role in the description of the Brownian motion of a free particle. Indeed, the velocity as well as the displacement of a free Brownian particle may be considered as Markov processes (but over time evolution intervals which are very different in the two cases).

• *Velocity*

When the random force is Gaussian and delta-correlated (Gaussian white noise), the velocity of the Brownian particle is a Markov process. The Markovian character of $v(t)$ allows us to write a Fokker–Planck equation for $f(v,t)$. The description of Brownian motion by the Fokker–Planck equation is equivalent to its description by the Langevin equation. In either case we are interested in the evolution over a time interval Δt much shorter than the relaxation time τ_r of the average velocity.

• *Displacement*

Over a time interval $\Delta t \ll \tau_r$, the displacement $x(t) - x_0$ of the Brownian particle from a given initial position is not a Markov process. However, over a time interval $\Delta t \gg \tau_r$, the particle's velocity undergoes a very large number of modifications between two successive observations of the particle's position. As a result, the quantity $x(t + \Delta t) - x(t)$ may then be considered as independent of the coordinate at times prior to t. Over such time intervals, the displacement of the Brownian particle is thus a Markov process.

Indeed, in the Langevin description, the displacement of the Brownian particle obeys a second-order differential equation and thus cannot be considered as a Markov

[10] This can be checked by carrying out the substitutions $t \to \Delta t$, $v \to v_0 + w$.

process, even if the random force corresponds to Gaussian white noise. However, in the viscous or overdamped limit, the inertia term of the Langevin equation is discarded, and the equation of motion takes the form of a first-order differential equation for the displacement:

$$\eta \frac{dx}{dt} = F(t). \tag{11.4.1}$$

Equation (11.4.1) is valid over an evolution time interval $\Delta t \gg \tau_r$. When the random force corresponds to white noise, we have the decorrelation property:

$$\langle x(t)F(t') \rangle = 0, \qquad t' > t, \tag{11.4.2}$$

which, in the Gaussian case, expresses a statistical independence. The displacement $x(t) - x_0 = (1/\eta)\int_0^t F(t')\,dt'$ is a non-stationary Markov process, called the *Wiener process*. The random force autocorrelation function being written in the form:

$$\langle F(t)F(t') \rangle = 2D\eta^2\delta(t - t'), \tag{11.4.3}$$

the probability density $p(x,t)$ obeys the *Einstein–Smoluchowski equation*:

$$\boxed{\frac{\partial p(x,t)}{\partial t} = D\frac{\partial^2 p(x,t)}{\partial x^2}.} \tag{11.4.4}$$

Equation (11.4.4) for $p(x,t)$, analogous to the Fokker–Planck equation for $f(v,t)$, is nothing but the usual diffusion equation.[11] Its fundamental solution for the initial condition $p(x,0) = \delta(x - x_0)$ is a Gaussian law of mean x_0 and of variance $\sigma_x^2(t) = 2Dt$:

$$\boxed{p(x,t) = (4\pi Dt)^{-1/2} \exp\left[-\frac{(x-x_0)^2}{4Dt}\right], \qquad t > 0.} \tag{11.4.5}$$

Otherwise stated, the *diffusion front* is Gaussian.

4.2. Brownian motion of a particle in a potential

If the Brownian motion takes place in a potential $\phi(x)$, the variation of the particle's velocity between times t and $t + \Delta t$ is not determined by the random force $F(t)$ alone, but it also depends on the force $-\partial\phi/\partial x$ acting on the particle, and thus on the particle's position at time t. This latter variable depending in turn on the velocity at times prior to t, the velocity is not a Markov process.[12]

---

[11] Equation (11.4.4) may also be derived in the framework of a *random walk* model (see Supplement 11A).

[12] The Brownian motion of a free particle therefore exhibits a very specific character due to the absence of a potential.

Brownian motion in a space-dependent potential cannot be described by a one-dimensional Markov process (that is, corresponding to a unique random process). Consider as an example the Brownian motion of a particle evolving in a harmonic oscillator potential. It may be described by the following Langevin equation,

$$m\frac{d^2x}{dt^2} + m\gamma\frac{dx}{dt} + m\omega_0^2 x = F(t), \tag{11.4.6}$$

where it is assumed that $F(t)$ corresponds to Gaussian white noise. Equation (11.4.6) being a second-order differential equation, neither the displacement $x(t) - x_0$ of the oscillator nor its velocity $v(t) = dx(t)/dt$ are Markov processes. On the other hand, the two-dimensional process $\{x(t), v(t)\}$ is Markovian. It can indeed be described by a set of two first-order differential equations,

$$\begin{cases} \dfrac{dx(t)}{dt} = v(t) \\ m\dfrac{dv(t)}{dt} + m\gamma v(t) + m\omega_0^2 x(t) = F(t), \end{cases} \tag{11.4.7}$$

and studied with the aid of arguments analogous to those developed for the velocity in the case of a free Brownian particle. We thus can show that the joint distribution function $f(x, v, t)$ obeys a Fokker–Planck equation.

Generally speaking, when a process is not Markovian, it may be considered as a 'projection' of a more complex Markov process (that is, a Markov process with a higher number of dimensions), obtained by introducing additional variables in the description. These additional variables serve to describe explicitly information which otherwise would be implicitly contained in the past values of the relevant variables. Obviously, this procedure is of practical interest only if the additional variables are in limited number. In the case of the Brownian motion of a harmonic oscillator, it is enough to consider both variables $x(t)$ and $v(t)$ to reduce the study to that of a two-dimensional Markov process.

Bibliography

P.M. Chaikin and T.C. Lubensky, *Principles of condensed matter physics*, Cambridge University Press, Cambridge, 1995.

C.W. Gardiner, *Handbook of stochastic methods*, Springer-Verlag, Berlin, third edition, 2004.

N.G. van Kampen, *Stochastic processes in physics and chemistry*, North-Holland, Amsterdam, third edition, 2007.

R. Kubo, M. Toda, and N. Hashitsume, *Statistical physics II: nonequilibrium statistical mechanics*, Springer-Verlag, Berlin, second edition, 1991.

R.M. Mazo, *Brownian motion: fluctuations, dynamics, and applications*, Oxford University Press, Oxford, 2002.

F. Reif, *Fundamentals of statistical and thermal physics*, McGraw-Hill, New York, 1965.

R. Zwanzig, *Nonequilibrium statistical mechanics*, Oxford University Press, Oxford, 2001.

References

A. Einstein, Über die von der molekularkinetischen Theorie der Wärme geforderte Bewegung von in ruhenden Flüssigkeiten suspendierten Teilchen, *Annalen der Physik* **17**, 549 (1905).

M. Smoluchowski, Zur kinetischen Theorie der Brownschen Molekularbewegung und der Suspensionen, *Annalen der Physik*, **21**, 756 (1906).

A.D. Fokker, Die mittlere Energie rotierender elektrischer Dipole im Strahlungsfeld, *Annalen der Physik* **43**, 810 (1914).

M. Planck, An essay on statistical dynamics and its amplification in the quantum theory (translated title), *Sitzungsberichte der preuss. Akademie der Wissenschaften* 324 (1917).

G.E. Uhlenbeck and L.S. Ornstein, On the theory of the Brownian motion, *Phys. Rev.* **36**, 823 (1930). Reprinted in *Selected papers on noise and stochastic processes* (N. Wax editor), Dover Publications, New York, 2003.

H.A. Kramers, Brownian motion in a field of force and the diffusion model of chemical reactions, *Physica* **7**, 284 (1940).

M.C. Wang and G.E. Uhlenbeck, On the theory of the Brownian motion II, *Rev. Mod. Phys.* **17**, 323 (1945). Reprinted in *Selected papers on noise and stochastic processes* (N. Wax editor), Dover Publications, New York, 2003.

J.E. MOYAL, Stochastic processes and statistical physics, *J. Royal Statist. Soc.* Ser. B **11**, 150 (1949).

R. KUBO, The fluctuation-dissipation theorem and Brownian motion, 1965, *Tokyo Summer Lectures in Theoretical Physics* (R. KUBO editor), Syokabo, Tokyo and Benjamin, New York, 1966.

R. KUBO, The fluctuation-dissipation theorem, *Rep. Prog. Phys.* **29**, 255 (1966).

Supplement 11A

Random walk

1. The drunken walker

Einstein's theory of Brownian motion amounts in fact to bringing back its study to a *random walk* problem. In the simplest description, the random walk takes place on an infinite one-dimensional lattice. A walker situated at a given site makes at the initial time a step likely to bring him, with equal probabilities, to one or the other of the two most neighboring sites. The procedure repeats itself after a time interval Δt, and so on. The problem consists in particular in determining the mean value and the variance of the walker's displacement as functions of time (this problem is also known as the *drunken man's walk*).

Random walk is widely used as a microscopic model for diffusion.[1] The problem to be solved may be formulated either in discrete time or in continuous time. To begin with, let us examine the discrete-time formulation.

1.1. Discrete-time formulation

The steps take place at times $k\Delta t$, where k is a non-negative integer. The probability $P_n(k)$ for the walker to be found on site n at time $k\Delta t$ obeys the following difference equation:

$$\boxed{P_n(k+1) = \frac{1}{2}\big[P_{n+1}(k) + P_{n-1}(k)\big].} \tag{11A.1.1}$$

This may also be written in the equivalent form:

$$P_n(k+1) - P_n(k) = \frac{1}{2}\big[P_{n+1}(k) + P_{n-1}(k) - 2P_n(k)\big]. \tag{11A.1.2}$$

Since there is no bias, equations (11A.1.1) and (11A.1.2) are invariant under the transformation $n-1 \longleftrightarrow n+1$.

---

[1] The model described here (*symmetric* random walk) may be extended to the biased case (*asymmetric* random walk). In this latter case, the probabilities that the walker at a given site makes a step towards one or the other of the two most neighboring sites are not equal. The asymmetric random walk is a microscopic model describing a *drift-diffusion* motion (the walker has in this case a finite average velocity).

1.2. Passage to a continuous-time description

We pass to a continuous-time description by letting Δt tend towards zero and the index k tend towards infinity, the product $k\Delta t$ becoming in this limit the time t. The probability $P_n(k)$ then tends towards the probability $p_n(t)$ for the walker to be found on site n at time t, and the quantity $P_n(k+1) - P_n(k)$ tends towards $\Delta t(dp_n(t)/dt)$. The difference equation (11A.1.2) thus takes, in the continuous-time limit, the form of a differential equation:[2]

$$\boxed{\frac{dp_n(t)}{dt} = w\big[p_{n+1}(t) + p_{n-1}(t) - 2p_n(t)\big].} \tag{11A.1.3}$$

In equation (11A.1.3), the quantity $w = (2\Delta t)^{-1}$ represents the transition probability per unit time between a given site and one or the other of its two nearest neighbors.

From the very definition of the model, it results that the displacement of the drunken walker is a Markov process. Equation (11A.1.3) is the master equation governing the evolution of the probabilities $p_n(t)$.

2. Diffusion of a drunken walker on a lattice

From now on we will use the continuous-time formulation. If the walker (whom we will henceforth call the particle) is situated on site $n = 0$ at the initial time $t = 0$, the initial condition for the probabilities reads:

$$p_n(0) = \delta_{n,0}. \tag{11A.2.1}$$

The solution $p_n(t)$ of the master equation (11A.1.3) for the initial condition (11A.2.1) can be expressed with the aid of the nth-order modified Bessel function of the first kind:

$$p_n(t) = e^{-2wt} I_{|n|}(2wt). \tag{11A.2.2}$$

On account of the properties of $I_{|n|}$, we can deduce from formula (11A.2.2) the asymptotic behavior of $p_n(t)$ ($t \to \infty$, $n \to \infty$ with n^2/t fixed):

$$p_n(t) \simeq (4\pi wt)^{-1/2} \exp\Big(-\frac{n^2}{4wt}\Big). \tag{11A.2.3}$$

We will be concerned with the diffusion properties of the particle. The two lowest-order moments of its displacement are defined by the formulas:

$$\langle x(t)\rangle = a \sum_{n=-\infty}^{\infty} n p_n(t), \qquad \langle x^2(t)\rangle = a^2 \sum_{n=-\infty}^{\infty} n^2 p_n(t), \tag{11A.2.4}$$

where a denotes the lattice spacing.

---

[2] Equation (11A.1.3) is a differential equation as far as time is concerned. It remains a difference equation as far as space is concerned, since the random walk takes place on a lattice.

In the symmetric random walk, the initial condition as well as the master equation being invariant under the transformation $n-1 \longleftrightarrow n+1$, we have, for any value of t:

$$\langle x(t)\rangle = 0. \tag{11A.2.5}$$

The second moment of the particle's displacement obeys the evolution equation:

$$\frac{d\langle x^2(t)\rangle}{dt} = wa^2 \sum_{n=-\infty}^{\infty} n^2 \big[p_{n+1}(t) + p_{n-1}(t) - 2p_n(t)\big], \tag{11A.2.6}$$

with the initial condition:

$$\langle x^2(0)\rangle = 0. \tag{11A.2.7}$$

The bracketed quantity on the right-hand side of equation (11A.2.6) is simply equal to $2\sum_{n=-\infty}^{\infty} p_n(t)$, that is, equal to 2 since the sum of probabilities is conserved. To determine $\langle x^2(t)\rangle$, we must therefore solve the differential equation:

$$\frac{d\langle x^2(t)\rangle}{dt} = 2wa^2, \tag{11A.2.8}$$

with the initial condition (11A.2.7). This gives, at any time t:

$$\langle x^2(t)\rangle = 2wa^2 t. \tag{11A.2.9}$$

The motion of the particle (random walker) is thus diffusive. The diffusion coefficient is:

$$\boxed{D = wa^2.} \tag{11A.2.10}$$

3. The diffusion equation

To pass to a spatially continuous description, we make the lattice spacing equal to Δx and we let this latter quantity tend towards zero. We set:

$$p_n(t) = p(x,t)\,\Delta x. \tag{11A.3.1}$$

The function $p(x,t)$ is the probability density of finding of the particle at point x at any given time t.

Expanding the master equation (11A.1.3) at lowest order in Δx, gives the partial differential equation obeyed by $p(x,t)$:

$$\Delta t \frac{\partial p(x,t)}{\partial t} = \frac{1}{2}(\Delta x)^2 \frac{\partial^2 p(x,t)}{\partial x^2}. \tag{11A.3.2}$$

With the scaling hypothesis:

$$\frac{(\Delta x)^2}{\Delta t} = 2D, \tag{11A.3.3}$$

equation (11A.3.2) identifies with a diffusion equation characterized by the diffusion coefficient D as given by formula (11A.2.10). The diffusion front $p(x,t)$ is Gaussian:

$$\boxed{p(x,t) = (4\pi Dt)^{-1/2} \exp\Big(-\frac{x^2}{4Dt}\Big), \qquad t>0.} \tag{11A.3.4}$$

This description of the particle's motion is equivalent to the Einstein–Smoluchowski theory of Brownian motion in the viscous limit.

Bibliography

S. CHANDRASEKHAR, *Stochastic problems in physics and astronomy*, Rev. Mod. Phys. **15**, 1 (1943). Reprinted in *Selected papers on noise and stochastic processes* (N. WAX editor), Dover Publications, New York, 2003.

C.W. GARDINER, *Handbook of stochastic methods*, Springer-Verlag, Berlin, third edition, 2004.

N.G. VAN KAMPEN, *Stochastic processes in physics and chemistry*, North-Holland, Amsterdam, third edition, 2007.

L. MANDEL and E. WOLF, *Optical coherence and quantum optics*, Cambridge University Press, Cambridge, 1995.

R.M. MAZO, *Brownian motion: fluctuations, dynamics, and applications*, Oxford University Press, Oxford, 2002.

G.H. WEISS, *Aspects and applications of the random walk*, North-Holland, Amsterdam, 1994.

R. ZWANZIG, *Nonequilibrium statistical mechanics*, Oxford University Press, Oxford, 2001.

References

A. EINSTEIN, Über die von der molekularkinetischen Theorie der Wärme geforderte Bewegung von in ruhenden Flüssigkeiten suspendierten Teilchen, *Annalen der Physik* **17**, 549 (1905).

M. KAC, Random walk and the theory of Brownian motion, *American Mathematical Monthly* **54**, 369 (1947). Reprinted in *Selected papers on noise and stochastic processes* (N. WAX editor), Dover Publications, New York, 2003.

Supplement 11B

Brownian motion: Gaussian processes

1. Harmonic analysis of stationary Gaussian processes

Consider a stochastic process $Y(t)$, assumed to be centered, and n arbitrary times $t_1, \ldots, t_n$. The set of values $\{y(t_1), \ldots, y(t_n)\}$ taken by the random function $Y(t)$ at these n times defines a n-dimensional random variable.

The process $Y(t)$ is Gaussian if, for any times $t_1, \ldots, t_n$, and whatever their number, the set $\{y(t_1), \ldots, y(t_n)\}$ is a Gaussian random variable. Each of these sets is entirely defined by the correlations $\langle y(t_i)y(t_j)\rangle$ $(1 \leq i \leq n,\ 1 \leq j \leq n)$. As regards the process $Y(t)$, it is fully characterized by its autocorrelation function $g(t, t') = \langle Y(t)Y(t')\rangle$. In particular, all the averages involving more than two times may be computed from the autocorrelation function.

If the process $Y(t)$ is stationary, its Fourier analysis can be carried out. We expand each realization $y(t)$, periodized with the period T, in Fourier series:

$$y(t) = \sum_{n=-\infty}^{\infty} a_n e^{-i\omega_n t}, \qquad \omega_n = \frac{2\pi n}{T}, \qquad 0 \leq t \leq T. \tag{11B.1.1}$$

The limit $T \to \infty$ is taken at the end of the calculations. Each Fourier coefficient a_n is a linear combination of the $\{y(t)\}$'s with $0 \leq t \leq T$:

$$a_n = \frac{1}{T} \int_0^T y(t) e^{i\omega_n t}\, dt. \tag{11B.1.2}$$

We also write:

$$Y(t) = \sum_{n=-\infty}^{\infty} A_n e^{-i\omega_n t}, \qquad 0 \leq t \leq T, \tag{11B.1.3}$$

with:

$$A_n = \frac{1}{T} \int_0^T Y(t) e^{i\omega_n t}\, dt. \tag{11B.1.4}$$

The coefficients A_n, which are linear superpositions of Gaussian random variables, are themselves Gaussian random variables.

The stationarity of the process $Y(t)$ implies that its Fourier coefficients of unequal angular frequencies are uncorrelated:

$$\langle A_n A_{n'}^* \rangle = \langle |A_n|^2 \rangle \delta_{n,n'}. \tag{11B.1.5}$$

Since they are Gaussian, they are statistically independent. The Fourier coefficients of a stationary Gaussian random process are thus independent Gaussian random variables.

2. Gaussian Markov stationary processes

A stationary Gaussian random process is, in addition, Markovian if it has an exponential autocorrelation function (and, accordingly, a Lorentzian spectral density). This result constitutes *Doob's theorem*. We provide here a demonstration of this theorem, assuming for the sake of simplicity that the $\{y(t_i)\}$'s are centered random variables of unit variance.

2.1. The conditional probability $p_{1|1}$ of a stationary Gaussian process

Consider the stationary Gaussian process $Y(t)$. Its one-time probability density reads:

$$p_1(y_1, t_1) = (2\pi)^{-1/2} e^{-y_1^2/2}. \tag{11B.2.1}$$

The law (11B.2.1) does not depend on t_1 (the process is stationary).

The two-time probability density $p_2(y_1, t_1; y_2, t_2)$, which depends only on $|t_2 - t_1|$, is the probability density of the two-dimensional Gaussian variable $\{y(t_1), y(t_2)\}$. To get its expression, we have to introduce the correlation coefficients, bounded by -1 and $+1$ (and here equal to the covariances):

$$\rho_{ij} = \langle y(t_i) y(t_j) \rangle = g(|t_j - t_i|), \qquad \{i, j\} = \{1, 2\}. \tag{11B.2.2}$$

We have $\rho_{ii} = 1$ and $\rho_{ij} = \rho_{ji}$. The covariance matrix is thus of the form:

$$\boldsymbol{M_2} = \begin{pmatrix} 1 & \rho_{12} \\ \rho_{12} & 1 \end{pmatrix}, \tag{11B.2.3}$$

where ρ_{12} is a function of $|t_2 - t_1|$. The two-time distribution $p_2(y_1, t_1; y_2, t_2)$ reads:

$$p_2(y_1, t_1; y_2, t_2) = (2\pi)^{-1} \big(\text{Det}\, \boldsymbol{A_2}\big)^{1/2} \exp\Big(-\frac{1}{2} \boldsymbol{y_2}.\boldsymbol{A_2}.\boldsymbol{y_2}\Big), \tag{11B.2.4}$$

where $\boldsymbol{y_2}$ denotes the vector of components y_1, y_2, and $\boldsymbol{A_2} = \boldsymbol{M_2}^{-1}$ the inverse matrix of $\boldsymbol{M_2}$, of determinant Det $\boldsymbol{A_2}$. Setting:

$$\boldsymbol{A_2} = \begin{pmatrix} a_{11} & a_{12} \\ a_{12} & a_{22} \end{pmatrix}, \tag{11B.2.5}$$

gives:

$$p_2(y_1,t_1;y_2,t_2) = (2\pi)^{-1}\big(\text{Det}\,\boldsymbol{A_2}\big)^{1/2}\exp\left[-\frac{1}{2}(a_{11}y_1^2+a_{22}y_2^2+2a_{12}y_1y_2)\right]. \quad (11B.2.6)$$

Since:

$$\boldsymbol{A_2} = \frac{1}{1-\rho_{12}^2}\begin{pmatrix} 1 & -\rho_{12} \\ -\rho_{12} & 1 \end{pmatrix} \quad (11B.2.7)$$

and:

$$\text{Det}\,\boldsymbol{A_2} = \frac{1}{1-\rho_{12}^2}, \quad (11B.2.8)$$

formula (11B.2.6) also reads:

$$p_2(y_1,t_1;y_2,t_2) = (2\pi)^{-1}(1-\rho_{12}^2)^{-1/2}\exp\left[-\frac{y_1^2+y_2^2-2\rho_{12}y_1y_2}{2(1-\rho_{12}^2)}\right]. \quad (11B.2.9)$$

For $t_1 < t_2$, the conditional probability $p_{1|1}(y_2,t_2|y_1,t_1)$ is obtained by dividing $p_2(y_1,t_1;y_2,t_2)$ (formula (11B.2.9)) by $p_1(y_1,t_1)$ (formula (11B.2.1)):

$$p_{1|1}(y_2,t_2|y_1,t_1) = [2\pi(1-\rho_{12}^2)]^{-1/2}\exp\left[-\frac{(y_2-y_1\rho_{12})^2}{2(1-\rho_{12}^2)}\right], \qquad t_1 < t_2. \quad (11B.2.10)$$

2.2. Necessary condition for the process to be Markovian

Similarly, the three-time distribution $p_3(y_1,t_1;y_2,t_2;y_3,t_3)$ reads:

$$p_3(y_1,t_1;y_2,t_2;y_3,t_3) = (2\pi)^{-3/2}\big(\text{Det}\,\boldsymbol{A_3}\big)^{1/2}\exp\Big(-\frac{1}{2}\boldsymbol{y_3}.\boldsymbol{A_3}.\boldsymbol{y_3}\Big), \quad (11B.2.11)$$

where $\boldsymbol{y_3}$ denotes the vector of components y_1, y_2, y_3, $\boldsymbol{M_3}$ the covariance matrix of dimensions 3×3, and $\boldsymbol{A_3} = \boldsymbol{M_3}^{-1}$ the inverse matrix of $\boldsymbol{M_3}$. The covariance matrix is of the form:

$$\boldsymbol{M_3} = \begin{pmatrix} 1 & \rho_{12} & \rho_{13} \\ \rho_{12} & 1 & \rho_{23} \\ \rho_{13} & \rho_{23} & 1 \end{pmatrix}. \quad (11B.2.12)$$

If we set:

$$\boldsymbol{A_3} = \begin{pmatrix} A_{11} & A_{12} & A_{13} \\ A_{12} & A_{22} & A_{23} \\ A_{13} & A_{23} & A_{33} \end{pmatrix}, \quad (11B.2.13)$$

formula (11B.2.11) can be rewritten as:

$$p_3(y_1,t_1;y_2,t_2;y_3,t_3) = (2\pi)^{-3/2}\big(\text{Det}\,\boldsymbol{A_3}\big)^{1/2}$$
$$\times\exp\left[-\frac{1}{2}\,(A_{11}y_1^2+A_{22}y_2^2+A_{33}y_3^2+2A_{12}y_1y_2+2A_{13}y_1y_3+2A_{23}y_2y_3)\right]. \quad (11B.2.14)$$

For $t_1 < t_2 < t_3$, the conditional probability $p_{1|2}(y_3, t_3|y_2, t_2; y_1, t_1)$ may be obtained by dividing $p_3(y_1, t_1; y_2, t_2; y_3, t_3)$ by $p_2(y_1, t_1; y_2, t_2)$.

For the process $Y(t)$ to be Markovian, it is necessary for the probability $p_{1|2}(y_3, t_3|y_2, t_2; y_1, t_1) = p_3(y_1, t_1; y_2, t_2; y_3, t_3)/p_2(y_1, t_1; y_2, t_2)$ to be independent of y_1. From the comparison between formulas (11B.2.6) and (11B.2.14) for the two-time and three-time probabilities, it follows that the absence of a term in $y_1 y_3$ in p_2 implies that we must have $A_{13} = 0$. As for the terms in y_1^2 and in $y_1 y_2$, they will cancel after the division of p_3 by p_2 provided that we have:

$$A_{11} = a_{11} = \frac{1}{1-\rho_{12}^2}, \qquad A_{12} = a_{12} = -\frac{\rho_{12}}{1-\rho_{12}^2}. \tag{11B.2.15}$$

Since A_{13} is proportional to the minor of ρ_{13} in $\boldsymbol{M_3}$, the condition $A_{13} = 0$ yields:

$$\rho_{23}\rho_{12} - \rho_{13} = 0. \tag{11B.2.16}$$

We can show that, if condition (11B.2.16) is satisfied, it is also the case of conditions (11B.2.15).

2.3. Doob's theorem

Equation (11B.2.16) expresses the condition for the Gaussian process $Y(t)$ to be, in addition, Markovian. This condition may be written in the form of a functional equation for the autocorrelation function:

$$\boxed{g(t_3 - t_2)g(t_2 - t_1) = g(t_3 - t_1), \qquad t_1 < t_2 < t_3.} \tag{11B.2.17}$$

The solution of equation (11B.2.17) is an exponential,

$$g(\tau) = e^{-\gamma|\tau|}, \qquad \gamma > 0, \tag{11B.2.18}$$

which demonstrates Doob's theorem.

3. Application to Brownian motion

Consider again the Langevin equation for the Brownian motion of a free particle,

$$m\frac{dv}{dt} + m\gamma v = F(t), \tag{11B.3.1}$$

where the Langevin force $F(t)$ is a random stationary process. We assume that the spectral density of $F(t)$ is independent of the angular frequency (white noise), which amounts to consider that the corresponding autocorrelation function is a delta function. The process $F(t)$ is then completely random. We make the supplementary hypothesis that it is a Gaussian process.

We assume that the Brownian particle has been in contact with the bath for a sufficiently long time to find itself in equilibrium at any finite time t. Its velocity $v(t)$ is then a random stationary process.

3.1. The Ornstein–Uhlenbeck process

The Fourier transforms of $F(t)$ and $v(t)$ are related by the formula:

$$v(\omega) = \frac{1}{m}\frac{1}{\gamma - i\omega}F(\omega). \tag{11B.3.2}$$

Since $F(t)$ is a Gaussian random process, $F(\omega)$ is a Gaussian random variable. The same is true of $v(\omega)$. Therefore, $v(t)$ is a Gaussian random process, termed the Ornstein–Uhlenbeck process, entirely characterized by its autocorrelation function $\langle v(t)v(t+\tau)\rangle$.

3.2. Verification of Doob's theorem

Since the autocorrelation function of $F(t)$ is a delta function, the solution $v(t)$ of the first-order differential equation (11B.3.1) is a Markov process. The spectral densities $S_v(\omega)$ and $S_F(\omega)$ are related by the formula:

$$S_v(\omega) = \frac{1}{m^2}\frac{1}{\gamma^2 + \omega^2}S_F(\omega). \tag{11B.3.3}$$

The noise is white. Its spectral density is independent of the angular frequency: $S_F(\omega) = S_F$. If the bath is at thermodynamic equilibrium at temperature T, we have $S_F = 2m\gamma kT$. We then deduce from equation (11B.3.3) the equilibrium velocity autocorrelation function:

$$\langle v(t)v\rangle = \frac{kT}{m}e^{-\gamma|t|}. \tag{11B.3.4}$$

In this example, we have verified Doob's theorem: the Ornstein–Uhlenbeck process, which is Gaussian and Markovian, has an exponential autocorrelation function.

3.3. Transition probability of the Ornstein–Uhlenbeck process

The process $v(t)$ being stationary, the one-time probability does not depend on time. Coming back to the dimensioned variables, we have:

$$p_1(v_1) = \Big(\frac{m}{2\pi kT}\Big)^{1/2}\exp\Big(-\frac{mv_1^2}{2kT}\Big). \tag{11B.3.5}$$

The two-time probability $p_2(v_1, t_1; v_2, t_2)$ depends only on $|t_2 - t_1|$. Since, at any time t_i, $v(t_i)$ is a Gaussian random variable of variance kT/m, we get from the general formula (11B.2.9) (re-establishing the dimensioned variables):

$$p_2(v_1, t_1; v_2, t_2) = \frac{m}{2\pi kT}\big(1 - e^{-2\gamma|t_2-t_1|}\big)^{-1/2}\exp\left(-\frac{m}{2kT}\frac{v_1^2 + v_2^2 - 2v_1v_2e^{-\gamma|t_2-t_1|}}{1 - e^{-2\gamma|t_2-t_1|}}\right). \tag{11B.3.6}$$

We deduce from the expressions (11B.3.5) and (11B.3.6) for $p_1(v_1)$ and $p_2(v_1, t_1; v_2, t_2)$ the conditional probability of the Ornstein–Uhlenbeck process:

$$p_{1|1}(v_2, t_2|v_1, t_1) = \left(\frac{m}{2\pi kT}\right)^{1/2} \left[1 - e^{-2\gamma(t_2-t_1)}\right]^{-1/2} \exp\left(-\frac{m}{2kT}\frac{\left[v_2 - v_1 e^{-\gamma(t_2-t_1)}\right]^2}{1 - e^{-2\gamma(t_2-t_1)}}\right), \tag{11B.3.7}$$

where $t_1 < t_2$.

The velocity distribution function at time t, obtained starting from an initial distribution $\delta(v - v_0)$, is thus:

$$f(v, t) = \left(\frac{m}{2\pi kT}\right)^{1/2} \left(1 - e^{-2\gamma t}\right)^{-1/2} \exp\left[-\frac{m}{2kT}\frac{\left(v - v_0 e^{-\gamma t}\right)^2}{1 - e^{-2\gamma t}}\right]. \tag{11B.3.8}$$

The distribution function (11B.3.8) is the fundamental solution of the Fokker–Planck equation.

Bibliography

N.G. VAN KAMPEN, *Stochastic processes in physics and chemistry*, North-Holland, Amsterdam, third edition, 2007.

R.M. MAZO, *Brownian motion: fluctuations, dynamics, and applications*, Oxford University Press, Oxford, 2002.

References

J.L. DOOB, The Brownian movement and stochastic equations, *Ann. Math.* **43**, 351 (1942). Reprinted in *Selected papers on noise and stochastic processes* (N. WAX editor), Dover Publications, New York, 2003.

M.C. WANG and G.E. UHLENBECK, On the theory of the Brownian motion II, *Rev. Mod. Phys.* **17**, 323 (1945). Reprinted in *Selected papers on noise and stochastic processes* (N. WAX editor), Dover Publications, New York, 2003.

Chapter 12

Linear responses and equilibrium correlations

This chapter constitutes an introduction to linear response theory. As it will be seen in the following chapter, this theory allows us to express the response properties of systems slightly departing from equilibrium in terms of the equilibrium correlation functions of the relevant dynamical variables. Adopting here an introductory approach, we limit ourselves to presenting in a general framework the linear response functions, on the one hand, and the equilibrium correlation functions, on the other hand. We do not aim, at this stage, to establish explicitly the link existing between these two types of quantities.

First, to introduce the linear response functions, we consider a system at equilibrium and we submit it to an applied field. When the perturbation is weak enough, its effect on the system may conveniently be described in the framework of a linear approximation. The response of the system to the perturbation is then studied with the aid of the linear response function or of its Fourier transform, the generalized susceptibility. These quantities depend only on the properties of the unperturbed system. The response function possesses the causality property (a physical effect cannot precede the cause which produces it). As a consequence, there exist formulas relating the real and imaginary parts of the generalized susceptibility, namely, the Kramers–Kronig relations. In the case of the response of a physical quantity to its own conjugate field, the rate of energy dissipation within the perturbed system is characterized by the imaginary part of the generalized susceptibility.

We then define the equilibrium autocorrelation functions of dynamical variables. In particular, we make precise the Fourier relation existing between the equilibrium autocorrelation functions and the dynamical structure factors (in terms of which many resonance and inelastic scattering experiments, either of radiation or of particles, are analyzed). The principal properties of the equilibrium autocorrelation functions are reviewed, a particular emphasis being put on the detailed balance relation obeyed by systems in canonical equilibrium.

1. Linear response functions

1.1. Definition

Consider a physical system in thermodynamic equilibrium, and submit it to an external field $a(t)$ assumed to be homogeneous, that is, spatially uniform.[1] The corresponding perturbation is described by the Hamiltonian:

$$H_1(t) = -a(t)A. \tag{12.1.1}$$

The field $a(t)$ is thus coupled to a conjugate physical quantity A. For instance, in the case of a perturbation induced by an electric field, the quantity conjugate to the field is the electric polarization,[2] whereas, in the case of a perturbation induced by a magnetic field, the quantity conjugate to the field is the magnetization.[3]

We are interested in a physical quantity B, of equilibrium average $\langle B\rangle$. We want to determine the response of B to the field $a(t)$, in other words, to compute at any time the modification $\delta\langle B(t)\rangle_a = \langle B(t)\rangle_a - \langle B\rangle$ of the average of B due to the applied field. For the sake of simplicity, we assume that B is centered ($\langle B\rangle = 0$), which allows us to identify $\delta\langle B(t)\rangle_a$ and $\langle B(t)\rangle_a$. In the linear range, the out-of-equilibrium average of B is written in the form:

$$\boxed{\big\langle B(t)\big\rangle_a = \int_{-\infty}^{\infty} \tilde{\chi}_{BA}(t,t')a(t')\,dt',} \tag{12.1.2}$$

where the real quantity $\tilde{\chi}_{BA}(t,t')$ is a *linear response function* depending only on the properties of the unperturbed system.

1.2. General properties

The linear response function $\tilde{\chi}_{BA}(t,t')$ is invariant under time-translation and is *causal*.

- *Time-translational invariance*

The unperturbed system being at equilibrium, the linear response function $\tilde{\chi}_{BA}(t,t')$ does not depend separately on the two arguments t and t', but on the difference $t-t'$ alone. Thus, the response function is invariant under time-translation.[4] Accordingly,

---

[1] The case of an inhomogeneous field will be treated in Section 5.

[2] See Supplement 12B about the polarization of an atom perturbed by an electric field and Supplement 13A about dielectric relaxation.

[3] See Supplement 13B on magnetic resonance.

[4] This property disappears when the unperturbed system is not in thermodynamic equilibrium. This is in particular the case with spin glasses and structural glasses, which display *aging properties* expressed possibly through the separate dependence of some response functions $\tilde{\chi}_{BA}(t,t')$ with respect to t and t'. These functions then depend, not only on the time difference, but also on the age of the system, that is, on the time elapsed since its preparation (see also Chapter 14 for further remarks about this question).

formula (12.1.2) has the structure of a convolution product:

$$\langle B(t) \rangle_a = \int_{-\infty}^{\infty} \tilde{\chi}_{BA}(t - t')a(t')\, dt'. \tag{12.1.3}$$

If the field is a delta function pulse ($a(t) = a\delta(t)$), then:

$$\langle B(t) \rangle_a = a\tilde{\chi}_{BA}(t). \tag{12.1.4}$$

The response function $\tilde{\chi}_{BA}(t)$ thus represents the *impulsional response*.[5]

• *Causal character*

The *causality principle* is a commonly admitted physical principle which states that any physical effect must follow in time the cause which produces it. A modification of the applied field taking place at a time t' may thus lead to a modification of $\langle B(t)\rangle_a$ only at times $t > t'$. In other words, the response function $\tilde{\chi}_{BA}(t, t')$ is causal, which means that it may be non-vanishing only for $t > t'$. The actual upper bound in the integral on the right-hand side of formulas (12.1.2) and (12.1.3) is thus t, and not $+\infty$.

1.3. Linear response of a quantity to its conjugate field

We will limit ourselves in this introduction to the study of the linear response of a physical quantity A to its own conjugate field $a(t)$. We denote simply by $\tilde{\chi}(t)$ (instead of $\tilde{\chi}_{AA}(t)$) the corresponding response function, which we will not try here to compute explicitly.[6]

2. Generalized susceptibilities

2.1. Linear response to a harmonic perturbation

The linear response of A to the harmonic field $a(t) = \Re e(ae^{-i\omega t})$ reads:

$$\langle A(t) \rangle_a = \int_{-\infty}^{\infty} \tilde{\chi}(t - t')\, \Re e(ae^{-i\omega t'})\, dt', \tag{12.2.1}$$

that is, the response function being real:

$$\langle A(t) \rangle_a = \Re e\left[\int_{-\infty}^{\infty} \tilde{\chi}(t - t')ae^{-i\omega t'}\, dt'\right]. \tag{12.2.2}$$

Since the argument of $\tilde{\chi}(t - t')$ must be positive, the upper bound of the integral on the right-hand side of equations (12.2.1) and (12.2.2) is in fact t. Setting $t - t' = \tau$, gives:

$$\langle A(t) \rangle_a = \Re e\left[ae^{-i\omega t}\int_{0}^{\infty} \tilde{\chi}(\tau)e^{i\omega\tau}\, d\tau\right]. \tag{12.2.3}$$

---

[5] It is also called the *retarded Green's function*.

[6] The general theory of linear response will be expounded in Chapter 13, where in particular the *Kubo formulas* enabling us to compute explicitly the linear response functions in terms of equilibrium correlation functions of the relevant dynamical variables will be established.

With the aid of the *generalized susceptibility* $\chi(\omega)$, defined as the Fourier transform of the causal function $\tilde{\chi}(t)$, that is, by the formula:

$$\chi(\omega) = \int_0^\infty \tilde{\chi}(t) e^{i\omega t}\, dt, \tag{12.2.4}$$

formula (12.2.3) reads:

$$\langle A(t) \rangle_a = \Re e \left[a e^{-i\omega t} \chi(\omega) \right]. \tag{12.2.5}$$

Setting $\chi(\omega) = \chi'(\omega) + i\chi''(\omega)$, we write the response $\langle A(t) \rangle_a$ to a harmonic field $a(t) = a \cos \omega t$ of real amplitude a in the form of a sum of two terms,

$$\langle A(t) \rangle_a = a \left[\chi'(\omega) \cos \omega t + \chi''(\omega) \sin \omega t \right], \tag{12.2.6}$$

in which $a\chi'(\omega) \cos \omega t$ represents the *in-phase* response and $a\chi''(\omega) \sin \omega t$ the *out-of-phase* response.

More generally, the response $\langle A(t) \rangle_a$ to an applied field $a(t)$ of Fourier transform $a(\omega)$ has the Fourier transform:[7]

$$\boxed{\langle A(\omega) \rangle_a = a(\omega) \chi(\omega).} \tag{12.2.7}$$

2.2. Definition of the generalized susceptibility $\chi(\omega)$

The definition (12.2.4) of the Fourier transform of $\tilde{\chi}(t)$ may pose a problem. The integral on the right-hand side of equation (12.2.4) may indeed not converge: in such a case, the generalized susceptibility does not exist as a function. It may however be defined in the distribution sense, that is, as the limit of a convenient sequence of functions. Such a limit may for instance be obtained by considering the sequence of functions $\chi(\omega + i\epsilon)$ defined by:

$$\chi(\omega + i\epsilon) = \int_0^\infty \tilde{\chi}(t) e^{i\omega t} e^{-\epsilon t}\, dt, \qquad \epsilon > 0, \tag{12.2.8}$$

and by letting ϵ tend towards zero at the end of the calculation. We then set:

$$\boxed{\chi(\omega) = \lim_{\epsilon \to 0^+} \chi(\omega + i\epsilon).} \tag{12.2.9}$$

The generalized susceptibility is thus defined as the Fourier transform in the distribution sense of the response function.

[7] For the sake of simplicity, we use the same notation $a(.)$ for the field $a(t)$ and its Fourier transform $a(\omega)$, as well as the same notation $\langle A(.) \rangle_a$ for the response $\langle A(t) \rangle_a$ and its Fourier transform $\langle A(\omega) \rangle_a$.

It is useful to introduce the function $\chi(z)$ defined for a complex argument z with $\Im m\, z > 0$ as the Fourier–Laplace transform of $\tilde{\chi}(t)$:

$$\boxed{\chi(z) = \int_0^\infty \tilde{\chi}(t) e^{izt}\, dt, \qquad \Im m\, z > 0.} \tag{12.2.10}$$

The function $\chi(z)$ is analytic in the upper complex half-plane. We assume in the rest of this chapter[8] that $\chi(z) \to 0$ as $z \to \infty$. As displayed by formula (12.2.9), the generalized susceptibility $\chi(\omega)$ is the limit of $\chi(z)$ as the point of affix $z = \omega + i\epsilon$ in the upper complex half-plane tends towards the point of abscissa ω on the real axis.

2.3. Spectral representations of $\chi(z)$

The response function being real, the real (*resp.* imaginary) part of the susceptibility is an even (*resp.* odd) function of ω. From formulas (12.2.8) and (12.2.9) yielding $\chi(\omega)$, we deduce the expressions for $\chi'(\omega)$ and $\chi''(\omega)$:

$$\left\{ \begin{aligned} \chi'(\omega) &= \lim_{\epsilon \to 0^+} \int_0^\infty \tilde{\chi}(t) \cos \omega t\, e^{-\epsilon t}\, dt \\ \chi''(\omega) &= \lim_{\epsilon \to 0^+} \int_0^\infty \tilde{\chi}(t) \sin \omega t\, e^{-\epsilon t}\, dt. \end{aligned} \right. \tag{12.2.11}$$

The function $\chi'(\omega)$ is the Fourier transform of the even part $\tilde{\chi}_p(t) = \frac{1}{2}[\tilde{\chi}(t) + \tilde{\chi}(-t)]$ of $\tilde{\chi}(t)$ and (up to a factor $1/i$) the function $\chi''(\omega)$ is the Fourier transform of its odd part $\tilde{\chi}_i(t) = \frac{1}{2}[\tilde{\chi}(t) - \tilde{\chi}(-t)]$. Setting $\tilde{\chi}_p(t) = \tilde{\chi}'(t)$ and $\tilde{\chi}_i(t) = i\tilde{\chi}''(t)$, we have $\chi'(\omega) = \int_{-\infty}^\infty \tilde{\chi}'(t) e^{i\omega t}\, dt$ and $\chi''(\omega) = \int_{-\infty}^\infty \tilde{\chi}''(t) e^{i\omega t}\, dt$.

• *Spectral representation of $\chi(z)$ in terms of $\chi''(\omega)$*

We can represent $\tilde{\chi}(t)$ with the aid of its odd part:

$$\tilde{\chi}(t) = 2\Theta(t)\tilde{\chi}_i(t) = 2i\Theta(t)\tilde{\chi}''(t), \tag{12.2.12}$$

where $\Theta(t)$ denotes the Heaviside function. Then, coming back to formula (12.2.10), valid for $\Im m\, z > 0$, we can write:

$$\chi(z) = 2i \int_0^\infty dt\, e^{izt} \int_{-\infty}^\infty \frac{d\omega}{2\pi} \chi''(\omega) e^{-i\omega t}, \qquad \Im m\, z > 0. \tag{12.2.13}$$

The integration over time leads to the *spectral representation* of $\chi(z)$ in terms of $\chi''(\omega)$:

$$\boxed{\chi(z) = \frac{1}{\pi} \int_{-\infty}^\infty \frac{\chi''(\omega)}{\omega - z}\, d\omega.} \tag{12.2.14}$$

---

[8] Supplementary information on the behavior of $\chi(z)$ when $z \to \infty$ is given in the appendix at the end of this chapter.

Formula (12.2.14) displays the fact that $\chi(z)$ exhibits singularities only on the real axis. It is thus an analytic function of z in the upper complex half-plane.[9]

• *Spectral representation of* $\chi(z)$ *in terms of* $\chi'(\omega)$

We can alternatively represent $\tilde{\chi}(t)$ with the aid of its even part:

$$\tilde{\chi}(t) = 2\Theta(t)\tilde{\chi}_p(t) = 2\Theta(t)\tilde{\chi}'(t). \tag{12.2.15}$$

Proceeding as above, we obtain a spectral representation of $\chi(z)$ in terms of $\chi'(\omega)$:

$$\chi(z) = \frac{1}{i\pi}\int_{-\infty}^{\infty} \frac{\chi'(\omega)}{\omega - z}\, d\omega. \tag{12.2.16}$$

Both representations (12.2.16) and (12.2.14) are equivalent. Formula (12.2.14) is however more frequently used, due to the fact that $\chi''(\omega)$ is a quantity often easier to measure than $\chi'(\omega)$, and is also directly related to the energy dissipation within the system.[10]

3. The Kramers–Kronig relations

The *Kramers–Kronig relations* are integral transformations relating the real part and the imaginary part of any generalized susceptibility. They were established in 1926 by R. Kronig and in 1927 by H.A. Kramers. Their existence relies on the fact that the response function $\tilde{\chi}(t)$ is causal, and therefore that the function $\chi(z)$ defined by the Fourier–Laplace integral (12.2.10) is analytic in the upper complex half-plane. This analyticity property follows from the causality alone, and it does not depend on the specific form of $\tilde{\chi}(t)$. This is why it is possible to derive the Kramers–Kronig relations without having at hand explicit expressions either for $\tilde{\chi}(t)$ or for $\chi(\omega)$. Here we will directly deduce[11] these relations from the spectral representations (12.2.14) and (12.2.16).

Writing these representations for $z = \omega + i\epsilon$ ($\epsilon > 0$), gives the formulas:

$$\chi(\omega + i\epsilon) = \frac{1}{\pi}\int_{-\infty}^{\infty} \frac{\chi''(\omega')}{\omega' - (\omega + i\epsilon)}\, d\omega' \tag{12.3.1}$$

and:

$$\chi(\omega + i\epsilon) = \frac{1}{i\pi}\int_{-\infty}^{\infty} \frac{\chi'(\omega')}{\omega' - (\omega + i\epsilon)}\, d\omega'. \tag{12.3.2}$$

[9] The spectral representation (12.2.14) may be used to define $\chi(z)$ at any point of affix z situated outside the real axis. It defines an analytic function of z either in the upper complex half-plane or in the lower complex half-plane (see Chapter 13), in contrast to the integral definition (12.2.10) valid only for $\Im m\, z > 0$.

[10] See Section 4.

[11] An alternative way, very commonly used, to establish the Kramers–Kronig relations, will be presented in appendix at the end of this chapter.

Taking the limit $\epsilon \to 0^+$, in which $\chi(\omega + i\epsilon)$ tends towards $\chi(\omega) = \chi'(\omega) + i\chi''(\omega)$, we get[12] from formulas (12.3.1) and (12.3.2) the Kramers–Kronig relations:

$$\begin{cases} \chi'(\omega) = \dfrac{1}{\pi}\,\mathrm{vp}\displaystyle\int_{-\infty}^{\infty} \frac{\chi''(\omega')}{\omega' - \omega}\,d\omega' \\[2ex] \chi''(\omega) = -\dfrac{1}{\pi}\,\mathrm{vp}\displaystyle\int_{-\infty}^{\infty} \frac{\chi'(\omega')}{\omega' - \omega}\,d\omega'. \end{cases} \tag{12.3.3}$$

Formulas (12.3.3) are a direct consequence of the causality principle.

Thus, the real and imaginary parts of $\chi(\omega)$ are not independent of each other, but they are related by an integral transformation: $\chi''(\omega)$ is the *Hilbert transform* of $\chi'(\omega)$, and $\chi'(\omega)$ is the *inverse Hilbert transform*[13] of $\chi''(\omega)$. The knowledge either of $\chi'(\omega)$ or of $\chi''(\omega)$ is thus enough to fully determine the generalized susceptibility. Most often, we measure $\chi''(\omega)$ (for instance, in an absorption experiment). If the measurement of $\chi''(\omega)$ is carried out over a sufficiently large angular frequencies interval, the Kramers–Kronig relations then allow us to get $\chi'(\omega)$.

4. Dissipation

The instantaneous power absorbed by the system submitted to the field $a(t)$ is:

$$\frac{dW}{dt} = a(t)\frac{d\langle A(t)\rangle_a}{dt}. \tag{12.4.1}$$

The linear response $\langle A(t)\rangle_a$ to the harmonic field $a(t) = a\cos\omega t$ is given by formula (12.2.5). The corresponding instantaneous absorbed power is thus:

$$\frac{dW}{dt} = a^2\omega\cos\omega t\big[-\chi'(\omega)\sin\omega t + \chi''(\omega)\cos\omega t\big]. \tag{12.4.2}$$

The average power absorbed over one period (or over an integer number of periods) is given by the formula:

$$\overline{\frac{dW}{dt}} = \frac{1}{2}a^2\omega\chi''(\omega). \tag{12.4.3}$$

---

[12] We make use of the relation:

$$\lim_{\epsilon\to 0^+}\frac{1}{x + i\epsilon} = \mathrm{vp}\,\frac{1}{x} - i\pi\delta(x),$$

where the symbol vp denotes the Cauchy principal value.

[13] The Hilbert transformation and the inverse Hilbert transformation are identical except for the sign.

The average energy absorption rate in the system is thus related to the imaginary part of the generalized susceptibility. The energy supplied by the external field is eventually dissipated irreversibly within the system, that is, transformed into heat.[14] For this reason, $\chi''(\omega)$ is qualified as the *dissipative part* of the generalized susceptibility.

At thermodynamic equilibrium, the average dissipated power is positive.[15] As a consequence, the quantity $\omega\chi''(\omega)$ is positive.

5. Non-uniform phenomena

5.1. Linear response functions

We are still considering here a physical system at thermodynamic equilibrium. We submit it to an inhomogeneous external field $a(\boldsymbol{r},t)$. The perturbation Hamiltonian may in principle be written in a form generalizing the Hamiltonian (12.1.1) of the homogeneous case:

$$H_1(t) = -\int d\boldsymbol{r}\, a(\boldsymbol{r},t)A(\boldsymbol{r}). \tag{12.5.1}$$

In formula (12.5.1), $A(\boldsymbol{r})$ is the physical quantity conjugate to the field $a(\boldsymbol{r},t)$ and the spatial integration is carried over the whole volume of the system. In some cases however, the perturbation Hamiltonian involves more naturally, instead of the applied field, the potential from which this latter derives. Let us take the example of a system of electrons perturbed by a non-uniform electric field $\boldsymbol{E}(\boldsymbol{r},t) = -\nabla\phi(\boldsymbol{r},t)$. The perturbation Hamiltonian reads $H_1(t) = \int d\boldsymbol{r}\,\phi(\boldsymbol{r},t)\rho(\boldsymbol{r})$, where $\rho(\boldsymbol{r}) = e\sum_i \delta(\boldsymbol{r}-\boldsymbol{r}_i)$ is the charge density at point $\boldsymbol{r}$ (e denotes the electron charge and the $\{\boldsymbol{r}_i\}$'s are the positions of the different electrons). The charge density $\rho(\boldsymbol{r})$ is thus coupled to the potential $\phi(\boldsymbol{r},t)$.

In an inhomogeneous situation, to determine the response of the physical quantity $B(\boldsymbol{r})$ to the field $a(\boldsymbol{r},t)$ amounts to computing the modification $\delta\langle B(\boldsymbol{r},t)\rangle_a$ of the average value of $B(\boldsymbol{r})$ due to the applied field. We assume that $B(\boldsymbol{r})$ is centered. In the linear range, its out-of-equilibrium average reads:

$$\boxed{\langle B(\boldsymbol{r},t)\rangle_a = \int d\boldsymbol{r}' \int_{-\infty}^{\infty} \tilde{\chi}_{BA}(\boldsymbol{r},t;\boldsymbol{r}',t')a(\boldsymbol{r}',t')\,dt'.} \tag{12.5.2}$$

The real quantity $\tilde{\chi}_{BA}(\boldsymbol{r},t;\boldsymbol{r}',t')$ is a non-local linear response function. The spatial integration on the right-hand side of formula (12.5.2) is carried out over the whole volume of the system.

As in the homogeneous case, the linear response function is invariant under time-translation and is also causal. Moreover, if the unperturbed system is invariant under

[14] A simple example of a dissipative system is a harmonic oscillator damped by viscous friction (see Supplement 12A). In this case, the energy supplied by the field is transferred to the incoherent degrees of freedom of the damping fluid.

[15] This property will be demonstrated in Chapter 14 with the aid of the microscopic expression for the average dissipated power.

space translation, which we assume to be the case in the following of this section, the response function $\tilde{\chi}_{BA}(\boldsymbol{r},t\,;\boldsymbol{r}',t')$ does not depend separately on the two arguments $\boldsymbol{r}$ and $\boldsymbol{r}'$, but only on the difference $\boldsymbol{r}-\boldsymbol{r}'$.

As in the uniform case, we limit ourselves here to the study of the linear response of a physical quantity $A(\boldsymbol{r})$ to its own conjugate field $a(\boldsymbol{r},t)$, denoting simply by $\tilde{\chi}(\boldsymbol{r},t)$ the corresponding response function.

5.2. Generalized susceptibilities

The linear response of $A(\boldsymbol{r})$ to the harmonic field $a(\boldsymbol{r},t)=\Re e[ae^{i(\boldsymbol{q}.\boldsymbol{r}-\omega t)}]$ reads:

$$\big\langle A(\boldsymbol{r},t)\big\rangle_a=\int d\boldsymbol{r}'\int_{-\infty}^{\infty}\tilde{\chi}(\boldsymbol{r}-\boldsymbol{r}',t-t')\,\Re e[ae^{i(\boldsymbol{q}.\boldsymbol{r}'-\omega t')}]\,dt', \tag{12.5.3}$$

that is:

$$\big\langle A(\boldsymbol{r},t)\big\rangle_a=\Re e\bigg[\int d\boldsymbol{r}'\int_{-\infty}^{\infty}\tilde{\chi}(\boldsymbol{r}-\boldsymbol{r}',t-t')ae^{i(\boldsymbol{q}.\boldsymbol{r}'-\omega t')}\,dt'\bigg]. \tag{12.5.4}$$

Setting $t-t'=\tau$ and $\boldsymbol{r}-\boldsymbol{r}'=\boldsymbol{\rho}$, gives:

$$\big\langle A(\boldsymbol{r},t)\big\rangle_a=\Re e\bigg[ae^{i(\boldsymbol{q}.\boldsymbol{r}-\omega t)}\int d\boldsymbol{\rho}\,e^{-i\boldsymbol{q}.\boldsymbol{\rho}}\int_0^{\infty}\tilde{\chi}(\boldsymbol{\rho},\tau)e^{i\omega\tau}\,d\tau\bigg]. \tag{12.5.5}$$

In terms of the generalized susceptibility $\chi(\boldsymbol{q},\omega)$, defined as the spatial and temporal Fourier transform[16] of $\tilde{\chi}(\boldsymbol{r},t)$, that is, by the formula:

$$\chi(\boldsymbol{q},\omega)=\int d\boldsymbol{r}\,e^{-i\boldsymbol{q}.\boldsymbol{r}}\lim_{\epsilon\to0^+}\int_0^{\infty}\tilde{\chi}(\boldsymbol{r},t)e^{i\omega t}e^{-\epsilon t}\,dt, \tag{12.5.6}$$

formula (12.5.5) reads:

$$\big\langle A(\boldsymbol{r},t)\big\rangle_a=\Re e\Big[ae^{i(\boldsymbol{q}.\boldsymbol{r}-\omega t)}\chi(\boldsymbol{q},\omega)\Big]. \tag{12.5.7}$$

More generally, the response $\langle A(\boldsymbol{r},t)\rangle_a$ to an applied field of Fourier transform $a(\boldsymbol{q},\omega)$ has the spatial and temporal Fourier transform:

$$\boxed{\big\langle A(\boldsymbol{q},\omega)\big\rangle_a=a(\boldsymbol{q},\omega)\chi(\boldsymbol{q},\omega).} \tag{12.5.8}$$

The study of the generalized susceptibility $\chi(\boldsymbol{q},\omega)$ at given wave vector $\boldsymbol{q}$ can be carried out in the same way as the study of $\chi(\omega)$. Using notations extending those of the homogeneous case, we can in particular write the spectral representation:[17]

$$\boxed{\chi(\boldsymbol{q},z)=\frac{1}{\pi}\int_{-\infty}^{\infty}\frac{\chi''(\boldsymbol{q},\omega)}{\omega-z}\,d\omega,} \tag{12.5.9}$$

[16] Note that, because of the causality principle, the spatial and temporal variables do not play equivalent roles as far as the response function is concerned. The spatial Fourier transform of the response function is a true Fourier transform, whereas the so-called temporal 'Fourier' transform is in fact a Fourier–Laplace transform (which must be taken in the distribution sense).

[17] As in the uniform case, this spectral representation can be extended to any point of affix z situated outside the real axis.

as well as Kramers–Kronig relations between $\chi'(\boldsymbol{q},\omega)$ and $\chi''(\boldsymbol{q},\omega)$.

In addition, the response function being real, we have the properties:

$$\begin{cases} \chi'(\boldsymbol{q},\omega) = \chi'(-\boldsymbol{q},-\omega) \\ \chi''(\boldsymbol{q},\omega) = -\chi''(-\boldsymbol{q},-\omega). \end{cases} \tag{12.5.10}$$

6. Equilibrium correlation functions

The dynamical properties of systems in thermodynamic equilibrium can be expressed in terms of equilibrium *correlation functions*. These functions are the appropriate quantities for interpreting several important experimental techniques, especially in condensed matter physics. Some of these techniques involve resonance methods. This is for instance the case of the nuclear magnetic resonance, the paramagnetic electronic resonance, or the Mössbauer spectroscopy. Others involve inelastic scattering either of radiation (acoustic waves, light, X-rays ...) or of neutral or charged particles (neutrons, electrons ...).

In the rest of this chapter, we will introduce the equilibrium correlation functions and study their main properties. We will directly adopt notations appropriate to the non-uniform case.

6.1. Definition

The system at equilibrium is described by a time-independent Hamiltonian H_0. A dynamical variable or observable $A(\boldsymbol{r})$ possibly depending on the space point $\boldsymbol{r}$ is associated with each physical quantity of the system. An example of such an observable is the particle density $n(\boldsymbol{r}) = \sum_i \delta(\boldsymbol{r}-\boldsymbol{r}_i)$.

In classical mechanics, a dynamical variable $A(\boldsymbol{r},t)$ is a function of the generalized coordinates and momenta, which evolve according to the Hamilton's equations governed by H_0. In quantum mechanics, we associate with an observable a Hermitean operator $A(\boldsymbol{r},t) = e^{iH_0t/\hbar}A(\boldsymbol{r})e^{-iH_0t/\hbar}$, which evolves according to the Heisenberg equation:

$$i\hbar\frac{dA(\boldsymbol{r},t)}{dt} = [A(\boldsymbol{r},t), H_0]. \tag{12.6.1}$$

Consider two classical operators[18] (dynamical variables) $A_i(\boldsymbol{r},t)$ and $A_j(\boldsymbol{r},t)$, assumed to be centered for the sake of simplicity. Their equilibrium correlation function

[18] In the quantum case, because of the non-commutativity of the operators, there are several non-equivalent definitions of the correlation function of two operators $A_i(\boldsymbol{r},t)$ and $A_j(\boldsymbol{r},t)$. It is even the case for the autocorrelation function of an operator $A(\boldsymbol{r},t)$, since $A(\boldsymbol{r},t)$ does not in general commute with $A(\boldsymbol{r}',t')$. We use in particular the *symmetric correlation function* $\tilde{S}_{A_iA_j}(\boldsymbol{r},t) = \frac{1}{2}\langle A_i(\boldsymbol{r},t)A_j + A_jA_i(\boldsymbol{r},t)\rangle$, as well as, for a system in canonical equilibrium, the Kubo *canonical correlation function* $\tilde{K}_{A_iA_j}(\boldsymbol{r},t) = \beta^{-1}\int_0^\beta \langle e^{\lambda H_0}A_je^{-\lambda H_0}A_i(\boldsymbol{r},t)\rangle\,d\lambda$, where $\beta = (kT)^{-1}$. In the classical limit, $\tilde{S}_{A_iA_j}(\boldsymbol{r},t)$ and $\tilde{K}_{A_iA_j}(\boldsymbol{r},t)$ both reduce to $\tilde{C}_{A_iA_j}(\boldsymbol{r},t)$ (see Chapter 14). In the present discussion, we use the notation $\tilde{C}$ for the correlation function (in the quantum case, $\tilde{C}$ stands in fact for either $\tilde{S}$ or $\tilde{K}$).

is defined by the formula:

$$\boxed{\tilde{C}_{A_iA_j}(\boldsymbol{r},t\,;\boldsymbol{r}',t') = \langle A_i(\boldsymbol{r},t)A_j(\boldsymbol{r}',t')\rangle,} \tag{12.6.2}$$

where the symbol $\langle\ldots\rangle$ denotes the thermodynamic equilibrium average. When $A_i = A_j = A$, the quantity:

$$\tilde{C}_{AA}(\boldsymbol{r},t\,;\boldsymbol{r}',t') = \langle A(\boldsymbol{r},t)A(\boldsymbol{r}',t')\rangle \tag{12.6.3}$$

is the *autocorrelation function* of the operator $A(\boldsymbol{r},t)$.

For a translationally invariant system (such as a homogeneous fluid), the equilibrium correlation function of two operators A_i and A_j depends on one spatial variable and one temporal variable. It is therefore denoted by $\tilde{C}_{A_iA_j}(\boldsymbol{r},t)$.

6.2. Power spectrum of an operator: generalization of the Wiener–Khintchine theorem

Consider a centered operator $A(\boldsymbol{r},t)$. To define its power spectrum, we first introduce the temporal Fourier transform[19] of $A(\boldsymbol{r},t)$:

$$A(\boldsymbol{r},\omega) = \int_{-\infty}^{\infty} A(\boldsymbol{r},t)e^{i\omega t}\,dt. \tag{12.6.4}$$

Inversely, we write:

$$A(\boldsymbol{r},t) = \frac{1}{2\pi}\int_{-\infty}^{\infty} A(\boldsymbol{r},\omega)e^{-i\omega t}\,d\omega. \tag{12.6.5}$$

Using the fact that the autocorrelation function $\tilde{C}_{AA}(\boldsymbol{r},t\,;\boldsymbol{r}',t')$ does not depend separately on t and on t', but only on the difference $t-t'$, we can show that:

$$\langle A(\boldsymbol{r},\omega)A(\boldsymbol{r}',\omega')\rangle = 2\pi\delta(\omega+\omega')C_{AA}(\boldsymbol{r},\omega\,;\boldsymbol{r}'). \tag{12.6.6}$$

In formula (12.6.6), $C_{AA}(\boldsymbol{r},\omega\,;\boldsymbol{r}')$ denotes the temporal Fourier transform of the autocorrelation function $\tilde{C}_{AA}(\boldsymbol{r},t;\boldsymbol{r}',0)$, defined by:[20]

$$C_{AA}(\boldsymbol{r},\omega\,;\boldsymbol{r}') = \int_{-\infty}^{\infty} \tilde{C}_{AA}(\boldsymbol{r},t\,;\boldsymbol{r}',0)e^{i\omega t}\,dt. \tag{12.6.7}$$

The function $C_{AA}(\boldsymbol{r},\omega\,;\boldsymbol{r})$ is called the *power spectrum* or the *spectral density*[21] of the fluctuations of $A(\boldsymbol{r},t)$. Formula (12.6.7) constitutes the generalization of the

[19] For the sake of simplicity, we use the same notation $A(\boldsymbol{r},.)$ for the operator $A(\boldsymbol{r},t)$ and its temporal Fourier transform $A(\boldsymbol{r},\omega)$.

[20] In contrast to the response functions, the correlation functions are not causal. The temporal Fourier transform involved in definition (12.6.7) is thus a true Fourier transform, and not a Fourier–Laplace transform.

[21] It remains to be proved that this quantity is actually positive. This property, which is a consequence of the fluctuation-dissipation theorem, will be established in Chapter 14.

Wiener–Khintchine theorem to a dynamical variable (classical case) or to an observable (quantum case).

6.3. Case of a homogeneous medium: spatial and temporal Fourier transform of the correlation function of two operators

We consider here an homogeneous medium. To compute the spatial Fourier transform of the correlation function $\tilde{C}_{A_iA_j}(\boldsymbol{r},t)$, we first take advantage of the space-translational invariance to rewrite $\tilde{C}_{A_iA_j}(\boldsymbol{r},t)$ in the following form:

$$\tilde{C}_{A_iA_j}(\boldsymbol{r},t) = \langle A_i(\boldsymbol{r}+\boldsymbol{r}',t)A_j(\boldsymbol{r}',0)\rangle. \tag{12.6.8}$$

We then compute its Fourier transform by introducing the supplementary integration $V^{-1}\int d\boldsymbol{r}' = 1$ (V denotes the volume of the system), which gives:

$$\int \tilde{C}_{A_iA_j}(\boldsymbol{r},t)e^{-i\boldsymbol{q}.\boldsymbol{r}}\,d\boldsymbol{r} = \frac{1}{V}\langle A_i(\boldsymbol{q},t)A_j(-\boldsymbol{q},0)\rangle. \tag{12.6.9}$$

In formula (12.6.9), we have introduced the spatial Fourier transforms[22] of an operator $A(\boldsymbol{r},t)$, defined by:

$$A(\boldsymbol{q},t) = \int A(\boldsymbol{r},t)e^{-i\boldsymbol{q}.\boldsymbol{r}}\,d\boldsymbol{r}. \tag{12.6.10}$$

The spatial and temporal Fourier transform $C_{A_iA_j}(\boldsymbol{q},\omega)$ of the correlation function $\tilde{C}_{A_iA_j}(\boldsymbol{r},t)$ is defined by:

$$C_{A_iA_j}(\boldsymbol{q},\omega) = \int d\boldsymbol{r}\,e^{-i\boldsymbol{q}.\boldsymbol{r}}\int_{-\infty}^{\infty} dt\,e^{i\omega t}\tilde{C}_{A_iA_j}(\boldsymbol{r},t). \tag{12.6.11}$$

Using formula (12.6.9), gives the relation (valid in a translationally invariant medium):

$$\boxed{C_{A_iA_j}(\boldsymbol{q},\omega) = \frac{1}{V}\int_{-\infty}^{\infty}\langle A_i(\boldsymbol{q},t)A_j(-\boldsymbol{q},0)\rangle e^{i\omega t}\,dt.} \tag{12.6.12}$$

6.4. Dynamical structure factor

We now consider an inelastic scattering process in the course of which, under the effect of the interaction with radiation, a system at equilibrium undergoes a transition from an initial state $|\lambda\rangle$ to a final state $|\lambda'\rangle$. The corresponding energy varies from ε_λ to $\varepsilon_{\lambda'}$, whereas the radiation energy varies from E to E'. The conservation of the global energy implies that $E+\varepsilon_\lambda = E'+\varepsilon_{\lambda'}$. The energy lost by the radiation is denoted by $\hbar\omega$ ($\hbar\omega = E - E' = \varepsilon_{\lambda'} - \varepsilon_\lambda$). We associate with the system–radiation

[22] We use the same notation $A(.,t)$ for the operator $A(\boldsymbol{r},t)$ and its spatial Fourier transform $A(\boldsymbol{q},t)$.

interaction a centered operator $A(\boldsymbol{r})$. For instance, in the case of scattering of light by a fluid in equilibrium,[23] the radiation is scattered by the density fluctuations of the fluid, and therefore the operator $A(\boldsymbol{r})$ is proportional to the density fluctuation $\delta n(\boldsymbol{r}) = n(\boldsymbol{r}) - \langle n \rangle$.

An incident plane-wave initial state $|\boldsymbol{k}\rangle$ is scattered to a final state which, in the framework of the Born approximation for scattering, is considered as being a plane-wave state[24] $|\boldsymbol{k}'\rangle$. The matrix element of the interaction operator between states $|\boldsymbol{k}\rangle$ and $|\boldsymbol{k}'\rangle$ is $\langle \boldsymbol{k}'|A(\boldsymbol{r})|\boldsymbol{k}\rangle = \int e^{-i\boldsymbol{k}'.\boldsymbol{r}} A(\boldsymbol{r}) e^{i\boldsymbol{k}.\boldsymbol{r}}\, d\boldsymbol{r}$: it is thus identical to the Fourier component $A(-\boldsymbol{q})$ of the operator $A(\boldsymbol{r})$ (having set $\boldsymbol{q} = \boldsymbol{k} - \boldsymbol{k}'$). At lowest perturbation order, the probability per unit time of the process $(|\boldsymbol{k}\rangle, \lambda) \to (|\boldsymbol{k}'\rangle, \lambda')$ is given by the Fermi golden rule:

$$W_{(\boldsymbol{k}',\lambda'),(\boldsymbol{k},\lambda)} = \frac{2\pi}{\hbar} |\langle \lambda'|A(-\boldsymbol{q})|\lambda\rangle|^2 \delta(\hbar\omega + \varepsilon_\lambda - \varepsilon_{\lambda'}). \tag{12.6.13}$$

The probability per unit time $W_{\boldsymbol{k}',\boldsymbol{k}}$ of the process $|\boldsymbol{k}\rangle \to |\boldsymbol{k}'\rangle$ is obtained by weighting $W_{(\boldsymbol{k}',\lambda'),(\boldsymbol{k},\lambda)}$ by the occupation probability p_λ of the initial state of the system at equilibrium, and by summing over all initial states and all final states:

$$W_{\boldsymbol{k}',\boldsymbol{k}} = \frac{2\pi}{\hbar} \sum_{\lambda,\lambda'} p_\lambda |\langle \lambda'|A(-\boldsymbol{q})|\lambda\rangle|^2 \delta(\hbar\omega + \varepsilon_\lambda - \varepsilon_{\lambda'}). \tag{12.6.14}$$

We associate with $W_{\boldsymbol{k}',\boldsymbol{k}}$ the *scattering function* $S(\boldsymbol{q},\omega) = \hbar^2 W_{\boldsymbol{k}',\boldsymbol{k}}$:

$$S(\boldsymbol{q},\omega) = 2\pi \sum_{\lambda,\lambda'} p_\lambda |\langle \lambda'|A(-\boldsymbol{q})|\lambda\rangle|^2 \delta\Big(\omega + \frac{\varepsilon_\lambda - \varepsilon_{\lambda'}}{\hbar}\Big). \tag{12.6.15}$$

When $A(\boldsymbol{r})$ is the density fluctuation $\delta n(\boldsymbol{r})$, the scattering function $S(\boldsymbol{q},\omega)$ is also designated as the *dynamical structure factor*. This latter denomination will be adopted generically in the rest of this chapter.

Introducing in formula (12.6.15) the Fourier representation of the delta function, $\delta(\omega) = (2\pi)^{-1} \int_{-\infty}^{\infty} e^{i\omega t}\, dt$, we can express $S(\boldsymbol{q},\omega)$ with the aid of an autocorrelation function. Since the operator $A(\boldsymbol{r})$ is Hermitean, we have:

$$\langle \lambda'|A(-\boldsymbol{q})|\lambda\rangle^* = \langle \lambda|A(\boldsymbol{q})|\lambda'\rangle. \tag{12.6.16}$$

This gives:

$$S(\boldsymbol{q},\omega) = \int_{-\infty}^{\infty} \sum_{\lambda,\lambda'} p_\lambda \langle \lambda| e^{i\varepsilon_\lambda t/\hbar} A(\boldsymbol{q}) e^{-i\varepsilon_{\lambda'} t/\hbar} |\lambda'\rangle\langle \lambda'|A(-\boldsymbol{q})|\lambda\rangle e^{i\omega t}\, dt, \tag{12.6.17}$$

---

[23] See Supplement 16B.

[24] This implies that the interaction of the radiation with the target has be weak enough in order to allow us to neglect multiple scattering. If the Born approximation is not appropriate, we replace the interaction potential $A(\boldsymbol{r})$ by a convenient pseudopotential.

that is:

$$S(\boldsymbol{q},\omega)=\int_{-\infty}^{\infty}\langle A(\boldsymbol{q},t)A(-\boldsymbol{q},0)\rangle e^{i\omega t}\,dt. \tag{12.6.18}$$

Equation (12.6.18) shows that the dynamical structure factor[25] is the Fourier transform of the autocorrelation function $\langle A(\boldsymbol{q},t)A(-\boldsymbol{q},0)\rangle$. This property was demonstrated by L. van Hove in 1954.

In the particular case of a translationally invariant medium, comparing formula (12.6.18) with the formula (12.6.12) written for $A_i = A_j = A$ gives:

$$S(\boldsymbol{q},\omega)=V C_{AA}(\boldsymbol{q},\omega). \tag{12.6.19}$$

It is then possible to deduce from the dynamical structure factor the autocorrelation function $\tilde{C}_{AA}(\boldsymbol{r},t)$.

7. Properties of the equilibrium autocorrelation functions

We will review some general properties of the equilibrium autocorrelation functions $\langle A(\boldsymbol{q},t)A(-\boldsymbol{q},0)\rangle$ of a centered operator $A(\boldsymbol{r},t)$ and of their Fourier transform $S(\boldsymbol{q},\omega)$. We will adopt the quantum framework. The passage to the classical limit will be carried out if necessary.

7.1. Stationarity

Any equilibrium correlation function depends only on the difference of its two temporal arguments. We therefore have the equality:

$$\big\langle A(\boldsymbol{q},t)A(-\boldsymbol{q},0)\big\rangle=\big\langle A(\boldsymbol{q},t+t_0)A(-\boldsymbol{q},t_0)\big\rangle, \tag{12.7.1}$$

the instant t_0 being arbitrary. Choosing $t_0=-t$ gives us, in particular:

$$\big\langle A(\boldsymbol{q},t)A(-\boldsymbol{q},0)\big\rangle=\big\langle A(\boldsymbol{q},0)A(-\boldsymbol{q},-t)\big\rangle. \tag{12.7.2}$$

7.2. Time-derivation

We have the relation:

$$\big\langle \dot{A}(\boldsymbol{q},0)A(-\boldsymbol{q},t)\big\rangle=-\big\langle A(\boldsymbol{q},0)\dot{A}(-\boldsymbol{q},t)\big\rangle. \tag{12.7.3}$$

---

[25] Simple examples of dynamical structure factors are presented in Supplement 12C. The dynamical structure factor in a fluid will be studied in Supplement 16B.

To derive formula (12.7.3), we can for instance[26] take the derivative of formula (12.7.1) with respect to t_0:

$$\left\langle \frac{\partial A(\boldsymbol{q}, t+t_0)}{\partial t_0} A(-\boldsymbol{q}, t_0) \right\rangle + \left\langle A(\boldsymbol{q}, t+t_0) \frac{\partial A(-\boldsymbol{q}, t_0)}{\partial t_0} \right\rangle = 0. \tag{12.7.4}$$

Making $t_0 = 0$ in equation (12.7.4), we first verify the property:

$$\langle \dot{A}(\boldsymbol{q}, t) A(-\boldsymbol{q}, 0) \rangle = -\langle A(\boldsymbol{q}, t) \dot{A}(-\boldsymbol{q}, 0) \rangle, \tag{12.7.5}$$

then, translating all times by $-t$ and changing then t into $-t$, formula (12.7.3).

7.3. Complex conjugation

We have the property:

$$\boxed{\langle A(\boldsymbol{q}, t) A(-\boldsymbol{q}, 0) \rangle^* = \langle A(\boldsymbol{q}, 0) A(-\boldsymbol{q}, t) \rangle.} \tag{12.7.6}$$

To demonstrate equality (12.7.6), we make explicit the definition of the equilibrium averages on both sides. Formula (12.7.6) thus reads:

$$\Big(\mathrm{Tr}\big[\rho_0 e^{iH_0 t/\hbar} A(\boldsymbol{q}, 0) e^{-iH_0 t/\hbar} A(-\boldsymbol{q}, 0)\big]\Big)^* = \mathrm{Tr}\big[\rho_0 A(\boldsymbol{q}, 0) e^{iH_0 t/\hbar} A(-\boldsymbol{q}, 0) e^{-iH_0 t/\hbar}\big], \tag{12.7.7}$$

where ρ_0 denotes the equilibrium density operator. Using the complex conjugation property $\big[\mathrm{Tr}(AB)\big]^* = \mathrm{Tr}(B^\dagger A^\dagger)$ as well as the invariance of the trace under a circular permutation of the operators, we verify formulas (12.7.7) and (12.7.6).

7.4. Real character of the function $S(\boldsymbol{q}, \omega)$

Formulas (12.7.2) and (12.7.6) allow us to demonstrate that $S(\boldsymbol{q}, \omega)$, as defined by formula (12.6.18), is a real quantity.[27]

Taking the complex conjugate of the definition formula (12.6.18), and making use of the complex conjugation formula (12.7.6), gives:

$$S^*(\boldsymbol{q}, \omega) = \int_{-\infty}^{\infty} \langle A(\boldsymbol{q}, 0) A(-\boldsymbol{q}, t) \rangle e^{-i\omega t}\, dt, \tag{12.7.8}$$

or, changing t into $-t$ in the integral on the right-hand side of equation (12.7.8):

$$S^*(\boldsymbol{q}, \omega) = \int_{-\infty}^{\infty} \langle A(\boldsymbol{q}, 0) A(-\boldsymbol{q}, -t) \rangle e^{i\omega t}\, dt. \tag{12.7.9}$$

Using the stationarity property (12.7.2), we verify that $S^*(\boldsymbol{q}, \omega) = S(\boldsymbol{q}, \omega)$.

---

[26] We can alternatively demonstrate formula (12.7.3) by writing the Heisenberg equations relative to the operators $A(\boldsymbol{q}, t)$ and $A(-\boldsymbol{q}, t)$, that is, $i\hbar \dot{A}(\boldsymbol{q}, t) = [A(\boldsymbol{q}, t), H_0]$ and $i\hbar \dot{A}(-\boldsymbol{q}, t) = [A(-\boldsymbol{q}, t), H_0]$, and then using the invariance of the trace under a circular permutation of the operators (together with the fact that the equilibrium density operator is a function of H_0).

[27] Note that the real character of $S(\boldsymbol{q}, \omega)$ is already apparent in formula (12.6.15).

7.5. Specific properties of the classical autocorrelation functions

The autocorrelation functions $\langle A(\boldsymbol{q},t)A(-\boldsymbol{q},0)\rangle$ are in general complex quantities. It is only for classical systems with an inversion symmetry (invariant under the transformation $\boldsymbol{r} \to -\boldsymbol{r}$) that the autocorrelation functions $\langle A(\boldsymbol{q},t)A(-\boldsymbol{q},0)\rangle$ are real. They are then even functions of time.

7.6. The Kubo–Martin–Schwinger condition

The correlation functions at canonical equilibrium obey the *Kubo–Martin–Schwinger condition*:

$$\boxed{\big\langle A(\boldsymbol{q},0)A(-\boldsymbol{q},t)\big\rangle = \big\langle A(-\boldsymbol{q},t-i\hbar\beta)A(\boldsymbol{q},0)\big\rangle.} \tag{12.7.10}$$

To demonstrate formula (12.7.10), we make explicit the definition of equilibrium averages and we use the explicit form $\rho_0 = Z^{-1}e^{-\beta H_0}$ of the canonical density operator ($Z = \operatorname{Tr} e^{-\beta H_0}$ denotes the partition function). The left-hand side of equation (12.7.10) reads:

$$\frac{1}{Z}\operatorname{Tr}\big[e^{-\beta H_0}A(\boldsymbol{q},0)\,e^{iH_0t/\hbar}A(-\boldsymbol{q},0)e^{-iH_0t/\hbar}\big], \tag{12.7.11}$$

that is, carrying out circular permutations of operators under the trace,

$$\frac{1}{Z}\operatorname{Tr}\big[e^{iH_0t/\hbar}A(-\boldsymbol{q},0)e^{-iH_0(t-i\hbar\beta)/\hbar}A(\boldsymbol{q},0)\big], \tag{12.7.12}$$

or:

$$\frac{1}{Z}\operatorname{Tr}\big[e^{-\beta H_0}e^{i(t-i\hbar\beta)H_0/\hbar}A(-\boldsymbol{q},0)e^{-i(t-i\hbar\beta)H_0/\hbar}A(\boldsymbol{q},0)\big]. \tag{12.7.13}$$

This last expression is simply $\langle A(-\boldsymbol{q},t-i\hbar\beta)A(\boldsymbol{q},0)\rangle$, which demonstrates formula (12.7.10).

7.7. The detailed balance relation

The *detailed balance relation* reads:

$$\boxed{S(\boldsymbol{q},\omega) = S(-\boldsymbol{q},-\omega)e^{\beta\hbar\omega}.} \tag{12.7.14}$$

Like the Kubo–Martin–Schwinger condition from which it derives, formula (12.7.14) is a specific property of systems at canonical equilibrium. It expresses the relation between dynamical structure factors relative to scattering processes which are inverse from one another, that is, which are respectively characterized by wave vector and angular frequency changes $(\boldsymbol{q},\omega)$, on the one hand, and $(-\boldsymbol{q},-\omega)$, on the other. If the system possesses the inversion symmetry $\boldsymbol{r} \to -\boldsymbol{r}$, the detailed balance relation takes the simpler form:

$$S(\boldsymbol{q},\omega) = S(\boldsymbol{q},-\omega)e^{\beta\hbar\omega}. \tag{12.7.15}$$

To demonstrate relation (12.7.14), we first rewrite formula (12.6.18) for $S(\boldsymbol{q},\omega)$ by making use of the properties (12.7.6) (complex conjugation) and (12.7.10) (Kubo–Martin–Schwinger condition), which gives:

$$S(\boldsymbol{q},\omega)=\int_{-\infty}^{\infty}\big\langle A(-\boldsymbol{q},t-i\hbar\beta)A(\boldsymbol{q},0)\big\rangle^{*}e^{i\omega t}\,dt, \tag{12.7.16}$$

or, since $S(\boldsymbol{q},\omega)$ is real:

$$S(\boldsymbol{q},\omega)=\int_{-\infty}^{\infty}\big\langle A(-\boldsymbol{q},t-i\hbar\beta)A(\boldsymbol{q},0)\big\rangle e^{-i\omega t}\,dt. \tag{12.7.17}$$

This expression for $S(\boldsymbol{q},\omega)$ must be compared with that for $S(-\boldsymbol{q},-\omega)$:

$$S(-\boldsymbol{q},-\omega)=\int_{-\infty}^{\infty}\big\langle A(-\boldsymbol{q},t)A(\boldsymbol{q},0)\big\rangle e^{-i\omega t}\,dt. \tag{12.7.18}$$

To relate $S(\boldsymbol{q},\omega)$ and $S(-\boldsymbol{q},-\omega)$, we can study the contour integral:

$$I=\oint_{\Gamma_I}\big\langle A(-\boldsymbol{q},\tau)A(\boldsymbol{q},0)\big\rangle e^{-i\omega\tau}\,d\tau, \tag{12.7.19}$$

where Γ_I is the rectangular contour of large sides equal to $2R$ represented in Fig. 12.1 (R will be made to tend towards infinity at the end of the calculation). The function to be integrated is analytic in the domain interior to the contour.[28] The integral I thus vanishes according to Cauchy's theorem.

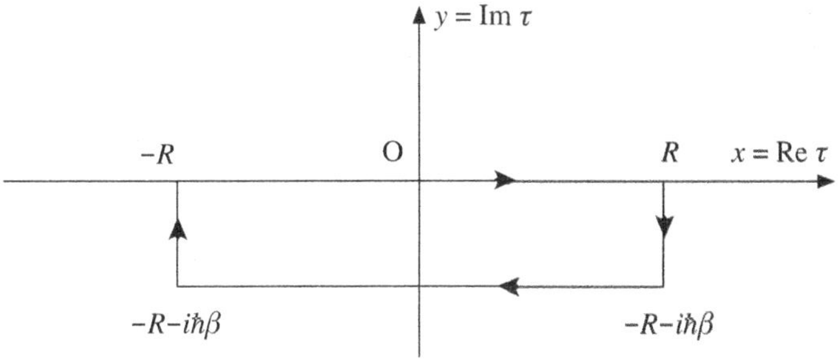

Fig. 12.1 The integration contour Γ_I.

[28] We have:

$$\langle A(-\boldsymbol{q},\tau)A(\boldsymbol{q},0)\rangle=\frac{1}{Z}\sum_{\lambda,\lambda'}|A(\boldsymbol{q},0)_{\lambda'\lambda}|^{2}e^{-\beta\varepsilon_\lambda+i\tau(\varepsilon_\lambda-\varepsilon_{\lambda'})}.$$

For $-\beta\leq\Im m\,\tau\leq 0$, the series on the right-hand side converges.

Detailing the various contributions to I, we obtain the equality:

$$\int_{-R}^{R} \langle A(-\boldsymbol{q},t)A(\boldsymbol{q},0)\rangle e^{-i\omega t}\,dt + \int_{0}^{\beta} \langle A(-\boldsymbol{q},R-i\hbar y)A(\boldsymbol{q},0)\rangle e^{-i\omega R}e^{-\hbar\omega y}(-i\hbar)\,dy + e^{-\beta\hbar\omega}\int_{R}^{-R} \langle A(-\boldsymbol{q},t-i\hbar\beta)A(\boldsymbol{q},0)\rangle e^{-i\omega t}\,dt + \int_{\beta}^{0} \langle A(-\boldsymbol{q},-R-i\hbar y)A(\boldsymbol{q},0)\rangle e^{i\omega R}e^{-\hbar\omega y}(-i\hbar)\,dy = 0. \tag{12.7.20}$$

In the limit $R \to \infty$, the contributions to I of the two vertical segments of abscissas R and $-R$ vanish, provided that the autocorrelation functions $\langle A(-\boldsymbol{q},R-i\hbar y)A(\boldsymbol{q},0)\rangle$ and $\langle A(-\boldsymbol{q},-R-i\hbar y)A(\boldsymbol{q},0)\rangle$ themselves vanish in this limit.[29] We then obtain the equality:

$$\int_{-\infty}^{\infty} \langle A(-\boldsymbol{q},t)A(\boldsymbol{q},0)\rangle e^{-i\omega t}\,dt = e^{-\beta\hbar\omega}\int_{-\infty}^{\infty} \langle A(-\boldsymbol{q},t-i\hbar\beta)A(\boldsymbol{q},0)\rangle e^{-i\omega t}\,dt. \tag{12.7.21}$$

The left-hand side of equation (12.7.21) is $S(-\boldsymbol{q},-\omega)$, whereas the integral on the right-hand side is just $S(\boldsymbol{q},\omega)$, which gives us:

$$S(-\boldsymbol{q},-\omega) = e^{-\beta\hbar\omega}S(\boldsymbol{q},\omega), \tag{12.7.22}$$

and this demonstrates the detailed balance relation (12.7.14).

[29] This hypothesis is physically meaningful in a system with an infinite number of degrees of freedom. However, in systems with a finite number of degrees of freedom, the correlation functions are oscillating functions, and this limit in principle does not exist. We can nevertheless define it by treating correlation functions as distributions, which allows us to extend the detailed balance relation to these systems.

Appendix

12A. An alternative derivation of the Kramers–Kronig relations

To establish the Kramers–Kronig relations, we can study the contour integral:

$$J = \oint_{\Gamma_J} \frac{\chi(z)}{z - \omega}\, dz, \tag{12A.1}$$

where ω denotes a real angular frequency. The function $\chi(z)$, analytic in the upper complex half-plane, is defined by the Fourier–Laplace integral (12.2.10). The integration contour Γ_J shown in Fig. 12.2 avoids, by using a semicircle of radius ϵ, the pole of abscissa ω of the function $\chi(z)/(z - \omega)$. The contour Γ_J is closed by a semicircle of large radius R.

The function $\chi(z)$ being analytic in the upper complex half-plane, the integral J of formula (12A.1) vanishes according to Cauchy's theorem. The function $\chi(z)$ generally vanishes as z tends towards infinity.[30] The contribution of the semicircle of radius R to the integral J thus vanishes in the limit $R \to \infty$. By detailing the other contributions to J, we get:

$$\int_{-\infty}^{\omega-\varepsilon} \frac{\chi(\omega')}{\omega' - \omega}\, d\omega' + \int_{\omega+\epsilon}^{\infty} \frac{\chi(\omega')}{\omega' - \omega}\, d\omega' + \int_{\pi}^{0} i\chi(\omega + \epsilon e^{i\theta})\, d\theta = 0. \tag{12A.2}$$

In the limit $\epsilon \to 0^+$, the third integral on the left-hand side of equation (12A.2) tends towards $-i\pi\chi(\omega)$. This gives the formula:

$$\chi(\omega) = -\frac{i}{\pi}\, \mathrm{vp} \int_{-\infty}^{\infty} \frac{\chi(\omega')}{\omega' - \omega}\, d\omega'. \tag{12A.3}$$

Identifying separately the real and imaginary parts of both sides of equation (12A.3), we get the Kramers–Kronig relations (formulas (12.3.3)).

---

[30] As a simple example, we can quote the generalized susceptibility $\chi_{xx}(z)$ of a harmonic oscillator, damped or not, which decreases proportionally to z^{-2} as z tends towards infinity, except in the viscous limit in which it decreases proportionally to z^{-1} (see Supplement 12A).

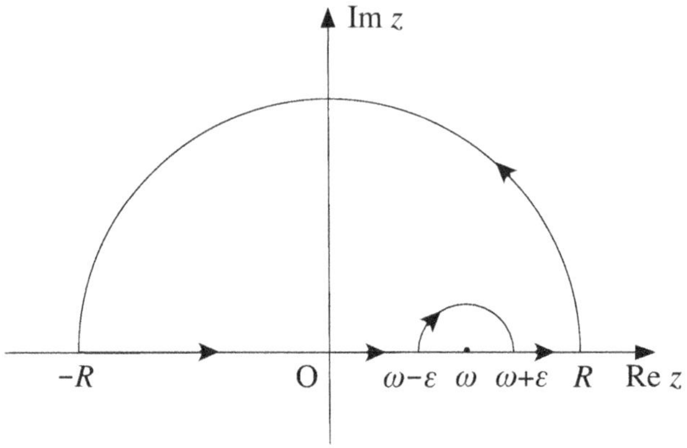

Fig. 12.2 Contour Γ_J used for the derivation of the Kramers–Kronig relations.

The Kramers–Kronig relations need to be modified if $\chi(z)$ does not tend towards zero but towards a finite constant χ_∞ as z tends towards infinity.[31] In such a case, the out-of-equilibrium average of the physical quantity A involves, besides the retarded term $\int_{-\infty}^{\infty}\tilde{\chi}(t-t')a(t')dt'$, an instantaneous contribution $\chi_\infty a(t)$ (due to a term $\chi_\infty\delta(t)$ in the response function). In this case,we can still write the Kramers–Kronig relations by working on the difference $\chi(z)-\chi_\infty$:

$$\begin{cases} \chi'(\omega)-\chi_\infty = \dfrac{1}{\pi}\,\mathrm{vp}\displaystyle\int_{-\infty}^{\infty}\dfrac{\chi''(\omega')}{\omega'-\omega}\,d\omega' \\[2ex] \chi''(\omega) = -\dfrac{1}{\pi}\,\mathrm{vp}\displaystyle\int_{-\infty}^{\infty}\dfrac{\chi'(\omega')-\chi_\infty}{\omega'-\omega}\,d\omega'. \end{cases} \tag{12A.4}$$

[31] An example of this type of behavior appears in dielectric relaxation (see Supplement 13A).

Bibliography

P.M. Chaikin and T.C. Lubensky, *Principles of condensed matter physics*, Cambridge University Press, Cambridge, 1995.

S. Dattagupta, *Relaxation phenomena in condensed matter physics*, Academic Press, Orlando, 1987.

D. Forster, *Hydrodynamic fluctuations, broken symmetries, and correlation functions*, Westview Press, Boulder, 1995.

C. Kittel, *Introduction to solid state physics*, Wiley, New York, eighth edition, 2005.

R. Kubo, M. Toda, and N. Hashitsume, *Statistical physics II: nonequilibrium statistical mechanics*, Springer-Verlag, Berlin, second edition, 1991.

L.D. Landau and E.M. Lifshitz, *Electrodynamics of continuous media*, Butterworth-Heinemann, Oxford, second edition, 1984.

S.W. Lovesey, *Condensed matter physics: dynamic correlations*, The Benjamin/ Cummings Publishing Company, Reading, second edition, 1986.

P.C. Martin, *Measurements and correlation functions*, Les Houches Lecture Notes 1967 (C. De Witt and R. Balian editors), Gordon and Breach, New York, 1968.

G. Parisi, *Statistical field theory*, Westview Press, Boulder, 1998.

M. Plischke and B. Bergersen, *Equilibrium statistical physics*, World Scientific, Singapore, third edition, 2006.

References

R. de L. Kronig, On the theory of dispersion of X-rays, *J. Opt. Soc. Am.* **12**, 547 (1926).

H.A. Kramers, La diffusion de la lumière par les atomes, *Atti del Congresso Internazionale dei Fisici (Como)*, **2**, 545, Zanichelli, Bologna, 1927.

Supplement 12A

Linear response of a damped oscillator

1. General interest of the study

The dynamical properties of many physical systems are controlled by oscillator modes. The information about the angular frequency and the damping of these modes is contained in the linear response functions and the generalized susceptibilities, as well as in the associated equilibrium correlation functions.[1]

We aim here to study the linear response properties of a classical harmonic oscillator damped by viscous friction. The extreme simplicity of this model allows us to compute directly –that is, without having recourse to the general linear response theory– the displacement response function and the corresponding generalized susceptibility. Their properties generalize to any system with modes at finite angular frequency.

2. The undamped oscillator

The Hamiltonian of a one-dimensional harmonic oscillator of mass m and spring constant k reads:

$$H_0 = \frac{p^2}{2m} + \frac{1}{2}kx^2. \qquad (12A.2.1)$$

Hamilton's equations for the displacement $x(t)$ and the momentum $p(t)$ are:

$$\dot{x} = \frac{p}{m}, \qquad \dot{p} = -kx. \qquad (12A.2.2)$$

To determine the modes, we assume that both $x(t)$ and $p(t)$ are proportional to $e^{-i\omega t}$. This gives the characteristic equation:

$$-\omega^2 + \frac{k}{m} = 0, \qquad (12A.2.3)$$

[1] See the general theory of linear response expounded in Chapter 13.

whose solutions are the angular frequencies of the modes:

$$\omega = \pm\omega_0, \qquad \omega_0 = \left(\frac{k}{m}\right)^{1/2}. \tag{12A.2.4}$$

The equation of motion of the oscillator reads:

$$m\frac{d^2x}{dt^2} + m\omega_0^2 x = 0. \tag{12A.2.5}$$

The Hamiltonian H_0 (formula (12A.2.1)), Hamilton's equations (12A.2.2), and the equation of motion (12A.2.5) of the undamped oscillator are invariant under time-reversal.

3. Oscillator damped by viscous friction

We can introduce damping in a phenomenological way by assuming that the oscillator is placed in a viscous fluid. The particle of mass m is then submitted, besides the restoring force $-kx$, to a viscous friction force $-m\gamma(dx/dt)$ characterized by the friction coefficient $\gamma > 0$.

The equation of motion of the oscillator then reads:

$$m\frac{d^2x}{dt^2} + m\gamma\frac{dx}{dt} + m\omega_0^2 x = 0. \tag{12A.3.1}$$

In contrast to equation (12A.2.5), equation (12A.3.1) is not time-reversal invariant. This equation, which has been introduced phenomenologically, does not follow directly (that is, without approximations) from a microscopic Hamiltonian.[2] The viscous force represents the mean effect produced on the oscillator by its interaction with the many incoherent degrees of freedom of the fluid which surrounds it. The energy of the damped oscillator evolving according to equation (12A.3.1) tends to flow irreversibly towards the modes of the fluid. This flow corresponds to a dissipation of the oscillator's energy.

3.1. Modes

The characteristic equation determining the angular frequencies of the modes of the damped oscillator reads:

$$-\omega^2 - i\gamma\omega + \omega_0^2 = 0. \tag{12A.3.2}$$

Its solutions are:

$$\omega = \pm\omega_1 - i\frac{\gamma}{2}, \qquad \omega_1 = \left(\omega_0^2 - \frac{\gamma^2}{4}\right)^{1/2}. \tag{12A.3.3}$$

[2] The equation of motion (12A.3.1) can be derived in the framework of the Caldeira–Leggett dissipation model. The Caldeira–Leggett Hamiltonian describes the global system made up of the oscillator and the modes of the bath with which it is coupled. If these modes form a continuum, we can obtain for the oscillator an irreversible equation of motion. In the Ohmic dissipation case, this equation takes the form (12A.3.1) (see also about this question Supplement 14A).

Two different types of damped motions may take place, depending on whether $\gamma < 2\omega_0$ or $\gamma > 2\omega_0$.

3.2. Case $\gamma < 2\omega_0$: underdamped motion

In this case, ω_1 is real. For initial conditions $x(0) = x_0$ and $\dot{x}(0) = v_0$, the solution of equation (12A.3.1) reads:

$$x(t) = \Big[x_0 \cos\omega_1 t + \big(v_0 + \frac{\gamma}{2}x_0\big)\frac{1}{\omega_1}\sin\omega_1 t\Big]e^{-\gamma t/2}, \qquad t > 0. \tag{12A.3.4}$$

The displacement $x(t)$ oscillates at the angular frequency ω_1. The amplitude of the oscillations decreases over the course of time with the decay time $\tau = 2\gamma^{-1}$.

3.3. Case $\gamma > 2\omega_0$: overdamped motion

Then ω_1 is purely imaginary. The displacement $x(t)$ is a linear combination of two real exponentials decreasing with the inverse decay times:

$$\begin{cases} \tau_1^{-1} = \dfrac{\gamma}{2}\Big[1 + (1 - 4\omega_0^2\gamma^{-2})^{1/2}\Big] \\ \tau_2^{-1} = \dfrac{\gamma}{2}\Big[1 - (1 - 4\omega_0^2\gamma^{-2})^{1/2}\Big]. \end{cases} \tag{12A.3.5}$$

In the viscous limit $\gamma \gg 2\omega_0$, the decay times $\tau_1 \simeq \gamma^{-1}$ and $\tau_2 \simeq \gamma\omega_0^{-2}$ are well separated: $\tau_1 \ll \tau_2$. At times $t \gg \tau_1$, we can then neglect the exponential of decay time τ_1 in the expression for $x(t)$, which amounts to discarding the inertia term in the equation of motion (12A.3.1). This latter equation then reduces to a first-order differential equation:

$$m\gamma\frac{dx}{dt} + m\omega_0^2 x = 0. \tag{12A.3.6}$$

The solution of equation (12A.3.6) for the initial condition $x(0) = x_0$ reads:

$$x(t) = x_0 e^{-\omega_0^2 t/\gamma}, \qquad t > 0. \tag{12A.3.7}$$

4. Generalized susceptibility

We apply to the system an external force $F_{\text{ext}}(t)$. The equation of motion now reads:

$$m\frac{d^2x}{dt^2} + m\gamma\frac{dx}{dt} + m\omega_0^2 x = F_{\text{ext}}(t). \tag{12A.4.1}$$

The generalized susceptibility $\chi_{xx}(\omega)$ corresponding to the response of the displacement is obtained by considering a stationary harmonic regime in which both the applied force and the displacement vary as $e^{-i\omega t}$. This gives:

$$\boxed{\chi_{xx}(\omega) = \frac{1}{m}\frac{1}{-\omega^2 + \omega_0^2 - i\gamma\omega}.} \tag{12A.4.2}$$

The real and imaginary parts of $\chi_{xx}(\omega)$ are:

$$\begin{cases} \chi'_{xx}(\omega) = \dfrac{1}{m}\dfrac{\omega_0^2-\omega^2}{\left(\omega^2-\omega_0^2\right)^2+\gamma^2\omega^2} \\ \chi''_{xx}(\omega) = \dfrac{1}{m}\dfrac{\gamma\omega}{\left(\omega^2-\omega_0^2\right)^2+\gamma^2\omega^2}. \end{cases} \tag{12A.4.3}$$

The average dissipated power, proportional to $\omega\chi''_{xx}(\omega)$, is actually positive.

Formula (12A.4.2) can be extended to a complex argument z of positive imaginary part. We thus define a function $\chi_{xx}(z)$ analytic in the upper complex half-plane:

$$\chi_{xx}(z) = \frac{1}{m}\frac{1}{-z^2+\omega_0^2-i\gamma z}, \qquad \Im m\, z > 0. \tag{12A.4.4}$$

The function $\chi_{xx}(z)$ decreases proportionally to z^{-2} as z tends towards infinity (except in the viscous limit[3]). The poles of the continuation of $\chi_{xx}(z)$ in the lower complex half-plane are the complex angular frequencies $\pm\omega_1 - i\gamma/2$ of the damped oscillator.

The characteristic features of the function $\chi_{xx}(\omega)$ depend on the value of the damping. We will describe them in both cases $\gamma < 2\omega_0$ and $\gamma > 2\omega_0$ (restricting ourselves in this latter case to the viscous limit).

4.1. Susceptibility in the case $\gamma < 2\omega_0$

The expression (12A.4.2) for $\chi_{xx}(\omega)$ can be decomposed into partial fractions:

$$\chi_{xx}(\omega) = \frac{1}{2m\omega_1}\left(-\frac{1}{\omega-\omega_1+\frac{i\gamma}{2}} + \frac{1}{\omega+\omega_1+\frac{i\gamma}{2}}\right). \tag{12A.4.5}$$

The decomposition (12A.4.5) is valid whatever the damping (that is, weak or not). However, this decomposition presents a practical interest only when ω_1 is real, that is for $\gamma < 2\omega_0$. In this case, it follows that $\chi''_{xx}(\omega)$ is the algebraic sum of two Lorentzians centered at $\omega = \pm\omega_1$ and of width γ (Fig. 12A.1):

$$\chi''_{xx}(\omega) = \frac{1}{2m\omega_1}\left[\frac{\frac{\gamma}{2}}{\left(\omega-\omega_1\right)^2+\frac{\gamma^2}{4}} - \frac{\frac{\gamma}{2}}{\left(\omega+\omega_1\right)^2+\frac{\gamma^2}{4}}\right]. \tag{12A.4.6}$$

---

[3] See Subsection 4.2.

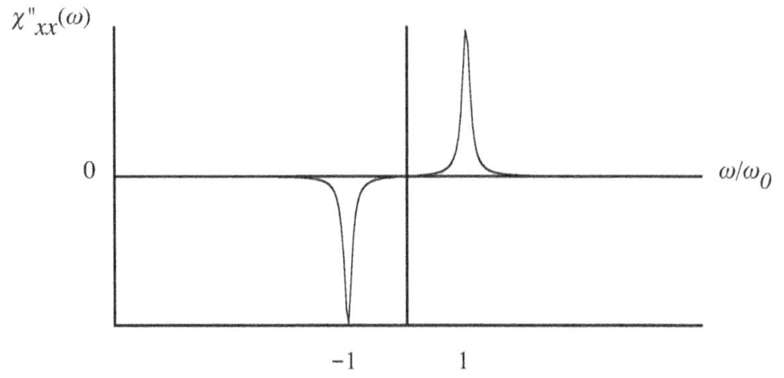

Fig. 12A.1 Imaginary part of the generalized susceptibility in the case $\gamma < 2\omega_0$.

In the limit of vanishing damping ($\gamma \ll \omega_0$), $\chi''_{xx}(\omega)$ appears as the algebraic sum of two delta function spikes centered at $\omega = \pm\omega_0$:

$$\chi''_{xx}(\omega) = \frac{\pi}{2m\omega_0}\big[\delta(\omega - \omega_0) - \delta(\omega + \omega_0)\big]. \tag{12A.4.7}$$

4.2. Susceptibility in the viscous limit

In this case, $\chi_{xx}(\omega)$ reads:

$$\chi_{xx}(\omega) = \frac{1}{m}\frac{1}{\omega_0^2 - i\gamma\omega}. \tag{12A.4.8}$$

The function:

$$\chi_{xx}(z) = \frac{1}{m}\frac{1}{\omega_0^2 - i\gamma z}, \qquad \Im m\, z > 0, \tag{12A.4.9}$$

decreases proportionally to z^{-1} as z tends towards infinity. We have:

$$\chi''_{xx}(\omega) = \frac{1}{m}\frac{\gamma\omega}{\omega_0^4 + \gamma^2\omega^2}. \tag{12A.4.10}$$

The quantity $\chi''_{xx}(\omega)/\omega$ is made up of a unique Lorentzian spike, centered at $\omega = 0$ and of width $2\omega_0^2\gamma^{-1}$ (Fig. 12A.2).

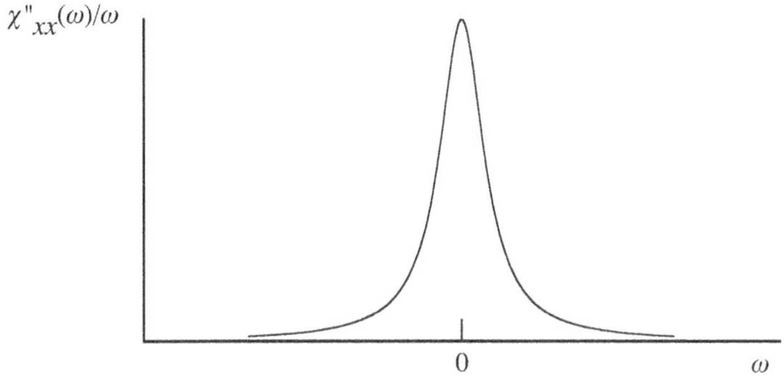

Fig. 12A.2 The function $\chi''_{xx}(\omega)/\omega$ in the viscous limit.

5. The displacement response function

The displacement response function $\tilde{\chi}_{xx}(t)$ is deduced from $\chi_{xx}(\omega)$ by inverse Fourier transformation:

$$\tilde{\chi}_{xx}(t) = \frac{1}{2\pi}\int_{-\infty}^{\infty} \chi_{xx}(\omega)e^{-i\omega t}\,d\omega. \tag{12A.5.1}$$

We will provide its expression in the two previously considered cases.

5.1. Response function in the underdamped case

Introducing the expression (12A.4.5) for $\chi_{xx}(\omega)$ in formula (12A.5.1), gives, after integration:

$$\tilde{\chi}_{xx}(t) = \Theta(t)\frac{\sin\omega_1 t}{m\omega_1}e^{-\gamma t/2}. \tag{12A.5.2}$$

The displacement response function oscillates and tends towards zero as $t \to \infty$. In the undamped case, $\tilde{\chi}_{xx}(t)$ oscillates indefinitely without decreasing:

$$\tilde{\chi}_{xx}(t) = \Theta(t)\frac{\sin\omega_0 t}{m\omega_0}\cdot \tag{12A.5.3}$$

5.2. Response function in the viscous limit

We now use the expression (12A.4.8) for $\chi_{xx}(\omega)$. The inverse Fourier integral (12A.5.1) then yields:

$$\tilde{\chi}_{xx}(t) = \Theta(t)\frac{1}{m\gamma}e^{-\omega_0^2 t/\gamma}. \tag{12A.5.4}$$

The displacement response function tends towards zero without oscillating as $t \to \infty$.

Bibliography

P.M. Chaikin and T.C. Lubensky, *Principles of condensed matter physics*, Cambridge University Press, Cambridge, 1995.

S.W. Lovesey, *Condensed matter physics: dynamic correlations*, The Benjamin/ Cummings Publishing Company, second edition, Reading, 1986.

P.C. Martin, *Measurements and correlation functions*, Les Houches Lecture Notes 1967 (C. De Witt and R. Balian editors), Gordon and Breach, New York, 1968.

Supplement 12B

Electronic polarization

1. Semiclassical model

In the presence of a non-resonant electromagnetic wave, an atomic system acquires an electric polarization. The response function allowing us to determine this polarization can be computed in the framework of a semiclassical model, in which the atom is quantified whereas the electromagnetic field of the wave is treated classically. This model is simple enough to enable us to directly obtain the response function and the corresponding generalized susceptibility. This semiclassical calculation allows us in particular to introduce the notion of *oscillator strength* associated with a transition, and to justify the Lorentz model (fully classical) of the elastically bound electron.

Consider an atomic system with a fundamental level of energy ε_0 and excited levels of energies ε_n, assumed non-degenerate for the sake of simplicity, to which correspond eigenstates $|\phi_0\rangle$ and $|\phi_n\rangle$. We assume that the atom, initially in its fundamental state $|\phi_0\rangle$, is excited by a non-resonant[1] plane wave of angular frequency ω. Under the effect of this excitation, there appears an induced dipolar electric moment, oscillating at the angular frequency ω and proportional to the electric field of the wave if this field is weak.

The electric field $E(t)$ of the wave, parallel to the Ox axis, is assumed to be spatially uniform. The perturbation is described by the dipolar electric Hamiltonian:

$$H_1(t) = -eE(t)x, \tag{12B.1.1}$$

in which e denotes the charge of the electron and x the component of its displacement along Ox. Our aim is to compute the out-of-equilibrium average of the induced dipolar electric moment $P = ex$. This average $\langle P(t)\rangle_a$ is defined by the formula:

$$\langle P(t)\rangle_a = e\langle\psi(t)|x|\psi(t)\rangle, \tag{12B.1.2}$$

where $|\psi(t)\rangle$ represents the state of the system at time t. In the linear range, we write:

$$\langle P(t)\rangle_a = \int_{-\infty}^{\infty} \tilde{\chi}(t-t')E(t')\,dt'. \tag{12B.1.3}$$

In equation (12B.1.3), $\tilde{\chi}(t) = e^2\tilde{\chi}_{xx}(t)$ denotes the linear response function of the polarization.

[1] The angular frequency ω thus coincides with none of the Bohr angular frequencies $\omega_{n0} = (\varepsilon_n - \varepsilon_0)/\hbar$ associated with the transitions taking place from $|\phi_0\rangle$.

2. Polarization response function

To compute $\tilde{\chi}(t)$, we take for the applied field a delta function pulse: $E(t) = E\delta(t)$. We assume that, prior to the application of the field, the atom is in its fundamental state. We thus have:

$$|\psi(t)\rangle = e^{-i\varepsilon_0 t/\hbar}|\phi_0\rangle, \qquad t < 0, \tag{12B.2.1}$$

and, just before the application of the field (for $t = 0^-$):

$$|\psi(0^-)\rangle = |\phi_0\rangle. \tag{12B.2.2}$$

The field pulse produces a discontinuity of the state of the system. To determine this discontinuity, we integrate the Schrödinger equation:

$$i\hbar\,\frac{d|\psi(t)\rangle}{dt} = \big[H_0 - eE\delta(t)x\big]\,|\psi(t)\rangle \tag{12B.2.3}$$

between times $t = 0^-$ and $t = 0^+$, which yields:

$$i\hbar\Big(|\psi(0^+)\rangle - |\psi(0^-)\rangle\Big) = -eEx\int_{0^-}^{0^+}\delta(t)|\psi(t)\rangle\,dt. \tag{12B.2.4}$$

At first perturbation order, $|\psi(t)\rangle$ must be replaced by $|\phi_0\rangle$ on the right-hand side of equation (12B.2.4). On account of the initial condition (12B.2.2), we get:

$$i\hbar\Big(|\psi(0^+)\rangle - |\phi_0\rangle\Big) = -eEx|\phi_0\rangle. \tag{12B.2.5}$$

The atomic state just after the application of the field is:[2]

$$|\psi(0^+)\rangle = |\phi_0\rangle - \frac{e}{i\hbar}E\sum_n|\phi_n\rangle\langle\phi_n|x|\phi_0\rangle. \tag{12B.2.6}$$

The atomic state at a subsequent time $t > 0$ is thus:

$$|\psi(t)\rangle = e^{-i\varepsilon_0 t/\hbar}|\phi_0\rangle - \frac{e}{i\hbar}E\sum_n e^{-i\varepsilon_n t/\hbar}|\phi_n\rangle\langle\phi_n|x|\phi_0\rangle. \tag{12B.2.7}$$

The average induced dipolar electric moment $\langle P(t)\rangle_a = e\langle\psi(t)|x|\psi(t)\rangle$ can be computed with the aid of the state vector $|\psi(t)\rangle$ given by formula (12B.2.7). The matrix element $\langle\phi_0|x|\phi_0\rangle$ vanishing for symmetry reasons, we get at first perturbation order:

$$\big\langle P(t)\big\rangle_a = \frac{2e^2E}{\hbar}\sum_n|\langle\phi_n|x|\phi_0\rangle|^2\sin\omega_{n0}t, \qquad t > 0. \tag{12B.2.8}$$

The linear response function of the atomic system is thus:

$$\boxed{\tilde{\chi}(t) = \Theta(t)\frac{2e^2}{\hbar}\sum_n|\langle\phi_n|x|\phi_0\rangle|^2\sin\omega_{n0}t.} \tag{12B.2.9}$$

[2] We have introduced the closure relation $\sum_n|\phi_n\rangle\langle\phi_n| = 1$.

3. Generalized susceptibility

For a harmonic applied field $E(t) = \Re e(Ee^{-i\omega t})$, the linear response of the polarization reads:

$$\big\langle P(t)\big\rangle_a = \Re e\big[Ee^{-i\omega t}\chi(\omega)\big]. \qquad (12B.3.1)$$

The generalized susceptibility is defined by the formula $\chi(\omega) = \lim_{\epsilon\to 0^+} \chi(\omega + i\epsilon)$, with:

$$\chi(\omega + i\epsilon) = \frac{2e^2}{\hbar}\sum_n |\langle\phi_n|x|\phi_0\rangle|^2 \int_0^\infty \sin\omega_{n0}t\, e^{i\omega t}e^{-\epsilon t}\, dt. \qquad (12B.3.2)$$

After the integration over time has been carried out, we get:

$$\boxed{\chi(\omega) = \frac{e^2}{\hbar}\lim_{\epsilon\to 0^+}\sum_n |\langle\phi_n|x|\phi_0\rangle|^2\left(-\frac{1}{\omega-\omega_{n0}+i\epsilon}+\frac{1}{\omega+\omega_{n0}+i\epsilon}\right).} \qquad (12B.3.3)$$

The real and imaginary parts of the generalized susceptibility are:

$$\begin{cases} \chi'(\omega) = \dfrac{e^2}{\hbar}\displaystyle\sum_n |\langle\phi_n|x|\phi_0\rangle|^2\left(-\mathrm{vp}\,\frac{1}{\omega-\omega_{n0}}+\mathrm{vp}\,\frac{1}{\omega+\omega_{n0}}\right) \\[2ex] \chi''(\omega) = \dfrac{\pi e^2}{\hbar}\displaystyle\sum_n |\langle\phi_n|x|\phi_0\rangle|^2\big[\delta(\omega-\omega_{n0})-\delta(\omega+\omega_{n0})\big]. \end{cases} \qquad (12B.3.4)$$

4. Comparison with the Lorentz model

The Lorentz model is a fully classical model, in which the motion of the electron is described as that of a bound charged particle.[3] Accordingly, in the presence of an electromagnetic wave, the electron is submitted to the electric field of the wave as well as to a restoring force proportional to its displacement. Its equation of motion is that of a harmonic oscillator (whose angular frequency is denoted by ω_0) submitted to an external force:

$$m\frac{d^2x}{dt^2} + m\omega_0^2 x = eE(t). \qquad (12B.4.1)$$

4.1. The susceptibility in the Lorentz model

If $E(t) = \Re e(Ee^{-i\omega t})$ is the electric field of a wave of angular frequency ω, equation (12B.4.1) has, in stationary regime, a solution of the form:

$$x(t) = \Re e(xe^{-i\omega t}). \qquad (12B.4.2)$$

We write:

$$ex = \chi_{\mathrm{cl}}(\omega)E, \qquad (12B.4.3)$$

[3] This model has historically played a very important role in the study of the optical properties of material media.

where $\chi_{\text{cl}}(\omega)$ denotes the generalized susceptibility of the Lorentz model (or classical susceptibility). To determine it, we first compute $\chi_{\text{cl}}(\omega + i\epsilon)$, and we then take the limit $\epsilon \to 0^+$. At finite ϵ, retaining in the denominator only terms of order lower than two in ϵ, we get:

$$\chi_{\text{cl}}(\omega + i\epsilon) \simeq \frac{e^2}{m} \frac{1}{-\omega^2 + \omega_0^2 - 2i\epsilon\omega}. \tag{12B.4.4}$$

In the limit $\epsilon \to 0^+$, we get the generalized susceptibility of the Lorentz model, whose real and imaginary parts are:

$$\begin{cases} \chi'_{\text{cl}}(\omega) = \dfrac{e^2}{2m\omega_0}\left(-\text{vp}\,\dfrac{1}{\omega - \omega_0} + \text{vp}\,\dfrac{1}{\omega + \omega_0}\right) \\[2ex] \chi''_{\text{cl}}(\omega) = \dfrac{\pi e^2}{2m\omega_0}\left[\delta(\omega - \omega_0) - \delta(\omega + \omega_0)\right]. \end{cases} \tag{12B.4.5}$$

4.2. Comparison with the semiclassical susceptibility: definition of the oscillator strength

Formulas (12B.4.5) may be directly compared with the semiclassical formulas (12B.3.4) written for the particular case of a system with two energy levels ε_0 and ε_1. To this end, we have to identify the angular frequency ω_0 of the oscillator of the Lorentz model with the Bohr angular frequency $\omega_{10} = (\varepsilon_1 - \varepsilon_0)/\hbar$. More precisely, introducing the quantity:

$$f_{10} = \frac{2m\omega_{10}}{\hbar} |\langle\phi_1|x|\phi_0\rangle|^2, \tag{12B.4.6}$$

we can write the identity:

$$\chi(\omega) = f_{10}\chi_{\text{cl}}(\omega)_{[\omega_0 \to \omega_{10}]}. \tag{12B.4.7}$$

The quantity f_{10} is a real dimensionless number characteristic of the transition $|\phi_0\rangle \to |\phi_1\rangle$, called the oscillator strength of this transition.

More generally, for a system with several energy levels, we introduce the oscillator strength associated with the transition $|\phi_0\rangle \to |\phi_n\rangle$, defined by the formula:

$$f_{n0} = \frac{2m\omega_{n0}}{\hbar} |\langle\phi_n|x|\phi_0\rangle|^2. \tag{12B.4.8}$$

For an unperturbed Hamiltonian of the form $H_0 = (p^2/2m) + \phi(\boldsymbol{r})$, where $\boldsymbol{r}$ and $\boldsymbol{p}$ are the electron position and momentum operators, it is possible to prove the following property, termed the *oscillator strength sum rule* (or the *Thomas–Reiche–Kuhn sum rule*):[4]

$$\sum_n f_{n0} = 1. \tag{12B.4.9}$$

[4] See Chapter 14 for a general discussion of sum rules.

Using formulas (12B.3.4), on the one hand, and formulas (12B.4.5), on the other hand, we then check the identities:

$$\begin{cases} \chi'(\omega) = \sum_n f_{n0} \chi'_{\text{cl}}(\omega)_{[\omega_0 \to \omega_{n0}]} \\ \chi''(\omega) = \sum_n f_{n0} \chi''_{\text{cl}}(\omega)_{[\omega_0 \to \omega_{n0}]}, \end{cases} \tag{12B.4.10}$$

from which we deduce the relation:

$$\chi(\omega) = \sum_n f_{n0} \chi_{\text{cl}}(\omega)_{[\omega_0 \to \omega_{n0}]}. \tag{12B.4.11}$$

Thus, the semiclassical calculation allows us, for a non-resonant wave, to justify the classical Lorentz model of the elastically bound electron. The semiclassical generalized susceptibility appears as a linear combination of generalized susceptibilities of the type of the Lorentz model's susceptibility. The angular frequencies of the different oscillators identify with the atomic Bohr angular frequencies. The proportion of oscillators with a given angular frequency is equal to the oscillator strength associated with the corresponding transition.

Bibliography

C. Cohen-Tannoudji, B. Diu, and F. Laloë, *Quantum mechanics*, Vol. 2, Hermann and Wiley, Paris, second edition, 1977.

C. Kittel, *Introduction to solid state physics*, Wiley, New York, eighth edition, 2005.

F. Wooten, *Optical properties of solids*, Academic Press, New York, 1972.

Supplement 12C

Some examples of dynamical structure factors

1. The examples

We present here two elementary examples of dynamical structure factors. The systems under consideration, made up of a unique atom, either free or submitted to a harmonic potential, are described by extremely simple equations of motion.[1]

2. Free atom

Consider a unique atom of mass m, whose position and momentum operators are respectively denoted by $\boldsymbol{r}_0$ and $\boldsymbol{p}_0$. This atom, which is free, is in thermal equilibrium at temperature T. For the sake of simplicity, we assume that the equilibrium statistics is the Maxwell–Boltzmann one. The quantum character of the equations of motion will however be taken into account.

2.1. The density and its Fourier transform

The operator associated with the interaction of the radiation with the system is the density $n(\boldsymbol{r},t) = \delta[\boldsymbol{r} - \boldsymbol{r}_0(t)]$. The spatial Fourier transform[2] of $n(\boldsymbol{r},t)$, defined by:

$$n(\boldsymbol{q},t) = \int n(\boldsymbol{r},t) e^{-i\boldsymbol{q}.\boldsymbol{r}}\, d\boldsymbol{r}, \tag{12C.2.1}$$

is:

$$n(\boldsymbol{q},t) = \exp\big[-i\boldsymbol{q}.\boldsymbol{r}_0(t)\big]. \tag{12C.2.2}$$

For a free atom, we simply have $\boldsymbol{r}_0(t) = \boldsymbol{r}_0 + (\boldsymbol{p}_0/m)t$, which gives:

$$n(\boldsymbol{q},t) = \exp\Big[-i\boldsymbol{q}.\big(\boldsymbol{r}_0 + \frac{\boldsymbol{p}_0}{m}t\big)\Big]. \tag{12C.2.3}$$

---

[1] In more complex systems such as fluids, the determination of the dynamical structure factor involves solving much more complicated equations of motion, but follows analogous lines. This subject is treated in Supplement 16B.

[2] We use the same notation $n(.\,,t)$ for the density $n(\boldsymbol{r},t)$ and its spatial Fourier transform $n(\boldsymbol{q},t)$.

The expression (12C.2.3) for $n(\boldsymbol{q},t)$ may be factorized into a product of exponentials. We must take into account the fact that the operators $\boldsymbol{r}_0$ and $\boldsymbol{p}_0$ do not commute. Using the *Glauber identity*,[3] we end up in one or the other of the two following equivalent factorized expressions:

$$\begin{cases} n(\boldsymbol{q},t) = \exp\Big(-\dfrac{i\hbar q^2 t}{2m}\Big)\exp\Big(-i\dfrac{\boldsymbol{p}_0.\boldsymbol{q}}{m}t\Big)\exp(-i\boldsymbol{q}.\boldsymbol{r}_0) \\ n(\boldsymbol{q},t) = \exp\Big(\dfrac{i\hbar q^2 t}{2m}\Big)\exp(-i\boldsymbol{q}.\boldsymbol{r}_0)\exp\Big(-i\dfrac{\boldsymbol{p}_0.\boldsymbol{q}}{m}t\Big). \end{cases} \tag{12C.2.4}$$

2.2. The density autocorrelation function

The equilibrium density autocorrelation function is defined as the equilibrium average of the product $n(\boldsymbol{q},t)n(-\boldsymbol{q},0)$. With the aid of the first of formulas (12C.2.4), this product reads:

$$n(\boldsymbol{q},t)n(-\boldsymbol{q},0) = \exp\Big(-\frac{i\hbar q^2 t}{2m}\Big)\exp\Big(-i\frac{\boldsymbol{p}_0.\boldsymbol{q}}{m}t\Big). \tag{12C.2.5}$$

Taking the average over the Maxwell–Boltzmann distribution (a Gaussian function of $\boldsymbol{p}_0$), we get:

$$\big\langle n(\boldsymbol{q},t)n(-\boldsymbol{q},0)\big\rangle = \exp\Big[-\frac{q^2 t(t+i\hbar\beta)}{2m\beta}\Big], \qquad \beta = (kT)^{-1}. \tag{12C.2.6}$$

In the same way, with the aid of the second of formulas (12C.2.4) applied to $n(-\boldsymbol{q},t)$, we can write:

$$n(\boldsymbol{q},0)n(-\boldsymbol{q},t) = \exp\Big(\frac{i\hbar q^2 t}{2m}\Big)\exp\Big(i\frac{\boldsymbol{p}_0.\boldsymbol{q}}{m}t\Big). \tag{12C.2.7}$$

On average, we have:

$$\big\langle n(\boldsymbol{q},0)n(-\boldsymbol{q},t)\big\rangle = \exp\Big[-\frac{q^2 t(t-i\hbar\beta)}{2m\beta}\Big]. \tag{12C.2.8}$$

Both $\langle n(\boldsymbol{q},t)n(-\boldsymbol{q},0)\rangle$ and $\langle n(\boldsymbol{q},0)n(-\boldsymbol{q},t)\rangle$ are invariant under the change $\boldsymbol{q}\to-\boldsymbol{q}$ (as a consequence of the fact that the system possesses the inversion symmetry $\boldsymbol{r}\to-\boldsymbol{r}$). We then verify the Kubo–Martin–Schwinger condition:

$$\big\langle n(\boldsymbol{q},0)n(-\boldsymbol{q},t)\big\rangle = \big\langle n(-\boldsymbol{q},t-i\hbar\beta)n(\boldsymbol{q},0)\big\rangle. \tag{12C.2.9}$$

---

[3] For two operators A and B which both commute with their commutator $[A,B]$, we have the Glauber identity:

$$e^{A+B} = e^A e^B e^{-\frac{1}{2}[A,B]}.$$

2.3. The dynamical structure factor

The dynamical structure factor is defined by the formula:

$$S(\boldsymbol{q},\omega)=\int_{-\infty}^{\infty}\langle n(\boldsymbol{q},t)n(-\boldsymbol{q},0)\rangle e^{i\omega t}\,dt. \tag{12C.2.10}$$

Expressing $\langle n(\boldsymbol{q},t)n(-\boldsymbol{q},0)\rangle$ with the aid of formula (12C.2.6) and carrying out the integration over time, gives:

$$\boxed{S(\boldsymbol{q},\omega)=\left(\frac{2\pi m\beta}{q^2}\right)^{1/2}\exp\left[-\frac{m\beta}{2q^2}\left(\omega-\hbar\frac{q^2}{2m}\right)^2\right].} \tag{12C.2.11}$$

At given wave vector $\boldsymbol{q}$, $S(\boldsymbol{q},\omega)$ is a Gaussian function of ω, centered at $\omega=\hbar q^2/2m$ and of variance $q^2\,kT/m$ (a quantity which increases with q and with the temperature). The dynamical structure factor given by formula (12C.2.11) verifies the detailed balance relation $S(\boldsymbol{q},\omega)=S(\boldsymbol{q},-\omega)e^{\beta\hbar\omega}$.

3. Atom in a harmonic potential

Assume now that the scattering atom evolves in a harmonic oscillator potential. In the formula for $n(\boldsymbol{r},t)$, the expression for $\boldsymbol{r}_0(t)$ must be appropriate to the potential in which the atom evolves. The quantum character of the equations of motion and of the equilibrium statistics will be taken into account in the calculation.

3.1. The density autocorrelation function

The expression for the product $n(\boldsymbol{q},t)n(-\boldsymbol{q},0)$ involves the components x_0 and $x_0(t)$ of $\boldsymbol{r}_0$ and $\boldsymbol{r}_0(t)$ along the wave vector $\boldsymbol{q}$:

$$n(\boldsymbol{q},t)n(-\boldsymbol{q},0)=\exp[-iqx_0(t)]\exp(iqx_0). \tag{12C.3.1}$$

We can group together the two factors on the right-hand side of equation (12C.3.1), taking into account the fact that $x_0(t)$ and x_0 do not commute. Using again the Glauber identity, gives:

$$\exp[-iqx_0(t)]\exp(iqx_0)=\exp\Big(\frac{1}{2}q^2[x_0(t),x_0]\Big)\exp[-iqx_0(t)+iqx_0]. \tag{12C.3.2}$$

For an oscillator of mass m and of angular frequency ω_0, we have $x_0(t)=x_0\cos\omega_0 t+(p_0/m\omega_0)\sin\omega_0 t$. The commutator $[x_0(t),x_0]$ is a scalar which needs no averaging. We thus have:

$$\langle n(\boldsymbol{q},t)n(-\boldsymbol{q},0)\rangle=\exp\Big(\frac{1}{2}q^2[x_0(t),x_0]\Big)\langle\exp[-iqx_0(t)+iqx_0]\rangle. \tag{12C.3.3}$$

To determine the density autocorrelation function, we are thus led to compute the equilibrium average of an exponentiated operator e^A, in which $A=-iq[x_0(t)-x_0]$ is a linear combination of the position and momentum operators of the oscillator. Using the identity:

$$\langle e^A \rangle = \exp\left\langle \frac{A^2}{2} \right\rangle, \tag{12C.3.4}$$

applicable to this type of operator,[4] we get:

$$\langle \exp[-iqx_0(t) + iqx_0] \rangle = \exp\Big(-\frac{1}{2}q^2\langle [x_0(t) - x_0]^2 \rangle\Big). \tag{12C.3.5}$$

This gives:

$$\langle n(\boldsymbol{q}, t)n(-\boldsymbol{q}, 0) \rangle = \exp\Big(-q^2\left[\langle x_0^2 \rangle - \langle x_0(t)x_0 \rangle\right]\Big). \tag{12C.3.6}$$

Formula (12C.3.6) involves the autocorrelation function $\langle x_0(t)x_0 \rangle$, as well as the average $\langle x_0^2 \rangle$. We have:

$$\begin{cases} \langle x_0(t)x_0 \rangle = \dfrac{\hbar}{2m\omega_0}[(1 + n_0)e^{-i\omega_0 t} + n_0 e^{i\omega_0 t}], \\[2ex] \langle x_0^2 \rangle = \dfrac{\hbar}{2m\omega_0} \coth \dfrac{\beta\hbar\omega_0}{2}, \end{cases} \tag{12C.3.7}$$

where $n_0 = (e^{\beta\hbar\omega_0} - 1)^{-1}$ denotes the Bose–Einstein distribution function at temperature $T = (k\beta)^{-1}$.

3.2. Analysis of the expression for $\langle n(\boldsymbol{q}, t)n(-\boldsymbol{q}, 0) \rangle$

The autocorrelation function $\langle n(\boldsymbol{q}, t)n(-\boldsymbol{q}, 0) \rangle$ given by formula (12C.3.6) appears as the product of two exponential factors. The first of them, time-independent,

$$\exp(-q^2\langle x_0^2 \rangle) = \exp\left(-\frac{\hbar q^2}{2m\omega_0} \coth \frac{\beta\hbar\omega_0}{2}\right), \tag{12C.3.8}$$

is called the *Debye–Waller factor*[5] and is denoted by $e^{-2W(\boldsymbol{q})}$. The second factor, which depends on time,

$$\exp(q^2\langle x_0(t)x_0 \rangle) = \exp\left(\frac{\hbar q^2}{2m\omega_0}\left[(1 + n_0)e^{-i\omega_0 t} + n_0 e^{i\omega_0 t}\right]\right), \tag{12C.3.9}$$

may be written equivalently as::

$$\exp(q^2\langle x_0(t)x_0 \rangle) = \exp\left(\frac{\hbar q^2}{4m\omega_0 \sinh\frac{\beta\hbar\omega_0}{2}}\left[e^{(-i\omega_0 t + \frac{\beta\hbar\omega_0}{2})} + e^{(i\omega_0 t - \frac{\beta\hbar\omega_0}{2})}\right]\right). \tag{12C.3.10}$$

---

[4] Note that identity (12C.3.4) is analogous to the formula of probability theory yielding the average of the exponential of a Gaussian centered random variable.

[5] The same type of factor, originating from the thermal motion of the atoms around their equilibrium positions, is involved in the scattering of neutrons or X-rays by a crystal.

Setting:

$$y = \frac{\hbar q^2}{2m\omega_0 \sinh \frac{\beta\hbar\omega_0}{2}}, \tag{12C.3.11}$$

we can rewrite the expression (12C.3.10) for $\exp\big(q^2\langle x_0(t)x_0\rangle\big)$ with the aid of the following series expansion:

$$\exp\Big[\frac{1}{2}y(x+\frac{1}{x})\Big] = \sum_{n=-\infty}^{\infty} x^n I_n(y), \qquad n = 0, \pm 1, \pm 2, \ldots, \tag{12C.3.12}$$

where the $I_n(y)$'s are the modified Bessel functions of the first kind. We finally obtain the formula:

$$\big\langle n(\boldsymbol{q},t)n(-\boldsymbol{q},0)\big\rangle = e^{-2W(\boldsymbol{q})} \sum_{n=-\infty}^{\infty} I_n(y) e^{\frac{1}{2}n\beta\hbar\omega_0 - in\omega_0 t}. \tag{12C.3.13}$$

3.3. The dynamical structure factor

The corresponding dynamical structure factor is obtained by Fourier transformation of $\langle n(\boldsymbol{q},t)n(-\boldsymbol{q},0)\rangle$ according to formula (12C.2.10). This gives:

$$S(\boldsymbol{q},\omega) = 2\pi e^{-2W(\boldsymbol{q})+\frac{1}{2}\beta\hbar\omega} \sum_{n=-\infty}^{\infty} I_n(y)\delta(\omega - n\omega_0). \tag{12C.3.14}$$

The dynamical structure factor given by formula (12C.3.14) verifies the detailed balance relation[6] $S(\boldsymbol{q},\omega) = S(\boldsymbol{q},-\omega)\, e^{\beta\hbar\omega}$ (the system possesses inversion symmetry).

In the sum on the right-hand side of equation (12C.3.14), the term $n = 0$ corresponds to an elastic scattering process with no energy exchange between the radiation and the target (here, the atom). The terms $n = \pm 1$ represent the one-quantum contributions to the dynamical structure factor, whereas the terms with higher $|n|$ correspond to contributions involving a larger number of quanta.

[6] To check this relation, we use the following property of the modified Bessel functions:

$$I_{-n}(y) = I_n(y).$$

Bibliography

S.W. Lovesey, *Condensed matter physics: dynamic correlations*, The Benjamin/Cummings Publishing Company, Reading, second edition, 1986.

S.W. Lovesey, *Theory of neutron scattering from condensed matter*, Oxford University Press, Oxford, 1984.

Chapter 13

General linear response theory

Linear response theory is a general formalism applicable to any physical system slightly departing from equilibrium. It allows us to express the linear response functions and the generalized susceptibilities in terms of the equilibrium correlation functions of the relevant dynamical variables. For instance, in the case of the response to an applied external field, one of the two relevant dynamical variables represents the physical quantity conjugate to the field whereas the other represents the physical quantity whose out-of-equilibrium average is measured. The computation of the linear response functions relies on a first-order perturbation expansion of the density operator (or of the phase space distribution function) of the system coupled to the field. The expressions thus obtained constitute the Kubo formulas of linear response theory.

In this chapter, after having introduced this general formalism and established the Kubo formulas, we examine the symmetry properties of the linear response functions and of the associated equilibrium correlation functions. Some of these properties depend on the way that the relevant dynamical variables behave under time-reversal. The Kubo formulas of linear response theory thus enable us in particular to demonstrate the Onsager reciprocity relations.

1. The object of linear response theory

A commonly used method of carrying out measurements on a physical system is to submit it to an external force and to observe the way it reacts. For the result of such an experiment to adequately reflect the properties of the system, the perturbation due to the applied force must be sufficiently weak.

In this very general framework, three distinct types of measurements can be carried out: actual response measurements, susceptibility measurements consisting in determining the response of the system to a harmonic force, and relaxation measurements in which, after having removed a force that had been applied for a very long time, we study the return of the system to equilibrium. The results of these three types of measurements are respectively expressed in terms of response functions, generalized susceptibilities, and *relaxation functions*. In the linear range, these quantities depend solely on the properties of the unperturbed system, and each of them interrelates with the others.

The object of *linear response theory* is to allow us, for any specific physical problem, to determine the response functions, the generalized susceptibilities, and the relaxation functions. In the linear range, all these quantities can be expressed with the aid of equilibrium correlation functions of the relevant dynamical variables of the unperturbed system. The corresponding expressions constitute the *Kubo formulas*. To derive them, we make use of a first-order perturbation expansion of the density operator of the system with respect to the perturbation created by the external field.

2. First-order evolution of the density operator

At the microscopic level, physical systems are generally described by quantum mechanics (however, in some cases, for instance to treat the translational degrees of freedom of the molecules of a gas or a liquid, we can use classical microscopic equations). We will adopt the Schrödinger picture (or its classical analog), in which the properties of a system are determined with the aid of its density operator (or of its phase space distribution function). To begin with, we will determine the evolution of these latter quantities at first perturbation order.

2.1. Response of an isolated system

Consider a physical system, initially at thermodynamic equilibrium and described by a time-independent Hamiltonian H_0. The corresponding density operator will be denoted by ρ_0. For a system in equilibrium with a thermostat at temperature T, ρ_0 is the canonical density operator:[1]

$$\rho_0 = \frac{1}{Z} e^{-\beta H_0}, \qquad \beta = (kT)^{-1}. \tag{13.2.1}$$

[1] This is the most frequent situation. This is why it is quite common practice to make this hypothesis. However, in principle, in the Kubo theory, the equilibrium distribution is not necessarily the canonical one.

From an initial time t_0 (the limit $t_0 \to -\infty$ will be taken at the end of the calculation), we isolate the system from the thermostat and we submit it to an external field $a(t)$, assumed to be spatially uniform.[2] The perturbation is described by the Hamiltonian:[3]

$$H_1(t) = -a(t)A, \tag{13.2.2}$$

where A is the Hermitean operator[4] associated with the physical quantity conjugate to the field. The Hamiltonian of the perturbed system is:

$$H = H_0 + H_1(t). \tag{13.2.3}$$

For $t > t_0$, the system is isolated. Its density operator $\rho(t)$ obeys the Liouville–von Neumann equation:

$$\frac{d\rho(t)}{dt} = -i\mathcal{L}\rho(t), \tag{13.2.4}$$

where $\mathcal{L}$ denotes the Liouville operator associated with the total Hamiltonian.[5] We aim to determine, at first perturbation order, the solution of equation (13.2.4) for the initial condition $\rho(t_0) = \rho_0$.

2.2. Evolution of the density operator in the linear regime

The Liouville operator is the sum of two terms:

$$\mathcal{L} = \mathcal{L}_0 + \mathcal{L}_1. \tag{13.2.5}$$

This decomposition corresponds to the decomposition (13.2.3) of the Hamiltonian. In the same way, the density operator may be written in the form:

$$\rho(t) = \rho_0 + \delta\rho(t). \tag{13.2.6}$$

Making use of formulas (13.2.5) and (13.2.6) in equation (13.2.4), we get, since $i\mathcal{L}_0\rho_0$ vanishes:

$$\frac{d\delta\rho(t)}{dt} = -i\mathcal{L}_1\rho_0 - i\mathcal{L}_0\delta\rho(t) - i\mathcal{L}_1\delta\rho(t). \tag{13.2.7}$$

---

[2] The case of a non-uniform field will be treated in Section 9.

[3] The forces whose effect may be described by a Hamiltonian of the type (13.2.2) are termed *mechanical forces*. There are other types of force, whose effect cannot be expresssed in this way. For instance, temperature or chemical potential inhomogeneities controlled from outside produce within a system generalized forces which give rise to a heat flux or to a particle flux. Such forces are termed *thermal forces*. The response to this latter type of force will be studied in Chapter 16.

[4] The perturbation Hamiltonian $H_1(t)$ may in general appear as a sum of the type $-\sum_i a_i(t)A_i$. The effect of each term of the sum can be studied separately in the linear regime. In such a situation, each operator A_i is not necessarily Hermitean (the overall hermiticity of $H_1(t)$ only requires that, if $A_i = A_j^\dagger$, we have $a_i = a_j^*$).

[5] In this first stage, we make use of the Liouville operator formalism which allows us to treat formally in a unified manner classical and quantum problems as well. We will use in all cases the denomination of density operator and the notation $\rho(t)$ (for a classical system, $\rho(t)$ will stand for the phase space distribution function).

The third term on the right-hand-side of equation (13.2.7) is of higher order. To get the first-order correction to ρ_0, it is enough to solve the evolution equation:

$$\frac{d\delta\rho(t)}{dt} = -i\mathcal{L}_1\rho_0 - i\mathcal{L}_0\delta\rho(t), \tag{13.2.8}$$

for the initial condition $\delta\rho(t_0) = 0$.

To this end, let us set $\delta\rho(t) = e^{-i\mathcal{L}_0 t}F(t)$. The evolution equation of $F(t)$ reads:

$$\frac{dF(t)}{dt} = -ie^{i\mathcal{L}_0 t}\mathcal{L}_1\rho_0. \tag{13.2.9}$$

The solution of equation (13.2.9) for the initial condition $F(t_0) = 0$ is:

$$F(t) = -i\int_{t_0}^{t} e^{i\mathcal{L}_0 t'}\mathcal{L}_1\rho_0\, dt'. \tag{13.2.10}$$

We deduce from equation (13.2.10) the expression for $\delta\rho(t)$:

$$\delta\rho(t) = -i\int_{t_0}^{t} e^{-i\mathcal{L}_0(t-t')}\mathcal{L}_1\rho_0\, dt'. \tag{13.2.11}$$

The major simplification arising from the linearity hypothesis is the presence of the unperturbed Liouville operator in the exponent of the evolution operator.

At this stage, it is convenient to make precise the meaning[6] of $\mathcal{L}_1\rho_0$:

$$\mathcal{L}_1\rho_0 = \frac{1}{\hbar}[H_1, \rho_0]. \tag{13.2.12}$$

Formula (13.2.11) thus reads:

$$\delta\rho(t) = -\frac{i}{\hbar}\int_{t_0}^{t} e^{-i\mathcal{L}_0(t-t')}[H_1, \rho_0]\, dt', \tag{13.2.13}$$

that is, taking into account the expression (13.2.2) for the perturbation Hamiltonian:

$$\delta\rho(t) = \frac{i}{\hbar}\int_{t_0}^{t} a(t')e^{-i\mathcal{L}_0(t-t')}[A, \rho_0]\, dt'. \tag{13.2.14}$$

In formula (13.2.14), the operator A, written in the Schrödinger picture, does not depend on time. The Liouville operator $\mathcal{L}_0$ acts only on A and not on ρ_0, since ρ_0 and H_0 commute. This gives:

$$\delta\rho(t) = \frac{i}{\hbar}\int_{t_0}^{t} a(t')[A^I(t'-t), \rho_0]\, dt', \tag{13.2.15}$$

where $A^I(t) = e^{i\mathcal{L}_0 t}A = e^{iH_0 t/\hbar}Ae^{-iH_0 t/\hbar}$ denotes the operator A in the interaction picture (that is, in the Heisenberg picture with respect to H_0).

The limit $t_0 \to -\infty$ once taken,we get:

$$\boxed{\delta\rho(t) = \frac{i}{\hbar}\int_{-\infty}^{t} a(t')[A^I(t'-t), \rho_0]\, dt'.} \tag{13.2.16}$$

[6] The notations of the quantum case being more familiar, it is this latter case that we choose to develop here, the classical case being treated in an appendix at the end of this chapter.

3. The linear response function

3.1. The Kubo formula

Consider a physical quantity represented by a Hermitean operator B. We want to determine at first order the effect of the perturbation described by $H_1(t)$ on the temporal evolution of the out-of-equilibrium average $\langle B(t)\rangle_a$. In the Schrödinger picture, $\langle B(t)\rangle_a$ reads:

$$\langle B(t)\rangle_a = \mathrm{Tr}\big[\rho(t)B\big], \tag{13.3.1}$$

that is, using the decomposition (13.2.6) of the density operator:

$$\langle B(t)\rangle_a = \langle B\rangle + \mathrm{Tr}\big[\delta\rho(t)B\big]. \tag{13.3.2}$$

In equation (13.2.12), $\langle B\rangle = \mathrm{Tr}\big(\rho_0 B\big)$ denotes the equilibrium average of B. For the sake of simplicity, we assume from now on that B is centered: $\langle B\rangle = 0$.

In these conditions, we deduce from formula (13.2.16) for $\delta\rho(t)$ the expression for $\langle B(t)\rangle_a$ at first perturbation order:

$$\langle B(t)\rangle_a = \frac{i}{\hbar}\int_{-\infty}^{t} a(t')\mathrm{Tr}\big([A^I(t'-t),\rho_0]B\big)\,dt'. \tag{13.3.3}$$

Using the invariance of the trace under a circular permutation of the operators gives:

$$\langle B(t)\rangle_a = \frac{i}{\hbar}\int_{-\infty}^{t} a(t')\mathrm{Tr}\big([B,A^I(t'-t)]\rho_0\big)\,dt'. \tag{13.3.4}$$

In equation (13.3.4), the quantity $\mathrm{Tr}\big([B,A^I(t'-t)]\rho_0\big)$ is simply the equilibrium average $\langle [B,A^I(t'-t)]\rangle$. It is thus possible to carry out a translation of its temporal arguments without modifying it and to write:[7]

$$\langle B(t)\rangle_a = \frac{i}{\hbar}\int_{-\infty}^{t} a(t')\langle [B^I(t-t'),A]\rangle\,dt'. \tag{13.3.5}$$

The response $\langle B(t)\rangle_a$ at first perturbation order is of the form:

$$\langle B(t)\rangle_a = \int_{-\infty}^{\infty} \tilde\chi_{BA}(t-t')a(t')\,dt'. \tag{13.3.6}$$

Formula (13.3.5) shows that the linear response function $\tilde\chi_{BA}(t)$ is expressed in terms of an equilibrium average of a commutator of unperturbed dynamical variables.

[7] Whereas $\langle B(t)\rangle_a$ denotes the out-of-equilibrium average of B, a quantity such as $\langle [B^I(t-t'),A]\rangle$ denotes an equilibrium average of a commutator of dynamical variables in the interaction picture. No misinterpretation being possible, we will suppress from now on the superscript label I for the operators in the interaction picture appearing in the response functions and the associated correlation functions.

The expression obtained for the response function,

$$\tilde{\chi}_{BA}(t) = \frac{i}{\hbar}\Theta(t)\langle[B(t), A]\rangle, \tag{13.3.7}$$

where $\Theta(t)$ denotes the Heaviside function, is the *Kubo formula*[8] for $\tilde{\chi}_{BA}(t)$. The average involved in formula (13.3.7) is an equilibrium average computed with the aid of the density operator ρ_0. Interestingly, if A or B commutes with H_0, $\tilde{\chi}_{BA}(t)$ vanishes.

3.2. Expression for $\tilde{\chi}_{BA}(t)$ with the aid of the eigenstates and eigenvalues of H_0

We denote by $\{|\phi_n\rangle\}$ a base of eigenstates of H_0, of energies ε_n (the levels are assumed non-degenerate). It is also a base of eigenstates of ρ_0, since ρ_0 and H_0 commute. We can rewrite formula (13.3.7) in the form:

$$\tilde{\chi}_{BA}(t) = \frac{i}{\hbar}\Theta(t)\sum_n \langle\phi_n|[B(t), A]\rho_0|\phi_n\rangle, \tag{13.3.8}$$

that is, denoting by $\Pi_n = \langle\phi_n|\rho_0|\phi_n\rangle$ the equilibrium population of state $|\phi_n\rangle$:

$$\tilde{\chi}_{BA}(t) = \frac{i}{\hbar}\Theta(t)\sum_{n,q} \Pi_n \left(B_{nq}A_{qn}e^{i\omega_{nq}t} - A_{nq}B_{qn}e^{i\omega_{qn}t}\right). \tag{13.3.9}$$

In formula (13.3.9), the quantities $\omega_{nq} = (\varepsilon_n - \varepsilon_q)/\hbar$ are the Bohr angular frequencies of the unperturbed system. We can also write, inverting the indices n and q in the second sum on the right-hand side of equation (13.3.9):

$$\tilde{\chi}_{BA}(t) = \frac{i}{\hbar}\Theta(t)\sum_{n,q}(\Pi_n - \Pi_q)B_{nq}A_{qn}e^{i\omega_{nq}t}. \tag{13.3.10}$$

The response function $\tilde{\chi}_{BA}(t)$ thus appears as a linear superposition of imaginary exponentials oscillating at the unperturbed Bohr angular frequencies. For a system with a finite number of degrees of freedom, the spectrum of H_0 is discrete, and the same is true of the Fourier spectrum of $\tilde{\chi}_{BA}(t)$. The response function is then a countable sum of periodic functions. Such a function does not vanish as $t \to \infty$: a finite system thus possesses in this sense an infinitely long 'memory'.[9] We will now illustrate this feature of the response function with two simple examples.

[8] The classical analog of formula (13.3.7) is demonstrated in an appendix at the end of this chapter.

[9] The system considered here is a finite system described by a Hamiltonian. This excludes a system such as the oscillator damped by viscous friction, whose displacement response function vanishes as $t \to \infty$.

• *Harmonic oscillator*

The response function $\tilde{\chi}_{xx}(t)$ of the displacement of a harmonic oscillator with mass m and angular frequency ω_0 is a sinusoidal function of angular frequency ω_0:

$$\tilde{\chi}_{xx}(t) = \Theta(t)\frac{\sin\omega_0 t}{m\omega_0}. \tag{13.3.11}$$

• *Atom perturbed by an electric field*

Let us re-examine the response function $\tilde{\chi}_{xx}(t)$ giving the polarization of an atomic system perturbed by an electric field when the unperturbed atom is in its fundamental state $|\phi_0\rangle$.

In this case, the density operator is the projector $\rho_0 = |\phi_0\rangle\langle\phi_0|$, and the determination of $\tilde{\chi}_{xx}(t)$ is a zero-temperature calculation. Formulas (13.3.7) and (13.3.9) simply read:

$$\tilde{\chi}_{xx}(t) = \frac{i}{\hbar}\Theta(t)\langle\phi_0|[x(t), x]|\phi_0\rangle, \tag{13.3.12}$$

and (the populations of the various states being $\Pi_0 = 1, \Pi_{n\neq 0} = 0$):

$$\tilde{\chi}_{xx}(t) = \frac{i}{\hbar}\Theta(t)\sum_n \left(x_{0n}x_{n0}e^{i\omega_{0n}t} - x_{0n}x_{n0}e^{i\omega_{n0}t}\right). \tag{13.3.13}$$

Thus, the response function $\chi_{xx}(t)$ is a sum of sinusoidal functions:

$$\tilde{\chi}_{xx}(t) = \Theta(t)\frac{2}{\hbar}\sum_n |\langle\phi_0|x|\phi_n\rangle|^2 \sin\omega_{n0}t. \tag{13.3.14}$$

Formula (13.3.14) is in accordance with the result of the direct calculation.

From the two examples above, we verify that in a finite system the response function does not vanish as $t \to \infty$. However, in an infinite system, the Fourier spectrum of the response function is continuous and the response function approaches zero as $t \to \infty$.

4. Relation with the canonical correlation function

We are interested here in a system initially at equilibrium with a thermostat at temperature T, in which case ρ_0 is the canonical density operator (13.2.1). The equilibrium population of the state $|\phi_n\rangle$ is thus:

$$\Pi_n = \frac{1}{Z}e^{-\beta\varepsilon_n}. \tag{13.4.1}$$

We will show that, in these conditions, the average value $\langle[B(t), A]\rangle$ involved in the Kubo formula (13.3.7) can be rewritten in a more simple way (that is, without involving a commutator).

To compute $\langle [B(t), A] \rangle = \mathrm{Tr}\big([A, \rho_0] B(t)\big)$, we use the following identity:[10]

$$[A, e^{-\beta H_0}] = e^{-\beta H_0} \int_0^\beta e^{\lambda H_0} [H_0, A] e^{-\lambda H_0} \, d\lambda. \tag{13.4.2}$$

Inserting in formula (13.4.2) the evolution equation for A ($i\hbar \dot{A} = [A, H_0]$), we get:

$$[A, \rho_0] = -i\hbar \rho_0 \int_0^\beta e^{\lambda H_0} \dot{A} e^{-\lambda H_0} \, d\lambda. \tag{13.4.3}$$

The response function (13.3.7) takes the form:

$$\tilde{\chi}_{BA}(t) = \Theta(t) \int_0^\beta \big\langle e^{\lambda H_0} \dot{A} e^{-\lambda H_0} B(t) \big\rangle \, d\lambda. \tag{13.4.4}$$

At this stage, it is convenient to introduce the Kubo *canonical correlation function* $\tilde{K}_{BA}(t)$, defined as:

$$\boxed{\tilde{K}_{BA}(t) = \frac{1}{\beta} \int_0^\beta \big\langle A(-i\hbar\lambda) B(t) \big\rangle \, d\lambda,} \tag{13.4.5}$$

where $A(-i\hbar\lambda) = e^{\lambda H_0} A e^{-\lambda H_0}$ denotes the operator A in the interaction picture at the imaginary time $-i\hbar\lambda$.

Formula (13.4.4) shows that the response function $\tilde{\chi}_{BA}(t)$ of a system initially at equilibrium with a thermostat at temperature T can be expressed in terms of the canonical correlation function[11] of B with $\dot{A}$:

$$\boxed{\tilde{\chi}_{BA}(t) = \beta \Theta(t) \tilde{K}_{B\dot{A}}(t).} \tag{13.4.6}$$

5. Generalized susceptibility

The linear response $\langle B(t) \rangle_a$ to an applied field $a(t)$ of Fourier transform $a(\omega)$, conjugate to a physical quantity A, has the Fourier transform:

$$\big\langle B(\omega) \big\rangle_a = \chi_{BA}(\omega) a(\omega). \tag{13.5.1}$$

The generalized susceptibility $\chi_{BA}(\omega)$ is the Fourier transform in the distribution sense of the response function $\tilde{\chi}_{BA}(t)$. To get it, we first compute:

$$\chi_{BA}(\omega + i\epsilon) = \int_0^\infty \tilde{\chi}_{BA}(t) e^{i\omega t} e^{-\epsilon t} \, dt, \qquad \epsilon > 0, \tag{13.5.2}$$

[10] We can check the identity (13.4.2) by computing the matrix elements of both sides on the base $\{|\phi_n\rangle\}$.

[11] The classical analog of formula (13.4.6) is demonstrated in an appendix at the end of this chapter.

and we then take the limit $\epsilon \to 0^+$:

$$\chi_{BA}(\omega) = \lim_{\epsilon \to 0^+} \chi_{BA}(\omega + i\epsilon). \tag{13.5.3}$$

If $\tilde{\chi}_{BA}(t)$ is expressed with the aid of formula (13.3.10), the above procedure yields the expression:

$$\boxed{\chi_{BA}(\omega) = \frac{1}{\hbar} \sum_{n,q} (\Pi_n - \Pi_q) B_{nq} A_{qn} \lim_{\epsilon \to 0^+} \frac{1}{\omega_{qn} - \omega - i\epsilon}.} \tag{13.5.4}$$

It is useful, z being the affix of a point of the upper complex half-plane, to introduce the Fourier–Laplace transform $\chi_{BA}(z)$ of $\tilde{\chi}_{BA}(t)$:

$$\chi_{BA}(z) = \int_0^\infty \tilde{\chi}_{BA}(t) e^{izt}\, dt, \qquad \Im m\, z > 0. \tag{13.5.5}$$

Using once again formula (13.3.10), we get:

$$\boxed{\chi_{BA}(z) = \frac{1}{\hbar} \sum_{n,q} (\Pi_n - \Pi_q) B_{nq} A_{qn} \frac{1}{\omega_{qn} - z}.} \tag{13.5.6}$$

Formula (13.5.6) clearly displays the singularities of $\chi_{BA}(z)$: a pole of $\chi_{BA}(z)$ on the real axis is associated with each Bohr angular frequency ω_{qn}. For a finite system described by a Hamiltonian, the poles of $\chi_{BA}(z)$ remain separated from each other by a minimal distance. It is then possible to define $\chi_{BA}(z)$ in the whole complex plane (except for the poles) by a direct analytical continuation of formula (13.5.6). Let us now apply formula (13.5.6) to the two previously considered examples.

- *Harmonic oscillator*

For any $z \neq \pm\omega_0$, we have:

$$\chi_{xx}(z) = \frac{1}{2m\omega_0} \left(-\frac{1}{z - \omega_0} + \frac{1}{z + \omega_0} \right). \tag{13.5.7}$$

- *Atom perturbed by an electric field*

Consider an atom initially in its fundamental state $|\phi_0\rangle$ and perturbed by an electric field. For any $z \neq \pm\,\omega_{n0}$, we have:

$$\chi_{xx}(z) = \frac{1}{\hbar} \sum_n |\langle\phi_0|x|\phi_n\rangle|^2 \left(-\frac{1}{z - \omega_{n0}} + \frac{1}{z + \omega_{n0}} \right). \tag{13.5.8}$$

In general, as the system's size tends towards infinity, the poles of $\chi_{BA}(z)$ push closer and closer together. In the limit of an infinite system, they form a continuum. The discrete set of poles situated on the real axis then becomes a branch cut and formula (13.5.6) is only valid for $\Im m\, z \neq 0$. It is then convenient to introduce a function characterizing the density of poles on the real axis as a function of ω: this is the *spectral function* $\xi_{BA}(\omega)$.

6. Spectral function

6.1. Definition

The spectral function $\xi_{BA}(\omega)$ is defined by:

$$\boxed{\xi_{BA}(\omega) = \frac{\pi}{\hbar} \sum_{n,q} (\Pi_n - \Pi_q) B_{nq} A_{qn} \delta(\omega_{qn} - \omega).} \tag{13.6.1}$$

Only the terms $n \neq q$ of the double sum on the right-hand side of formula (13.6.1) yield a non-vanishing contribution to $\xi_{BA}(\omega)$. Like the response function, the spectral function vanishes if A or B commutes with H_0.

Since the operators A and B have been assumed Hermitean, we have the property:

$$\xi^*_{BA}(\omega) = \xi_{AB}(\omega). \tag{13.6.2}$$

6.2. Inverse Fourier transform

The inverse Fourier transform of $\xi_{BA}(\omega)$ is, by definition:

$$\tilde{\xi}_{BA}(t) = \frac{1}{2\pi} \int_{-\infty}^{\infty} \xi_{BA}(\omega) e^{-i\omega t} \, d\omega. \tag{13.6.3}$$

Importing the expression (13.6.1) for $\xi_{BA}(\omega)$ into formula (13.6.3), we demonstrate the formula:

$$\tilde{\xi}_{BA}(t) = \frac{1}{2\hbar} \langle [B(t), A] \rangle. \tag{13.6.4}$$

Comparing formula (13.6.4) with the Kubo formula (13.3.7) for the response function, we verify that the relation:

$$\boxed{\tilde{\chi}_{BA}(t) = 2i\Theta(t)\tilde{\xi}_{BA}(t)} \tag{13.6.5}$$

holds between the response function and the inverse Fourier transform of the spectral function.

6.3. Spectral representation of $\chi_{BA}(z)$

Formulas (13.5.6) and (13.6.1) show that we can write a spectral representation of $\chi_{BA}(z)$ in terms of $\xi_{BA}(\omega)$:

$$\boxed{\chi_{BA}(z) = \frac{1}{\pi} \int_{-\infty}^{\infty} \frac{\xi_{BA}(\omega)}{\omega - z} \, d\omega.} \tag{13.6.6}$$

In contrast to the integral definition (13.5.5), which is only valid for $\Im m\, z > 0$, the spectral representation (13.6.6) allows us to define $\chi_{BA}(z)$ at any point of affix z outside the real axis.

We have the property (valid for any z outside the real axis):

$$\chi^*_{BA}(z) = \chi_{AB}(z^*). \tag{13.6.7}$$

On the real axis, according to formula (13.5.2) and to formula (13.6.6) (written for $z = \omega + i\epsilon$), we have:

$$\chi_{BA}(\omega) = \frac{1}{\pi} \lim_{\epsilon \to 0^+} \int \frac{\xi_{BA}(\omega')}{\omega' - \omega - i\epsilon}\, d\omega'. \tag{13.6.8}$$

The function $\chi_{BA}(z)$ defined by the spectral representation (13.6.6) reaches different values on opposite sides of the cut:

$$\lim_{\epsilon \to 0^+} \chi_{BA}(\omega + i\epsilon) \neq \lim_{\epsilon \to 0^+} \chi_{BA}(\omega - i\epsilon). \tag{13.6.9}$$

We have:

$$\lim_{\epsilon \to 0^+} \chi_{BA}(\omega + i\epsilon) - \lim_{\epsilon \to 0^+} \chi_{BA}(\omega - i\epsilon) = 2i\, \xi_{BA}(\omega). \tag{13.6.10}$$

The spectral function $\xi_{BA}(\omega)$ thus represents, up to the factor $(2i)^{-1}$, the difference between the values taken by $\chi_{BA}(z)$ at the upper edge and at the lower edge of the cut at the point of abscissa $\omega = \Re e\, z$. On the left-hand side of equation (13.6.10), the first term is simply $\chi_{BA}(\omega)$, whereas the second term, which involves $\chi_{BA}(\omega - i\epsilon) = \chi^*_{AB}(\omega + i\epsilon)$, is equal to $\lim_{\epsilon \to 0^+} \chi_{BA}(\omega - i\epsilon) = \chi^*_{AB}(\omega)$. Formula (13.6.10) thus reads:

$$\xi_{BA}(\omega) = \frac{1}{2i}\big[\chi_{BA}(\omega) - \chi^*_{AB}(\omega)\big]. \tag{13.6.11}$$

In general, the susceptibilities $\chi_{BA}(\omega)$ and $\chi_{AB}(\omega)$ are unequal, and the spectral function $\xi_{BA}(\omega)$ does not identify with the imaginary part of $\chi_{BA}(\omega)$. However, in the particular case $B = A$, we have:

$$\boxed{\xi_{AA}(\omega) = \Im m\, \chi_{AA}(\omega) = \chi''_{AA}(\omega).} \tag{13.6.12}$$

In the case of a finite system, $\xi_{BA}(\omega)$ is a countable sum of Dirac distributions. Thus, it is not a function, but a distribution. We provide below its expression in the two simple examples already considered.

- *Harmonic oscillator*

We have:

$$\xi_{xx}(\omega) = \chi''_{xx}(\omega) = \frac{\pi}{2m\omega_0}\big[\delta(\omega - \omega_0) - \delta(\omega + \omega_0)\big]. \tag{13.6.13}$$

- *Atom perturbed by an electric field*

The spectral function $\xi_{xx}(\omega)$ for an atom, initially in its fundamental state $|\phi_0\rangle$ and perturbed by an electric field, reads:

$$\xi_{xx}(\omega) = \chi''_{xx}(\omega) = \frac{\pi}{\hbar} \sum_n |\langle \phi_0 | x | \phi_n \rangle|^2 \,\big[\delta(\omega - \omega_{n0}) - \delta(\omega + \omega_{n0})\big]. \tag{13.6.14}$$

7. Relaxation

In the course of a relaxation measurement, the system is first submitted to a field $a(t)$ for a sufficiently long time. This field is then suddenly removed. Our aim is to study the ensuing relaxation of a physical quantity B, that is, the way that the out-of-equilibrium average $\langle B(t)\rangle_a$ tends towards the equilibrium average[12] $\langle B\rangle$. The deviation $\langle B(t)\rangle_a - \langle B\rangle$ will be denoted for short by $\delta\langle B(t)\rangle_a$.

7.1. Preparation of an out-of-equilibrium state

At the initial time (which will be taken equal to $-\infty$), the system is in equilibrium with a thermostat. We then isolate it from the thermostat and we submit it to a field $a(t)$ of the form:

$$a(t) = ae^{\eta t}\Theta(-t), \qquad \eta > 0. \tag{13.7.1}$$

The field reaches its final value a over a characteristic time of order η^{-1} (Fig. 13.1). The system thus progressively departs from its initial equilibrium state. In the limit $\eta \to 0^+$, the perturbation is established adiabatically.

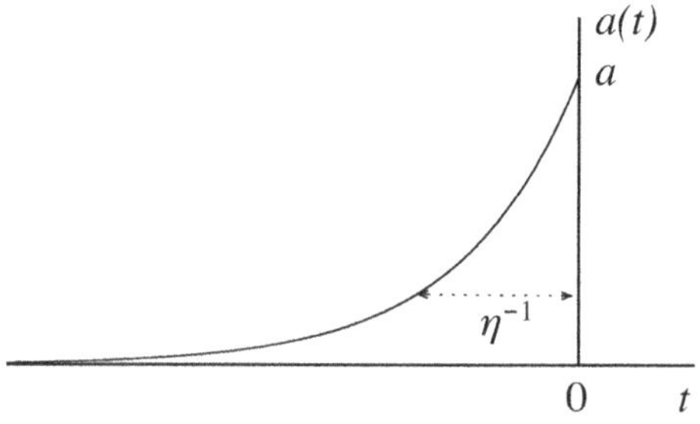

Fig. 13.1 The field $a(t)$.

At time $t = 0$, the field is suddenly suppressed. The system then evolves freely under the effect of its unperturbed Hamiltonian H_0 alone. The behavior of $\delta\langle B(t)\rangle_a$ for $t > 0$ describes the relaxation of B from the out-of-equilibrium state attained at $t = 0$. The applied field is assumed to be weak enough for the relaxation to be properly described by a linear theory.

7.2. Determination of $\delta\langle B(t)\rangle_a$

In the linear regime, we have:

$$\delta\big\langle B(t)\big\rangle_a = a\int_{-\infty}^{\infty} e^{\eta t'}\Theta(-t')\tilde{\chi}_{BA}(t-t')\,dt', \tag{13.7.2}$$

[12] Here, the physical quantity B is not assumed to be centered.

that is:

$$\delta\langle B(t)\rangle_a = a e^{\eta t} \int_t^\infty \tilde{\chi}_{BA}(t') e^{-\eta t'}\, dt'. \tag{13.7.3}$$

In the limit $\eta \to 0^+$, formula (13.7.3) reduces to:

$$\delta\langle B(t)\rangle_a = a \int_t^\infty \tilde{\chi}_{BA}(t')\, dt', \tag{13.7.4}$$

where the integral $\int_t^\infty \tilde{\chi}_{BA}(t')dt'$ is defined as $\lim_{\epsilon\to 0^+} \int_t^\infty \tilde{\chi}_{BA}(t')e^{-\epsilon t'}dt'$. The general formula (13.7.4) is valid for $t \leq 0$ as well as for $t \geq 0$. We will now analyze these two cases successively.

- *Case* $t \leq 0$

For $t \leq 0$, the effective lower bound of the integral on the right-hand side of equation (13.7.4) is 0. This gives:

$$\delta\langle B(t \leq 0)\rangle_a = a \lim_{\epsilon\to 0^+} \int_0^\infty \tilde{\chi}_{BA}(t) e^{-\epsilon t}\, dt. \tag{13.7.5}$$

Thus, the deviation from equilibrium $\langle \delta B(t)\rangle_a$ for $t \leq 0$ does not depend on time. It is expressed as the product of the amplitude a of the applied field by the static susceptibility $\chi_{BA}(\omega = 0) = \lim_{\epsilon\to 0^+} \int_0^\infty \tilde{\chi}_{BA}(t)e^{-\epsilon t}\, dt$:

$$\delta\langle B(t \leq 0)\rangle_a = a\chi_{BA}(\omega = 0). \tag{13.7.6}$$

- *Case* $t \geq 0$

For $t \geq 0$, we start from formula (13.7.4), rewritten in the form:

$$\delta\langle B(t \geq 0)\rangle_a = a\chi_{BA}(\omega = 0) - a \int_0^t \tilde{\chi}_{BA}(t')\, dt'. \tag{13.7.7}$$

We compute the second term on the right-hand side of equation (13.7.7) using the Kubo formula (13.4.6), rewritten as:

$$\tilde{\chi}_{BA}(t) = -\Theta(t) \int_0^\beta \langle A(-i\hbar\lambda)\dot{B}(t)\rangle\, d\lambda. \tag{13.7.8}$$

The integration over time once carried out, we deduce from formula (13.7.7) the expression for $\delta\langle B(t)\rangle_a$ for $t \geq 0$:

$$\delta\langle B(t \geq 0)\rangle_0 = a\chi_{BA}(\omega = 0) + a \int_0^\beta \langle A(-i\hbar\lambda)B(t)\rangle\, d\lambda - a \int_0^\beta \langle A(-i\hbar\lambda)B\rangle\, d\lambda. \tag{13.7.9}$$

7.3. The Kubo formula for the static susceptibility

The static susceptibility plays a central role in the computation of $\delta\langle B(t)\rangle_a$ for $t \leq 0$ (formula (13.7.6)) and for $t \geq 0$ as well (formula (13.7.9)). On account of the expression (13.7.8) for the response function, the static susceptibility reads:

$$\chi_{BA}(\omega = 0) = -\lim_{\epsilon \to 0^+} \int_0^\infty dt\, e^{-\epsilon t} \int_0^\beta \langle A(-i\hbar\lambda)\dot{B}(t)\rangle\, d\lambda. \tag{13.7.10}$$

To deduce from formula (13.7.10) a Kubo formula for $\chi_{BA}(\omega = 0)$, we carry out an integration by parts at finite $\epsilon > 0$ of the integral over time on the right-hand side. This gives:

$$\chi_{BA}(\omega = 0) = \int_0^\beta \langle A(-i\hbar\lambda)B\rangle\, d\lambda - \lim_{\epsilon \to 0^+} \epsilon \int_0^\infty dt\, e^{-\epsilon t} \int_0^\beta \langle A(-i\hbar\lambda)B(t)\rangle\, d\lambda. \tag{13.7.11}$$

The matrix elements of $B(t)$ oscillate at the Bohr angular frequencies of the unperturbed system. Therefore we have:[13]

$$\lim_{\epsilon \to 0^+} \epsilon \int_0^\infty dt\, e^{-\epsilon t} \int_0^\beta \langle A(-i\hbar\lambda)B(t)\rangle\, d\lambda = \beta\langle A^0 B^0\rangle. \tag{13.7.12}$$

In formula (13.7.12), the operators A^0 and B^0 are defined[14] by their matrix elements on the eigenbase of H_0:

$$\langle\phi_n|A^0|\phi_q\rangle = \begin{cases} \langle\phi_n|A|\phi_q\rangle, & \varepsilon_n = \varepsilon_q \\ 0, & \varepsilon_n \neq \varepsilon_q. \end{cases} \tag{13.7.13}$$

We deduce from these results an expression for the static susceptibility:[15]

$$\chi_{BA}(\omega = 0) = \int_0^\beta \langle A(-i\hbar\lambda)B\rangle\, d\lambda - \beta\langle A^0 B^0\rangle, \tag{13.7.14}$$

which involves the equal time canonical correlation function $\tilde{K}_{B-B^0,A-A^0}(t = 0)$:

$$\boxed{\chi_{BA}(\omega = 0) = \beta\tilde{K}_{B-B^0,A-A^0}(t = 0).} \tag{13.7.15}$$

If A or B commutes with H_0, $\chi_{BA}(\omega = 0)$ vanishes.

---

[13] We use the formula:

$$\lim_{\epsilon \to 0^+} \epsilon \int_0^\infty e^{-\epsilon t + i\nu t}\, dt = \begin{cases} 0, & \nu \neq 0 \\ 1, & \nu = 0. \end{cases}$$

[14] If the energy levels of H_0 are non-degenerate, the operators A^0 and B^0 possess only diagonal elements on the eigenbase of H_0. Then they may be called the *diagonal parts* of A and B with respect to H_0.

[15] A more thorough discussion of the static susceptibility is presented in an appendix at the end of this chapter. In particular, the static susceptibility $\chi_{BA}(\omega = 0)$ is compared with the isothermal susceptibility χ_{BA}^T.

7.4. Relaxation function

For $t \leq 0$, $\langle \delta B(t) \rangle_a$ is proportional to $\chi_{BA}(\omega = 0)$ (formula (13.7.6)). We thus have, using the Kubo formula (13.7.15):

$$\boxed{\delta \langle B(t \leq 0) \rangle_a = a\beta \tilde{K}_{B-B^0, A-A^0}(t = 0).} \tag{13.7.16}$$

For $t \geq 0$, the quantity B relaxes. Introducing the Kubo formula (13.7.15) for $\chi_{BA}(\omega = 0)$ into formula (13.7.9) gives:

$$\delta \big\langle B(t \geq 0) \big\rangle_a = a \int_0^\beta \langle A(-i\hbar\lambda) B(t) \rangle \, d\lambda - a\beta \big\langle A^0 B^0 \big\rangle, \tag{13.7.17}$$

that is:

$$\boxed{\delta \big\langle B(t \geq 0) \big\rangle_a = a\beta \tilde{K}_{B-B^0, A-A^0}(t).} \tag{13.7.18}$$

Thus, the relaxation of B for $t \geq 0$ involves the canonical correlation function $\tilde{K}_{B-B^0, A-A^0}(t)$. Formula (13.7.18) is the *Kubo formula* for the relaxation.

In terms of the relaxation function $\Phi_{BA}(t)$, defined for $t \geq 0$ by:

$$\Phi_{BA}(t) = \beta \tilde{K}_{B-B^0, A-A^0}(t), \qquad t \geq 0, \tag{13.7.19}$$

we have:

$$\delta \big\langle B(t \geq 0) \big\rangle_a = a \Phi_{BA}(t). \tag{13.7.20}$$

According to formula (13.7.4), the response and relaxation functions are related by:

$$\boxed{\Phi_{BA}(t) = \int_t^\infty \tilde{\chi}_{BA}(t') \, dt', \qquad t \geq 0.} \tag{13.7.21}$$

We have, inversely:

$$\boxed{\tilde{\chi}_{BA}(t) = -\frac{d}{dt} \Phi_{BA}(t), \qquad t \geq 0.} \tag{13.7.22}$$

7.5. Usual relaxation laws

Most relaxation processes may be described in terms of an exponential relaxation function:

$$\Phi(t) = \Phi(0) e^{-t/\tau}, \qquad t \geq 0. \tag{13.7.23}$$

This is the *Debye relaxation law*.[16] The relaxation function (13.7.23) is fully characterized by the relaxation time τ and the initial value $\Phi(0)$.

In many systems however,[17] the relaxation dynamics depart from exponential laws (and are generally much slower). The relaxation processes may then be described by means of various other functions, including the *Kohlrausch–Williams–Watts* or *stretched exponential* law:

$$\Phi(t) = \Phi(0)e^{-(t/\tau)^{\beta}}, \qquad 0 < \beta < 1, \qquad t \geq 0, \tag{13.7.24}$$

and a law of the type:

$$\Phi(t) = \Phi(0)\frac{1}{1+(t/\tau)^{\delta}}, \qquad \delta > 0, \qquad t \geq 0, \tag{13.7.25}$$

decreasing for $t \gg \tau$ as a power law.

7.6. Determination of the equilibrium correlation functions from the linear response functions

This problematics is the inverse of the one which was developed till now in the chapter: owing to the Kubo formulas, we can deduce the equilibrium correlation functions from the associated response functions in a weakly out-of-equilibrium situation. Such a procedure is of practical interest when these response functions can be determined independently.

For instance, some relaxation problems in linear regime can be treated by directly computing from the linearized equations of motion the Fourier–Laplace transform $\delta\langle B(z)\rangle_a$ of $\delta\langle B(t)\rangle_a$, defined by:

$$\delta\big\langle B(z)\big\rangle_a = \int_0^\infty \delta\big\langle B(t)\big\rangle_a e^{izt}\,dt, \qquad \Im m\, z > 0. \tag{13.7.26}$$

In linear regime, $\delta\langle B(z)\rangle_a$ is related to $\chi_{BA}(z)$. From formulas (13.7.7) and (13.7.8), we get the relation:[18]

$$\delta\big\langle B(z)\big\rangle_a = \frac{\delta\big\langle B(t=0)\big\rangle_a}{iz}\left[\frac{\chi_{BA}(z)}{\chi_{BA}(z=0)} - 1\right]. \tag{13.7.27}$$

Introducing into formula (13.7.27) the expression for $\delta\langle B(z)\rangle_a$ directly deduced from the equations of motion, we can obtain the expression for $\chi_{BA}(z)$ and deduce from it the equilibrium correlation function involved in the appropriate Kubo formula. This method can be used for the computation of transport coefficients.[19]

---

[16] The Debye model of dielectric relaxation is described in Supplement 13A.

[17] This is, for instance, the case in spin glasses and structural glasses, as well as in other complex systems.

[18] Formula (13.7.27) contains the same information as formula (13.7.7). It may be viewed as more aesthetic than this latter formula. However it gets complicated when there are several input operators, otherwise stated, when the perturbation Hamiltonian is of the form $-\sum_i a_i(t)A_i$.

[19] The *Green–Kubo formulas* allowing us to compute the transport coefficients from the equilibrium correlation functions will be established in Chapters 15 and 16.

8. Symmetries of the response and correlation functions

The symmetries of the problem under consideration allow us to obtain some relations obeyed by the linear response functions and the associated equilibrium correlation functions.

As all the functions involved in linear response theory are interrelated, it is enough to study the symmetry properties of one of them, for instance the function $\tilde{\xi}_{BA}(t)$ (proportional to the equilibrium average $\langle [B(t), A] \rangle$).

8.1. Stationarity

The unperturbed system being at equilibrium, the function $\tilde{\xi}_{BA}(t)$ is stationary, that is, invariant under time translation. We therefore have:

$$\frac{1}{2\hbar}\langle [B(t), A] \rangle = -\frac{1}{2\hbar}\langle [A(-t), B] \rangle, \tag{13.8.1}$$

that is:

$$\boxed{\tilde{\xi}_{BA}(t) = -\tilde{\xi}_{AB}(-t).} \tag{13.8.2}$$

By Fourier transformation, we get:

$$\boxed{\xi_{BA}(\omega) = -\xi_{AB}(-\omega).} \tag{13.8.3}$$

8.2. Complex conjugation

Since:

$$\frac{1}{2\hbar}\langle [B(t), A] \rangle^* = -\frac{1}{2\hbar}\langle [B(t), A] \rangle, \tag{13.8.4}$$

we have:

$$\boxed{\tilde{\xi}^*_{BA}(t) = -\tilde{\xi}_{BA}(t),} \tag{13.8.5}$$

and:

$$\boxed{\xi^*_{BA}(\omega) = -\xi_{BA}(-\omega) = \xi_{AB}(\omega).} \tag{13.8.6}$$

Therefore, if the spectral function $\xi_{AB}(\omega)$ is invariant when we exchange A and B, it is a real and odd function of ω. If $\xi_{AB}(\omega)$ changes sign in this permutation, it is a purely imaginary and even function of ω.

8.3. Properties related to time-reversal

The way the operators A and B as well as the density operator ρ_0 behave under time-reversal determines additional symmetry properties of the response and correlation functions.

Before analyzing these properties, we have to characterize the effect of time-reversal on A and B, as well as on H_0 and ρ_0. In quantum mechanics, time-reversal is described by an anti-unitary operator τ. If $|\tau\phi_n\rangle$ and $|\tau\phi_q\rangle$ are the states deduced from the states $|\phi_n\rangle$ and $|\phi_q\rangle$ by time-reversal, we have the equality $\langle\tau\phi_n|\tau\phi_q\rangle = \langle\phi_q|\phi_n\rangle$. Time-reversal applied to an operator A leads to a new operator $\tau A\tau^\dagger$. Time-reversal applied to a product of operators implies a reversal of the order of the operators.

• *Signature of an operator*

Most often, an operator A transforms under time-reversal into:

$$\tau A(t)\tau^\dagger = \epsilon_A A(-t), \tag{13.8.7}$$

where $\epsilon_A = \pm 1$. The operator A is then said to have a well-defined signature ϵ_A under time-reversal. For instance, time-reversal does not change the position operator of a particle, but it reverses its velocity operator:

$$\tau x(t)\tau^\dagger = x(-t), \qquad \tau v(t)\tau^\dagger = -v(-t). \tag{13.8.8}$$

The signature of the position is thus $\epsilon_x = +1$, while that of the velocity is $\epsilon_v = -1$.

• *Effect of a magnetic field*

In the absence of a magnetic field, H_0 and ρ_0 are time-reversal invariant. However, H_0 and ρ_0 may depend on an external magnetic field $\boldsymbol{H}$ breaking this invariance. Such a field is coupled to physical quantities of signature -1. We have:

$$\tau H_0(\boldsymbol{H})\tau^\dagger = H_0(-\boldsymbol{H}), \qquad \tau\rho_0(\boldsymbol{H})\tau^\dagger = \rho_0(-\boldsymbol{H}). \tag{13.8.9}$$

The operators deduced of H_0 and of ρ_0 by time-reversal correspond to the Hamiltonian and to the density operator of the system placed in the field $-\boldsymbol{H}$.

8.4. Reciprocity relations

We first assume that there is no magnetic field and that H_0 and ρ_0 are time-reversal invariant:

$$\tau H_0\tau^\dagger = H_0, \qquad \tau\rho_0\tau^\dagger = \rho_0. \tag{13.8.10}$$

We consider operators A and B possessing well-defined signatures ϵ_A and ϵ_B. Applying the rules about the behavior of the different operators under time-reversal, we can derive the following two identities:

$$\begin{cases} \langle B(t)A\rangle = \epsilon_A\epsilon_B\langle A(t)B\rangle \\ \langle AB(t)\rangle = \epsilon_A\epsilon_B\langle BA(t)\rangle. \end{cases} \tag{13.8.11}$$

Hence, the reciprocity relations verified by the function $\tilde{\xi}_{BA}(t)$,

$$\boxed{\tilde{\xi}_{BA}(t) = \epsilon_A\epsilon_B\tilde{\xi}_{AB}(t),} \tag{13.8.12}$$

and, by Fourier transformation, those verified by the spectral function:

$$\boxed{\xi_{BA}(\omega) = \epsilon_A\epsilon_B\xi_{AB}(\omega),} \tag{13.8.13}$$

From formula (13.8.6), it follows that, if A and B have identical signatures, the spectral function $\xi_{AB}(\omega) = \xi_{BA}(\omega)$ is a real and odd function of ω, whereas, if A and B have opposite signatures, the spectral function $\xi_{AB}(\omega) = -\xi_{BA}(\omega)$ is a purely imaginary and even function of ω.

As for the response function and the generalized susceptibility, the reciprocity relations read:

$$\tilde{\chi}_{BA}(t) = \epsilon_A\epsilon_B\tilde{\chi}_{AB}(t), \tag{13.8.14}$$

and:

$$\chi_{BA}(\omega) = \epsilon_A\epsilon_B\chi_{AB}(\omega). \tag{13.8.15}$$

In the presence of a magnetic field $\boldsymbol{H}$, we have:

$$\tilde{\chi}_{BA}(t, \boldsymbol{H}) = \epsilon_A\epsilon_B\tilde{\chi}_{AB}(t, -\boldsymbol{H}), \tag{13.8.16}$$

and:

$$\chi_{BA}(\omega, \boldsymbol{H}) = \epsilon_A\epsilon_B\chi_{AB}(\omega, -\boldsymbol{H}). \tag{13.8.17}$$

Formula (13.8.17) is verified by the electrical conductivity tensor in the presence of an external magnetic field, as well as by the magnetic susceptibility tensor.[20]

The Kubo formulas of linear response theory thus provide a microscopic way to compute the kinetic coefficients. These coefficients are shown to obey the Onsager or Onsager–Casimir reciprocity relations, which are thus demonstrated.

9. Non-uniform phenomena

The Kubo formula for the response function generalizes to the case in which the system is perturbed by a non-homogeneous external field $a(\boldsymbol{r}, t)$. Then, the perturbation is described by a Hamiltonian of the form:

$$H_1(t) = -\int d\boldsymbol{r}\, a(\boldsymbol{r}, t)A(\boldsymbol{r}). \tag{13.9.1}$$

In the linear range, the out-of-equilibrium average of a centered physical quantity $B(\boldsymbol{r})$ reads:

$$\langle B(\boldsymbol{r}, t)\rangle_a = \int d\boldsymbol{r}' \int_{-\infty}^{\infty} \tilde{\chi}_{BA}(\boldsymbol{r}, t\,; \boldsymbol{r}', t')a(\boldsymbol{r}', t')\, dt'. \tag{13.9.2}$$

The response is in this case both non-local and retarded.

[20] See Supplement 13B.

The Kubo formula generalizing formula (13.3.7) for the homogeneous case reads:

$$\boxed{\tilde{\chi}_{BA}(\boldsymbol{r},t\,;\boldsymbol{r}',t') = \frac{i}{\hbar}\Theta(t-t')\big\langle[B(\boldsymbol{r},t),A(\boldsymbol{r}',t')]\big\rangle.} \tag{13.9.3}$$

Since the unperturbed system is at equilibrium, the response function $\tilde{\chi}_{BA}(\boldsymbol{r},t\,;\boldsymbol{r}',t')$ only depends on $t-t'$. In addition, when the unperturbed system is space-translational invariant, $\tilde{\chi}_{BA}(\boldsymbol{r},t\,;\boldsymbol{r}',t')$ does not depend separately on $\boldsymbol{r}$ and $\boldsymbol{r}'$, but on the difference $\boldsymbol{r}-\boldsymbol{r}'$ alone.

Thus, in linear response theory, the response of the system to an external perturbation, describing an out-of-equilibrium situation, is expressed in terms of a two-time average in the equilibrium state, that is, of an equilibrium correlation function. The out-of-equilibrium average $\langle B(\boldsymbol{r},t)\rangle_a$ of a centered physical quantity $B(\boldsymbol{r})$, linear for weak excitations, is computed with the aid of a response function $\tilde{\chi}_{BA}(\boldsymbol{r},t\,;\boldsymbol{r}',t')$ causal and time-translational invariant (as well as, if the case arises, space-translational invariant).

Appendices

13A. Classical linear response

The classical formulas of linear response theory may be derived from the analogous quantum formulas via the correspondence:

$$\{\ ,\ \} \longleftrightarrow \frac{i}{\hbar}[\ ,\ ] \tag{13A.1}$$

between commutators and Poisson brackets.

However, without recourse to this correspondence, we will present here a direct derivation of the classical formulas for the first-order correction to the distribution function and the linear response function. We will limit ourselves to the homogeneous case.

13A.1. Expression for $\delta\rho(t)$ in classical mechanics

We have:

$$\mathcal{L}_1\rho_0 = -i\{H_1, \rho_0\}. \tag{13A.2}$$

Formula (13.2.11) thus reads:

$$\delta\rho(t) = -\int_{t_0}^{t} e^{-i\mathcal{L}_0(t-t')}\{H_1, \rho_0\}\, dt', \tag{13A.3}$$

that is, taking into account the expression (13.2.2) for the perturbation Hamiltonian:

$$\delta\rho(t) = \int_{t_0}^{t} a(t')e^{-i\mathcal{L}_0(t-t')}\{A, \rho_0\}\, dt'. \tag{13A.4}$$

The Liouville operator $\mathcal{L}_0$ acts only on A. This gives:

$$\delta\rho(t) = \int_{t_0}^{t} a(t')\{A(t'-t), \rho_0\}\, dt', \tag{13A.5}$$

where $A(t) = e^{i\mathcal{L}_0 t}A$. The limit $t_0 \to -\infty$ once taken, gives:

$$\delta\rho(t) = \int_{-\infty}^{t} a(t')\{A(t'-t), \rho_0\}\, dt'. \tag{13A.6}$$

13A.2. The Kubo formula for the classical linear response function

The physical quantity B is assumed to be centered. Its average in the presence of the external field is:

$$\langle B(t)\rangle_a = \frac{1}{N!h^{3N}} \int \delta\rho(t) B \, dqdp. \tag{13A.7}$$

Importing into formula (13A.7) the expression (13A.6) for $\delta\rho(t)$, we obtain:

$$\langle B(t)\rangle_a = \frac{1}{N!h^{3N}} \int_{-\infty}^{t} a(t')\, dt' \int \{A(t'-t), \rho_0\} B \, dqdp. \tag{13A.8}$$

Using the property:[21]

$$\int \{A, \rho_0\} B \, dqdp = -\int \{A, B\} \rho_0 \, dqdp, \tag{13A.9}$$

we get:

$$\langle B(t)\rangle_a = \frac{1}{N!h^{3N}} \int_{-\infty}^{t} a(t')\, dt' \int \{B, A(t'-t)\} \rho_0 \, dqdp, \tag{13A.10}$$

that is, after a translation of the temporal arguments of the equilibrium correlation function $\langle\{B, A(t'-t)\}\rangle = (N!h^{3N})^{-1} \int \{B, A(t'-t)\}\rho_0 \, dqdp$:

$$\langle B(t)\rangle_a = \int_{-\infty}^{t} a(t') \langle\{B(t-t'), A\}\rangle \, dt'. \tag{13A.11}$$

Formula (13A.11) shows that the classical linear response function is expressed in terms of an equilibrium average of a Poisson bracket of dynamical variables (Kubo formula):

$$\boxed{\tilde{\chi}_{BA}(t) = \Theta(t)\langle\{B(t), A\}\rangle.} \tag{13A.12}$$

A1.3. Expression for $\tilde{\chi}_{BA}(t)$ in terms of the correlation function $\langle B(t)\dot{A}\rangle$

When ρ_0 corresponds to canonical equilibrium, the response function given by the general formula (13A.12) may be recast in the following simpler form:

$$\boxed{\tilde{\chi}_{BA}(t) = \beta\Theta(t)\langle B(t)\dot{A}\rangle.} \tag{13A.13}$$

To demonstrate formula (13A.13), we rewrite the average of the Poisson bracket $\{B(t), A\}$ involved in formula (13A1.12) as:

$$\langle\{B(t), A\}\rangle = \frac{1}{N!h^{3N}} \int \{B(t), A\}\rho_0 \, dqdp = \frac{1}{N!h^{3N}} \int \{A, \rho_0\}\, B(t) \, dqdp. \tag{13A.14}$$

[21] The identity (13A.9) can be demonstrated using integration by parts and the fact that the distribution function ρ_0 vanishes for $q_i = \pm\infty$ or $p_i = \pm\infty$.

By definition of the Poisson bracket, we have:

$$\{A, \rho_0\} = \sum_i \left(\frac{\partial A}{\partial p_i} \frac{\partial \rho_0}{\partial q_i} - \frac{\partial A}{\partial q_i} \frac{\partial \rho_0}{\partial p_i} \right), \tag{13A.15}$$

where the index i denotes the different degrees of freedom of the system of Hamiltonian H_0. Since ρ_0 is the canonical distribution function, we have:

$$\frac{\partial \rho_0}{\partial q_i} = -\beta \rho_0 \frac{\partial H_0}{\partial q_i}, \qquad \frac{\partial \rho_0}{\partial p_i} = -\beta \rho_0 \frac{\partial H_0}{\partial p_i}, \tag{13A.16}$$

that is:

$$\{A, \rho_0\} = -\beta \rho_0 \{A, H_0\}. \tag{13A.17}$$

On account of the evolution equation $\dot{A} = \{H_0, A\}$, formula (13A.17) reads:

$$\{A, \rho_0\} = \beta \rho_0 \dot{A}. \tag{13A.18}$$

We finally obtain the equality:

$$\langle \{B(t), A\} \rangle = \beta \langle B(t) \dot{A} \rangle, \tag{13A.19}$$

which demonstrates formula[22] (13A.13).

13B. Static susceptibility of an isolated system and isothermal susceptibility

13B.1. Static susceptibility of an isolated system

Let us come back to the derivation of a Kubo formula for $\chi_{BA}(\omega = 0)$. The static susceptibility is defined as the Fourier transform of $\tilde{\chi}_{BA}(t)$ at vanishing angular frequency:

$$\chi_{BA}(\omega = 0) = \lim_{\epsilon \to 0^+} \int_0^\infty \tilde{\chi}_{BA}(t) e^{-\epsilon t}\, dt. \tag{13B.1}$$

To obtain $\chi_{BA}(\omega = 0)$, we use the Kubo formula which allows us to express $\tilde{\chi}_{BA}(t)$ in terms of the canonical correlation function[23] $\tilde{K}_{B\dot{A}}(t)$:

$$\tilde{\chi}_{BA}(t) = -\Theta(t) \int_0^\beta \langle A(-i\hbar\lambda) \dot{B}(t) \rangle \, d\lambda. \tag{13B.2}$$

Importing the expression (13B.2) for $\tilde{\chi}_{BA}(t)$ into formula (13B.1), the integration over time having been carried out and the limit $\epsilon \to 0^+$ taken, gives:

$$\chi_{BA}(\omega = 0) = \int_0^\beta \langle A(-i\hbar\lambda) B \rangle \, d\lambda - \lim_{t \to \infty} \int_0^\beta \langle A(-i\hbar\lambda) B(t) \rangle \, d\lambda. \tag{13B.3}$$

---

[22] This formula can also be deduced very directly from its quantum analog (formula (13.4.6)): indeed, in the classical case, the different operators commute and the Kubo canonical correlation function $\tilde{K}_{B\dot{A}}(t)$ identifies with $\langle B(t) \dot{A} \rangle$.

[23] We assume that the system is initially at equilibrium with a thermostat at temperature T.

Formula (13B.3) will be especially useful when comparing $\chi_{BA}(\omega = 0)$ with the isothermal susceptibility.[24]

Now, to recover the result previously obtained in the main text, we can show[25] that the matrix elements between eigenstates of same energy of the operators A and B are the only ones which are left in the second term on the right-hand side of equation (13B.3) in the limit $t \to \infty$. We therefore have:

$$\chi_{BA}(\omega = 0) = \int_0^\beta \langle A(-i\hbar\lambda)B\rangle\, d\lambda - \beta\langle A^0 B^0\rangle, \tag{13B.4}$$

that is, in terms of the equal time canonical correlation function of the operators $A - A^0$ and $B - B^0$:

$$\boxed{\chi_{BA}(\omega = 0) = \beta\tilde{K}_{B-B^0,A-A^0}(t = 0).} \tag{13B.5}$$

The static susceptibility is not necessarily equal to a thermodynamic susceptibility. In the following, we will define and determine the isothermal susceptibility χ_{BA}^T, and examine in which physical conditions it can safely be identified with $\chi_{BA}(\omega = 0)$.

13B.2. Isothermal susceptibility

The isothermal susceptibility corresponds to a physical situation differing from the preceding one: the system submitted to the applied field is not isolated from the outside, but remains permanently in contact with a thermostat at temperature T.

To compute χ_{BA}^T, we assume that the system, while being in contact with the thermostat, is submitted to an external time-independent field a. Its Hamiltonian thus reads:

$$H = H_0 - aA. \tag{13B.6}$$

In the linear regime, the isothermal susceptibility corresponding to the response of the physical quantity B to the perturbation induced by the field a is defined as follows:

$$\chi_{BA}^T = \lim_{a\to 0} \frac{\langle B\rangle_a - \langle B\rangle}{a}. \tag{13B.7}$$

[24] See Subsection 13B.3.

[25] We use Abel's theorem:

$$\lim_{t\to\infty} \int_0^\beta \langle A(-i\hbar\lambda)B(t)\rangle\, d\lambda = \lim_{\epsilon\to 0^+} \epsilon \int_0^\infty e^{-\epsilon t} \int_0^\beta \langle A(-i\hbar\lambda)B(t)\rangle\, d\lambda\, dt,$$

and the identity:

$$\lim_{\epsilon\to 0^+} \epsilon \int_0^\infty e^{-\epsilon t} \int_0^\beta \langle A(-i\hbar\lambda)B(t)\rangle\, d\lambda\, dt = \beta\langle A^0 B^0\rangle.$$

In formula (13B.7), $\langle B\rangle_a$ denotes the equilibrium average of B in the presence of the applied field:

$$\langle B\rangle_a = \frac{\mathrm{Tr}\big[Be^{-\beta(H_0-aA)}\big]}{\mathrm{Tr}\big[e^{-\beta(H_0-aA)}\big]}, \tag{13B.8}$$

whereas $\langle B\rangle$ is the equilibrium average of B in the absence of a field:

$$\langle B\rangle = \frac{\mathrm{Tr}\big(Be^{-\beta H_0}\big)}{\mathrm{Tr}\big(e^{-\beta H_0}\big)}. \tag{13B.9}$$

To compute $\langle B\rangle_a$ at first order, we write the first-order expansion for the exponential $e^{-\beta(H_0-aA)}$:

$$e^{-\beta(H_0-aA)} = e^{-\beta H_0}\Big[1 + a\int_0^\beta A(-i\hbar\lambda)\,d\lambda + O(a^2)\Big], \tag{13B.10}$$

and we deduce from it the first-order expansion for the density operator at equilibrium in the presence of the field:

$$\frac{e^{-\beta(H_0-aA)}}{\mathrm{Tr}\big[e^{-\beta(H_0-aA)}\big]} \simeq \frac{e^{-\beta H_0}}{\mathrm{Tr}\big[e^{-\beta H_0}\big]}\Big(1 + a\int_0^\beta \big[A(-i\hbar\lambda) - \langle A\rangle\big]\,d\lambda\Big). \tag{13B.11}$$

In formula (13B.11), $\langle A\rangle$ denotes the equilibrium average of A in the absence of a field. We then deduce from formula (13B.7) the expression for the isothermal susceptibility:

$$\chi_{BA}^T = \int_0^\beta \langle A(-i\hbar\lambda)B\rangle\,d\lambda - \beta\langle A\rangle\langle B\rangle. \tag{13B.12}$$

Formula (13B.12) involves the equal time canonical correlation function of the operators $A - \langle A\rangle$ and $B - \langle B\rangle$:

$$\boxed{\chi_{BA}^T = \beta\tilde{K}_{B-\langle B\rangle,A-\langle A\rangle}(t=0).} \tag{13B.13}$$

In the particular case $B = A$, the isothermal susceptibility is related to the equal time canonical correlation function of the fluctuations of A:

$$\chi_{AA}^T = \beta\tilde{K}_{A-\langle A\rangle,A-\langle A\rangle}(t=0). \tag{13B.15}$$

In the classical case, formula (13B.14) allows us to relate χ_{AA}^T to the variance of the equilibrium fluctuations of A in the absence of a field:

$$\chi_{AA}^T = \beta\langle(A-\langle A\rangle)^2\rangle. \tag{13B.16}$$

13B.3. Discussion

The comparison between the static susceptibility of an isolated system computed via the Kubo formula, on the one hand, and the isothermal susceptibility, on the other hand, displays the fact that these two susceptibilities are in principle different.

As shown by formulas (13B.4) and (13B.12), $\chi_{BA}(\omega = 0)$ and χ^T_{BA} are equal only if we have the identity:

$$\langle A^0 B^0 \rangle = \langle A \rangle \langle B \rangle. \tag{13B.16}$$

Condition (13B.16) also reads:

$$\lim_{t\to\infty} \int_0^\beta \langle A(-i\hbar\lambda)B(t)\rangle \, d\lambda = \beta \langle A \rangle \langle B \rangle. \tag{13B.17}$$

This equality expresses a physically reasonable hypothesis, related to the ergodicity properties of the concerned dynamical variables. Although this equality does not hold, a priori, in general, it is verified in systems with a very large number of degrees of freedom.

Bibliography

P.M. CHAIKIN and T.C. LUBENSKY, *Principles of condensed matter physics*, Cambridge University Press, Cambridge, 1995.

S. DATTAGUPTA, *Relaxation phenomena in condensed matter physics*, Academic Press, Orlando, 1987.

D. FORSTER, *Hydrodynamic fluctuations, broken symmetries, and correlation functions*, Westview Press, Boulder, 1995.

R. KUBO, M. TODA, and N. HASHITSUME, *Statistical physics II: nonequilibrium statistical mechanics*, Springer-Verlag, Berlin, second edition, 1991.

S.W. LOVESEY, *Condensed matter physics: dynamic correlations*, The Benjamin/Cummings Publishing Company, Reading, second edition, 1986.

P.C. MARTIN, *Measurements and correlation functions*, Les Houches Lecture Notes 1967 (C. DE WITT and R. BALIAN editors), Gordon and Breach, New York, 1968.

M. PLISCHKE and B. BERGERSEN, *Equilibrium statistical physics*, World Scientific, Singapore, third edition, 2006.

D. ZUBAREV, V. MOROZOV, and G. RÖPKE, *Statistical mechanics of nonequilibrium processes,* Vol. 2: *Relaxation and hydrodynamic processes*, Akademie Verlag, Berlin, 1997.

R. ZWANZIG, *Nonequilibrium statistical mechanics*, Oxford University Press, Oxford, 2001.

References

R. KOHLRAUSCH, *Annalen der Physik* **12**, 393 (1847).

P. DEBYE, *Polare Molekeln*, Hirzel, Leipzig, 1929.

R. KUBO, Statistical-mechanical theory of irreversible processes. I. General theory and simple applications to magnetic and conduction problems, *J. Phys. Soc. Japan* **12**, 570 (1957).

R. KUBO, The fluctuation-dissipation theorem, *Rep. Prog. Phys.* **29**, 255 (1966).

G. WILLIAMS and D.C. WATTS, Non-symmetrical dielectric relaxation behavior arising from a simple empirical decay function, *Trans. Faraday Soc.* **66**, 80 (1970).

Supplement 13A

Dielectric relaxation

1. Dielectric permittivity and polarizability

Consider a homogeneous dielectric, which we assume to be either isotropic or of cubic symmetry. When it is submitted to an external electric field $\boldsymbol{E}_{\text{ext}}$, the dielectric sample, of volume V, acquires a dipolar electric moment $\boldsymbol{M}$ and a polarization $\boldsymbol{P} = \boldsymbol{M}/V$.

If the applied field is weak enough, the response of the dielectric material is linear. It may be described either by a macroscopic quantity, the dielectric permittivity $\varepsilon(\omega)$, or by a microscopic one, the polarizability $\alpha(\omega)$. To establish the relation between $\varepsilon(\omega)$ and $\alpha(\omega)$, we have to define precisely the different fields which come into play: at the macroscopic level, the *Maxwell field* $\boldsymbol{E}_{\text{Max}}$ and, at the microscopic level, the *local field* $\boldsymbol{E}_{\text{loc}}$.

1.1. The Maxwell field and the local field

The Maxwell field is the macroscopic field involved in the Maxwell equations. It represents the average of the microscopic field $\boldsymbol{e}$ over a small region surrounding each point:

$$\boldsymbol{E}_{\text{Max}} = \langle \boldsymbol{e} \rangle. \tag{13A.1.1}$$

The theory of dielectric media relies also on the notion of local field. The local field $\boldsymbol{E}_{\text{loc}}$ at a point (for instance at a site of a crystal) is the sum of the macroscopic field $\boldsymbol{E}_{\text{Max}}$ and the field created by the dipoles inside the sample, except for the dipole present at the considered site.[1]

To establish the relation between the Maxwell field and the local field, we consider an ellipsoidal sample with one axis parallel to the applied field (in such an ellipsoid, if the local field is uniform, the polarization is also uniform). We generally write the field created by the dipoles in the form of a sum of three terms denoted by $\boldsymbol{E}_1$, $\boldsymbol{E}_2$, and $\boldsymbol{E}_3$. The field $\boldsymbol{E}_1$ is the *depolarizing field* created by the charges situated on the external surface of the sample. To define the fields $\boldsymbol{E}_2$ and $\boldsymbol{E}_3$, we imagine a fictitious spherical cavity dug in the dielectric and centered at the considered point. The polarization charges on the surface of this cavity are at the origin of the field

[1] The local field $\boldsymbol{E}_{\text{loc}}$ must not be confused with the microscopic field $\boldsymbol{e}$ which is the sum of the external field and the field created by all the dipoles inside the sample.

$\boldsymbol{E}_2$, termed the *Lorentz cavity field*, whereas the dipoles inside this cavity create a field $\boldsymbol{E}_3$. On account of these various contributions, the local field reads:

$$\boldsymbol{E}_{\text{loc}} = \boldsymbol{E}_{\text{ext}} + \boldsymbol{E}_1 + \boldsymbol{E}_2 + \boldsymbol{E}_3. \tag{13A.1.2}$$

The Maxwell field is the sum of the external field and the depolarizing field:

$$\boldsymbol{E}_{\text{Max}} = \boldsymbol{E}_{\text{ext}} + \boldsymbol{E}_1. \tag{13A.1.3}$$

Between the Maxwell field and the local field, we have the relation:

$$\boldsymbol{E}_{\text{loc}} = \boldsymbol{E}_{\text{Max}} + \boldsymbol{E}_2 + \boldsymbol{E}_3. \tag{13A.1.4}$$

- *Depolarizing field*

We generally write the depolarizing field as:

$$E_{1x} = -\mathcal{N}_x P_x, \qquad E_{1y} = -\mathcal{N}_y P_y, \qquad E_{1z} = -\mathcal{N}_z P_z, \tag{13A.1.5}$$

where $\mathcal{N}_x$, $\mathcal{N}_y$, and $\mathcal{N}_z$ are the *depolarizing factors.*[2]

- *Lorentz field*

The Lorentz field $\boldsymbol{E}_2$ due to the polarization charges on the surface of the fictitious spherical cavity is:

$$\boldsymbol{E}_2 = \frac{4\pi}{3}\boldsymbol{P}. \tag{13A.1.6}$$

- *Field of the dipoles inside the cavity*

Among the fields contributing to the local field, the field $\boldsymbol{E}_3$ is the only one which depends on the structure of the material. In a medium that is either isotropic or of cubic symmetry this field vanishes:

$$\boldsymbol{E}_3 = 0. \tag{13A.1.7}$$

On account of formulas (13A.1.6) and (13A.1.7), the relation (13A.1.4) reads:

$$\boxed{\boldsymbol{E}_{\text{loc}} = \boldsymbol{E}_{\text{Max}} + \frac{4\pi}{3}\boldsymbol{P},} \tag{13A.1.8}$$

independently of the form (spherical or not) of the sample. Formula (13A.1.8), called the *Lorentz relation*, is always valid (provided that the considered medium is either isotropic or of cubic symmetry).

In the case of a spherical sample for which $\mathcal{N}_x = \mathcal{N}_y = \mathcal{N}_z = 4\pi/3$, the relation (13A.1.2) simply becomes:

$$\boldsymbol{E}_{\text{loc}} = \boldsymbol{E}_{\text{ext}}. \tag{13A.1.9}$$

---

[2] The values of the depolarizing factors depend on the ratios between the principal axes of the ellipsoid. In the case of a spherical sample, we have $\mathcal{N} = 4\pi/3$ for any axis.

1.2. The Clausius–Mossotti relation

In the linear response regime, the dielectric permittivity ε of the medium is defined by the relation:

$$\boldsymbol{E}_{\text{Max}} + 4\pi \boldsymbol{P} = \varepsilon \boldsymbol{E}_{\text{Max}}. \tag{13A.1.10}$$

In the case of a medium that is either isotropic or of cubic symmetry, we deduce from formulas (13A.1.8) and (13A.1.10) the following relation between the local field and the Maxwell field:

$$\boldsymbol{E}_{\text{loc}} = \frac{\varepsilon + 2}{3} \boldsymbol{E}_{\text{Max}}. \tag{13A.1.11}$$

Adopting now a microscopic point of view, we introduce the polarizability α linked to each dipole $\boldsymbol{p}$ by writing the following proportionality relation:

$$\boldsymbol{p} = \alpha \boldsymbol{E}_{\text{loc}}. \tag{13A.1.12}$$

If N is the number of dipoles per unit volume, then:

$$\boldsymbol{P} = N\alpha \boldsymbol{E}_{\text{loc}}. \tag{13A.1.13}$$

The polarizability characterizes an atomic or ionic property, whereas the dielectric permittivity also depends on the way the atoms or the ions are assembled within the material. Using the Lorentz relation (13A.1.8), as well as formulas (13A.1.10) and (13A.1.13), we derive the *Clausius–Mossotti relation* between ε and α:

$$\boxed{\frac{\varepsilon - 1}{\varepsilon + 2} = \frac{4\pi}{3} N\alpha.} \tag{13A.1.14}$$

1.3. Absorption of an electromagnetic wave

The complex index of refraction $\hat{n}$ of the dielectric medium is defined by:

$$\hat{n}^2 = \varepsilon. \tag{13A.1.15}$$

It is of the form $\hat{n} = n + i\kappa$, where n the usual refractive index and κ the extinction coefficient characterizing the damping of the wave. Denoting respectively by ε' and ε'' the real and imaginary parts of ε, we have:

$$\varepsilon' = n^2 - \kappa^2, \qquad \varepsilon'' = 2n\kappa. \tag{13A.1.16}$$

The electric field of an electromagnetic wave of angular frequency ω propagating in the medium along the Oz axis is of the form:

$$E(z,t) = E_0 e^{-i\omega(t - nz/c)} e^{-\omega\kappa z/c}. \tag{13A.1.17}$$

The absorption of the wave by the medium is characterized by the ratio:

$$\frac{|E(z,t)|^2}{|E(z=0,t)|^2} = e^{-Kz}. \tag{13A.1.18}$$

The absorption coefficient K, which represents the inverse of an absorption length, is related to the imaginary part of the dielectric permittivity:

$$K = \frac{\omega \varepsilon''}{nc}. \tag{13A.1.19}$$

2. Microscopic polarization mechanisms

The calculation of the permittivity of a dielectric material requires a microscopic study. In practice, there are three main polarization mechanisms, whose relative importance depends of the angular frequency of the external field.

• *Electronic polarization*

The electronic polarization results from the displacement of the electrons with respect to the nucleus, that is from the deformation of the electronic layer. In the optical range, the electronic polarization is the mechanism providing the most important contribution to the dielectric permittivity. We can compute the electronic polarizability either by using the classical model of the elastically bound electron or, quantum-mechanically, by having recourse to the oscillator strengths.

• *Ionic polarization*

The ionic polarization comes from the relative displacement of ions of opposite signs in the presence of an applied electric field. The corresponding resonances are situated in the far infrared.

• *Orientational or dipolar polarization*

This polarization mechanism takes place at even lower frequencies, in substances composed of permanent electric moments more or less free to change their orientation in the field. This situation is commonly encountered in gases as well as in liquids.

Our aim here is to study the orientational polarization of a dielectric liquid composed of polar molecules, that is, of molecules possessing a permanent dipole moment. We will compute the dielectric permittivity of this liquid, first by following the phenomenological theory proposed by P. Debye, then by elaborating a microscopic model treated in the framework of linear response theory.

3. The Debye theory of dielectric relaxation

In some polar liquids, for instance water, alcohol ..., the static dielectric permittivities take important values. For instance, for water at room temperature, the static dielectric permittivity ε_s equals 81, whereas the dielectric permittivity at optical frequencies[3] ε_∞ equals 1.77. The difference between ε_s and ε_∞ is mainly due to the orientational polarization of the dipolar moments, effective at low frequencies but negligible at frequencies higher than $\sim 10^{10}$ s^{-1}.

3.1. The Debye model for $\alpha(\omega)$

In 1929, P. Debye explained the important values of ε_s in some liquids by assuming that the molecules carry permanent dipolar moments likely to depart from their equilibrium orientation, their return towards equilibrium being characterized by a relaxation time[4] τ.

[3] The fact that $\varepsilon_\infty \neq 1$ is due to the other polarization mechanisms which come into play at frequencies higher than those we are interested in.

[4] The relaxation times may depend strongly on the temperature. For instance, in water at room temperature we find $\tau^{-1} \sim 3 \times 10^{10}$ s^{-1}, whereas at $-20°$C we find $\tau^{-1} \sim 10^3$ s^{-1}.

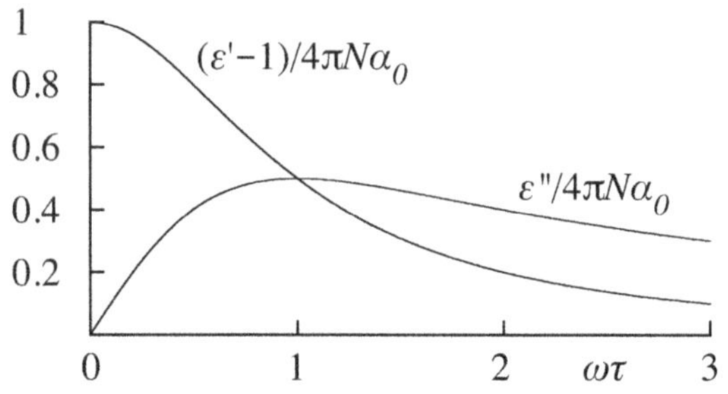

Fig. 13A.1 Debye model: $[\varepsilon'(\omega) - 1]/4\pi N\alpha_0$ and $\varepsilon''(\omega)/4\pi N\alpha_0$ as functions of $\omega\tau$.

If the angular frequency ω of the applied field is much larger than τ^{-1}, the molecule becomes unable to 'follow' the field. Debye proposed accordingly to write the polarizability in the form:

$$\alpha(\omega) = \frac{\alpha_0}{1 - i\omega\tau}, \tag{13A.3.1}$$

where α_0 denotes the static orientational polarizability. In a low density liquid, we have $\varepsilon(\omega) \simeq 1$. We can therefore assimilate the local field and the Maxwell field, and simplify the Clausius–Mossotti formula (13A.1.14) into:

$$\varepsilon(\omega) - 1 = 4\pi N\alpha(\omega). \tag{13A.3.2}$$

This gives,[5] on account of the modelization (13A.3.1):

$$\boxed{\varepsilon(\omega) = 1 + \frac{4\pi N\alpha_0}{1 - i\omega\tau}.} \tag{13A.3.3}$$

The real and imaginary parts of $\varepsilon(\omega)$ are given by the formulas:

$$\begin{cases} \varepsilon'(\omega) - 1 = (\varepsilon_s - 1)\dfrac{1}{1 + \omega^2\tau^2} \\ \varepsilon''(\omega) = (\varepsilon_s - 1)\dfrac{\omega\tau}{1 + \omega^2\tau^2}, \end{cases} \tag{13A.3.4}$$

with $\varepsilon_s - 1 = 4\pi N\alpha_0$. The corresponding curves are represented in Fig. 13A.1.

[5] We have here $\varepsilon_\infty = 1$, since the other polarization mechanisms are not taken into account.

3.2. The Cole–Cole diagram

Another representation widely used in dielectric relaxation problems is the *Cole–Cole diagram*, in which $\varepsilon'(\omega)$ is put as abscissa and $\varepsilon''(\omega)$ as ordinate. Eliminating ω between $\varepsilon'(\omega)$ and $\varepsilon''(\omega)$ (formulas (13A.3.4)) gives:

$$\left(\varepsilon' - \frac{\varepsilon_s + 1}{2}\right)^2 + \varepsilon''^2 = \frac{(\varepsilon_s - 1)^2}{4}. \tag{13A.3.5}$$

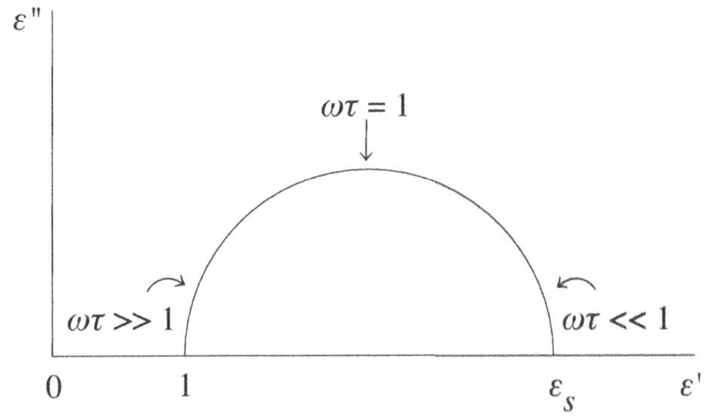

Fig. 13A.2 Debye model: the Cole–Cole diagram.

Thus, the Cole–Cole diagram associated with the Debye model is a semicircle ($\varepsilon'' > 0$). The relaxation time may be directly read on this diagram, since $\varepsilon''(\omega)$ takes its maximum value at $\omega = \tau^{-1}$ (Fig. 13A.2).

3.3. Absorption of an electromagnetic wave

Since the density of the liquid is low, we have $n \simeq 1$, and the absorption coefficient given by formula (13A.1.19) simply reads:

$$K \simeq \frac{\omega\varepsilon''}{c}. \tag{13A.3.6}$$

From the Debye model, we get:

$$K = \frac{\varepsilon_s - 1}{c} \frac{\omega^2\tau}{1 + \omega^2\tau^2}. \tag{13A.3.7}$$

The absorption coefficient as given by formula (13A.3.7) first increases with ω, then it saturates for $\omega\tau \gg 1$: this is the *Debye plateau*.

3.4. Discussion

The Debye theory is actually valid only for $\omega\tau \ll 1$, as displayed by the fact that the part $\omega\tau \gg 1$ of the experimental Cole–Cole diagram exhibits a kind of protuberance, that is, a zone in which $\varepsilon' < 1$, absent from the semicircular Cole–Cole diagram corresponding to the Debye model.

Similarly, for $\omega\tau \gg 1$, the measured absorption coefficient does not exhibit a plateau, but instead decreases. The Debye model of orientational polarization thus proves to be insufficient to explain the shape of the dielectric relaxation curves actually observed in polar liquids. We shall now see what improvements can be brought by a microscopic description based on linear response theory.

4. A microscopic model of orientational polarization

We consider here a liquid made of 'rigid' molecules.[6] Their orientational motion can properly be described in the framework of classical mechanics (in an energy interval of order kT, there are, at room temperature, many rotational levels).

4.1. The Kubo formula for $\varepsilon(\omega)$

We assume that the external field $\boldsymbol{E}_{\text{ext}}(t)$, parallel to the Ox axis, produces only a small perturbation. The Hamiltonian of the perturbed system reads:[7]

$$H = H_0 - ME_{\text{ext}}(t). \tag{13A.4.1}$$

The liquid being isotropic on average, the induced average dipolar moment is parallel to Ox. In the linear response regime, it is of the form:

$$\langle M(t)\rangle_a = \int_{-\infty}^{\infty} \tilde{\chi}_{MM}(t-t')E_{\text{ext}}(t')\,dt'. \tag{13A.4.2}$$

By Fourier transformation, we get (using standard notations):

$$\langle M(\omega)\rangle_a = \chi_{MM}(\omega)E_{\text{ext}}(\omega). \tag{13A.4.3}$$

The response function $\tilde{\chi}_{MM}(t)$ is given by the Kubo formula,

$$\tilde{\chi}_{MM}(t) = \beta\Theta(t)\langle M(t)\dot{M}\rangle, \tag{13A.4.4}$$

which may conveniently be rewritten in the form:

$$\tilde{\chi}_{MM}(t) = -\beta\Theta(t)\langle \dot{M}(t)M\rangle. \tag{13A.4.5}$$

We assume that the relation between the local field and the external field is simply given by formula (13A.1.9). We can then identify $N\alpha(\omega)$ and $V^{-1}\chi_{MM}(\omega)$. On

[6] We exclude molecules with free or nearly free internal degrees of freedom.

[7] For the sake of simplicity, we simply denote by M the component M_x of the induced dipolar moment.

account of the Clausius–Mossotti relation (13A.1.14), $\chi_{MM}(\omega)$ and $\varepsilon(\omega)$ are related by the formula:

$$\frac{\varepsilon(\omega)-1}{\varepsilon(\omega)+2}=\frac{4\pi}{3}\frac{1}{V}\chi_{MM}(\omega), \tag{13A.4.6}$$

that is:

$$\frac{\varepsilon(\omega)-1}{\varepsilon(\omega)+2}=\frac{4\pi}{3}\frac{1}{V}\int_0^\infty \tilde{\chi}_{MM}(t)e^{i\omega t}\,dt. \tag{13A.4.7}$$

Importing the expression (13A.4.5) for $\tilde{\chi}_{MM}(t)$ into equation (13A.4.7), we get the formula:

$$\frac{\varepsilon(\omega)-1}{\varepsilon(\omega)+2}=-\frac{4\pi}{3}\frac{1}{V}\beta\int_0^\infty \frac{d\langle M(t)M\rangle}{dt}e^{i\omega t}\,dt, \tag{13A.4.8}$$

which, due to the average isotropy of the liquid, may be rewritten as:

$$\frac{\varepsilon(\omega)-1}{\varepsilon(\omega)+2}=-\frac{4\pi}{9kTV}\int_0^\infty \frac{d\langle \boldsymbol{M}(t).\boldsymbol{M}\rangle}{dt}e^{i\omega t}\,dt. \tag{13A.4.9}$$

Formula (13A.4.9) is valid whatever the origin of the dipolar moment of the sample (ionic, orientational ...). We will now apply it to the determination of the orientational polarization in a polar fluid of low density.

4.2. Dielectric permittivity of a low density fluid made up of polar molecules with low polarizability

In a low density fluid, we have $\varepsilon(\omega)\simeq 1$, and formula (13A.4.9) simplifies to:

$$\varepsilon(\omega)-1=-\frac{4\pi}{3kTV}\int_0^\infty \frac{d\langle \boldsymbol{M}(t).\boldsymbol{M}\rangle}{dt}e^{i\omega t}\,dt. \tag{13A.4.10}$$

When the molecules are of low polarizability, the electric moment of the system originates essentially from the orientation of the molecular dipoles. We therefore write:

$$\varepsilon(\omega)-1=-\frac{4\pi}{3kTV}\int_0^\infty \frac{d\langle \boldsymbol{M}_0(t).\boldsymbol{M}_0\rangle}{dt}e^{i\omega t}\,dt, \tag{13A.4.11}$$

where $\boldsymbol{M}_0$ is the orientational contribution to the dipolar moment.[8]

---

[8] As a general rule, the dipolar moment $\boldsymbol{M}$ depends on the coordinates of the centers of mass of the molecules (symbolically denoted below by $\boldsymbol{r}$), on the set of Euler angles in the laboratory frame (denoted by Ω), and on the internal coordinates fixing the position of the nuclei in the molecules (denoted by n_i). We can write, if the internal displacements remain of small amplitude,

$$\boldsymbol{M}=\boldsymbol{M}(\boldsymbol{r},\Omega,n_i=0)+\sum_i\frac{\partial \boldsymbol{M}}{\partial n_i}n_i+\cdots,$$

that is:

$$\boldsymbol{M}=\boldsymbol{M}_0+\boldsymbol{M}_1,$$

where $\boldsymbol{M}_0$ is the contribution to the dipolar moment due to the orientation of the molecules (considered as rigid rods) and $\boldsymbol{M}_1$ the contribution of the vibrations. The vibrations being assumed to be of weak amplitude, the most significant contribution to $\boldsymbol{M}$ is $\boldsymbol{M}_0$. It is the only contribution which we will take into account here, which amounts to setting $\varepsilon_\infty=1$.

4.3. Autocorrelation function $\langle \boldsymbol{M}_0(t).\boldsymbol{M}_0 \rangle$

The dipolar moment $\boldsymbol{M}_0$ is the sum of the individual dipole moments of the different polar molecules:

$$\boldsymbol{M}_0 = \sum_i \boldsymbol{\mu}_i. \tag{13A.4.12}$$

The fluid being of low density, the orientational correlations between different molecules may be neglected. Accordingly, we write:

$$\langle \boldsymbol{M}_0(t).\boldsymbol{M}_0 \rangle \simeq NV \langle \boldsymbol{\mu}_i(t).\boldsymbol{\mu}_i \rangle \tag{13A.4.13}$$

(NV is the total number of molecules). This gives:

$$\varepsilon(\omega) - 1 = -\frac{4\pi N}{3kT} \int_0^\infty \frac{d\langle \boldsymbol{\mu}_i(t).\boldsymbol{\mu}_i \rangle}{dt} e^{i\omega t}\, dt. \tag{13A.4.14}$$

At vanishing angular frequency, we are left with:

$$\varepsilon_s - 1 = \frac{4\pi N}{3kT} \langle \boldsymbol{\mu}_i^2 \rangle. \tag{13A.4.15}$$

From equations (13A.4.14) and (13A.4.15), we get:

$$\frac{\varepsilon(\omega) - 1}{\varepsilon_s - 1} = -\int_0^\infty \frac{d}{dt} \left[\frac{\langle \boldsymbol{\mu}_i(t).\boldsymbol{\mu}_i \rangle}{\langle \boldsymbol{\mu}_i^2 \rangle} \right] e^{i\omega t}\, dt. \tag{13A.4.16}$$

If $\boldsymbol{u}_i$ denotes the unit vector parallel to $\boldsymbol{\mu}_i$, we can rewrite equation (13A.4.16) in the form:

$$\boxed{\frac{\varepsilon(\omega) - 1}{\varepsilon_s - 1} = -\int_0^\infty \frac{dS(t)}{dt} e^{i\omega t}\, dt,} \tag{13A.4.17}$$

where the function $S(t) = \langle \boldsymbol{u}_i(t).\boldsymbol{u}_i \rangle$ is proportional to the autocorrelation function of an individual dipole. At initial time, we have $S(0) = u_i^2 = 1$. Due to the collisions undergone by the molecule, $S(t)$ decreases as time increases (and it eventually vanishes).

4.4. Autocorrelation function of an individual dipole: interpretation of the experimental Cole–Cole diagram

The explicit calculation of $S(t)$ requires a model for the microscopic dynamics of a rigid dipole. If we assume that $S(t)$ decreases exponentially, that is, if we write:

$$S(t) = e^{-t/\tau}, \qquad t > 0, \tag{13A.4.18}$$

the time τ being a measure of the duration of the correlation of $\boldsymbol{u}_i(t)$ with $\boldsymbol{u}_i(0)$, we recover the Debye expression for the dielectric permittivity:

$$\frac{\varepsilon(\omega) - 1}{\varepsilon_s - 1} = \frac{1}{1 - i\omega\tau}. \tag{13A.4.19}$$

However, as shown by the experiments, the result (13A.4.19) is not correct at the angular frequencies which involve times much shorter than τ. In other words, the approximate form (13A.4.18) of $S(t)$ is not valid at times $t \ll \tau$. The autocorrelation function $S(t)$ (evenly prolongated for $t < 0$) must indeed be analytic at $t = 0$, and its first derivative must vanish at the origin (whereas its second derivative at the origin must be negative for physical reasons). To obtain the behavior of $\varepsilon(\omega)$ at angular frequencies $\omega \gg \tau^{-1}$, we can integrate by parts the integral on the right-hand side of equation (13A.4.17):

$$\frac{\varepsilon(\omega) - 1}{\varepsilon_s - 1} = -\frac{1}{i\omega} e^{i\omega t} \dot{S}(t) \Big|_0^\infty + \frac{1}{i\omega} \int_0^\infty \ddot{S}(t) e^{i\omega t}\, dt. \tag{13A.4.20}$$

Repeating the integration by parts, we generate an expansion of $\varepsilon(\omega)$ in inverse powers of ω:

$$\frac{\varepsilon(\omega) - 1}{\varepsilon_s - 1} = \frac{\dot{S}(0)}{i\omega} - \frac{\ddot{S}(0)}{(i\omega)^2} + \cdots \tag{13A.4.21}$$

As $\dot{S}(0) = 0$, the first non-vanishing term of the expansion (13A.4.21) is the term in $\ddot{S}(0)$. Therefore, for $\omega\tau \gg 1$, we can write the approximate formula:

$$\frac{\varepsilon(\omega) - 1}{\varepsilon_s - 1} \simeq \frac{\ddot{S}(0)}{\omega^2}, \tag{13A.4.22}$$

in which $\ddot{S}(0) < 0$. Equating the real and imaginary parts of both sides of the approximate equation (13A.4.22) gives:

$$\begin{cases} \varepsilon'(\omega) - 1 \simeq (\varepsilon_s - 1) \dfrac{\ddot{S}(0)}{\omega^2} \\ \varepsilon''(\omega) \simeq 0. \end{cases} \tag{13A.4.23}$$

Formulas (13A.4.23) allow us to understand the origin[9] of the protuberance $\varepsilon' < 1$ observed on the experimental Cole–Cole diagram. They show in addition that the absorption coefficient $K = \omega\varepsilon''/nc$ vanishes as $\omega \to \infty$, which explains the lowering of the curve $K(\omega)$ observed for $\omega\tau \gg 1$.

[9] The precise microscopic mechanism accounting for the analyticity of $S(t)$ at $t = 0$ is not described here. By analogy with the description of the velocity autocorrelation function of a Brownian particle, we can think of this mechanism as involving some microscopic time much shorter than the relaxation time.

Bibliography

N.W. Ashcroft and N.D. Mermin, *Solid state physics*, W.B. Saunders Company, Philadelphia, 1976.

W. Jones and N.H. March, *Theoretical solid-state physics: non-equilibrium and disorder*, Vol. 2, Wiley, New York, 1973. Reprinted, Dover Publications, New York, 1985.

C. Kittel, *Introduction to solid state physics*, Wiley, New York, eighth edition, 2005.

R. Kubo, M. Toda, and N. Hashitsume, *Statistical physics II: nonequilibrium statistical mechanics*, Springer-Verlag, Berlin, second edition, 1991.

D.A. McQuarrie, *Statistical mechanics*, University Science Books, Sausalito, second edition, 2000.

R. Zwanzig, *Nonequilibrium statistical mechanics*, Oxford University Press, Oxford, 2001.

References

P. Debye, *Polare Molekeln*, Hirzel, Leipzig, 1929.

K.S. Cole and R.H. Cole, Dispersion and absorption in dielectrics I. Alternating current characteristics, *J. Chem. Phys.* **9**, 341 (1941).

S.H. Glarum, Dielectric relaxation of polar liquids, *J. Chem. Phys.* **33**, 1371 (1960).

R.H. Cole, Correlation function theory of dielectric relaxation, *J. Chem. Phys.* **42**, 637 (1965).

Supplement 13B

Magnetic resonance

1. Formulation of the problem

We propose here, as an example of application of the quantum theory of linear response, a study of the principle of magnetic resonance.[1] Consider a system of spins placed in an external magnetic field $\boldsymbol{H}(t)$. Each spin $\boldsymbol{S}$ carries a magnetic moment $\boldsymbol{M} = \gamma\boldsymbol{S}$, where γ is the gyromagnetic ratio.[2] In magnetic resonance experiments, the applied magnetic field is the sum of a static field $\boldsymbol{H}_0$ and a small oscillating field $\boldsymbol{H}_1(t)$ perpendicular to $\boldsymbol{H}_0$:

$$\boldsymbol{H}(t) = \boldsymbol{H}_0 + \boldsymbol{H}_1(t). \tag{13B.1.1}$$

In the absence of any other interactions, the average magnetic moment $\langle \boldsymbol{M}(t)\rangle_a$ induced by the applied field evolves according to the equation:

$$\frac{d\langle \boldsymbol{M}(t)\rangle_a}{dt} = \gamma \langle \boldsymbol{M}(t)\rangle_a \times \boldsymbol{H}(t). \tag{13B.1.2}$$

In the presence of interactions inducing a relaxation of the average magnetic moment, we write, instead of equation (13B.1.2), an evolution equation of the form:

$$\frac{d\langle \boldsymbol{M}(t)\rangle_a}{dt} = \gamma \langle \boldsymbol{M}(t)\rangle_a \times \boldsymbol{H}(t) + \left.\frac{d\langle \boldsymbol{M}(t)\rangle_a}{dt}\right|_{\text{relax}}, \tag{13B.1.3}$$

in which the relaxation term $d\langle \boldsymbol{M}(t)\rangle_a/dt|_{\text{relax}}$ accounts for the different phenomena producing the relaxation of $\langle \boldsymbol{M}(t)\rangle_a$ towards its equilibrium value.

We will first present the phenomenological theory of magnetic resonance relying on the *Bloch equations*. Then we will solve the same problem microscopically via the quantum theory of linear response.[3]

---

[1] There are several types of magnetic resonance phenomena, such as nuclear magnetic resonance (NMR), electronic paramagnetic resonance (EPR) ..., thus denominated depending on whether the spins are carried by nuclei, electrons ... Here, we only provide a general description of the principle of the magnetic resonance phenomenon.

[2] The gyromagnetic ratio is negative for electronic spins, but most often positive for nuclear spins.

[3] This calculation will be restricted to the simple case in which damping is neglected.

2. Phenomenological theory

2.1. The Bloch equations

At equilibrium in the presence of the static field $\boldsymbol{H}_0$ assumed to be applied along the Oz axis, the only non-vanishing component of the average magnetic moment is $\langle M_z\rangle_a$, as expressed by the formulas:

$$\langle M_x\rangle_a = \langle M_y\rangle_a = 0, \qquad \langle M_z\rangle_a = M_0. \tag{13B.2.1}$$

We introduce for convenience the static magnetic susceptibility, defined by the relation:

$$M_0 = \chi_0 H_0. \tag{13B.2.2}$$

• *Relaxation of* $\langle M_z(t)\rangle_a$

The relaxation of $\langle M_z(t)\rangle_a$ towards its equilibrium value M_0 is described by the relaxation term:

$$\boxed{\left.\frac{d\big\langle M_z(t)\big\rangle_a}{dt}\right|_{\text{relax}} = -\frac{\big\langle M_z(t)\big\rangle_a - M_0}{T_1}.} \tag{13B.2.3}$$

The characteristic time T_1 in formula (13B.2.3) is called the *longitudinal relaxation time*. The component along Oz of the average magnetic moment is related to the average energy of the spin system. Its evolution is accordingly determined by the inelastic processes that are likely to modify this average energy. Let us take the example of the nuclear magnetic resonance of a magnetic crystalline sample. The spins are carried by atoms situated on the sites of a lattice. The inelastic processes mainly consist, in this case, of the interactions of the spins with phonons at thermal equilibrium.

• *Relaxation of* $\langle M_x(t)\rangle_a$ *and* $\langle M_y(t)\rangle_a$

The transverse components $\langle M_x(t)\rangle_a$ and $\langle M_y(t)\rangle_a$ of the average magnetic moment relax towards their vanishing equilibrium value with a characteristic time T_2 called the *transverse relaxation time*:

$$\boxed{\left.\frac{d\big\langle M_{x,y}(t)\big\rangle_a}{dt}\right|_{\text{relax}} = -\frac{\big\langle M_{x,y}(t)\big\rangle_a}{T_2}.} \tag{13B.2.4}$$

The relaxation of $\langle M_x(t)\rangle_a$ and $\langle M_y(t)\rangle_a$ is mainly due to spin–spin interactions.

Clearly, the microscopic determination of T_1 and T_2 requires a knowledge of the interaction mechanisms of a spin with its environment. Depending on the particular conditions of the system under study, we can have either $T_1 \simeq T_2$ or $T_1 \gg T_2$.

The evolution equations deduced from equation (13B.1.3) by projection on the three axes, with the relaxation terms as given by formulas (13B.2.3) and (13B.2.4), constitute the Bloch equations.

2.2. Linear response to a transverse rotating field: resonance

We first assume that $\boldsymbol{H}_1(t)$ is a rotating field at the angular frequency ω, of components $H_{1x}(t) = H_1 \cos\omega t$, $H_{1y}(t) = H_1 \sin\omega t$. The evolution equations of $\langle M_x(t)\rangle_a$ and $\langle M_y(t)\rangle_a$ read in this case:

$$\left\{\begin{aligned} \frac{d\langle M_x(t)\rangle_a}{dt} &= \gamma\langle M_y(t)\rangle_a H_0 - \gamma\langle M_z(t)\rangle_a H_1 \sin\omega t - \frac{\langle M_x(t)\rangle_a}{T_2} \\ \frac{d\langle M_y(t)\rangle_a}{dt} &= -\gamma\langle M_x(t)\rangle_a H_0 + \gamma\langle M_z(t)\rangle_a H_1 \cos\omega t - \frac{\langle M_y(t)\rangle_a}{T_2}. \end{aligned}\right. \tag{13B.2.5}$$

If the amplitude of $\boldsymbol{H}_1(t)$ is weak enough, we can, in the framework of a first-order expansion of equations (13B.2.5), replace $\langle M_z(t)\rangle_a$ by its equilibrium value M_0. This gives, for $\langle M_x(t)\rangle_a$ and $\langle M_y(t)\rangle_a$, the following linear evolution equations:[4]

$$\left\{\begin{aligned} \frac{d\langle M_x(t)\rangle_a}{dt} &= \gamma\langle M_y(t)\rangle_a H_0 - \gamma M_0 H_1 \sin\omega t - \frac{\langle M_x(t)\rangle_a}{T_2} \\ \frac{d\langle M_y(t)\rangle_a}{dt} &= -\gamma\langle M_x(t)\rangle_a H_0 + \gamma M_0 H_1 \cos\omega t - \frac{\langle M_y(t)\rangle_a}{T_2}. \end{aligned}\right. \tag{13B.2.6}$$

To solve the system of coupled differential equations (13B.2.6), we set:

$$\langle M_\pm(t)\rangle_a = \langle M_x(t)\rangle_a \pm i\langle M_y(t)\rangle_a. \tag{13B.2.7}$$

This gives a set of two uncoupled differential equations:

$$\left\{\begin{aligned} \frac{d\langle M_+(t)\rangle_a}{dt} &= i\omega_0\langle M_+(t)\rangle_a + i\gamma M_0 H_1 e^{i\omega t} - \frac{\langle M_+(t)\rangle_a}{T_2} \\ \frac{d\langle M_-(t)\rangle_a}{dt} &= -i\omega_0\langle M_-(t)\rangle_a - i\gamma M_0 H_1 e^{-i\omega t} - \frac{\langle M_-(t)\rangle_a}{T_2}. \end{aligned}\right. \tag{13B.2.8}$$

In equations (13B.2.8), we have introduced the angular frequency $\omega_0 = -\gamma H_0$ (Larmor angular frequency in the field $\boldsymbol{H}_0$).

In a stationary regime of angular frequency ω, we look for a solution of equations (13B.2.8) of the form:

$$\left\{\begin{aligned} \langle M_+(t)\rangle_a &= M_+(\omega)e^{i\omega t} \\ \langle M_-(t)\rangle_a &= M_-(\omega)e^{-i\omega t}. \end{aligned}\right. \tag{13B.2.9}$$

The amplitudes $M_+(\omega)$ and $M_-(\omega)$ obey the equations:

$$\left\{\begin{aligned} i\omega M_+(\omega) &= i\omega_0 M_+(\omega) + i\gamma M_0 H_1 - \frac{M_+(\omega)}{T_2} \\ -i\omega M_-(\omega) &= -i\omega_0 M_-(\omega) - i\gamma M_0 H_1 - \frac{M_-(\omega)}{T_2}, \end{aligned}\right. \tag{13B.2.10}$$

[4] Note that the longitudinal relaxation time implicitly contained in $\langle M_z(t)\rangle_a$ disappears from the evolution equations of $\langle M_x(t)\rangle_a$ and $\langle M_y(t)\rangle_a$ once these equations are linearized. To measure T_1, we thus have either to observe a non-linear phenomenon or to study a transitory phenomenon such as the relaxation of $\langle M_z(t)\rangle_a$.

from which we get:

$$\begin{cases} M_+(\omega) = \dfrac{\gamma M_0 H_1}{(\omega - \omega_0) - iT_2^{-1}} \\ \\ M_-(\omega) = \dfrac{\gamma M_0 H_1}{(\omega - \omega_0) + iT_2^{-1}}. \end{cases} \tag{13B.2.11}$$

The common modulus $M(\omega)$ of $M_+(\omega)$ and $M_-(\omega)$, given by:

$$M(\omega) = |\gamma| M_0 H_1 \left[(\omega - \omega_0)^2 + T_2^{-2}\right]^{-1/2}, \tag{13B.2.12}$$

exhibits a maximum for $\omega = \omega_0$ (resonance).

2.3. Transverse susceptibility

Let us come back to $\langle M_x(t)\rangle_a$ and $\langle M_y(t)\rangle_a$. We have:

$$\langle M_x(t)\rangle_a = \frac{1}{2}\gamma M_0 H_1 \left[\frac{e^{i\omega t}}{(\omega - \omega_0) - iT_2^{-1}} + \frac{e^{-i\omega t}}{(\omega - \omega_0) + iT_2^{-1}}\right] \tag{13B.2.13}$$

(as well as an analogous formula for $\langle M_y(t)\rangle_a$). The average magnetic moment described by $\langle M_x(t)\rangle_a$ and $\langle M_y(t)\rangle_a$ is a response to the rotating field $\boldsymbol{H}_1(t)$. We can write $\langle M_x(t)\rangle_a$ in the form:

$$\langle M_x(t)\rangle_a = H_1\left[\chi_T'(\omega)\cos\omega t + \chi_T''(\omega)\sin\omega t\right], \tag{13B.2.14}$$

where:

$$\begin{cases} \chi_T'(\omega) = \dfrac{\omega_0 \chi_0 (\omega_0 - \omega)}{(\omega - \omega_0)^2 + T_2^{-2}} \\ \\ \chi_T''(\omega) = \dfrac{\omega_0 \chi_0 T_2^{-1}}{(\omega - \omega_0)^2 + T_2^{-2}} \end{cases} \tag{13B.2.15}$$

are the real and imaginary parts of the *transverse susceptibility* $\chi_T(\omega)$.

2.4. Response to a linearly polarized field

Consider now a field $\boldsymbol{H}_1(t)$ linearly polarized along Ox with $H_{1x}(t) = H_1\cos\omega t$. The linear response to this field is of the form:[5]

$$\langle M_x(t)\rangle_a = H_1\left[\chi_{xx}'(\omega)\cos\omega t + \chi_{xx}''(\omega)\sin\omega t\right] \tag{13B.2.16}$$

(together with an analogous formula for $\langle M_y(t)\rangle_a$). As $\boldsymbol{H}_1(t)$ may be considered as the half-sum of two rotating fields of angular frequencies ω and $-\omega$, we can write:

$$\begin{cases} \chi_{xx}'(\omega) = \dfrac{1}{2}\left[\chi_T'(\omega) + \chi_T'(-\omega)\right] \\ \\ \chi_{xx}''(\omega) = \dfrac{1}{2}\left[\chi_T''(\omega) - \chi_T''(-\omega)\right]. \end{cases} \tag{13B.2.17}$$

[5] The condensed notation $\chi_{xx}(\omega)$ stands for the susceptibility $\chi_{M_x M_x}(\omega)$.

The generalized susceptibility $\chi_{xx}(\omega) = \chi'_{xx}(\omega) + i\chi''_{xx}(\omega)$ thus reads:

$$\chi_{xx}(\omega) = -\frac{1}{2}\omega_0\chi_0\left(\frac{1}{\omega - \omega_0 + iT_2^{-1}} - \frac{1}{\omega + \omega_0 + iT_2^{-1}}\right). \tag{13B.2.18}$$

2.5. Average absorbed power in a linearly polarized field

The instantaneous power absorbed by the system submitted to the field $\boldsymbol{H}_1(t)$ is given by:

$$\frac{dW}{dt} = \boldsymbol{H}_1(t) \cdot \frac{d\langle \boldsymbol{M}(t)\rangle_a}{dt}. \tag{13B.2.19}$$

In the case of the linearly polarized field under consideration, we have, on average over one period,

$$\overline{\frac{dW}{dt}} = \frac{1}{2}H_1^2\omega\chi''_{xx}(\omega), \tag{13B.2.20}$$

with:

$$\chi''_{xx}(\omega) = \frac{1}{2}\omega_0\chi_0 T_2^{-1}\left[\frac{1}{(\omega-\omega_0)^2 + T_2^{-2}} - \frac{1}{(\omega+\omega_0)^2 + T_2^{-2}}\right]. \tag{13B.2.21}$$

For $\omega \simeq \omega_0$, the leading contribution to $\chi''_{xx}(\omega)$ is that of the first term of formula (13B.2.21). The average absorbed power takes approximately the form of a Lorentzian centered at ω_0 and of width $2T_2^{-1}$:

$$\boxed{\overline{\frac{dW}{dt}} \simeq \frac{1}{4}H_1^2\omega_0^2\chi_0 T_2^{-1}\frac{1}{(\omega-\omega_0)^2 + T_2^{-2}}.} \tag{13B.2.22}$$

The width of the resonance line is thus determined by T_2. Inversely, the measurement of the linewidth provides information about the microscopic processes governing T_2.

3. A microscopic model

Consider a spin $\boldsymbol{S}$ placed in a static magnetic field $\boldsymbol{H}_0$. Submit it in addition to a transverse magnetic field $\boldsymbol{H}_1(t)$ of small amplitude. We are looking for the average magnetic moment $\langle \boldsymbol{M}(t)\rangle_a$ at first perturbation order.

The unperturbed Hamiltonian is:[6]

$$h_0 = -\boldsymbol{M}.\boldsymbol{H}_0. \tag{13B.3.1}$$

[6] In order to avoid any confusion with the magnetic field, the unperturbed Hamiltonian is denoted here by a lower case letter (the same is done below for the perturbed Hamiltonian). The expression (13B.3.1) for h_0 corresponds to the spin part of the unperturbed Hamiltonian (this latter may also have an orbital part). The spin–lattice and spin–spin interactions are not taken into account here.

The eigenvalues of h_0 are $\varepsilon_m = -m\gamma\hbar H_0 = m\hbar\omega_0$, the magnetic quantum number m being an integer or a half-integer. The corresponding eigenstates are denoted by $|m\rangle$. The perturbation Hamiltonian is:

$$h_1(t) = -\boldsymbol{M}.\boldsymbol{H}_1(t) = -\big[M_x H_{1x}(t) + M_y H_{1y}(t)\big]. \tag{13B.3.2}$$

Depending on the choices made for the functions $H_{1x}(t)$ and $H_{1y}(t)$, the form (13B.3.2) of $h_1(t)$ may allow us in particular to describe either a rotating field or a linearly polarized field.

According to linear response theory, the average induced magnetic moment $\langle M_x(t)\rangle_a$ is given by:

$$\big\langle M_x(t)\big\rangle_a = \int_{-\infty}^{\infty} \big[\tilde{\chi}_{xx}(t-t')H_{1x}(t') + \tilde{\chi}_{xy}(t-t')H_{1y}(t')\big]\, dt'. \tag{13B.3.3}$$

The response functions $\tilde{\chi}_{M_x M_x}(t)$ and $\tilde{\chi}_{M_x M_y}(t)$, respectively denoted here by $\tilde{\chi}_{xx}(t)$ and $\tilde{\chi}_{xy}(t)$ for the sake of simplicity, may be determined by using the Kubo formulas, which read in this case:

$$\tilde{\chi}_{xx}(t) = \frac{i}{\hbar}\Theta(t)\gamma^2\big\langle[S_x(t), S_x(0)]\big\rangle, \tag{13B.3.4}$$

and:

$$\tilde{\chi}_{xy}(t) = \frac{i}{\hbar}\Theta(t)\gamma^2\big\langle[S_x(t), S_y(0)]\big\rangle. \tag{13B.3.5}$$

We can also deduce $\tilde{\chi}_{xx}(t)$ and $\tilde{\chi}_{xy}(t)$ from the generalized susceptibilities $\chi_{xx}(\omega)$ and $\chi_{xy}(\omega)$ by inverse Fourier transformation. We will adopt here this latter method.[7]

3.1. Generalized susceptibilities and response functions

To begin with, we consider the zero-temperature case. Then only the fundamental state $|-s\rangle$ (spin antiparallel to $\boldsymbol{H}_0$) has a non-vanishing population. This gives:

$$\chi_{xx}(\omega) = \gamma^2\hbar^{-1} \lim_{\epsilon\to 0^+} \sum_{m\neq -s} \left(\frac{|\langle -s|S_x|m\rangle|^2}{\omega_{m,-s} - \omega - i\epsilon} - \frac{|\langle -s|S_x|m\rangle|^2}{\omega_{-s,m} - \omega - i\epsilon} \right) \tag{13B.3.6}$$

and:

$$\chi_{xy}(\omega) = \gamma^2\hbar^{-1} \lim_{\epsilon\to 0^+} \sum_{m\neq -s} \left(\frac{\langle -s|S_x|m\rangle\langle m|S_y|-s\rangle}{\omega_{m,-s} - \omega - i\epsilon} - \frac{\langle m|S_x|-s\rangle\langle -s|S_y|m\rangle}{\omega_{-s,m} - \omega - i\epsilon} \right). \tag{13B.3.7}$$

[7] We compute the susceptibilities by using the general formula:

$$\chi_{BA}(\omega) = \frac{1}{\hbar}\sum_{n,q}(\Pi_n - \Pi_q)B_{nq}A_{qn} \lim_{\epsilon\to 0^+} \frac{1}{\omega_{qn} - \omega - i\epsilon}.$$

To make explicit the expressions (13B.3.6) and (13B.3.7) for $\chi_{xx}(\omega)$ and $\chi_{xy}(\omega)$, we compute the matrix elements $\langle -s|S_x|m\rangle$ and $\langle -s|S_y|m\rangle$. For a spin s (m then varying from $-s$ to $+s$), we have:

$$\langle -s|S_x|m\rangle = \frac{\hbar}{2}(2s)^{1/2}\delta_{m,-s+1}, \qquad \langle -s|S_y|m\rangle = \frac{i\hbar}{2}(2s)^{1/2}\delta_{m,-s+1}, \tag{13B.3.8}$$

and[8] $\omega_{m,-s} = -\omega_{-s,m} = \omega_0$. This finally gives:

$$\chi_{xx}(\omega) = -\frac{\gamma^2 s\hbar}{2}\lim_{\epsilon\to 0^+}\left(\frac{1}{\omega-\omega_0+i\epsilon} - \frac{1}{\omega+\omega_0+i\epsilon}\right) \tag{13B.3.9}$$

and:

$$\chi_{xy}(\omega) = i\frac{\gamma^2 s\hbar}{2}\lim_{\epsilon\to 0^+}\left(\frac{1}{\omega-\omega_0+i\epsilon} + \frac{1}{\omega+\omega_0+i\epsilon}\right). \tag{13B.3.10}$$

These results may be generalized at finite temperature. In this case, we have:

$$\chi_{xx}(\omega) = \frac{\gamma^2}{2}\langle S_z\rangle \lim_{\epsilon\to 0^+}\left(\frac{1}{\omega-\omega_0+i\epsilon} - \frac{1}{\omega+\omega_0+i\epsilon}\right) \tag{13B.3.11}$$

and:

$$\chi_{xy}(\omega) = -\frac{i\gamma^2}{2}\langle S_z\rangle \lim_{\epsilon\to 0^+}\left(\frac{1}{\omega-\omega_0+i\epsilon} + \frac{1}{\omega+\omega_0+i\epsilon}\right). \tag{13B.3.12}$$

In formulas (13B.3.11) and (13B.3.12), $\langle S_z\rangle$ denotes the equilibrium average of the component of the spin along Oz. At $T=0$, we have $\langle S_z\rangle = -s\hbar$. Formulas (13B.3.11) and (13B.3.12) then respectively coincide with formulas (13B.3.9) and (13B.3.10).

The response functions $\tilde{\chi}_{xx}(t)$ and $\tilde{\chi}_{xy}(t)$ are obtained by inverse Fourier transformation. The limit $\epsilon \to 0^+$ having been taken, we obtain the formulas:

$$\tilde{\chi}_{xx}(t) = -\gamma^2\langle S_z\rangle\hbar\Theta(t)\sin\omega_0 t \tag{13B.3.13}$$

and:

$$\tilde{\chi}_{xy}(t) = -\gamma^2\langle S_z\rangle\hbar\Theta(t)\cos\omega_0 t. \tag{13B.3.14}$$

3.2. The Onsager relations

We can, in the same way, compute $\chi_{yx}(\omega)$ and $\tilde{\chi}_{yx}(t)$. It can be verified that the susceptibilities and the response functions obey the Onsager relations in the presence of the magnetic field $\boldsymbol{H}_0$:

$$\begin{cases} \chi_{yx}(\omega, \boldsymbol{H}_0) = \chi_{xy}(\omega, -\boldsymbol{H}_0), \\ \tilde{\chi}_{yx}(t, \boldsymbol{H}_0) = \tilde{\chi}_{xy}(t, -\boldsymbol{H}_0). \end{cases} \tag{13B.3.15}$$

[8] We assume here $\gamma < 0$, that is, $\omega_0 > 0$.

3.3. Response to a transverse rotating field: resonance

Let us examine again the case of a transverse rotating field of components $H_{1x}(t) = H_1 \cos\omega t$, $H_{1y}(t) = H_1 \sin\omega t$. The corresponding average magnetic moment can be computed with the aid of formula (13B.3.3). Re-establishing $\epsilon > 0$ finite in the expressions of the response functions, we have:

$$\big\langle M_x(t)\big\rangle_a = -\gamma^2\langle S_z\rangle H_1 \lim_{\epsilon\to 0^+}\int_{-\infty}^{t}\big[\cos\omega t' \sin\omega_0(t-t') + \sin\omega t' \cos\omega_0(t-t')\big]e^{-\epsilon(t-t')}\,dt', \tag{13B.3.16}$$

that is, after integration:

$$\big\langle M_x(t)\big\rangle_a = -\frac{1}{2}\gamma^2\langle S_z\rangle H_1 \lim_{\epsilon\to 0^+}\left[\frac{e^{i\omega t}}{-(\omega-\omega_0)+i\epsilon} - \frac{e^{-i\omega t}}{\omega-\omega_0+i\epsilon}\right]. \tag{13B.3.17}$$

The limit $\epsilon \to 0^+$ once taken, we get:

$$\big\langle M_x(t)\big\rangle_a = -\gamma^2\langle S_z\rangle H_1\left[-\mathrm{vp}\,\frac{1}{\omega-\omega_0}\cos\omega t + \pi\delta(\omega-\omega_0)\sin\omega t\right]. \tag{13B.3.18}$$

Formula (13B.3.18) displays a resonance at $\omega = \omega_0$. The resonance peak is infinitely narrow, which is unphysical. Actually, a realistic calculation should take into account the spin–lattice and spin–spin interactions in the expression of the unperturbed Hamiltonian. We would then obtain a broadened resonance peak.

3.4. Spin correlation functions

We can also compute the symmetric correlation functions of the transverse spins, denoted by $\tilde{S}_{xx}(t)$ and $\tilde{S}_{xy}(t)$, defined by the formulas:

$$\tilde{S}_{xx}(t) = \frac{1}{2}\big\langle S_x(t)S_x(0) + S_x(0)S_x(t)\big\rangle, \tag{13B.3.19}$$

and:

$$\tilde{S}_{xy}(t) = \frac{1}{2}\big\langle S_x(t)S_y(0) + S_y(0)S_x(t)\big\rangle. \tag{13B.3.20}$$

We will present here the computation of $\tilde{S}_{xx}(t)$ (that of $\tilde{S}_{xy}(t)$ can be carried out along similar lines). We deduce $\tilde{S}_{xx}(t)$ from its Fourier transform $S_{xx}(\omega)$, as given by the formula:[9]

$$S_{xx}(\omega) = -\frac{\pi}{2}\langle S_z\rangle\hbar\coth\frac{\beta\hbar\omega}{2}\big[\delta(\omega_0-\omega) - \delta(\omega_0+\omega)\big], \qquad \beta = (kT)^{-1}. \tag{13B.3.21}$$

[9] We compute $S_{xx}(\omega)$ with the aid of the following general formula, which will be established in Chapter 14:

$$S_{BA}(\omega) = \pi\sum_{n,q}(\Pi_n + \Pi_q)B_{nq}A_{qn}\delta(\omega_{qn}-\omega)$$

(the notations are those of the general linear response theory).

From the expression (13B.3.11) for $\chi_{xx}(\omega) = \chi_{M_x M_x}(\omega)$, we deduce, the limit $\epsilon \to 0^+$ once taken:

$$\chi''_{S_x S_x}(\omega) = \frac{\pi}{2}\langle S_z\rangle\big[\delta(\omega+\omega_0) - \delta(\omega-\omega_0)\big]. \qquad (13B.3.22)$$

This verifies the relation:[10]

$$S_{xx}(\omega) = \hbar \coth\frac{\beta\hbar\omega}{2}\chi''_{S_x S_x}(\omega). \qquad (13B.3.23)$$

We deduce from equation (13B.3.21) the correlation function $\tilde{S}_{xx}(t)$:

$$\tilde{S}_{xx}(t) = -\frac{1}{2}\langle S_z\rangle\hbar\coth\frac{\beta\hbar\omega_0}{2}\cos\omega_0 t. \qquad (13B.3.24)$$

3.5. Comparison with the phenomenological approach

As for the response to a transverse rotating field, the expressions of the average magnetic moment $\langle M_x(t)\rangle_a$ obtained by solving the linearized Bloch equations (formula (13B.2.13)), on the one hand, and by applying the linear response theory (formula (13B.3.17)), on the other hand, are closely similar.

The expressions of $\chi_{xx}(\omega)$ obtained in both approaches, respectively given by formulas (13B.2.18) and (13B.3.11), may even be fully identified.[11] To this end, it is enough to attribute to the parameter ϵ the physical meaning of ${T_2}^{-1}$.

---

[10] The relation (13B.3.23) between $S_{xx}(\omega)$ and $\chi''_{xx}(\omega)$ constitutes the expression for this problem of the fluctuation-dissipation theorem (see Chapter 14).

[11] We have the identity:

$$-\gamma^2\langle S_z\rangle = \omega_0\chi_0,$$

equivalent to the relation $\langle M_z\rangle_a = M_0$.

Bibliography

A. ABRAGAM, *The principles of nuclear magnetism*, Clarendon Press, Oxford, 1961.

R. BALIAN, *From microphysics to macrophysics*, Vol. 1, Springer-Verlag, Berlin, 1991; Vol. 2, Springer-Verlag, Berlin, 1992.

C. COHEN-TANNOUDJI, B. DIU, and F. LALOË, *Quantum mechanics*, Vol. 1, Hermann and Wiley, Paris, second edition, 1977.

C. KITTEL, *Introduction to solid state physics*, Wiley, New York, eighth edition, 2005.

R. KUBO, M. TODA, and N. HASHITSUME, *Statistical physics II: nonequilibrium statistical mechanics*, Springer-Verlag, second edition, Berlin, 1991.

C.P SLICHTER, *Principles of magnetic resonance*, Springer-Verlag, third edition, Berlin, 1996.

Chapter 14

The fluctuation-dissipation theorem

This chapter deals with the fluctuation-dissipation theorem, which constitutes one of the fundamental statements of the statistical physics of linear irreversible processes.

The fluctuation-dissipation theorem expresses a relation between the dissipative part of a generalized susceptibility and the associated equilibrium correlation function. In practice, this relation is used either to describe the intrinsic fluctuations of a dynamical variable using the characteristics of the susceptibility or, inversely, to deduce the susceptibility from the relevant equilibrium fluctuations.

The fluctuation-dissipation theorem is derived from the Kubo formulas of linear response theory. As a general rule, the energy absorbed on average by a dissipative system coupled to an external field $a(t)$ *is larger than the energy restored to the field by the system. The energy supplied by the field is eventually dissipated irreversibly within the system. In the case of a linear coupling described by the perturbation Hamiltonian* $-a(t)A$, *the average dissipated power is related to the imaginary part of the generalized susceptibility* $\chi_{AA}(\omega)$. *This susceptibility is expressed in terms of the equilibrium autocorrelation function of the relevant dynamical variable (namely, the physical quantity* A*). The autocorrelation function of an operator characterizing the fluctuations of the associated physical quantity, the Kubo formula for* $\chi''_{AA}(\omega)$ *establishes a relation between the dissipation and the equilibrium fluctuations of* A. *This relation constitutes the fluctuation-dissipation theorem.*

1. Dissipation

The energy received by a system of infinite size coupled to an external field is eventually transformed into heat. It is thus dissipated irreversibly within the system. In a harmonic linear regime, the average dissipated power is related to the generalized susceptibility. Consequently, one of the means of determining experimentally the susceptibility consists in measuring the average dissipated power.

The relation between the average dissipated power and the susceptibility can be established in several ways, either in the general framework of the linear response theory, or by comparing the expression for the average dissipated power deduced from the Fermi golden rule with the Kubo formula for the susceptibility.

1.1. Computation of the average dissipated power via the linear response theory

Consider a system of unperturbed Hamiltonian H_0, submitted to a spatially uniform applied field $a(t)$. The corresponding perturbation is described by the Hamiltonian:[1]

$$H_1(t) = -a(t)A, \tag{14.1.1}$$

in which the Hermitean operator A represents the physical quantity coupled to the external field. We assume that the latter is harmonic ($a(t) = \Re e(ae^{-i\omega t})$). Our aim is to calculate the average dissipated power $\overline{dW/dt}$. This quantity may be obtained, either by calculating the average power received by the system and eventually dissipated, or from the average evolution rate of the total energy of the system coupled to the field.

- *Average power received by the system*

The instantaneous power received by the system submitted to the field $a(t)$ is:

$$\frac{dW}{dt} = a(t)\frac{d\langle A(t)\rangle_a}{dt}. \tag{14.1.2}$$

In a stationary harmonic regime of angular frequency ω, we have:

$$\langle A(t)\rangle_a = \Re e\big[ae^{-i\omega t}\chi_{AA}(\omega)\big]. \tag{14.1.3}$$

For a real amplitude a, the instantaneous power received by the system is thus:

$$\frac{dW}{dt} = a^2\omega\cos\omega t\big[-\chi'_{AA}(\omega)\sin\omega t + \chi''_{AA}(\omega)\cos\omega t\big]. \tag{14.1.4}$$

On average, the power dissipated within the system is equal to the power it receives. The average dissipated power is thus proportional to $\omega\chi''_{AA}(\omega)$:

$$\boxed{\overline{\frac{dW}{dt}} = \frac{1}{2}a^2\omega\chi''_{AA}(\omega).} \tag{14.1.5}$$

[1] The study may be generalized to the case in which $H_1(t)$ is a sum of the type $-\sum_j a_j(t)A_j$. To compute the dissipation, it is then necessary to take into account the linear response of the different observables A_i, that is, the set of generalized susceptibilities $\chi_{A_iA_j}(\omega)$.

• *Average evolution rate of the energy of the system coupled to the field*

We can also obtain the average dissipated power by considering as the 'system' (in the thermodynamic sense) the system of unperturbed Hamiltonian H_0 placed in the field $a(t)$, and by determining the average evolution rate of the energy $E_{\rm tot}$ of this global ensemble. This latter energy is given by:

$$E_{\rm tot} = \mathrm{Tr}\big[\rho(t)H\big], \tag{14.1.6}$$

where $H = H_0 - a(t)A$ is the Hamiltonian of the system coupled to the field and $\rho(t)$ the density operator of the system. We have:

$$\frac{dE_{\rm tot}}{dt} = \frac{d}{dt}\Big(\mathrm{Tr}\big[\rho(t)H\big]\Big), \tag{14.1.7}$$

that is:

$$\frac{dE_{\rm tot}}{dt} = \mathrm{Tr}\Big[\frac{d\rho(t)}{dt}H\Big] - \frac{da(t)}{dt}\mathrm{Tr}\big[\rho(t)A\big]. \tag{14.1.8}$$

The first term on the right-hand side of equation (14.1.8) vanishes on account of the evolution equation of $\rho(t)$. We are left with:

$$\frac{dE_{\rm tot}}{dt} = -\frac{da(t)}{dt}\langle A(t)\rangle_a. \tag{14.1.9}$$

In the considered harmonic regime, we therefore have, at any time,

$$\frac{dE_{\rm tot}}{dt} = a^2\omega\sin\omega t\big[\chi'_{AA}(\omega)\cos\omega t + \chi''_{AA}(\omega)\sin\omega t\big], \tag{14.1.10}$$

and, on average:

$$\boxed{\overline{\frac{dE_{\rm tot}}{dt}} = \frac{1}{2}a^2\omega\chi''_{AA}(\omega).} \tag{14.1.11}$$

As shown by formulas (14.1.4) and (14.1.10), the instantaneous power received by the system is not equal to the instantaneous evolution rate of the energy of the system coupled to the field. However, as displayed by formulas (14.1.5) and (14.1.11), these quantities are equal on average. This is why the average power dissipated within the system may also be obtained from the average evolution rate of the total energy of the system coupled to the field.

1.2. Computation of the average dissipated power via the Fermi golden rule

The Hamiltonian H_0 has a base of eigenstates $\{|\phi_n\rangle\}$, of energies ε_n. The system is submitted to the perturbation $-aA\cos\omega t$. We associate with this perturbation a transition rate between an initial state $|\phi_n\rangle$ and final states $|\phi_q\rangle$ given, according to the Fermi golden rule, by:

$$\frac{\pi a^2}{2\hbar^2}\sum_q |\langle\phi_q|A|\phi_n\rangle|^2\big[\delta(\omega_{qn}-\omega)+\delta(\omega_{nq}-\omega)\big]. \tag{14.1.12}$$

The transition $|\phi_n\rangle \to |\phi_q\rangle$ corresponds to an absorption process if $\varepsilon_q > \varepsilon_n$, and to an induced emission process if $\varepsilon_q < \varepsilon_n$.

We denote by $dW/dt|_{\rm abs}$ the energy absorbed per unit time by the system in thermal equilibrium at temperature T. We compute it from the first term of expression (14.1.12), summed over all initial states $|\phi_n\rangle$ weighted by their average equilibrium occupation probabilities $\Pi_n \propto e^{-\beta\varepsilon_n}$:

$$\frac{dW}{dt}\bigg|_{\rm abs} = \frac{\pi a^2}{2\hbar^2}\sum_{n,q}\Pi_n \hbar\omega |A_{qn}|^2 \delta(\omega_{qn} - \omega). \qquad (14.1.13)$$

Similarly, we denote by $dW/dt|_{\rm em}$ the energy emitted per unit time by the system. We compute it in an analogous way from the second term of expression (14.1.12). This gives:

$$\frac{dW}{dt}\bigg|_{\rm em} = \frac{\pi a^2}{2\hbar^2}\sum_{n,q}\Pi_n \hbar\omega |A_{qn}|^2 \delta(\omega_{nq} - \omega), \qquad (14.1.14)$$

or, the operator A being Hermitean:

$$\frac{dW}{dt}\bigg|_{\rm em} = \frac{\pi a^2}{2\hbar^2}\sum_{n,q}\Pi_q \hbar\omega |A_{qn}|^2 \delta(\omega_{qn} - \omega). \qquad (14.1.15)$$

The energy actually received per unit time by the system is equal to the difference $dW/dt|_{\rm abs} - dW/dt|_{\rm em}$:

$$\frac{dW}{dt}\bigg|_{\rm abs} - \frac{dW}{dt}\bigg|_{\rm em} = \frac{\pi a^2}{2\hbar}\sum_{n,q}(\Pi_n - \Pi_q)\omega |A_{qn}|^2 \delta(\omega_{qn} - \omega), \qquad (14.1.16)$$

which is a non-negative quantity,[2] as it should. Formula (14.1.16) is a microscopic expression for the total energy received per unit time by the system, that is, for the average dissipated power $\overline{dW/dt}$. Comparing formula (14.1.16) with the definition of the spectral function[3] $\xi_{AA}(\omega)$, we verify the relation:

$$\boxed{\overline{\frac{dW}{dt}} = \frac{1}{2}a^2\omega\xi_{AA}(\omega).} \qquad (14.1.17)$$

The spectral function $\xi_{AA}(\omega)$ being just the imaginary part $\chi''_{AA}(\omega)$ of the generalized susceptibility, the expressions (14.1.5) and (14.1.17) for $\overline{dW/dt}$ are effectively identical.

---

[2] Indeed, for $\omega > 0$, the non-vanishing terms of the sum on the right-hand side of equation (14.1.16) correspond to $\varepsilon_q > \varepsilon_n$, that is, to $\Pi_n > \Pi_q$.

[3] The spectral function $\xi_{AA}(\omega)$ is defined by the formula:

$$\xi_{AA}(\omega) = \frac{\pi}{\hbar}\sum_{n,q}(\Pi_n - \Pi_q)|A_{qn}|^2 \delta(\omega_{qn} - \omega).$$

The close relation between the calculations carried out via the linear response theory, on the one hand, and via the Fermi golden rule, on the other hand, comes out of the fact that both calculations are first perturbation order treatments.

1.3. Determination of the generalized susceptibility

The real and imaginary parts of the generalized susceptibility may be deduced from one another by means of the Kramers–Kronig relations. Therefore, it results from formulas (14.1.5) or (14.1.17) that the measurement of the average dissipation as a function of the angular frequency of the external field suffices in principle to determine the generalized susceptibility and the linear response function.

2. Equilibrium fluctuations

At equilibrium, the correlation between the fluctuations of two physical quantities is characterized by the correlation function of the associated operators. In the classical case, the correlation function $\tilde{C}_{BA}(t)$ of two operators A and B is defined unambiguously (for centered operators, we have $\tilde{C}_{BA}(t) = \langle B(t)A \rangle$). However, in the quantum case, because of the non-commutativity of operators, we generally use one or the other of two non-equivalent definitions for the correlation function of two operators A and B. It is even so for the autocorrelation function of an operator A, since $A(t)$ does not in general commute with $A(t')$.

2.1. Symmetric and canonical correlation functions

We limit ourselves here to the study of the correlation functions of two Hermitean operators A and B whose diagonal parts with respect to H_0 vanish ($A^0 = 0$, $B^0 = 0$). This hypothesis implies in particular that A and B are centered:

$$\langle A \rangle = \mathrm{Tr}(\rho_0 A) = 0, \qquad \langle B \rangle = \mathrm{Tr}(\rho_0 B) = 0. \tag{14.2.1}$$

• *The symmetric correlation function*

The symmetric correlation function $\tilde{S}_{BA}(t)$ of two centered operators A and B is defined by the formula:

$$\boxed{\tilde{S}_{BA}(t) = \frac{1}{2}\langle \{A, B(t)\}_+ \rangle,} \tag{14.2.2}$$

where $\{A, B(t)\}_+ = AB(t) + B(t)A$ denotes the anticommutator of A and $B(t)$. The quantity:

$$\tilde{S}_{AA}(t) = \frac{1}{2}\langle \{A, A(t)\}_+ \rangle \tag{14.2.3}$$

is the symmetric autocorrelation function of the operator A.

• *The canonical correlation function*

For systems in canonical equilibrium at temperature T, we also make use of the canonical correlation function of the operators A and B, defined by:

$$\tilde{K}_{BA}(t) = \frac{1}{\beta}\int_0^\beta \langle e^{\lambda H_0} A e^{-\lambda H_0} B(t)\rangle \, d\lambda, \qquad \beta = (kT)^{-1}. \tag{14.2.4}$$

2.2. Expression for the Fourier transforms $S_{BA}(\omega)$ and $K_{BA}(\omega)$ over an eigenbase of H_0

Let $S_{BA}(\omega)$ and $K_{BA}(\omega)$ be the respective Fourier transforms of $\tilde{S}_{BA}(t)$ and $\tilde{K}_{BA}(t)$:

$$S_{BA}(\omega) = \int_{-\infty}^{\infty} \tilde{S}_{BA}(t) e^{i\omega t}\, dt, \qquad K_{BA}(\omega) = \int_{-\infty}^{\infty} \tilde{K}_{BA}(t) e^{i\omega t}\, dt. \tag{14.2.5}$$

• *Expression for* $S_{BA}(\omega)$

From the formulas:

$$\langle B(t)A\rangle = \sum_{n,q} \Pi_n B_{nq} A_{qn} e^{-i\omega_{qn} t} \tag{14.2.6}$$

and:

$$\langle AB(t)\rangle = \sum_{n,q} \Pi_q A_{qn} B_{nq} e^{-i\omega_{qn} t}, \tag{14.2.7}$$

we deduce the expression for $S_{BA}(\omega)$:

$$S_{BA}(\omega) = \pi \sum_{n,q} (\Pi_n + \Pi_q) B_{nq} A_{qn} \delta(\omega_{qn} - \omega). \tag{14.2.8}$$

• *Expression for* $K_{BA}(\omega)$

To obtain $K_{BA}(\omega)$, we first rewrite formula (14.2.4) for $\tilde{K}_{BA}(t)$ in the form:

$$\tilde{K}_{BA}(t) = \frac{1}{\beta\hbar} \sum_{n,q} (\Pi_n - \Pi_q) \frac{B_{nq} A_{qn}}{\omega_{qn}} e^{i\omega_{nq} t}. \tag{14.2.9}$$

Hence the expression for $K_{BA}(\omega)$ is:

$$K_{BA}(\omega) = \frac{2\pi}{\beta\hbar} \sum_{n,q} (\Pi_n - \Pi_q) \frac{B_{nq} A_{qn}}{\omega_{qn}} \delta(\omega_{qn} - \omega). \tag{14.2.10}$$

2.3. Relation between $S_{BA}(\omega)$ and $K_{BA}(\omega)$

On account of the identity:

$$\Pi_n - \Pi_q = (\Pi_n + \Pi_q) \frac{\Pi_n - \Pi_q}{\Pi_n + \Pi_q} = (\Pi_n + \Pi_q) \frac{1 - e^{-\beta(\varepsilon_q - \varepsilon_n)}}{1 + e^{-\beta(\varepsilon_q - \varepsilon_n)}}, \tag{14.2.11}$$

we deduce from formulas (14.2.8) and (14.2.10) a direct relation (not involving the eigenstates of H_0) between $S_{BA}(\omega)$ and $K_{BA}(\omega)$:

$$S_{BA}(\omega) = \frac{\beta\hbar\omega}{2} \coth \frac{\beta\hbar\omega}{2} K_{BA}(\omega). \tag{14.2.12}$$

2.4. The classical limit

In the classical limit, the different operators commute and we get:

$$\tilde{S}_{BA}(t) = \tilde{K}_{BA}(t) = \tilde{C}_{BA}(t). \tag{14.2.13}$$

Both symmetric and canonical correlation functions are thus identical in this limit.

Accordingly, we have, in the classical limit $\beta\hbar\omega \ll 1$:

$$S_{BA}(\omega) = K_{BA}(\omega) = C_{BA}(\omega). \tag{14.2.14}$$

3. The fluctuation-dissipation theorem

The *fluctuation-dissipation theorem* constitutes the central kernel of the linear response theory. It expresses a general relation between the response or the relaxation in linear regime and the equilibrium fluctuations.[4] It allows us in particular to establish a link between the energy dissipation in the course of linear irreversible processes and the spectral density of the equilibrium fluctuations. The fluctuation-dissipation theorem is used to predict the characteristics of the fluctuations or of the thermal noise given those of the admittance, or, inversely, to deduce the admittance from the thermal fluctuations. The Nyquist theorem constitutes an example of the first procedure, whereas the Onsager's derivation of the reciprocity relations between kinetic coefficients is the oldest example of the second one.

The first formulation of this theorem was given by A. Einstein in 1905 in the context of his study of Brownian motion. This result was extended by H. Nyquist in 1928 to the thermal noise in an electrical circuit at equilibrium, and later generalized by L. Onsager in 1931 in the form of a hypothesis about the regression of fluctuations. The fluctuation-dissipation theorem was demonstrated in the framework of quantum linear response theory by H.B. Callen and T.A. Welton in 1951, and by R. Kubo in 1957.

[4] Violations of this theorem are only observed in out-of-equilibrium systems such as spin glasses or structural glasses. These systems relax very slowly and display aging properties: the time scale of the response to an external perturbation or of a correlation function between two physical quantities increases with the age of the system, that is, with the time elapsed since its preparation. In other words, these functions decrease all the more slowly when the system is older. The aging manifests itself in particular by the separate dependance of response and/or correlation functions of some dynamical variables with respect to their two temporal arguments. In such a case, the fluctuation-dissipation theorem does not hold.

3.1. Relation between $S_{BA}(\omega)$ or $K_{BA}(\omega)$ and the spectral function $\xi_{BA}(\omega)$

Using the expressions (14.2.8) and (14.2.10) for $S_{BA}(\omega)$ and $K_{BA}(\omega)$, together with the definition of the spectral function[5] $\xi_{BA}(\omega)$, we verify the relations:

$$\boxed{S_{BA}(\omega) = \hbar \coth \frac{\beta\hbar\omega}{2} \xi_{BA}(\omega)} \tag{14.3.1}$$

and:

$$\boxed{K_{BA}(\omega) = \frac{2}{\beta\omega} \xi_{BA}(\omega).} \tag{14.3.2}$$

3.2. Formulations of the fluctuation-dissipation theorem

Both formulas (14.3.1) and (14.3.2) lead to formulations of the fluctuation-dissipation theorem for a system in thermodynamic equilibrium at temperature $T = (k\beta)^{-1}$.

For instance, in the case of a perturbation Hamiltonian of the form (14.1.1), the dissipation is related to $\xi_{AA}(\omega) = \chi''_{AA}(\omega)$, whereas $S_{AA}(\omega)$ or $K_{AA}(\omega)$ may be interpreted as the spectral density of the fluctuations[6] of A. We can write the relations:

$$\chi''_{AA}(\omega) = \left(\hbar \coth \frac{\beta\hbar\omega}{2}\right)^{-1} \int_{-\infty}^{\infty} \tilde{S}_{AA}(t) e^{i\omega t}\, dt \tag{14.3.3}$$

and:

$$\chi''_{AA}(\omega) = \frac{\beta\omega}{2} \int_{-\infty}^{\infty} \tilde{K}_{AA}(t) e^{i\omega t}\, dt, \tag{14.3.4}$$

which show that the knowledge of the energy dissipation in the system perturbed by the field $a(t)$ coupled to the physical quantity A is equivalent to that of the dynamics of the equilibrium fluctuations of A.

In the classical limit, $\tilde{S}_{AA}(t)$ and $\tilde{K}_{AA}(t)$ identify with $\tilde{C}_{AA}(t)$ (formula (14.2.14)), and formulas (14.3.3) and (14.3.4) simply read:

$$\chi''_{AA}(\omega) = \frac{\beta\omega}{2} \int_{-\infty}^{\infty} \langle A(t)A \rangle e^{i\omega t}\, dt. \tag{14.3.5}$$

3.3. A few examples

There are a wide variety of applications of the fluctuation-dissipation theorem. Two examples are quoted below.[7]

---

[5] The spectral function $\xi_{BA}(\omega)$ is defined by the formula:

$$\xi_{BA}(\omega) = \frac{\pi}{\hbar} \sum_{n,q} (\Pi_n - \Pi_q)\, B_{nq} A_{qn} \delta(\omega_{qn} - \omega).$$

[6] See Subsection 4.2.

[7] The examples presented here are classical. A quantum example, concerning the displacement of a harmonic oscillator coupled with a phonon bath, will be treated in Supplement 14A.

• *Motion described by the generalized Langevin equation*

Consider, within a fluid in equilibrium, a particle whose motion is described by the generalized Langevin equation: due to its fluid environment, the particle is submitted to a fluctuating force and a retarded friction force. During the relaxation of a velocity fluctuation, energy is dissipated within the fluid.

The admittance $\mathcal{A}(\omega)$ identifies with the Fourier–Laplace transform $\chi_{vx}(\omega)$ of the linear response function $\tilde{\chi}_{vx}(t)$. We have $\mathcal{A}(\omega) = 1/m[\gamma(\omega) - i\omega]$, where $\gamma(\omega)$ denotes the generalized friction coefficient. The dissipative part[8] of $\mathcal{A}(\omega)$ is related to the equilibrium velocity autocorrelation function:

$$\Re e\,\mathcal{A}(\omega) = \frac{1}{2kT}\int_{-\infty}^{\infty}\langle v(t)v\rangle e^{i\omega t}\,dt. \tag{14.3.6}$$

Formula (14.3.6), called the first fluctuation-dissipation theorem in the terminology of R. Kubo, can be derived by applying the linear response theory to the isolated system constituted by the particle coupled with the bath.

The real part of the generalized friction coefficient is the Fourier transform of the Langevin force autocorrelation function:

$$\Re e\,\gamma(\omega) = \frac{1}{2mkT}\int_{-\infty}^{\infty}\langle F(t)F\rangle e^{i\omega t}\,dt. \tag{14.3.7}$$

Formula (14.3.7), called the second fluctuation-dissipation theorem, can be derived from formula (14.3.6) together with the generalized Langevin equation.[9]

• *Light scattering by a fluid*

Another application of the fluctuation-dissipation theorem comes into play in the study of light scattering by the density fluctuations of a fluid. From the measurement of the spectrum of scattered light, we deduce the spectrum of the density fluctuations. These fluctuations are of two types: thermal fluctuations, due to fluctuations of the local entropy, and mechanical fluctuations, due to damped sound waves. For fluctuations of small amplitude, low angular frequency, and large wavelength, the spectrum of the scattered light can be deduced from the response functions as obtained from the linearized equations of hydrodynamics. Light scattering experiments thus provide a way to measure the dissipative coefficients of a fluid.[10]

---

[8] The velocity and position operators having opposite signatures with respect to time-reversal, the dissipative part of $\mathcal{A}(\omega)$ is its real part $\chi'_{vx}(\omega)$. We have the relation $\xi_{vx}(\omega) = -i\chi'_{vx}(\omega)$. An analogous discussion, concerning electrical conductivity, is carried out in Chapter 15.

[9] In the classical case, the spectral density of the random force is thus proportional to the real part of the generalized friction coefficient:

$$S_F(\omega) = 2mkT\,\Re e\,\gamma(\omega).$$

In the quantum case, we have instead:

$$S_F(\omega) = \hbar\omega\coth\frac{\beta\hbar\omega}{2}m\,\Re e\,\gamma(\omega).$$

[10] See Supplement 16B.

4. Positivity of $\omega\chi''_{AA}(\omega)$

We consider, as previously, a system of unperturbed Hamiltonian H_0, submitted to a perturbation $-a(t)A$, where $a(t)$ is a harmonic field of angular frequency ω.

The microscopic expression (14.1.16) for the average dissipated power $\overline{dW/dt}$ shows that this quantity is positive: a stable dissipative system absorbs on average, from the field to which it is coupled, more energy than it restores to it. Hence, using the expression (14.1.5) for $\overline{dW/dt}$, the positivity of $\omega\chi''_{AA}(\omega)$:

$$\boxed{\omega\chi''_{AA}(\omega) \geq 0.} \tag{14.4.1}$$

We deduce from formulas (14.3.3) and (14.3.4) the equalities:

$$\omega\chi''_{AA}(\omega) = \omega\left(\hbar\coth\frac{\beta\hbar\omega}{2}\right)^{-1} S_{AA}(\omega) \tag{14.4.2}$$

and:

$$\omega\chi''_{AA}(\omega) = \frac{\beta\omega^2}{2} K_{AA}(\omega). \tag{14.4.3}$$

Formulas (14.4.2) and (14.4.3) show that the positivity of $\omega\chi''_{AA}(\omega)$ implies the positivity of $S_{AA}(\omega)$ as well as the positivity of $K_{AA}(\omega)$:

$$\boxed{S_{AA}(\omega) \geq 0, \qquad K_{AA}(\omega) \geq 0.} \tag{14.4.4}$$

The inequalities (14.4.4) ensure a posteriori that it is actually possible to interpret $S_{AA}(\omega)$ or $K_{AA}(\omega)$ as the (positive) spectral density of the fluctuations of A. The Fourier relations (14.2.5) (written for $B = A$) thus represent the generalization of the Wiener–Khintchine theorem to a dynamical variable or an observable.

5. Static susceptibility

5.1. Expression for $\chi_{AA}(\omega = 0)$

The imaginary part $\chi''_{AA}(\omega)$ of the generalized susceptibility is an odd function of ω. Taking advantage of this property, we can deduce from the fluctuation-dissipation theorem the expression for $\chi_{AA}(\omega = 0)$ in terms of an equilibrium correlation function.

The function $\chi''_{AA}(\omega)$ being odd, $\chi'_{AA}(\omega = 0)$ identifies with the static susceptibility. The Kramers–Kronig relation for $\chi'_{AA}(\omega = 0)$,

$$\chi'_{AA}(\omega = 0) = \frac{1}{\pi}\int_{-\infty}^{\infty} \frac{\chi''_{AA}(\omega)}{\omega}\, d\omega, \tag{14.5.1}$$

thus yields an integral representation of the static susceptibility:

$$\chi_{AA}(\omega = 0) = \frac{1}{\pi} \int_{-\infty}^{\infty} \frac{\chi''_{AA}(\omega)}{\omega}\, d\omega. \tag{14.5.2}$$

According to the fluctuation-dissipation theorem (14.3.4), we then get:

$$\chi_{AA}(\omega = 0) = \frac{\beta}{2\pi} \int_{-\infty}^{\infty} K_{AA}(\omega)\, d\omega. \tag{14.5.3}$$

We thus retrieve the proportionality relation between the static susceptibility and the equal time canonical correlation function:[11]

$$\boxed{\chi_{AA}(\omega = 0) = \beta \tilde{K}_{AA}(t = 0).} \tag{14.5.4}$$

In the classical case, formula (14.5.4) simply reads:

$$\boxed{\chi_{AA}(\omega = 0) = \beta \langle A^2 \rangle.} \tag{14.5.5}$$

5.2. Thermodynamic sum rule

Since it has been assumed that $A^0 = 0$ (and thus that $\langle A \rangle = 0$), the static susceptibility is identical to the isothermal susceptibility:

$$\chi_{AA}(\omega = 0) = \chi_{AA}^T = \left.\frac{\partial A}{\partial a}\right|_T \cdot \tag{14.5.6}$$

In formula (14.5.6), a denotes a static external field applied to the system.

Formula (14.5.4) (which takes the form (14.5.5) in the classical case) thus allows us to express the thermodynamic derivative $\partial A/\partial a|_T$ in terms of the correlation function $\tilde{K}_{AA}(t = 0)$ (equal to $\langle A^2 \rangle$ in the classical case):

$$\left.\frac{\partial A}{\partial a}\right|_T = \beta \tilde{K}_{AA}(t = 0). \tag{14.5.7}$$

Formula (14.5.7) constitutes the *thermodynamic sum rule*.

---

[11] The expression for $\chi_{BA}(\omega = 0)$ involves in principle the equal time correlation function $\tilde{K}_{B-B^0, A-A^0}(t = 0)$, which reduces to $\tilde{K}_{BA}(t = 0)$ since we have assumed $A^0 = 0$, $B^0 = 0$.

6. Sum rules

Generally speaking, the *sum rules* of the linear response theory are exact integral relations that any generalized susceptibility must verify. They stipulate that each moment of $\chi_{BA}(\omega)$ must be given by an equal time equilibrium correlation function of certain derivatives of the operators $A(t)$ and $B(t)$. The thermodynamic sum rule and the sum rule of the oscillator strengths are well-known examples of sum rules. In practice, the sum rules impose constraints on the phenomenological models likely to be proposed for the generalized susceptibilities (in particular concerning their behavior at large angular frequencies).

We will establish the sum rules, limiting ourselves to the simple case of the response of a physical quantity A to its own conjugate field.

6.1. Derivation

To study the behavior of the susceptibility $\chi_{AA}(\omega)$ at large angular frequencies, we introduce the function $\chi_{AA}(z)$, defined by its spectral representation in terms of $\chi''_{AA}(\omega)$:

$$\chi_{AA}(z) = \frac{1}{\pi}\int_{-\infty}^{\infty} \frac{\chi''_{AA}(\omega)}{\omega - z}\, d\omega, \qquad \Im m\, z \neq 0. \tag{14.6.1}$$

We expand the right-hand side of equation (14.6.1) in powers of $1/z$:

$$\chi_{AA}(z) = -\frac{1}{z}\frac{1}{\pi}\int_{-\infty}^{\infty} \omega \frac{\chi''_{AA}(\omega)}{\omega}\, d\omega - \frac{1}{z^2}\frac{1}{\pi}\int_{-\infty}^{\infty} \omega^2 \frac{\chi''_{AA}(\omega)}{\omega}\, d\omega - \cdots. \tag{14.6.2}$$

The coefficients of the expansion (14.6.2) are proportional to the moments of order $n \geq 1$ of $\chi''_{AA}(\omega)/\omega$.

According to the fluctuation-dissipation theorem (14.3.4), we have the equality:

$$\frac{1}{\pi}\int_{-\infty}^{\infty} \omega^n \frac{\chi''_{AA}(\omega)}{\omega}\, d\omega = \frac{\beta}{2\pi}\int_{-\infty}^{\infty} \omega^n K_{AA}(\omega)\, d\omega, \qquad n \geq 0. \tag{14.6.3}$$

The expansion (14.6.2) thus involve the moments of order $n \geq 1$ of $K_{AA}(\omega)$. According to the properties of the Fourier transformation, each moment of $K_{AA}(\omega)$ (including the zeroth-order moment, which does not appear in the expansion (14.6.2)) is proportional to the derivative of corresponding order,[12] computed at $t = 0$, of the autocorrelation

[12] We have, by definition:

$$\tilde{K}_{AA}(t) = \frac{1}{2\pi}\int_{-\infty}^{\infty} K_{AA}(\omega) e^{-i\omega t}\, d\omega.$$

Hence the expression for the moment of order n of $K_{AA}(\omega)$ is:

$$\frac{1}{2\pi}\int_{-\infty}^{\infty} \omega^n K_{AA}(\omega)\, d\omega = \left(i\frac{d}{dt}\right)^n \tilde{K}_{AA}(t)\Big|_{t=0}, \qquad n \geq 0.$$

function $\tilde{K}_{AA}(t)$. This gives the formula:

$$\boxed{\frac{1}{\pi}\int_{-\infty}^{\infty}\omega^n\frac{\chi''_{AA}(\omega)}{\omega}\,d\omega=\beta\Big(i\frac{d}{dt}\Big)^n\tilde{K}_{AA}(t)\Big|_{t=0},\qquad n\geq 0.}\tag{14.6.4}$$

The expression appearing on the right-hand side of equation (14.6.4) is a finite quantity. The relations obtained in this way for the various values of $n \geq 0$ constitute the sum rules that $\chi''_{AA}(\omega)$ must satisfy.

Formulas (14.6.4) are of interest for even n only. Indeed, since $\chi''_{AA}(\omega)$ is an odd function of ω, the moments of odd order of $\chi''_{AA}(\omega)/\omega$ vanish. Accordingly, $\tilde{K}_{AA}(t)$ is an even function of t analytic at $t=0$. Its derivatives of odd order at the origin vanish.

In the classical case, we deduce from equation (14.6.4), written for the moment of order $2p$ of $\chi''_{AA}(\omega)/\omega$, the relation:

$$\boxed{\frac{1}{\pi}\int_{-\infty}^{\infty}\omega^{2p-1}\chi''_{AA}(\omega)\,d\omega=\beta\langle[A^{(p)}]^2\rangle,\qquad p\geq 0.}\tag{14.6.5}$$

6.2. The thermodynamic sum rule revisited

Formula (14.6.4) reads for $n=0$:

$$\frac{1}{\pi}\int_{-\infty}^{\infty}\frac{\chi''_{AA}(\omega)}{\omega}\,d\omega=\beta\tilde{K}_{AA}(t=0).\tag{14.6.6}$$

The equality (14.6.6) may be rewritten in the equivalent form:

$$\chi_{AA}(\omega=0)=\beta\tilde{K}_{AA}(t=0),\tag{14.6.7}$$

a formula which is simply the thermodynamic sum rule (14.5.7).

6.3. *f*-sum rule

Let us examine in more detail the sum rule (14.6.5) in the case $p=1$:

$$\frac{1}{\pi}\int_{-\infty}^{\infty}\omega\chi''_{AA}(\omega)\,d\omega=\beta\langle\dot{A}^2\rangle.\tag{14.6.8}$$

This property is known as the *f-sum rule*. We will check it for the example of the polarization of an atom perturbed by an electric field.

The generalized susceptibility associated with the polarization of an atom, initially in its fundamental state $|\phi_0\rangle$ and perturbed by an electric field parallel to the Ox axis, is:

$$\chi(\omega)=\frac{e^2}{\hbar}\lim_{\epsilon\to 0^+}\sum_n|\langle\phi_n|x|\phi_0\rangle|^2\Big(-\frac{1}{\omega-\omega_{n0}+i\epsilon}+\frac{1}{\omega+\omega_{n0}+i\epsilon}\Big).\tag{14.6.9}$$

In formula (14.6.9), the states $\{|\phi_n\rangle\}$ ($n > 0$) are the excited unperturbed eigenstates and the ω_{n0}'s are the Bohr angular frequencies. In particular, we have:

$$\chi''(\omega) = \frac{\pi e^2}{\hbar} \sum_n |\langle\phi_n|x|\phi_0\rangle|^2 \big[\delta(\omega - \omega_{n0}) - \delta(\omega + \omega_{n0})\big]. \tag{14.6.10}$$

The f-sum rule (14.6.8) reads, in this case:

$$\frac{2m}{\hbar} \sum_n |\langle\phi_n|x|\phi_0\rangle|^2 \omega_{n0} = 1. \tag{14.6.11}$$

Formula (14.6.11) is simply the Thomas–Reiche–Kuhn (oscillator strength) sum rule.

6.4. Phenomenological models and sum rules

The various sum rules constitute a set of constraints which the phenomenological models obey only partially.

Let us come back to the example of the susceptibility $\chi_{xx}(\omega)$ of an oscillator damped by viscous friction, as given by the formula:

$$\chi_{xx}(\omega) = \frac{1}{m} \frac{1}{-\omega^2 + \omega_0^2 - i\gamma\omega}. \tag{14.6.12}$$

We have:

$$\chi''_{xx}(\omega) = \frac{1}{m} \frac{\gamma\omega}{(\omega^2 - \omega_0^2)^2 + \gamma^2\omega^2}. \tag{14.6.13}$$

The odd order moments of $\chi''_{xx}(\omega)/\omega$ actually vanish. The first two even order moments of $\chi''_{xx}(\omega)/\omega$ are finite. The thermodynamic sum rule and the f-sum rule are verified for any γ. We actually have the equalities:

$$\frac{1}{m\pi} \int_{-\infty}^{\infty} \frac{\gamma}{(\omega^2 - \omega_0^2)^2 + \gamma^2\omega^2}\, d\omega = \frac{1}{m\omega_0^2} \tag{14.6.14}$$

(thermodynamic sum rule) and:

$$\frac{1}{m\pi} \int_{-\infty}^{\infty} \frac{\gamma\omega^2}{(\omega^2 - \omega_0^2)^2 + \gamma^2\omega^2}\, d\omega = \frac{1}{m} \tag{14.6.15}$$

(f-sum rule). Formulas (14.6.14) and (14.6.15) express the fact that the oscillator in a viscous fluid reaches thermal equilibrium as a consequence of collisions with the molecules of the fluid: $m\omega_0^2\langle x^2\rangle/2 = kT/2$, $m\langle\dot{x}^2\rangle = kT/2$.

However, the moments of even order $n \geq 4$ of $\chi''_{xx}(\omega)/\omega$ diverge. This divergence displays the fact that the viscous damping model with an ω-independent friction coefficient is not satisfactory at large angular frequencies. This model can be improved by the introduction of a generalized friction coefficient $\gamma(\omega)$ (the equation of motion of the oscillator then involves a retarded friction force). For instance, with the generalized friction coefficient:

$$\gamma(\omega) = \gamma \frac{\omega_c}{\omega_c - i\omega}, \tag{14.6.16}$$

in which ω_c denotes an angular frequency characteristic of the damping fluid, the moment of order 4 of $\chi''_{xx}(\omega)/\omega$ becomes finite, the moments of even order $n > 4$ of this quantity being, however, still divergent.

Bibliography

P.M. CHAIKIN and T.C. LUBENSKY, *Principles of condensed matter physics*, Cambridge University Press, Cambridge, 1995.

C. COHEN-TANNOUDJI, B. DIU, and F. LALOË, *Quantum mechanics*, Vol. 2, Hermann and Wiley, Paris, second edition, 1977.

S. DATTAGUPTA, *Relaxation phenomena in condensed matter physics*, Academic Press, Orlando, 1987.

D. FORSTER, *Hydrodynamic fluctuations, broken symmetries, and correlation functions*, Westview Press, Boulder, 1995.

R. KUBO, M. TODA, and N. HASHITSUME, *Statistical physics II: nonequilibrium statistical mechanics*, Springer-Verlag, Berlin, second edition, 1991.

P.C. MARTIN, *Measurements and correlation functions*, Les Houches Lecture Notes 1967 (C. DE WITT and R. BALIAN editors), Gordon and Breach, New York, 1968.

A. MESSIAH, *Quantum mechanics*, North-Holland, Amsterdam, 1970.

D. ZUBAREV, V. MOROZOV, and G. RÖPKE, *Statistical mechanics of nonequilibrium processes*, Vol. 2: *Relaxation and hydrodynamic processes*, Akademie Verlag, Berlin, 1997.

R. ZWANZIG, *Nonequilibrium statistical mechanics*, Oxford University Press, Oxford, 2001.

References

H. NYQUIST, Thermal agitation of electric charge in conductors, *Phys. Rev.* **32**, 110 (1928).

L. ONSAGER, Reciprocal relations in irreversible processes. I., *Phys. Rev.* **37**, 405 (1931); II., *Phys. Rev.* **38**, 2265 (1931).

H.B. CALLEN and T.A. WELTON, Irreversibility and generalized noise, *Phys. Rev.* **83**, 34 (1951).

H.B. CALLEN and R.F. GREENE, On a theorem of irreversible thermodynamics, *Phys. Rev.* **86**, 702 (1952).

R. KUBO, The fluctuation-dissipation theorem and Brownian motion, 1965, *Tokyo Summer Lectures in Theoretical Physics* (R. KUBO editor), Syokabo, Tokyo and Benjamin, New York, 1966.

R. KUBO, The fluctuation-dissipation theorem, *Rep. Prog. Phys.* **29**, 255 (1966).

Supplement 14A

Dissipative dynamics of a harmonic oscillator

1. Oscillator coupled with a thermal bath

The classical or quantum dissipative dynamics of a harmonic oscillator may be studied in the framework of the Caldeira–Leggett model, in which the oscillator under study is linearly coupled with an environment made up of an infinite number of independent oscillators in thermal equilibrium. This coupling gives rise to a damping. Since the model is linear, it is possible to determine exactly the response function of the damped oscillator's displacement, and then, owing to the fluctuation-dissipation theorem, the associated correlation function.

We will first recall the expressions for the response function and the susceptibility associated with the displacement of the uncoupled oscillator, and compute the corresponding autocorrelation function. We will then examine how the oscillator's dynamics is modified by the coupling.

2. Dynamics of the uncoupled oscillator

The Hamiltonian of a one-dimensional harmonic oscillator of mass m and angular frequency ω_0 reads, in terms of the displacement x and the momentum p:

$$H_0 = \frac{p^2}{2m} + \frac{1}{2} m\omega_0^2 x^2, \tag{14A.2.1}$$

or, in terms of the annihilation and creation operators[1] a and $a^\dagger$:

$$H_0 = \hbar\omega_0\Big(a^\dagger a + \frac{1}{2}\Big). \tag{14A.2.2}$$

[1] Recall the formulas:

$$x = \Big(\frac{\hbar}{2m\omega_0}\Big)^{1/2}(a + a^\dagger), \qquad p = -i\Big(\frac{m\hbar\omega_0}{2}\Big)^{1/2}(a - a^\dagger).$$

The oscillator is submitted to an external force $F(t)$. The corresponding perturbation Hamiltonian is:

$$H_1(t) = -F(t)x. \tag{14A.2.3}$$

2.1. Displacement response function

The displacement response function $\tilde{\chi}_{xx}(t)$ is a linear combination of the four response functions $\tilde{\chi}_{aa^\dagger}(t)$, $\tilde{\chi}_{a^\dagger a}(t)$, $\tilde{\chi}_{aa}(t)$, and $\tilde{\chi}_{a^\dagger a^\dagger}(t)$. Given the expressions for $a(t)$ and $a^\dagger(t)$,

$$a(t) = ae^{-i\omega_0 t}, \qquad a^\dagger(t) = a^\dagger e^{i\omega_0 t}, \tag{14A.2.4}$$

we have, according to the Kubo formulas,

$$\tilde{\chi}_{aa^\dagger}(t) = \frac{i}{\hbar}\Theta(t)e^{-i\omega_0 t}, \qquad \tilde{\chi}_{a^\dagger a}(t) = -\frac{i}{\hbar}\Theta(t)e^{i\omega_0 t}, \tag{14A.2.5}$$

and:

$$\tilde{\chi}_{aa}(t) = 0, \qquad \tilde{\chi}_{a^\dagger a^\dagger}(t) = 0. \tag{14A.2.6}$$

The displacement response function thus reduces to the sum of two terms:

$$\tilde{\chi}_{xx}(t) = \frac{\hbar}{2m\omega_0}\big[\tilde{\chi}_{aa^\dagger}(t) + \tilde{\chi}_{a^\dagger a}(t)\big]. \tag{14A.2.7}$$

On account of formulas (14A.2.5), we have:

$$\boxed{\tilde{\chi}_{xx}(t) = \Theta(t)\frac{\sin\omega_0 t}{m\omega_0}.} \tag{14A.2.8}$$

2.2. Generalized susceptibility

The generalized susceptibility $\chi_{xx}(\omega)$ is the Fourier transform of $\tilde{\chi}_{xx}(t)$:

$$\chi_{xx}(\omega) = \lim_{\epsilon\to 0^+}\int_0^\infty \tilde{\chi}_{xx}(t)e^{i\omega t}e^{-\epsilon t}\,dt. \tag{14A.2.9}$$

From formula (14A.2.8), we get:

$$\boxed{\chi_{xx}(\omega) = \frac{1}{2m\omega_0}\lim_{\epsilon\to 0^+}\left(-\frac{1}{\omega-\omega_0+i\epsilon} + \frac{1}{\omega+\omega_0+i\epsilon}\right).} \tag{14A.2.10}$$

The real and imaginary parts of $\chi_{xx}(\omega)$ are:

$$\begin{cases} \chi'_{xx}(\omega) = \dfrac{1}{2m\omega_0}\left(-\mathrm{vp}\dfrac{1}{\omega-\omega_0} + \mathrm{vp}\dfrac{1}{\omega+\omega_0}\right) \\[2ex] \chi''_{xx}(\omega) = \dfrac{\pi}{2m\omega_0}\big[\delta(\omega-\omega_0) - \delta(\omega+\omega_0)\big]. \end{cases} \tag{14A.2.11}$$

2.3. Displacement autocorrelation function

The symmetric autocorrelation function of the displacement of the oscillator is:

$$\tilde{S}_{xx}(t) = \frac{1}{2}\langle xx(t) + x(t)x\rangle. \tag{14A.2.12}$$

Since the equilibrium averages $\langle aa\rangle$ and $\langle a^\dagger a^\dagger\rangle$ vanish, we simply have:

$$\tilde{S}_{xx}(t) = \frac{\hbar}{4m\omega_0}\big[\tilde{S}_{aa^\dagger}(t) + \tilde{S}_{a^\dagger a}(t)\big], \tag{14A.2.13}$$

that is, on account of the expressions (14A.2.4) for $a(t)$ and $a^\dagger(t)$:

$$\tilde{S}_{xx}(t) = \frac{\hbar}{2m\omega_0}\langle aa^\dagger + a^\dagger a\rangle \cos\omega_0 t. \tag{14A.2.14}$$

The equilibrium averages $\langle a^\dagger a\rangle$ and $\langle aa^\dagger\rangle$ are given by the formulas:

$$\langle a^\dagger a\rangle = n_0, \qquad \langle aa^\dagger\rangle = 1 + n_0, \tag{14A.2.15}$$

where $n_0 = (e^{\beta\hbar\omega_0} - 1)^{-1}$ denotes the Bose–Einstein distribution function at temperature $T = (k\beta)^{-1}$. This gives us the symmetric displacement autocorrelation function,

$$\boxed{\tilde{S}_{xx}(t) = \frac{\hbar}{2m\omega_0}\coth\frac{\beta\hbar\omega_0}{2}\cos\omega_0 t,} \tag{14A.2.16}$$

and, by Fourier transformation, the associated spectral density:

$$\boxed{S_{xx}(\omega) = \frac{\pi\hbar}{2m\omega_0}\coth\frac{\beta\hbar\omega_0}{2}\big[\delta(\omega-\omega_0) + \delta(\omega+\omega_0)\big].} \tag{14A.2.17}$$

In the absence of coupling, the oscillator is not damped. As shown by formulas (14A.2.8) and (14A.2.16), both the displacement response function and the displacement autocorrelation function oscillate indefinitely without decreasing.

2.4. The fluctuation-dissipation theorem

We can rewrite formula (14A.2.17) as:

$$S_{xx}(\omega) = \frac{\pi\hbar}{2m\omega_0}\coth\frac{\beta\hbar\omega}{2}\big[\delta(\omega-\omega_0) - \delta(\omega+\omega_0)\big]. \tag{14A.2.18}$$

Comparing the expression (14A.2.18) for $S_{xx}(\omega)$ with $\chi''_{xx}(\omega)$ as given by formula (14A.2.11), we verify the fluctuation-dissipation theorem:

$$\boxed{S_{xx}(\omega) = \hbar\coth\frac{\beta\hbar\omega}{2}\chi''_{xx}(\omega).} \tag{14A.2.19}$$

In fact, the uncoupled oscillator, which is not damped, does not constitute a genuine dissipative system. According to expression (14A.2.11) for $\chi''_{xx}(\omega)$, the oscillator absorbs energy only at its own angular frequency ω_0. As shown by expression (14A.2.17) for $S_{xx}(\omega)$, the displacement fluctuations show up exclusively at this angular frequency. The fluctuation-dissipation theorem treats this limiting case consistently.

3. Response functions and susceptibilities of the coupled oscillator

The coupling of the oscillator with a bath allows us to generate a truly dissipative system. For this to be the case, the number of the bath's oscillators has to tend towards infinity, their angular frequencies forming a continuum in this limit.

3.1. The Caldeira–Leggett Hamiltonian

The Caldeira–Leggett Hamiltonian reads, for the present problem,

$$H_{\mathrm{C-L}} = \frac{p^2}{2m} + \frac{1}{2}m\omega_0^2 x^2 + \frac{1}{2}\sum_{n=1}^{N}\left[\frac{p_n^2}{m_n} + m_n\omega_n^2\Big(x_n - \frac{c_n}{m_n\omega_n^2}x\Big)^2\right], \qquad (14A.3.1)$$

where the index $n = 1, \dots, N$ denotes each of the bath's oscillators (we can consider that these oscillators represent phonon modes). The bilinear coupling term between the oscillator under study and the bath is:

$$-\sum_{n=1}^{N} c_n x_n x = \hbar\sum_{n=1}^{N} g_n(b_n + b_n^\dagger)(a + a^\dagger), \qquad (14A.3.2)$$

where b_n and $b_n^\dagger$ are the annihilation and creation operators relative to the mode n. The quantities c_n or $g_n = -c_n(4mm_n\omega_0\omega_n)^{-1/2}$ are coupling constants. We generally neglect the terms $b_n a + b_n^\dagger a^\dagger$, which correspond to processes in which two quanta are either annihilated or created.[2] We then write, up to additive constants:

$$H_{\mathrm{C-L}} \simeq \hbar\omega_0 a^\dagger a + \sum_{n=1}^{N}\hbar\omega_n b_n^\dagger b_n + \sum_{n=1}^{N}\hbar g_n(ab_n^\dagger + a^\dagger b_n). \qquad (14A.3.3)$$

The coupling between the system and the bath is performed through processes in which the oscillator gains a quantum to the detriment of mode n, and conversely.

3.2. General formula for the time-derivative of a response function

To begin with, we will establish a general formula[3] for the time-derivative of a response function. Taking the derivative of the response function $\tilde{\chi}_{BA}(t)$ as given by the Kubo formula, that is:

$$\tilde{\chi}_{BA}(t) = \frac{i}{\hbar}\Theta(t)\langle[B(t), A]\rangle, \qquad (14A.3.4)$$

we obtain for $d\tilde{\chi}_{BA}(t)/dt$ the following expression:

$$\frac{d\tilde{\chi}_{BA}(t)}{dt} = \frac{i}{\hbar}\delta(t)\langle[B, A]\rangle + \frac{i}{\hbar}\Theta(t)\langle[\dot{B}(t), A]\rangle. \qquad (14A.3.5)$$

[2] These terms correspond to rapidly oscillating contributions whose effect may be neglected. Such an approximation is commonly used in optics, where it takes the name of *rotating wave approximation*.

[3] We use here the notations of the general linear response theory.

The second term on the right-hand side of equation (14A.3.5) identifies with $\tilde{\chi}_{\dot{B}A}(t)$. We thus have:

$$\frac{d\tilde{\chi}_{BA}(t)}{dt} = \frac{i}{\hbar}\delta(t)\langle[B,A]\rangle + \tilde{\chi}_{\dot{B}A}(t). \tag{14A.3.6}$$

3.3. Coupled equations for the response functions

Our aim is to compute the response function $\tilde{\chi}_{xx}(t)$ of the displacement of the coupled oscillator, the evolution of the global system being governed by the Caldeira–Leggett Hamiltonian.

- *Response functions* $\tilde{\chi}_{aa^\dagger}(t)$ *and* $\tilde{\chi}_{a^\dagger a}(t)$

Applying the general formula (14A.3.6) to $d\tilde{\chi}_{aa^\dagger}(t)/dt$, and taking into account the structure of the Hamiltonian (14A.3.3), we obtain:

$$\frac{d\tilde{\chi}_{aa^\dagger}(t)}{dt} = \frac{i}{\hbar}\delta(t) - i\omega_0\tilde{\chi}_{aa^\dagger}(t) - i\sum_n g_n\tilde{\chi}_{b_n a^\dagger}(t). \tag{14A.3.7}$$

We then apply formula (14A.3.6) to $d\tilde{\chi}_{b_n a^\dagger}(t)/dt$ $(n = 1, \ldots, N)$:

$$\frac{d\tilde{\chi}_{b_n a^\dagger}(t)}{dt} = -i\omega_n\tilde{\chi}_{b_n a^\dagger}(t) - ig_n\tilde{\chi}_{aa^\dagger}(t). \tag{14A.3.8}$$

Equation (14A.3.7) together with the set of equations (14A.3.8) for $n = 1, \ldots, N$ form a system of coupled differential equations for the response functions $\tilde{\chi}_{aa^\dagger}(t)$ and $\tilde{\chi}_{b_n a^\dagger}(t)$. We proceed in an analogous way as far as $\tilde{\chi}_{a^\dagger a}(t)$ is concerned.

- *Response functions* $\tilde{\chi}_{aa}(t)$ *and* $\tilde{\chi}_{a^\dagger a^\dagger}(t)$

As for the response functions $\tilde{\chi}_{aa}(t)$ and $\tilde{\chi}_{a^\dagger a^\dagger}(t)$, they vanish even in the presence of the coupling. Indeed, if we apply the general formula (14A.3.6) to $d\tilde{\chi}_{aa}(t)/dt$, the inhomogeneous term in $\delta(t)$ yields no contribution. The previous method allows us to write a system of linear and homogeneous coupled differential equations for the response functions $\tilde{\chi}_{aa}(t)$ and $\tilde{\chi}_{b_n a}(t)$, whose unique solution is $\tilde{\chi}_{aa}(t) = \tilde{\chi}_{b_n a}(t) = 0$. The same argument applies to $\tilde{\chi}_{a^\dagger a^\dagger}(t)$.

3.4. The generalized susceptibility $\chi_{xx}(\omega)$

Coupled equations for the generalized susceptibilities $\chi_{aa^\dagger}(\omega + i\epsilon)$ and $\chi_{b_n a^\dagger}(\omega + i\epsilon)$ at finite $\epsilon > 0$ can be deduced from the coupled differential equations (14A.3.7) and (14A.3.8) for the response functions $\tilde{\chi}_{aa^\dagger}(t)$ and $\tilde{\chi}_{b_n a^\dagger}(t)$:

$$\begin{cases} (\omega + i\epsilon - \omega_0)\chi_{aa^\dagger}(\omega + i\epsilon) = -\dfrac{1}{\hbar} + \displaystyle\sum_n g_n\chi_{b_n a^\dagger}(\omega + i\epsilon) \\ (\omega + i\epsilon - \omega_n)\chi_{b_n a^\dagger}(\omega + i\epsilon) = g_n\chi_{aa^\dagger}(\omega + i\epsilon), \qquad n = 1, \ldots, N. \end{cases} \tag{14A.3.9}$$

Hence, the susceptibilities $\chi_{b_n a^\dagger}(\omega + i\epsilon)$ once eliminated and the limit $\epsilon \to 0^+$ once taken, we have:

$$\chi_{aa^\dagger}(\omega) = -\frac{1}{\hbar}\lim_{\epsilon\to 0^+}\frac{1}{\omega + i\epsilon - \omega_0 - \sum_n g_n^2(\omega + i\epsilon - \omega_n)^{-1}}. \tag{14A.3.10}$$

The expression for $\chi_{a^\dagger a}(\omega)$ can be obtained in an analogous way:

$$\chi_{a^\dagger a}(\omega) = \frac{1}{\hbar} \lim_{\epsilon \to 0^+} \frac{1}{\omega + i\epsilon + \omega_0 - \sum_n g_n^2 (\omega + i\epsilon + \omega_n)^{-1}}. \tag{14A.3.11}$$

Since $\chi_{aa}(\omega)$ and $\chi_{a^\dagger a^\dagger}$ vanish even in the presence of the coupling, the generalized susceptibility $\chi_{xx}(\omega)$ of the coupled oscillator can be deduced from $\chi_{aa^\dagger}(\omega)$ and $\chi_{a^\dagger a}(\omega)$:

$$\chi_{xx}(\omega) = \frac{\hbar}{2m\omega_0} \left[\chi_{aa^\dagger}(\omega) + \chi_{a^\dagger a}(\omega)\right]. \tag{14A.3.12}$$

To study $\chi_{xx}(\omega)$, we thus have to analyze formulas (14A.3.10) and (14A.3.11).

4. Analysis of $\chi_{xx}(\omega)$

4.1. The function $\Sigma(\omega)$

The central quantity for the study of $\chi_{aa^\dagger}(\omega)$ is the function $\Sigma(\omega)$ as defined by:

$$\Sigma(\omega) = \lim_{\epsilon \to 0^+} \sum_n g_n^2 (\omega + i\epsilon - \omega_n)^{-1}. \tag{14A.4.1}$$

It has the form:

$$\Sigma(\omega) = g^2 \left[\Delta(\omega) - i\Gamma(\omega)\right], \tag{14A.4.2}$$

with:

$$\begin{cases} g^2 \Delta(\omega) = \sum_n g_n^2 \,\mathrm{vp}\, \dfrac{1}{\omega - \omega_n} \\[2ex] g^2 \Gamma(\omega) = \pi \sum_n g_n^2 \delta(\omega - \omega_n). \end{cases} \tag{14A.4.3}$$

To describe an irreversible dynamics, we have to consider the limit in which the number N of bath modes tends towards infinity, their angular frequencies forming a continuum in this limit. The real and imaginary parts $g^2\Delta(\omega)$ and $g^2\Gamma(\omega)$ of $\Sigma(\omega)$ can then be considered as continuous functions of ω. As shown by formula (14A.4.1), the continuation $\Sigma(z)$ of the function $\Sigma(\omega)$ to a complex argument z is analytic in the upper complex half-plane. Accordingly, the functions $\Delta(\omega)$ and $\Gamma(\omega)$ are not independent, but instead related to one another by Kramers–Kronig type formulas. In particular, we have:

$$\Delta(\omega) = \frac{1}{\pi} \,\mathrm{vp} \int_{-\infty}^{\infty} \frac{\Gamma(\omega')}{\omega - \omega'} \, d\omega'. \tag{14A.4.4}$$

4.2. Modelization of $\Sigma(\omega)$

We have to choose a function $\Sigma(\omega)$ conveniently suited to the description of an environment made up of phonon modes. If we assume for the sake of simplicity that the coupling constant $g_n = g_{\text{eff}}$ is independent of n, the function $\Gamma(\omega)$ reads:

$$\Gamma(\omega) = \frac{\pi g_{\text{eff}}^2}{g^2} \sum_n \delta(\omega - \omega_n). \tag{14A.4.5}$$

It is proportional to the density of phonon modes $Z(\omega) = N^{-1}\sum_n \delta(\omega - \omega_n)$. Setting $g_{\text{eff}}^2 = N^{-1}g^2$, gives:

$$\Gamma(\omega) = \pi Z(\omega). \tag{14A.4.6}$$

The modelization of $\Sigma(\omega)$ thus amounts to that of the density of modes.

• *Density of phonon modes*

We can choose to describe it simply by a Debye model:

$$Z(\omega) = \begin{cases} 3\omega^2/\omega_D^3, & 0 < \omega < \omega_D \\ 0, & \omega > \omega_D. \end{cases} \tag{14A.4.7}$$

The Debye angular frequency ω_D is a measure of the phonon bandwidth.

• *The function $\Delta(\omega)$*

In the Debye model, we have, according to formulas (14A.4.4), (14A.4.6) and (14A.4.7):

$$\Delta(\omega) = \text{vp} \int_0^{\omega_D} \frac{3\omega'^2}{\omega_D^3} \frac{1}{\omega - \omega'} \, d\omega'. \tag{14A.4.8}$$

In particular, we have:

$$\Delta(\omega) \simeq \frac{1}{\omega}, \qquad |\omega| \gg \omega_D, \tag{14A.4.9}$$

and:[4]

$$\Delta(\omega = 0) = -\frac{3}{2\omega_D}. \tag{14A.4.10}$$

The integration on the right-hand side of equation (14A.4.8) once carried out, we get:

$$\Delta(\omega) = -\frac{3}{\omega_D}\left[\frac{1}{2} + \frac{\omega}{\omega_D} + \left(\frac{\omega}{\omega_D}\right)^2 \ln\left|\frac{\omega_D - \omega}{\omega}\right|\right]. \tag{14A.4.11}$$

The function $\Delta(\omega)$ is represented in Fig. 14A.1.

---

[4] As displayed by the general formula (14A.4.4), the behavior $\Delta(\omega) \simeq \omega^{-1}$ at large angular frequencies of the function $\Delta(\omega)$ is not specific to the Debye model, but it is realized in all cases (this behavior is solely linked to the normalization of the density of modes $Z(\omega)$). Besides, the property $\Delta(\omega = 0) < 0$ also holds independently of the modelization chosen for the density of modes, since we have $\Delta(\omega = 0) = -\text{vp}\int_{-\infty}^{\infty} [Z(\omega)/\omega]\, d\omega$ with $Z(\omega)$ a positive quantity.

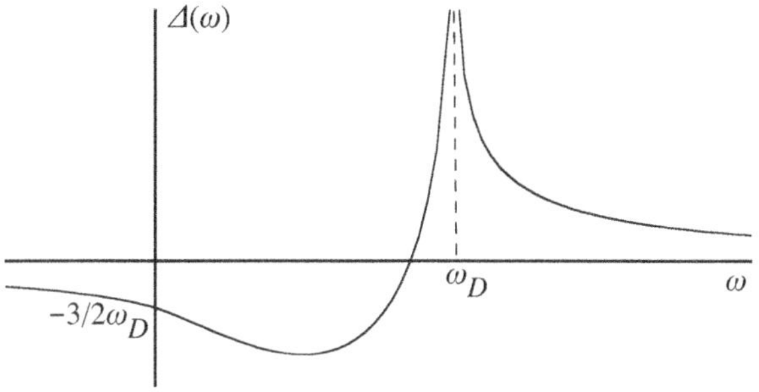

Fig. 14A.1 The function $\Delta(\omega)$ in the Debye model.

4.3. Stability condition

In terms of the function $\Sigma(\omega)$, the susceptibility $\chi_{aa^\dagger}(\omega)$ reads:

$$\chi_{aa^\dagger}(\omega) = -\frac{1}{\hbar}\frac{1}{\omega + i\epsilon - \omega_0 - \Sigma(\omega)}. \tag{14A.4.12}$$

Its real and imaginary parts are, on account of formula (14A.4.2):

$$\begin{cases} \chi'_{aa^\dagger}(\omega) = -\dfrac{1}{\hbar}\dfrac{\omega - \omega_0 - g^2\Delta(\omega)}{\left[\omega - \omega_0 - g^2\Delta(\omega)\right]^2 + \left[g^2\Gamma(\omega)\right]^2} \\ \chi''_{aa^\dagger}(\omega) = \dfrac{1}{\hbar}\dfrac{g^2\Gamma(\omega)}{\left[\omega - \omega_0 - g^2\Delta(\omega)\right]^2 + \left[g^2\Gamma(\omega)\right]^2}. \end{cases} \tag{14A.4.13}$$

At vanishing angular frequency, we have:

$$\chi'_{aa^\dagger}(\omega = 0) = \frac{1}{\hbar}\frac{1}{\omega_0 + g^2\Delta(\omega = 0)}. \tag{14A.4.14}$$

Now, according to the Kramers–Kronig relation for the susceptibility $\chi_{aa^\dagger}(\omega)$, the quantity $\chi'_{aa^\dagger}(\omega = 0)$ reads:

$$\chi'_{aa^\dagger}(\omega = 0) = \frac{1}{\pi}\int_{-\infty}^{\infty}\frac{\chi''_{aa^\dagger}(\omega)}{\omega}\,d\omega. \tag{14A.4.15}$$

In a stable dissipative system, the integrand $\chi''_{aa^\dagger}(\omega)/\omega$ is positive.[5] We thus necessarily have $\chi'_{a^\dagger a}(\omega = 0) > 0$, that is, according to formula (14A.4.14):

$$\omega_0 + g^2\Delta(\omega = 0) > 0. \tag{14A.4.16}$$

[5] The input and output operators (in other words, the operators A and B of the general linear response theory) are here Hermitean conjugate of one another. The result is that $\xi_{aa^\dagger}(\omega)$ is simply $\chi''_{aa^\dagger}(\omega)$. The average dissipated power, positive, is proportional to $\omega\xi_{aa^\dagger}(\omega)$, that is, to $\omega\chi''_{aa^\dagger}(\omega)$. This insures the positivity of the integrand on the right-hand side of formula (14A.4.15).

As $\Delta(\omega = 0) < 0$, the stability condition (14A.4.16) implies that the coupling constant g cannot exceed some value $g_{\max}$. In the particular case of the Debye model, in which $\Delta(\omega = 0)$ is given by formula (14A.4.10), we have $g_{\max} = (2\omega_0\omega_D/3)^{1/2}$.

4.4. Graphical determination of the maxima of $\chi''_{aa^\dagger}(\omega)$

As shown by formula (14A.4.13), $\chi''_{aa^\dagger}(\omega)$ exhibits in general a maximum in the vicinity of a solution ω_m of the equation:

$$\omega - \omega_0 - g^2\Delta(\omega) = 0. \tag{14A.4.17}$$

The solutions of equation (14A.4.17) can be obtained by looking for the intersections of the curve $\Delta(\omega)$ with the straight line of equation $f(\omega) = g^{-2}(\omega - \omega_0)$. Fig. 14A.2 presents, for two different values of ω_0 (one larger than ω_D, the other one smaller) the appropriate graphical constructions in the case of both weak and intermediate coupling.

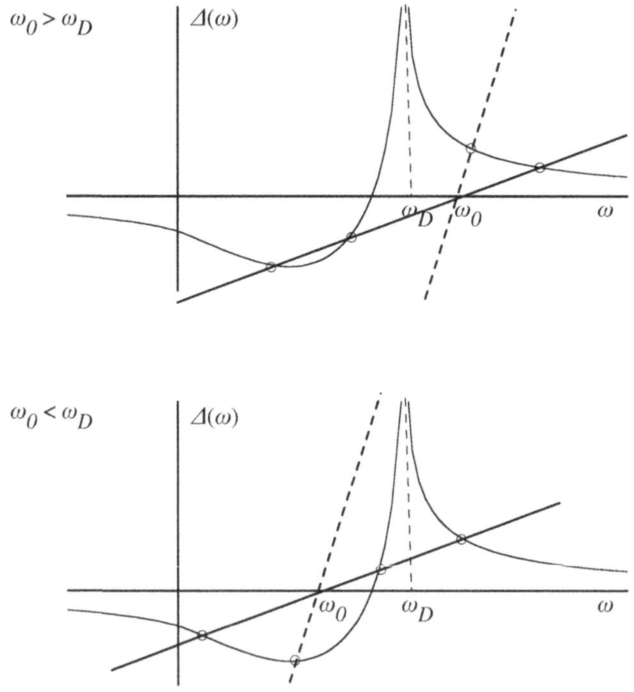

Fig. 14A.2 Determination of the maxima of $\chi''_{aa^\dagger}(\omega)$ for $\omega_0 > \omega_D$ and $\omega_0 < \omega_D$.

When the stability condition (14A.4.16) is fulfilled, the ordinate at the origin of the straight line of equation $f(\omega) = g^{-2}(\omega - \omega_0)$ is smaller than $\Delta(\omega = 0) = -3/2\omega_D$. All solutions of equation (14A.4.17) then correspond to angular frequencies $\omega_m > 0$. The number of solutions depends on the coupling strength. There is a unique solution in weak coupling,[6] shown by the intersection of the curve $\Delta(\omega)$ with the straight line $f(\omega)$ (dotted line), and three solutions at intermediate coupling, shown by the intersection of the curve $\Delta(\omega)$ with the straight line $f(\omega)$ (full line).

In a general manner, near a zero ω_m of equation (14A.4.17), we can write the following expansions:

$$\begin{cases} \omega - \omega_0 - g^2\Delta(\omega) \simeq (\omega - \omega_m)\left[1 - g^2\Delta'(\omega_m)\right] + \cdots \\ \Gamma(\omega) \simeq \Gamma(\omega_m) + \cdots \end{cases} \tag{14A.4.18}$$

For $\omega \simeq \omega_m$, $\chi''_{aa^\dagger}(\omega)$ thus takes approximately the form of a Lorentzian centered at ω_m, of width $2g^2\Gamma(\omega_m)[1 - g^2\Delta'(\omega_m)]^{-1}$ and of weight $[1 - g^2\Delta'(\omega_m)]^{-1}$:

$$\chi''_{aa^\dagger}(\omega) \simeq \frac{1}{\hbar}\frac{1}{1 - g^2\Delta'(\omega_m)}\frac{\dfrac{g^2\Gamma(\omega_m)}{1 - g^2\Delta'(\omega_m)}}{(\omega - \omega_m)^2 + \left[\dfrac{g^2\Gamma(\omega_m)}{1 - g^2\Delta'(\omega_m)}\right]^2}, \qquad \omega \simeq \omega_m. \tag{14A.4.19}$$

The corresponding expression for $\chi'_{aa^\dagger}(\omega)$ is:

$$\chi'_{aa^\dagger}(\omega) \simeq -\frac{1}{\hbar}\frac{1}{1 - g^2\Delta'(\omega_m)}\frac{\omega - \omega_m}{(\omega - \omega_m)^2 + \left[\dfrac{g^2\Gamma(\omega_m)}{1 - g^2\Delta'(\omega_m)}\right]^2}, \qquad \omega \simeq \omega_m. \tag{14A.4.20}$$

• *Weak-coupling solution*

When the coupling is weak, it is possible to obtain for $\chi'_{aa^\dagger}(\omega)$ and $\chi''_{aa^\dagger}(\omega)$ approximate formulas valid for the whole set of values of ω. In weak coupling, the term $[\omega - \omega_0 - g^2\Delta(\omega)]^2$ in the denominator of formulas (14A.4.13) is large as compared to the term in $[g^2\Gamma(\omega)]^2$ (except near ω_m, where it vanishes). It is thus possible, in the very small terms $g^2\Gamma(\omega)$ and $g^2\Delta(\omega)$, to replace ω by ω_0. The function $\chi''_{aa^\dagger}(\omega)$ is then correctly represented, for the whole set of values of ω, by a Lorentzian centered at $\omega_m \simeq \omega_0 + g^2\Delta(\omega_0)$, and of width $2g^2\Gamma(\omega_0)$:

$$\chi''_{aa^\dagger}(\omega) = \frac{1}{\hbar}\frac{\gamma_m}{(\omega - \omega_m)^2 + \gamma_m^2}, \qquad \gamma_m = g^2\Gamma(\omega_0). \tag{14A.4.21}$$

The corresponding expression of $\chi'_{aa^\dagger}(\omega)$ is:

$$\chi'_{aa^\dagger}(\omega) = -\frac{1}{\hbar}\frac{\omega - \omega_m}{(\omega - \omega_m)^2 + \gamma_m^2}. \tag{14A.4.22}$$

---

[6] There is a unique solution in weak coupling, whatever the value of ω_0. The logarithmic divergence of the function $\Delta(\omega)$ at $\omega = \omega_D$ is an artefact of the Debye model (in a more realistic description of the density of modes, the function $\Delta(\omega)$ would be negative for small values of ω and positive for large ones, the curve $\Delta(\omega)$ having the shape of a dispersion curve with a maximum close to the maximal value of the phonon angular frequency).

- *Resonance*

When $\omega_0 > \omega_D$, and whatever the coupling strength, one of the solutions of equation (14A.4.17) is larger than ω_0, that is, larger than ω_D. For this solution, denoted by ω_{m1}, we have $Z(\omega_{m_1}) = 0$ and $\Gamma(\omega_{m_1}) = 0$. The corresponding response is not damped. In the vicinity of ω_{m_1}, we have:

$$\chi''_{aa^\dagger}(\omega) \simeq \frac{1}{\hbar}\frac{1}{1 - g^2\Delta'(\omega_{m_1})}\pi\delta(\omega - \omega_{m_1}). \tag{14A.4.23}$$

This response is called a *resonance* of the model.

The other possible solutions, which appear at intermediate coupling, are such that $\omega_m < \omega_D$. We then have $Z(\omega_m) \neq 0$ and $\Gamma(\omega_m) \neq 0$. The corresponding response is actually damped.

4.5. The generalized susceptibility $\chi_{xx}(\omega)$ in weak coupling

Let us now discuss the weak coupling solution in the case $\omega_0 < \omega_D$. This solution, which then does not correspond to a resonance, is effectively damped.

Formulas (14A.3.10) and (14A.3.11) allow us to verify the properties:

$$\begin{cases} \chi'_{aa^\dagger}(\omega) = \chi'_{a^\dagger a}(-\omega) \\ \chi''_{aa^\dagger}(\omega) = -\chi''_{a^\dagger a}(-\omega). \end{cases} \tag{14A.4.24}$$

To deduce the generalized susceptibility $\chi_{xx}(\omega)$ of the oscillator weakly coupled to the phonons from the weak-coupling expressions of $\chi_{aa^\dagger}(\omega)$ and $\chi_{a^\dagger a}(\omega)$ (equations (14A.4.21) and (14A.4.22)), it is consistent to modify formula (14A.3.12) by taking into account the fact that the angular frequency of the oscillator, once shifted by the coupling, equals $\omega_m \simeq \omega_0 + g^2\Delta(\omega_0)$ instead of ω_0, and to write accordingly $\chi_{xx}(\omega)$ in the following approximate form:

$$\chi_{xx}(\omega) \simeq \frac{\hbar}{2m\omega_m}\left[\chi_{aa^\dagger}(\omega) + \chi_{a^\dagger a}(\omega)\right]. \tag{14A.4.25}$$

On account of the second of properties (14A.4.24), we have:

$$\chi''_{xx}(\omega) \simeq \frac{\hbar}{2m\omega_m}\left[\chi''_{aa^\dagger}(\omega) - \chi''_{aa^\dagger}(-\omega)\right]. \tag{14A.4.26}$$

Using then the weak-coupling expression (14A.4.21) for $\chi''_{aa^\dagger}(\omega)$ gives:

$$\chi''_{xx}(\omega) \simeq \frac{1}{2m\omega_m}\left[\frac{\gamma_m}{(\omega - \omega_m)^2 + \gamma_m^2} - \frac{\gamma_m}{(\omega + \omega_m)^2 + \gamma_m^2}\right]. \tag{14A.4.27}$$

Similarly, we write $\chi'_{xx}(\omega)$ in the form:

$$\chi'_{xx}(\omega) \simeq \frac{\hbar}{2m\omega_m}\left[\chi'_{aa^\dagger}(\omega) + \chi'_{aa^\dagger}(-\omega)\right]. \tag{14A.4.28}$$

Using then the weak-coupling expression (14A.4.22) for $\chi'_{aa^\dagger}(\omega)$ gives:

$$\chi'_{xx}(\omega) \simeq \frac{1}{2m\omega_m}\left[-\frac{\omega-\omega_m}{(\omega-\omega_m)^2+\gamma_m^2}+\frac{\omega+\omega_m}{(\omega+\omega_m)^2+\gamma_m^2}\right]. \qquad (14A.4.29)$$

In the limit $(\gamma_m \to 0, \omega_m \to \omega_0)$ in which the coupling with the phonons vanishes, formulas (14A.4.27) and (14A.4.29) identify with the corresponding formulas for the undamped oscillator (equations (14A.2.11)).

We deduce from formulas (14A.4.27) and (14A.4.29) the generalized susceptibility $\chi_{xx}(\omega)$ of the oscillator of angular frequency $\omega_0 < \omega_D$, weakly coupled to a bath of phonons of bandwidth ω_D:

$$\boxed{\chi_{xx}(\omega) \simeq \frac{1}{2m\omega_m}\left(-\frac{1}{\omega-\omega_m+i\gamma_m}+\frac{1}{\omega+\omega_m+i\gamma_m}\right).} \qquad (14A.4.30)$$

The damping coefficient $\gamma_m = g^2\Gamma(\omega_0)$ is determined by the density of modes at the angular frequency ω_0. The angular frequency shift $\omega_m - \omega_0 = g^2\Delta(\omega_0)$ is determined by the whole density of modes, as displayed by the integral expression (14A.4.4) for $\Delta(\omega)$.

5. Dynamics of the weakly coupled oscillator

According to formula (14A.4.30), the generalized susceptibility $\chi_{xx}(\omega)$ of the oscillator weakly coupled to the phonons is formally similar to that of a classical damped oscillator with angular frequency ω_m and friction coefficient $2\gamma_m$. The displacement response function is, accordingly:

$$\boxed{\tilde{\chi}_{xx}(t) \simeq \Theta(t)\frac{\sin\omega_m t}{m\omega_m}e^{-\gamma_m t}.} \qquad (14A.5.1)$$

To obtain the associated autocorrelation function, we make use of the fluctuation-dissipation theorem, which allows us to deduce $S_{xx}(\omega)$ from $\chi''_{xx}(\omega)$. This gives:

$$\boxed{S_{xx}(\omega) = \frac{\hbar}{2m\omega_m}\coth\frac{\beta\hbar\omega}{2}\left[\frac{\gamma_m}{(\omega-\omega_m)^2+\gamma_m^2}-\frac{\gamma_m}{(\omega+\omega_m)^2+\gamma_m^2}\right].} \qquad (14A.5.2)$$

We can compute $\tilde{S}_{xx}(t)$ by inverse Fourier transformation. The function $\tilde{S}_{xx}(t)$ has in general no simple analytic expression.[7] However, in the classical limit $\beta\hbar\omega \ll 1$ in which $\tilde{S}_{xx}(t)$ is given by the formula:

$$\tilde{S}_{xx}(t) = \frac{2kT\gamma_m}{\pi m}\int_{-\infty}^{\infty}\frac{1}{\left[(\omega-\omega_m)^2+\gamma_m^2\right]\left[(\omega+\omega_m)^2+\gamma_m^2\right]}e^{-i\omega t}\,d\omega, \qquad (14A.5.3)$$

[7] The poles of the function $\coth\beta\hbar z/2$ indeed play a role in the computation of the Fourier integral giving $\tilde{S}_{xx}(t)$.

we get, at second order in coupling:

$$\boxed{\tilde{S}_{xx}(t) \simeq \frac{kT}{m\omega_m^2}\left(\cos\omega_m t + \frac{\gamma_m}{\omega_m}\sin\omega_m|t|\right)e^{-\gamma_m|t|}.} \tag{14A.5.4}$$

Formulas (14A.5.1) and (14A.5.4) show that, in the presence of a weak coupling with the phonon bath, the motion of the oscillator in classical regime consists of a damped oscillation whose angular frequency ω_m is shifted with respect to ω_0 by a value proportional to the squared coupling constant. The correlation time $\tau_c = \gamma_m^{-1}$ varies as the inverse of the squared coupling constant. It diverges as the latter vanishes, the oscillation then persisting indefinitely without decreasing.

Bibliography

P.M. CHAIKIN and T.C. LUBENSKY, *Principles of condensed matter physics*, Cambridge University Press, Cambridge, 1995.

C. COHEN-TANNOUDJI, J. DUPONT-ROC, and G. GRYNBERG, *Atom–photon interactions: basic processes and interactions*, Wiley, New York, 1992.

H. HAKEN, *Synergetics*, Springer-Verlag, Berlin, third edition, 1983.

W.H. LOUISELL, *Quantum statistical properties of radiation*, Wiley, New York, 1973.

S.W. LOVESEY, *Condensed matter physics: dynamic correlations*, The Benjamin/Cummings Publishing Company, Reading, second edition, 1986.

P.C. MARTIN, *Measurements and correlation functions*, Les Houches Lecture Notes 1967 (C. DE WITT and R. BALIAN editors), Gordon and Breach, New York, 1968.

H.L. PÉCSELI, *Fluctuations in physical systems*, Cambridge University Press, Cambridge, 2000.

U. WEISS, *Quantum dissipative systems*, World Scientific, Singapore, third edition, 2008.

References

G.W. FORD, M. KAC, and P. MAZUR, Statistical mechanics of assemblies of coupled oscillators, *J. Math. Phys.* **6**, 504 (1965).

A.O. CALDEIRA and A.J. LEGGETT, Quantum tunnelling in a dissipative system, *Ann. Phys.* **149**, 374 (1983).

A.J. LEGGETT, S. CHAKRAVARTY, A.T. DORSEY, M.P.A. FISHER, A. GARG, and W. ZWERGER, Dynamics of the dissipative two-state system, *Rev. Mod. Phys.* **59**, 1 (1987).

Chapter 15

Quantum theory of electronic transport

According to the semiclassical theories of electrical conduction, in a solid, be it either a metal or a semiconductor, the electrons of a given band obey, between collisions, the semiclassical equations of motion, this fact being counterbalanced by the scattering due to phonons and lattice defects. In this approach, the scattering cross-sections as well as the band structures are computed quantum-mechanically, whereas the balance equations only take into account average occupation probabilities. The scatterers situated at different places are assumed to act incoherently.

The first fully quantum approach to the theory of electronic transport is that of the Kubo's theory of linear response. It enables us to write a microscopic expression for the conductivity tensor involving correlation functions of the relevant components of the electric current (Kubo–Nakano formula). In the homogeneous case, we can deduce from this formula an expression for the real part of the conductivity of a non-interacting electron gas in terms of matrix elements of one-particle currents (Kubo–Greenwood formula). At lowest order in $(k_F\ell)^{-1}$ *(*k_F *being the Fermi wave vector and* ℓ *the elastic mean free path relative to electron–impurity collisions), we thus recover for the conductivity the result of the semiclassical calculation relying on the Boltzmann equation. The corrections with respect to this latter result are due to quantum interference effects coming into play at length scales much larger than* ℓ*.*

1. The Kubo–Nakano formula

Generally speaking, linear response theory enables us to express each coefficient of the linear phenomenological laws of transport in terms of an equilibrium correlation function of the appropriate currents. The formulas thus obtained are generically termed the *Green–Kubo formulas.*

We will here establish the Green–Kubo formula for the electrical conductivity tensor. This formula, which allows us to express the components of the conductivity tensor in terms of a correlation function of the corresponding components of the electric current, is also called more specifically the *Kubo–Nakano formula.* The system of charge carriers being described at the microscopic level by quantum mechanics, the Kubo–Nakano formula is at the basis of the quantum theory of electronic transport. We will first assume that the applied electric field is spatially uniform. We will then generalize the study to the non-uniform case.

1.1. Conductivity in uniform electric field

Consider a conducting material to which a spatially uniform electric field $\boldsymbol{E}(t)$ is applied. If the field is parallel to the direction β, the perturbation Hamiltonian may be written as:

$$H_1(t) = -e\sum_i r_{i,\beta}E_\beta(t). \tag{15.1.1}$$

In formula (15.1.1), the $\{\boldsymbol{r}_i\}$'s are the position operators of the different electrons of the considered sample.

The electric current density operator at point $\boldsymbol{r}$ is defined by:

$$\boldsymbol{J}(\boldsymbol{r}) = \frac{e}{2}\sum_i\big[\boldsymbol{v}_i\delta(\boldsymbol{r}-\boldsymbol{r}_i)+\delta(\boldsymbol{r}-\boldsymbol{r}_i)\boldsymbol{v}_i\big], \tag{15.1.2}$$

where $\boldsymbol{v_i} = \dot{\boldsymbol{r}}_i$ is the velocity of the electron i. Since the field is uniform, the current density is also uniform, and it can therefore be written as:

$$\boldsymbol{J} = \frac{1}{V}\int \boldsymbol{J}(\boldsymbol{r})\,d\boldsymbol{r} \tag{15.1.3}$$

(V denotes the volume of the sample). Using formula (15.1.2), this gives:

$$\boldsymbol{J} = \frac{e}{V}\sum_i \boldsymbol{v}_i. \tag{15.1.4}$$

In the linear response regime, the average value $\langle J_\alpha(t)\rangle$ of the component of $\boldsymbol{J}$ parallel to the direction α reads:

$$\langle J_\alpha(t)\rangle = \int_{-\infty}^{\infty}\tilde{\chi}_{BA}(t-t')E_\beta(t')\,dt', \tag{15.1.5}$$

with $A = e\sum_i r_{i,\beta}$ and $B = J_\alpha$. According to the general linear response theory, the response function $\tilde{\chi}_{BA}(t)$ is given by the formula:

$$\tilde{\chi}_{BA}(t) = \frac{i}{\hbar}\Theta(t)\big\langle[J_\alpha(t), e\sum_i r_{i,\beta}]\big\rangle. \qquad (15.1.6)$$

The unperturbed system is assumed to be in canonical equilibrium at temperature T. We can therefore write $\tilde{\chi}_{BA}(t)$ with the aid of the Kubo canonical correlation function $\tilde{K}_{B\dot{A}}(t) = V\tilde{K}_{J_\alpha J_\beta}(t)$:

$$\tilde{\chi}_{BA}(t) = \Theta(t)V\int_0^\beta \big\langle J_\beta(-i\hbar\lambda)J_\alpha(t)\big\rangle\,d\lambda, \qquad \beta = (kT)^{-1}. \qquad (15.1.7)$$

The components of the associated generalized susceptibility, namely, the electrical conductivity tensor are:

$$\boxed{\sigma_{\alpha\beta}(\omega) = V\lim_{\epsilon\to 0^+}\int_0^\infty dt\, e^{(i\omega-\epsilon)t}\int_0^\beta \big\langle J_\beta(-i\hbar\lambda)J_\alpha(t)\big\rangle\,d\lambda.} \qquad (15.1.8)$$

The conductivity is thus expressed in terms of an equilibrium current–current correlation function. Formula (15.1.8) is the Kubo–Nakano formula for the particular case of a uniform electric field.[1]

1.2. Generalization to the non-uniform case

The previous study can be extended to the case in which the perturbation depends not only on time but also on the point $\boldsymbol{r}$ of space. The conductivity tensor is then a function not only of the angular frequency but also of the wave vector: $\underline{\sigma} = \underline{\sigma}(\boldsymbol{q},\omega)$.

For an inhomogeneous imposed external potential $\phi(\boldsymbol{r},t)$, the perturbation is described by the Hamiltonian:

$$H_1(t) = \int \phi(\boldsymbol{r},t)\rho(\boldsymbol{r})\,d\boldsymbol{r}, \qquad (15.1.9)$$

where $\rho(\boldsymbol{r}) = e\sum_i \delta(\boldsymbol{r}-\boldsymbol{r}_i)$ is the charge density at point $\boldsymbol{r}$.

To make the field $E_\beta(\boldsymbol{r},t) = -\nabla_\beta\phi(\boldsymbol{r},t)$ explicitly appear in the expression for the perturbation, we introduce the derivative $\dot{H}_1$ defined by $i\hbar\dot{H}_1 = [H_1, H_0]$ (the derivation corresponds to the unperturbed evolution):

$$\dot{H}_1 = \int \phi(\boldsymbol{r},t)\dot{\rho}(\boldsymbol{r})\,d\boldsymbol{r}. \qquad (15.1.10)$$

[1] Using the fact that the response function $\tilde{\chi}_{BA}(t)$ is real, we can also write the components of the conductivity tensor in uniform electric field in the equivalent form:

$$\sigma_{\alpha\beta}(\omega) = V\lim_{\epsilon\to 0^+}\int_0^\infty dt\, e^{(i\omega-\epsilon)t}\int_0^\beta \langle J_\alpha J_\beta(-t+i\hbar\lambda)\rangle\,d\lambda,$$

where J_α stands for $J_\alpha(t=0)$.

Taking into account the continuity equation $\dot{\rho}(\boldsymbol{r}) = -\nabla.\boldsymbol{J}(\boldsymbol{r})$, we get:

$$\dot{H}_1 = -\int \phi(\boldsymbol{r},t)\, \nabla.\boldsymbol{J}(\boldsymbol{r})\, d\boldsymbol{r}, \tag{15.1.11}$$

that is:

$$\dot{H}_1 = -\int \nabla.\big[\phi(\boldsymbol{r},t)\boldsymbol{J}(\boldsymbol{r})\big]\, d\boldsymbol{r} + \int \boldsymbol{J}(\boldsymbol{r}).\nabla\phi(\boldsymbol{r},t)\, d\boldsymbol{r}. \tag{15.1.12}$$

The first term on the right-hand side of equation (15.1.12) can be transformed into a surface integral of the flux $\phi(\boldsymbol{r},t)\boldsymbol{J}(\boldsymbol{r})$, and then cancelled owing to a proper choice of boundary conditions. Formula (15.1.12) then reduces to:

$$\dot{H}_1 = -\int E_\beta(\boldsymbol{r},t) J_\beta(\boldsymbol{r})\, d\boldsymbol{r}. \tag{15.1.13}$$

Formula (15.1.13) is of the general form:

$$\dot{H}_1 = -\int a(\boldsymbol{r},t)\dot{A}(\boldsymbol{r})\, d\boldsymbol{r}, \tag{15.1.14}$$

with $a(\boldsymbol{r},t) = E_\beta(\boldsymbol{r},t)$ and $\dot{A}(\boldsymbol{r}) = J_\beta(\boldsymbol{r})$.

In the linear response regime, the average value $\langle J_\alpha(\boldsymbol{r},t)\rangle$ reads:

$$\big\langle J_\alpha(\boldsymbol{r},t)\big\rangle = \int d\boldsymbol{r}' \int_{-\infty}^{\infty} \tilde{\chi}_{BA}(\boldsymbol{r}-\boldsymbol{r}',t-t') E_\beta(\boldsymbol{r}',t')\, dt', \tag{15.1.15}$$

with $B(\boldsymbol{r}) = J_\alpha(\boldsymbol{r})$. For an unperturbed system in canonical equilibrium at temperature T, the response function $\tilde{\chi}_{BA}(\boldsymbol{r}-\boldsymbol{r}',t-t')$ is given by the formula:

$$\tilde{\chi}_{BA}(\boldsymbol{r}-\boldsymbol{r}',t-t') = \Theta(t-t') \int_0^\beta \big\langle J_\beta(\boldsymbol{r}',-i\hbar\lambda) J_\alpha(\boldsymbol{r},t-t')\big\rangle\, d\lambda. \tag{15.1.16}$$

To determine the conductivity tensor, we introduce the spatial and temporal Fourier transforms $E_\beta(\boldsymbol{q},\omega)$ and $\langle J_\alpha(\boldsymbol{q},\omega)\rangle$ of $E_\beta(\boldsymbol{r},t)$ and $\langle J_\alpha(\boldsymbol{r},t)\rangle$, defined respectively by the formulas:

$$E_\beta(\boldsymbol{q},\omega) = \int d\boldsymbol{r} \int E_\beta(\boldsymbol{r},t) e^{i(\omega t-\boldsymbol{q}.\boldsymbol{r})}\, dt \tag{15.1.17}$$

and:

$$\big\langle J_\alpha(\boldsymbol{q},\omega)\big\rangle = \int d\boldsymbol{r} \int \big\langle J_\alpha(\boldsymbol{r},t)\big\rangle e^{i(\omega t-\boldsymbol{q}.\boldsymbol{r})}\, dt. \tag{15.1.18}$$

By Fourier transforming equation (15.1.15), we obtain a relation of the form:

$$\big\langle J_\alpha(\boldsymbol{q},\omega)\big\rangle = \sigma_{\alpha\beta}(\boldsymbol{q},\omega) E_\beta(\boldsymbol{q},\omega), \tag{15.1.19}$$

with, taking account of formula (15.1.16):

$$\sigma_{\alpha\beta}(\boldsymbol{q},\omega)=\int d(\boldsymbol{r}-\boldsymbol{r}')\,e^{-i\boldsymbol{q}.(\boldsymbol{r}-\boldsymbol{r}')}$$

$$\times\lim_{\epsilon\to0^+}\int_0^\infty d(t-t')\,e^{(i\omega-\epsilon)(t-t')}\int_0^\beta\langle J_\beta(\boldsymbol{r}',-i\hbar\lambda)J_\alpha(\boldsymbol{r},t-t')\rangle\,d\lambda. \tag{15.1.20}$$

Introducing the supplementary integration $\frac{1}{V}\int d\boldsymbol{r}=1$, and carrying out the change of variables $(\boldsymbol{r},\boldsymbol{r}-\boldsymbol{r}')\to(\boldsymbol{r},\boldsymbol{r}')$ in the double integral over space variables thus obtained, we can recast formula (15.1.20) in a form involving a correlation function of the spatial Fourier transforms of the current densities:[2]

$$\boxed{\sigma_{\alpha\beta}(\boldsymbol{q},\omega)=\frac{1}{V}\lim_{\epsilon\to0^+}\int_0^\infty dt\,e^{(i\omega-\epsilon)t}\int_0^\beta\langle J_\beta(-\boldsymbol{q},-i\hbar\lambda)J_\alpha(\boldsymbol{q},t)\rangle\,d\lambda,} \tag{15.1.21}$$

Formula (15.1.21) is the general Kubo–Nakano formula.

2. The Kubo–Greenwood formula

In some cases, we can write for the conductivity a more explicit formula. In particular, it is possible to express the real part of the conductivity at vanishing wave vector of a non-interacting electron gas in terms of matrix elements of one-particle currents. This expression for $\Re e\,\sigma_{\alpha\beta}(\boldsymbol{q}=0,\omega)$ constitutes the *Kubo–Greenwood formula*, which we will here derive from the Kubo–Nakano formula.[3]

If the electrons are free, the one-particle eigenstates are plane waves $|\boldsymbol{k}_i\rangle$ ($\langle\boldsymbol{r}|\boldsymbol{k}_i\rangle=V^{-1/2}e^{i\boldsymbol{k}_i.\boldsymbol{r}}$). More generally, especially in the presence of impurities, the eigenstates of wave vector $\boldsymbol{k}_i$, denoted then by $|\phi_{\boldsymbol{k}_i}\rangle$, may be other functions $\phi_{\boldsymbol{k}_i}(\boldsymbol{r})$. The set of electrons is described in an occupation number representation $\{n_i\}$, in which n_i is the occupation number[4] of the state of wave vector $\boldsymbol{k}_i$. For a non-interacting electron gas, of Hamiltonian $H_0=\sum_{\boldsymbol{k}_i}\varepsilon_{\boldsymbol{k}_i}a^\dagger_{\boldsymbol{k}_i}a_{\boldsymbol{k}_i}$, the annihilation and creation operators in the state of wave vector $\boldsymbol{k}_i$ at time t are given by:

$$a_{\boldsymbol{k}_i}(t)=a_{\boldsymbol{k}_i}e^{-i\varepsilon_{\boldsymbol{k}_i}t/\hbar},\qquad a^\dagger_{\boldsymbol{k}_i}(t)=a^\dagger_{\boldsymbol{k}_i}e^{i\varepsilon_{\boldsymbol{k}_i}t/\hbar}. \tag{15.2.1}$$

---

[2] We can write equivalently the components of the conductivity tensor as:

$$\sigma_{\alpha\beta}(\boldsymbol{q},\omega)=\frac{1}{V}\lim_{\epsilon\to0^+}\int_0^\infty dt\,e^{(i\omega-\epsilon)t}\int_0^\beta\langle J_\alpha(\boldsymbol{q})J_\beta(-\boldsymbol{q},-t+i\hbar\lambda)\rangle\,d\lambda,$$

where $J_\alpha(\boldsymbol{q})$ stands for $J_\alpha(\boldsymbol{q},t=0)$.

[3] An alternative derivation of the Kubo–Greenwood formula, relying on the relation between the real part of the conductivity at vanishing wave vector and the electromagnetic absorption properties, is presented in Supplement 15A.

[4] Since electrons are fermions, n_i may be equal either to 0 or to 1.

If the one-particle eigenstates are plane waves, the field operators (annihilation and creation of particles at point $\boldsymbol{r}$) are defined by the formulas:

$$\begin{cases} \psi(\boldsymbol{r},t) = V^{-1/2}\sum_i a_{\boldsymbol{k}_i}(t)e^{i\boldsymbol{k}_i.\boldsymbol{r}} \\ \psi^\dagger(\boldsymbol{r},t) = V^{-1/2}\sum_i a^\dagger_{\boldsymbol{k}_i}(t)e^{-i\boldsymbol{k}_i.\boldsymbol{r}}. \end{cases} \tag{15.2.2}$$

More generally, if the one-particle eigenstates are functions $\phi_{\boldsymbol{k}_i}(\boldsymbol{r})$, we define the field operators as follows:

$$\begin{cases} \psi(\boldsymbol{r},t) = \sum_i a_{\boldsymbol{k}_i}(t)\phi_{\boldsymbol{k}_i}(\boldsymbol{r}) \\ \psi^\dagger(\boldsymbol{r},t) = \sum_i a^\dagger_{\boldsymbol{k}_i}(t)\phi^*_{\boldsymbol{k}_i}(\boldsymbol{r}). \end{cases} \tag{15.2.3}$$

2.1. Expression for the currents $J_\alpha(\boldsymbol{q}=\boldsymbol{0},t)$ and $J_\beta(\boldsymbol{q}=\boldsymbol{0},-i\hbar\boldsymbol{\lambda})$ with the aid of the one-particle eigenstates

To express the current density with the aid of the field operators, we start from the continuity equation $\dot\rho(\boldsymbol{r}) = -\nabla.\boldsymbol{J}(\boldsymbol{r})$. The charge density being expressed, in terms of the field operators, as $\rho(\boldsymbol{r},t) = e\psi^\dagger(\boldsymbol{r},t)\psi(\boldsymbol{r},t)$, we can show that $J_\alpha(\boldsymbol{r},t)$ reads:

$$J_\alpha(\boldsymbol{r},t) = \frac{e\hbar}{2mi}\left[\psi^\dagger(\boldsymbol{r},t)\frac{\partial\psi(\boldsymbol{r},t)}{\partial x_\alpha} - \frac{\partial\psi^\dagger(\boldsymbol{r},t)}{\partial x_\alpha}\psi(\boldsymbol{r},t)\right]. \tag{15.2.4}$$

The spatial Fourier transform of $J_\alpha(\boldsymbol{r},t)$ is:[5]

$$J_\alpha(\boldsymbol{q},t) = \int d\boldsymbol{r}\, e^{-i\boldsymbol{q}.\boldsymbol{r}}\frac{e\hbar}{2mi}\sum_{\boldsymbol{k}_1,\boldsymbol{k}_2}\left[\phi^*_{\boldsymbol{k}_1}(\boldsymbol{r})\frac{\partial\phi_{\boldsymbol{k}_2}(\boldsymbol{r})}{\partial x_\alpha}a^\dagger_{\boldsymbol{k}_1}a_{\boldsymbol{k}_2} - \frac{\partial\phi^*_{\boldsymbol{k}_1}(\boldsymbol{r})}{\partial x_\alpha}\phi_{\boldsymbol{k}_2}(\boldsymbol{r})a^\dagger_{\boldsymbol{k}_1}a_{\boldsymbol{k}_2}\right]. \tag{15.2.5}$$

At vanishing wave vector, we get:

$$J_\alpha(\boldsymbol{q}=0,t) = \int d\boldsymbol{r}\,\frac{e\hbar}{2mi}\sum_{\boldsymbol{k}_1,\boldsymbol{k}_2}\left[\phi^*_{\boldsymbol{k}_1}(\boldsymbol{r})\frac{\partial\phi_{\boldsymbol{k}_2}(\boldsymbol{r})}{\partial x_\alpha}a^\dagger_{\boldsymbol{k}_1}a_{\boldsymbol{k}_2} - \frac{\partial\phi^*_{\boldsymbol{k}_1}(\boldsymbol{r})}{\partial x_\alpha}\phi_{\boldsymbol{k}_2}(\boldsymbol{r})a^\dagger_{\boldsymbol{k}_1}a_{\boldsymbol{k}_2}\right]. \tag{15.2.6}$$

The second term on the right-hand side of formula (15.2.6) can be integrated by parts: the integrated term does not contribute, whereas the other contribution of the integration by parts of this second term is seen to be equal to the first one. This gives[6]:

$$J_\alpha(\boldsymbol{q}=0,t) = \frac{e}{m}\sum_{\boldsymbol{k}_1,\boldsymbol{k}_2}\langle\boldsymbol{k}_1|p_\alpha|\boldsymbol{k}_2\rangle a^\dagger_{\boldsymbol{k}_1}(t)a_{\boldsymbol{k}_2}(t). \tag{15.2.7}$$

[5] In formulas (15.2.5) and (15.2.6), the operators $a^\dagger_{\boldsymbol{k}_1}$ and $a_{\boldsymbol{k}_2}$ are taken at time t (for the sake of simplicity, this time-dependence is not explicitly displayed).

[6] The eigenstates $|\phi_{\boldsymbol{k}_i}\rangle$ are denoted for short by $|\boldsymbol{k}_i\rangle$ in the remainder of this section even if they are not plane waves.

In equation (15.2.7), $p_\alpha = (\hbar/i)\partial/\partial x_\alpha$ denotes the component parallel to the direction α of the one-particle momentum. Using formulas (15.2.1), we obtain:

$$\left\{\begin{aligned} J_\alpha(\boldsymbol{q}=0,t) &= \frac{e}{m}\sum_{\boldsymbol{k}_1,\boldsymbol{k}_2}\langle\boldsymbol{k}_1|p_\alpha|\boldsymbol{k}_2\rangle e^{i(\varepsilon_{\boldsymbol{k}_1}-\varepsilon_{\boldsymbol{k}_2})t/\hbar}a^\dagger_{\boldsymbol{k}_1}a_{\boldsymbol{k}_2} \\ J_\beta(\boldsymbol{q}=0,-i\hbar\lambda) &= \frac{e}{m}\sum_{\boldsymbol{k}_3,\boldsymbol{k}_4}\langle\boldsymbol{k}_3|p_\beta|\boldsymbol{k}_4\rangle e^{(\varepsilon_{\boldsymbol{k}_3}-\varepsilon_{\boldsymbol{k}_4})\lambda}a^\dagger_{\boldsymbol{k}_3}a_{\boldsymbol{k}_4}.\end{aligned}\right. \tag{15.2.8}$$

2.2. Real part of the conductivity at vanishing wave vector

Importing the expressions (15.2.8) for $J_\alpha(\boldsymbol{q}=0,t)$ and $J_\beta(\boldsymbol{q}=0,-i\hbar\lambda)$ into the Kubo–Nakano formula (15.1.21) written for $\boldsymbol{q}=0$, gives:

$$\begin{aligned}\sigma_{\alpha\beta}(\boldsymbol{q}=0,\omega) = \frac{1}{V}\lim_{\epsilon\to0^+}\int_0^\infty dt\, e^{(i\omega-\epsilon)t}\sum_{\boldsymbol{k}_1,\boldsymbol{k}_2,\boldsymbol{k}_3,\boldsymbol{k}_4}\frac{e^2}{m^2}\langle\boldsymbol{k}_1|p_\alpha|\boldsymbol{k}_2\rangle\langle\boldsymbol{k}_3|p_\beta|\boldsymbol{k}_4\rangle \\ \times\, e^{i(\varepsilon_{\boldsymbol{k}_1}-\varepsilon_{\boldsymbol{k}_2})t/\hbar}\langle a^\dagger_{\boldsymbol{k}_3}a_{\boldsymbol{k}_4}a^\dagger_{\boldsymbol{k}_1}a_{\boldsymbol{k}_2}\rangle\int_0^\beta e^{(\varepsilon_{\boldsymbol{k}_3}-\varepsilon_{\boldsymbol{k}_4})\lambda}\,d\lambda.\end{aligned} \tag{15.2.9}$$

To make explicit $\sigma_{\alpha\beta}(\boldsymbol{q}=0,\omega)$, it remains to calculate the equilibrium average $\langle a^\dagger_{\boldsymbol{k}_3}a_{\boldsymbol{k}_4}a^\dagger_{\boldsymbol{k}_1}a_{\boldsymbol{k}_2}\rangle$. According to the general rules about equilibrium averages of products of creation and annihilation operators, the only two possibilities to get a non-vanishing result are $\boldsymbol{k}_1=\boldsymbol{k}_2, \boldsymbol{k}_3=\boldsymbol{k}_4$, and $\boldsymbol{k}_1=\boldsymbol{k}_4$, $\boldsymbol{k}_2=\boldsymbol{k}_3$. The contribution to $\sigma_{\alpha\beta}(\boldsymbol{q}=0,\omega)$ corresponding to $\boldsymbol{k}_1=\boldsymbol{k}_2, \boldsymbol{k}_3=\boldsymbol{k}_4$ involves the sum:

$$\sum_{\boldsymbol{k}_1,\boldsymbol{k}_3}\langle\boldsymbol{k}_1|p_\alpha|\boldsymbol{k}_1\rangle\langle\boldsymbol{k}_3|p_\beta|\boldsymbol{k}_3\rangle n_{\boldsymbol{k}_1}n_{\boldsymbol{k}_3}, \tag{15.2.10}$$

where $n_{\boldsymbol{k}_i}=\langle a^\dagger_{\boldsymbol{k}_i}a_{\boldsymbol{k}_i}\rangle$ is the average number of electrons in state $|\boldsymbol{k}_i\rangle$ at thermodynamic equilibrium. Since there is no current at equilibrium, we have:

$$\sum_{\boldsymbol{k}}\langle\boldsymbol{k}|p|\boldsymbol{k}\rangle\, n_{\boldsymbol{k}}=0. \tag{15.2.11}$$

The contribution $\boldsymbol{k}_1=\boldsymbol{k}_2, \boldsymbol{k}_3=\boldsymbol{k}_4$ to $\sigma_{\alpha\beta}(\boldsymbol{q}=0,\omega)$ thus vanishes, and we have only to take into account the contribution $\boldsymbol{k}_1=\boldsymbol{k}_4$, $\boldsymbol{k}_2=\boldsymbol{k}_3$. The component $\sigma_{\alpha\beta}(\boldsymbol{q}=0,\omega)$ of the conductivity tensor reads:

$$\begin{aligned}\sigma_{\alpha\beta}(\boldsymbol{q}=0,\omega) = \frac{1}{V}\lim_{\epsilon\to0^+}\int_0^\infty dt\, e^{(i\omega-\epsilon)t}\sum_{\boldsymbol{k}_1,\boldsymbol{k}_2}\frac{e^2}{m^2}\langle\boldsymbol{k}_1|p_\alpha|\boldsymbol{k}_2\rangle\langle\boldsymbol{k}_2|p_\beta|\boldsymbol{k}_1\rangle \\ \times\,(1-n_{\boldsymbol{k}_1})n_{\boldsymbol{k}_2}e^{i(\varepsilon_{\boldsymbol{k}_1}-\varepsilon_{\boldsymbol{k}_2})t/\hbar}\int_0^\beta e^{(\varepsilon_{\boldsymbol{k}_2}-\varepsilon_{\boldsymbol{k}_1})\lambda}\,d\lambda.\end{aligned} \tag{15.2.12}$$

The integration over t yields:

$$\lim_{\epsilon\to 0^+}\int_0^\infty dt\, e^{(i\omega-\epsilon)t} e^{i(\varepsilon_{\boldsymbol{k}_1}-\varepsilon_{\boldsymbol{k}_2})t/\hbar} = \mathrm{vp}\,\frac{i}{\omega-\dfrac{\varepsilon_{\boldsymbol{k}_2}-\varepsilon_{\boldsymbol{k}_1}}{\hbar}} + \pi\delta\Big(\omega-\frac{\varepsilon_{\boldsymbol{k}_2}-\varepsilon_{\boldsymbol{k}_1}}{\hbar}\Big). \quad (15.2.13)$$

In addition, we have:

$$\int_0^\beta e^{(\varepsilon_{\boldsymbol{k}_2}-\varepsilon_{\boldsymbol{k}_1})\lambda}\, d\lambda = \frac{e^{-\beta(\varepsilon_{\boldsymbol{k}_1}-\varepsilon_{\boldsymbol{k}_2})}-1}{\varepsilon_{\boldsymbol{k}_2}-\varepsilon_{\boldsymbol{k}_1}}, \quad (15.2.14)$$

and we can verify the formula:

$$(1-n_{\boldsymbol{k}_1})n_{\boldsymbol{k}_2}\big[e^{-\beta(\varepsilon_{\boldsymbol{k}_1}-\varepsilon_{\boldsymbol{k}_2})}-1\big] = n_{\boldsymbol{k}_1}-n_{\boldsymbol{k}_2}. \quad (15.2.15)$$

Taking into account equations (15.2.13), (15.2.14), and (15.2.15), we finally obtain the Kubo–Greenwood formula for the real part of the conductivity tensor at vanishing wave vector:[7]

$$\Re e\,\sigma_{\alpha\beta}(\omega) = \frac{\pi e^2}{m^2\omega V}\sum_{\boldsymbol{k}_1,\boldsymbol{k}_2}\langle\boldsymbol{k}_1|p_\alpha|\boldsymbol{k}_2\rangle\langle\boldsymbol{k}_2|p_\beta|\boldsymbol{k}_1\rangle(n_{\boldsymbol{k}_1}-n_{\boldsymbol{k}_2})\delta[\hbar\omega-(\varepsilon_{\boldsymbol{k}_2}-\varepsilon_{\boldsymbol{k}_1})]. \quad (15.2.16)$$

Formula (15.2.16) allows us to relate the real part of the conductivity at vanishing wave vector of a non-interacting electron gas to transitions between one-particle stationary states.

2.3. Relation with the electromagnetic absorption properties

The real part of the conductivity at vanishing wave vector corresponds to the dissipative part of this susceptibility, a property which can be proved by coming back to the general Kubo–Nakano formula (15.1.21).

To begin with, let us note that this latter formula was derived from the general formulas of linear response theory with $\dot{A}(\boldsymbol{r}) = J(\boldsymbol{r})$ and $B(\boldsymbol{r}) = J(\boldsymbol{r})$. Since the spatial Fourier transform at vanishing wave vector of $\boldsymbol{J}(\boldsymbol{r})$ is $\boldsymbol{J}(\boldsymbol{q}=0) = V\boldsymbol{J}$, with $\boldsymbol{J}$ the current density operator of the homogeneous system, we thus recover as it should for the conductivity at vanishing wave vector, the Kubo–Nakano formula for the homogeneous case (formula (15.1.8)).

In other words, the conductivity at vanishing wave vector directly follows from the general linear response theory as a generalized susceptibility $\chi_{BA}(\omega)$ with $\dot{A} = \boldsymbol{J}$ and $B = \boldsymbol{J}$. The signatures under time-reversal of the A and B operators coming into play are respectively $\epsilon_A = +1$ and $\epsilon_B = -1$. The associated Onsager reciprocity relation reads:

$$\chi_{BA}(\omega) = -\chi_{AB}(\omega). \quad (15.2.17)$$

[7] From now on, the conductivity tensor at vanishing wave vector will be simply designated for short by $\underline{\sigma}(\omega)$.

Therefore, for the dissipative part of this susceptibility, that is, for the spectral function $\xi_{BA}(\omega)$, we get:

$$2i\xi_{BA}(\omega) = \chi_{BA}(\omega) - \chi^*_{AB}(\omega) = \chi_{BA}(\omega) + \chi^*_{BA}(\omega) = 2\,\Re e\,\chi_{BA}(\omega). \qquad (15.2.18)$$

Formula (15.2.18) displays the fact that, as stated above, the real part of the conductivity tensor corresponds to the dissipative part of this susceptibility.

In a conductor, the real part of the conductivity tensor and the imaginary part of the dielectric permittivity tensor are related to one another by the formula:[8]

$$\Im m\,\underline{\varepsilon}(\omega) = \frac{4\pi}{\omega}\,\Re e\,\underline{\sigma}(\omega). \qquad (15.2.19)$$

In the framework of an independent electron model, the Kubo–Greenwood formula for $\Re e\,\underline{\sigma}(\omega)$ thus also gives access to the electromagnetic absorption properties as characterized by $\Im m\,\underline{\varepsilon}(\omega)$.

3. Conductivity of an electron gas in the presence of impurities

Let us now come back to the conductivity of a degenerate electron gas in the presence of impurities, a quantity previously obtained in the framework of a semiclassical calculation relying on the linearized Boltzmann equation.

3.1. Relaxation time

If the potential $V_i(\boldsymbol{r})$ created by the impurities remains weak enough, the electronic states in the presence of the impurities may be obtained from the electronic states in their absence via a perturbation calculation. The eigenvalue equation reads, in the presence of the impurities,

$$(H_0 + V_i)|\phi_{\boldsymbol{k}_i}\rangle = E_{\boldsymbol{k}_i}|\phi_{\boldsymbol{k}_i}\rangle, \qquad (15.3.1)$$

whereas in the absence of the impurities, it simply reads:

$$H_0|\boldsymbol{k}_i\rangle = \varepsilon_{\boldsymbol{k}_i}|\boldsymbol{k}_i\rangle. \qquad (15.3.2)$$

At first perturbation order, we have:

$$|\phi_{\boldsymbol{k}_i}\rangle = |\boldsymbol{k}_i\rangle + \sum_{\boldsymbol{k}_j \neq \boldsymbol{k}_i} \frac{\langle \boldsymbol{k}_j|V_i|\boldsymbol{k}_i\rangle}{\varepsilon_{\boldsymbol{k}_i} - \varepsilon_{\boldsymbol{k}_j}}|\boldsymbol{k}_j\rangle. \qquad (15.3.3)$$

Thus, for $\boldsymbol{k}_1 \neq \boldsymbol{k}_2$, the matrix element of p_x between states $|\phi_{\boldsymbol{k}_1}\rangle$ and $|\phi_{\boldsymbol{k}_2}\rangle$ reads:

$$\langle \phi_{\boldsymbol{k}_1}|p_x|\phi_{\boldsymbol{k}_2}\rangle = \hbar(k_{1x} - k_{2x})\frac{\langle \boldsymbol{k}_1|V_i|\boldsymbol{k}_2\rangle}{\varepsilon_{\boldsymbol{k}_2} - \varepsilon_{\boldsymbol{k}_1}}. \qquad (15.3.4)$$

---

[8] See Supplement 15A.

The matrix elements which have to be taken into account in the Kubo–Greenwood formula (15.2.16) are these matrix elements between perturbed states. The system, disordered, can be considered as isotropic: $\sigma_{\alpha\beta} = \sigma\delta_{\alpha\beta}$. We obtain:

$$\Re e\,\sigma(\omega) = \frac{\pi e^2}{m^2\omega V}\sum_{\boldsymbol{k}_1,\boldsymbol{k}_2}\hbar^2\frac{1}{3}|\boldsymbol{k}_1-\boldsymbol{k}_2|^2\frac{|\langle\boldsymbol{k}_1|V_i|\boldsymbol{k}_2\rangle|^2}{(\hbar\omega)^2}(n_{\boldsymbol{k}_1}-n_{\boldsymbol{k}_2})\delta\big[\hbar\omega-(\epsilon_{\boldsymbol{k}_2}-\epsilon_{\boldsymbol{k}_1})\big]. \tag{15.3.5}$$

This perturbative calculation is only valid for $V_i \ll \hbar\omega$, where V_i represents a typical value of the modulus of the matrix elements $\langle\boldsymbol{k}_1|V_i|\boldsymbol{k}_2\rangle$. Therefore, it cannot be applied at vanishing angular frequency.

In a metal, both $|\boldsymbol{k}_1|$ and $|\boldsymbol{k}_2|$ are close to the modulus k_F of the Fermi wave vector. We therefore have, θ denoting the angle between $\boldsymbol{k}_1$ and $\boldsymbol{k}_2$:

$$|\boldsymbol{k}_1-\boldsymbol{k}_2|^2 \simeq 2k_F^2(1-\cos\theta). \tag{15.3.6}$$

Also, we have:

$$n_{\boldsymbol{k}_1}-n_{\boldsymbol{k}_2} = f_0(\varepsilon_{\boldsymbol{k}_1}) - f_0(\varepsilon_{\boldsymbol{k}_2}), \tag{15.3.7}$$

where f_0 denotes the Fermi–Dirac function. Since, in the corresponding term of the Kubo–Greenwood formula, $\varepsilon_{\boldsymbol{k}_2}$ and $\varepsilon_{\boldsymbol{k}_1}$ are related by $\varepsilon_{\boldsymbol{k}_2}-\varepsilon_{\boldsymbol{k}_1}=\hbar\omega$, we can write, for $\hbar\omega \ll \varepsilon_F$, where ε_F is the Fermi energy,

$$f_0(\varepsilon_{\boldsymbol{k}_1}) - f_0(\varepsilon_{\boldsymbol{k}_2}) \simeq -\hbar\omega\frac{\partial f_0}{\partial\varepsilon_{\boldsymbol{k}_1}}, \tag{15.3.8}$$

that is, the electron gas being degenerate:

$$f_0(\varepsilon_{\boldsymbol{k}_1}) - f_0(\varepsilon_{\boldsymbol{k}_2}) \simeq \hbar\omega\,\delta(\varepsilon_{\boldsymbol{k}_1}-\varepsilon_F). \tag{15.3.9}$$

We finally obtain, in an intermediate angular frequencies range characterized by the double inequality $V_i \ll \hbar\omega \ll \varepsilon_F$:

$$\Re e\,\sigma(\omega) = \frac{e^2}{Vm^2\omega^2}\sum_{\boldsymbol{k}_1}\frac{\hbar^2k_1^2}{3}\frac{1}{\tau(\boldsymbol{k}_1)}\delta(\varepsilon_{\boldsymbol{k}_1}-\varepsilon_F). \tag{15.3.10}$$

In formula (15.3.10), the inverse relaxation time $[\tau(\boldsymbol{k}_1)]^{-1}$ is defined by the usual formula for electron–impurity collisions:

$$\frac{1}{\tau(\boldsymbol{k}_1)} = \frac{2\pi}{\hbar}\sum_{\boldsymbol{k}_2}|\langle\boldsymbol{k}_1|V_i|\boldsymbol{k}_2\rangle|^2(1-\cos\theta)\delta(\varepsilon_{\boldsymbol{k}_2}-\varepsilon_{\boldsymbol{k}_1}). \tag{15.3.11}$$

3.2. Conductivity

The real part of the conductivity may be easily computed due to the presence of the delta function in the sum on the right-hand side of equation (15.3.10). Assuming that the relaxation time is a function of the electron energy, this latter equation can be rewritten as:

$$\Re e\, \sigma(\omega) = \frac{e^2}{Vm^2\omega^2} \frac{1}{\tau(\varepsilon_F)} \sum_{\boldsymbol{k}_1} \frac{\hbar^2 k_1^2}{3} \delta\Big[\frac{\hbar^2}{2m}(k_1^2 - k_F^2)\Big], \tag{15.3.12}$$

where $\tau(\varepsilon_F)$ denotes the relaxation time at the Fermi level.

We have:

$$\frac{1}{V} \sum_{\boldsymbol{k}_1} \frac{\hbar^2 k_1^2}{3} \delta\Big[\frac{\hbar^2}{2m}(k_1^2 - k_F^2)\Big] = mn, \tag{15.3.13}$$

where $n = k_F^3/3\pi^2$ is the density of the electron gas. This gives:

$$\boxed{\Re e\, \sigma(\omega) = \frac{ne^2}{m\omega^2 \tau(\varepsilon_F)}.} \tag{15.3.14}$$

Note that, if we rewrite formula (15.3.14) in the form:

$$\Re e\, \sigma(\omega) = \frac{ne^2 \tau(\varepsilon_F)}{m\omega^2 \tau^2(\varepsilon_F)}, \tag{15.3.15}$$

it appears as the limit for $\omega\tau(\varepsilon_F) \gg 1$ of a generalized Drude–Lorentz formula[9] written with the relaxation time of the average velocity equal to $\tau(\varepsilon_F)$:

$$\Re e\, \sigma(\omega) = \frac{ne^2 \tau(\varepsilon_F)}{m} \frac{1}{1 + \omega^2 \tau^2(\varepsilon_F)}. \tag{15.3.16}$$

3.3. Beyond Drude's result: Ioffe–Regel criterion and quantum transport

The calculation carried out here, valid in an intermediate range of angular frequencies characterized by the double inequality $\tau^{-1}(\varepsilon_F) \ll \omega \ll \varepsilon_F/\hbar$, is compatible with the classical Drude's result taken in the limit[10] $\omega\tau(\varepsilon_F) \gg 1$. It is only when the condition

---

[9] The *generalized Drude–Lorentz formula* is obtained by solving, in harmonic regime of angular frequency ω, the evolution equation of the average electron velocity in uniform electric field:

$$m\frac{d\langle v\rangle}{dt} + m\frac{\langle v\rangle}{\tau} = eE(t).$$

For $E(t)$ and $\langle v(t)\rangle$ varying as $e^{-i\omega t}$, we get:

$$\sigma(\omega) = \frac{ne^2\tau}{m} \frac{1}{1 - i\omega\tau}.$$

[10] An alternative derivation of $\Re e\, \sigma(\omega)$ from the Kubo–Greenwood formula, in which the condition $\omega\tau(\varepsilon_F) \gg 1$ is not assumed to be fulfilled, is presented in Supplement 15A.

$\varepsilon_F \tau(\varepsilon_F) \gg \hbar$ is not fulfilled that specific quantum effects are likely to appear. This latter condition also reads, more simply, in the form of the Ioffe–Regel criterion,

$$k_F \ell \gg 1, \tag{15.3.17}$$

where $\ell \sim \hbar k_F \tau(\varepsilon_F)/m$ is the elastic mean free path of the electrons.

When disorder increases in such a way that the Ioffe–Regel criterion is violated, corrections to Drude's theory must be accounted for. These corrections, which can be shown to involve terms in inverse powers of $k_F \ell$, are due to quantum interference effects involving length scales much larger than ℓ. The elastic scattering by impurities does not destroy the phase coherence of the scattered waves. This coherence disappears over a *coherence length* $\ell_\phi(T) \gg \ell$, which generally depends on dynamical mechanisms (impurity motion, phonon scattering ...).

In mesoscopic systems of size $L \leq \ell_\phi(T)$, quantum interference effects, likely to lead to *electronic localization* (that is, to the absence of conduction) may modify the transport properties.[11] For their study, we have to devise a convenient method of calculation of the electrical conductivity. Such a method was first proposed by R. Landauer in 1957.[12]

[11] The coherence length $\ell_\phi(T)$ is a decreasing function of temperature. Any system at sufficiently low temperature is thus mesoscopic.

[12] See Supplement 15A.

Bibliography

N.W. ASHCROFT and N.D. MERMIN, *Solid state physics*, Holt-Saunders, Philadelphia, 1976.

A.L. FETTER and J.D. WALECKA, *Quantum theory of many-particle systems*, McGraw-Hill, New York, 1971. Reprinted, Dover Publications, New York, 2003.

D. FORSTER, *Hydrodynamic fluctuations, broken symmetries, and correlation functions*, Westview Press, Boulder, 1995.

Y. IMRY, *Introduction to mesoscopic physics*, Oxford University Press, Oxford, second edition, 2002.

W. JONES and N.H. MARCH, *Theoretical solid-state physics: non-equilibrium and disorder*, Vol. 2, Wiley, New York, 1973. Reprinted, Dover Publications, New York, 1985.

C. KITTEL, *Quantum theory of solids*, Wiley, New York, second edition, 1967.

R. KUBO, M. TODA, and N. HASHITSUME, *Statistical physics II: nonequilibrium statistical mechanics*, Springer-Verlag, Berlin, second edition, 1991.

M. PLISCHKE and B. BERGERSEN, *Equilibrium statistical physics*, World Scientific, Singapore, third edition, 2006.

H. SMITH and H.H. JENSEN, *Transport phenomena*, Oxford Science Publications, Oxford, 1989.

R. ZWANZIG, *Nonequilibrium statistical mechanics*, Oxford University Press, Oxford, 2001.

References

M.S. GREEN, Markoff random processes and the statistical mechanics of time-dependent phenomena, *J. Chem. Phys.* **20**, 1281 (1952).

M.S. GREEN, Markoff random processes and the statistical mechanics of time-dependent phenomena. II. Irreversible processes in fluids, *J. Chem. Phys.* **22**, 398 (1954).

R. KUBO, Statistical-mechanical theory of irreversible processes. I. General theory and simple applications to magnetic and conduction problems, *J. Phys. Soc. Japan* **12**, 570 (1957).

H. NAKANO, A method of calculation of electrical conductivity, *Prog. Theor. Phys.* **17**, 145 (1957).

R. LANDAUER, Spatial variation of currents and fields due to localized scatterers in metallic conduction, *IBM J. Res. Dev.* **1**, 223 (1957); Electrical resistance of disordered one-dimensional lattices, *Phil. Mag.* **21**, 863 (1970).

D.A. GREENWOOD, The Boltzmann equation in the theory of electrical conduction in metals, *Proc. Phys. Soc. London* **71**, 585 (1958).

R. ZWANZIG, Time-correlation functions and transport coefficients in statistical mechanics, *Ann. Rev. Phys. Chem.* **16**, 67 (1965).

J.S. LANGER and T. NEAL, Breakdown of the concentration expansion for the impurity resistivity of metals, *Phys. Rev. Lett.* **16**, 984 (1966).

D.J. THOULESS, Relation between the Kubo–Greenwood formula and the Boltzmann equation for electrical conductivity, *Phil. Mag.* **32**, 877 (1975).

Supplement 15A

Conductivity of a weakly disordered metal

1. Introduction

This supplement deals with the electrical conductivity in uniform electric field of a weakly disordered conductor at zero temperature. In the case of a macroscopic sample, the conductivity can be computed with the aid of the Kubo–Greenwood formula. This way to determine the conductivity, which makes use of the relation between its real part and the electromagnetic absorption properties, corresponds to an experimental procedure in which, after having placed the conducting sample devoid of contacts in an electromagnetic cavity, we measure the supplementary absorption due to its presence.

There is another approach, due to R. Landauer, allowing us to compute the conductivity of a disordered conductor. Landauer's approach is better suited to the case of systems of small dimensions (mesoscopic) and to the discussion of the *localization* phenomenon (absence of conduction) in one-dimensional disordered systems.

2. The Kubo–Greenwood formula

The Maxwell equations in a material medium read:

$$\begin{cases} \nabla \times \boldsymbol{H} = (1/c)(4\pi \boldsymbol{J} + \partial \boldsymbol{D}/\partial t) & \nabla . \boldsymbol{B} = 0 \\ \nabla . \boldsymbol{D} = 4\pi\rho & \nabla \times \boldsymbol{E} = -(1/c)\partial \boldsymbol{B}/\partial t. \end{cases} \tag{15A.2.1}$$

In equations (15A.2.1), $\boldsymbol{J}$ denotes the current density and ρ the density of free charges. Defining as usual the spatial and temporal Fourier transforms of the various fields involved, we write in linear regime the relations $\boldsymbol{D}(\boldsymbol{q},\omega) = \underline{\varepsilon}^{\circ}(\boldsymbol{q},\omega).\boldsymbol{E}(\boldsymbol{q},\omega)$ and $\boldsymbol{B}(\boldsymbol{q},\omega) = \underline{\mu}(\boldsymbol{q},\omega).\boldsymbol{H}(\boldsymbol{q},\omega)$, as well as Ohm's law $\boldsymbol{J}(\boldsymbol{q},\omega) = \underline{\sigma}(\boldsymbol{q},\omega).\boldsymbol{E}(\boldsymbol{q},\omega)$. The tensors $\underline{\varepsilon}^{\circ}$ and $\underline{\mu}$ are respectively the dielectric permittivity tensor and the magnetic permeability tensor of the medium,[1] whereas $\underline{\sigma}$ is the electrical conductivity tensor.

---

[1] Remember that we are using Gauss units, in which the dielectric permittivity of vacuum equals unity.

We are interested here in the propagation of an electromagnetic wave in a non-magnetic medium ($\mu_{\alpha\beta} = \delta_{\alpha\beta}$), assumed to be simple enough for the tensors $\underline{\varepsilon}^\circ$ and $\underline{\sigma}$ to be proportional to the unit matrix: $\varepsilon^\circ_{\alpha\beta} = \varepsilon^\circ \delta_{\alpha\beta}$, $\sigma_{\alpha\beta} = \sigma \delta_{\alpha\beta}$. We assume that the fields of the wave do not vary very much over a distance of the order of the electronic mean free path. Then ε° and σ do not depend on the wave vector $\boldsymbol{q}$, but only on the angular frequency ω of the wave.

2.1. Global dielectric permittivity

At vanishing angular frequency, the permittivity and the conductivity correspond to distinct physical processes: σ describes the motion of free charges (conduction electrons), whereas ε° describes the polarization of bound charges (electrons in completely filled bands).

At finite angular frequency, the distinction between ε° and σ is purely conventional. We can consider that both the response of the electrons and the electromagnetic properties are in fact determined by the sole global dielectric permittivity $\varepsilon(\omega)$, related to $\varepsilon^0(\omega)$ and $\sigma(\omega)$ by the formula:

$$\boxed{\varepsilon(\omega) = \varepsilon^\circ(\omega) + i\frac{4\pi}{\omega}\sigma(\omega).} \tag{15A.2.2}$$

The global permittivity $\varepsilon(\omega)$ includes the contribution of electrons of all bands, even of those which are only partially filled. This can be shown by analyzing in two different ways the Maxwell equation (15A.2.1) for $\nabla \times \boldsymbol{H}$.

If all the effects of the response of the electrons are described through a global dielectric permittivity $\varepsilon(\omega)$, we consider that there is no current in the medium and we write $\boldsymbol{D} = \varepsilon \boldsymbol{E}$. This gives:

$$\nabla \times \boldsymbol{H} = \frac{1}{c}\big[-i\omega\varepsilon(\omega)\big]\boldsymbol{E}. \tag{15A.2.3}$$

If we adopt the point of view according to which the response of the electrons is described both by a dielectric permittivity $\varepsilon^\circ(\omega)$ and an electrical conductivity $\sigma(\omega)$, the medium is crossed through by a current $\boldsymbol{J} = \sigma \boldsymbol{E}$ and we have:

$$\nabla \times \boldsymbol{H} = \frac{1}{c}\big[4\pi\sigma(\omega) - i\omega\varepsilon^\circ(\omega)\big]\boldsymbol{E}. \tag{15A.2.4}$$

Identifying the expressions (15A.2.3) and (15A.2.4) for $\nabla\times\boldsymbol{H}$, we get formula (15A.2.2).

2.2. Electromagnetic absorption and conductivity

Consider a conducting sample of volume V, whose electrons are treated as independent. The one-electron Hamiltonian being H_0, we denote by $\{|\phi_n\rangle\}$ a base of eigenstates of H_0, of energies ε_n. The energy levels are assumed non-degenerate for the sake of simplicity. We write $\boldsymbol{D} = \varepsilon \boldsymbol{E} = \boldsymbol{E} + 4\pi \boldsymbol{P}$, where $\boldsymbol{P} = \chi \boldsymbol{E}$ is the polarization[2] induced by the field $\boldsymbol{E}$ (χ is the electrical susceptibility of the medium).

The general theory of linear response allows us to obtain an expression for $\chi(\omega)$, then for $\varepsilon(\omega)$, in terms of the eigenstates and eigenvalues of H_0. The one-electron perturbation Hamiltonian reads, for an applied electric field[3] $\boldsymbol{E}(t)$:

$$H_1(t) = -e\boldsymbol{E}(t).\boldsymbol{r}. \tag{15A.2.5}$$

As a general rule, the Kubo formula for the electrical susceptibility tensor reads:

$$\chi_{\alpha\beta}(\omega) = \frac{e^2}{V\hbar} \lim_{\epsilon\to 0^+} \sum_{n,q} (\Pi_n - \Pi_q) \frac{\langle\phi_n|x_\alpha|\phi_q\rangle\langle\phi_q|x_\beta|\phi_n\rangle}{\omega_{qn} - \omega - i\epsilon}, \tag{15A.2.6}$$

where, at fixed temperature and chemical potential, the average occupation probability at equilibrium Π_n of the state $|\phi_n\rangle$ is the Fermi–Dirac function $f_0(\varepsilon_n)$. Like the tensors $\underline{\varepsilon}^\circ(\omega)$ and $\underline{\sigma}(\omega)$, the tensor $\underline{\chi}(\omega)$ is assumed here diagonal: $\chi_{\alpha\beta}(\omega) = \chi(\omega)\delta_{\alpha\beta}$. We have:

$$\chi(\omega) = \frac{e^2}{V\hbar} \lim_{\epsilon\to 0^+} \sum_{n,q} \big[f_0(\varepsilon_n) - f_0(\varepsilon_q)\big] \frac{|\langle\phi_q|x|\phi_n\rangle|^2}{\omega_{qn} - \omega - i\epsilon}. \tag{15A.2.7}$$

We can then deduce from $\chi(\omega)$ the dielectric permittivity $\varepsilon(\omega) = 1 + 4\pi\chi(\omega)$.

The polarization of the electrons of the completely filled bands generally leads to $\varepsilon^\circ(\omega)$ real. According to formula (15A.2.2), this gives:

$$\Im m\, \varepsilon(\omega) = \frac{4\pi}{\omega} \Re e\, \sigma(\omega), \tag{15A.2.8}$$

that is:

$$\Re e\, \sigma(\omega) = \omega\, \Im m\, \chi(\omega). \tag{15A.2.9}$$

Making use of formula (15A.2.7) for $\chi(\omega)$, we obtain an expression for the real part of the conductivity at vanishing wave vector:

$$\Re e\, \sigma(\omega) = \frac{\pi e^2 \omega}{V\hbar} \sum_{n,q} \big[f_0(\varepsilon_n) - f_0(\varepsilon_q)\big] |\langle\phi_q|x|\phi_n\rangle|^2 \delta(\omega_{qn} - \omega). \tag{15A.2.10}$$

Formula (15A.2.10) involves the matrix elements of the operator x. It is common practice to rewrite $\Re e\, \sigma(\omega)$ in terms of the matrix elements of the operator $v_x = (i\hbar)^{-1}[x, H_0]$:

$$\boxed{\Re e\, \sigma(\omega) = \frac{\pi e^2}{\omega V} \sum_n \sum_{q\neq n} |\langle\phi_q|v_x|\phi_n\rangle|^2 \big[f_0(\varepsilon_n) - f_0(\varepsilon_q)\big] \delta(\varepsilon_q - \varepsilon_n - \hbar\omega).} \tag{15A.2.11}$$

---

[2] We assume here that the polarization of the medium is uniquely of electronic origin.

[3] The fields varying little over a distance of the order of the electronic mean free path, they may be considered as spatially uniform.

We thus retrieve the Kubo–Greenwood formula, previously deduced from the general Kubo–Nakano formula giving the conductivity in terms of a correlation function of currents.

The average power absorbed per unit volume by the conductor placed in an electromagnetic wave of electric field $\mathbf{E}(t) = \Re e[\mathbf{E}_0 e^{-i\omega t}]$ is:

$$\overline{\frac{dW}{dt}} = \frac{1}{2} E_0^2 \omega \chi''(\omega), \tag{15A.2.12}$$

that is:

$$\overline{\frac{dW}{dt}} = \frac{1}{2} E_0^2 \, \Re e \, \sigma(\omega). \tag{15A.2.13}$$

3. Conductivity of a macroscopic system

Consider a three-dimensional macroscopic sample of a metal containing impurities. As the system is macroscopic, the eigenenergies of H_0 form a continuum. To compute $\Re e \, \sigma(\omega)$ using the Kubo–Greenwood formula (15A.2.11), we therefore introduce the density of states in energy $n(\varepsilon)$ per unit volume and per spin direction. Since the only non-vanishing matrix elements of v_x are its matrix elements between states of the same spin, we write:

$$\Re e \, \sigma(\omega) = \frac{2\pi e^2 V}{\omega} \iint n(\varepsilon_n) n(\varepsilon_q) |\langle \phi_q | v_x | \phi_n \rangle|^2 \left[f_0(\varepsilon_n) - f_0(\varepsilon_q) \right] \delta(\varepsilon_q - \varepsilon_n - \hbar\omega) \, d\varepsilon_n d\varepsilon_q. \tag{15A.3.1}$$

At zero temperature, the excitations of the system are electron–hole pairs: owing to the absorption of a quantum of energy $\hbar\omega$, an electron in a state of energy $\varepsilon_n < \varepsilon_F$ is put in a state of energy $\varepsilon_q (= \varepsilon_n + \hbar\omega) > \varepsilon_F$, leaving behind a hole (ε_F is the Fermi energy). Owing to the presence of the delta function on the right-hand side of equation (15A.3.1), we get:

$$\Re e \, \sigma(\omega) = 2\pi e^2 V \hbar \int_{\varepsilon_F - \hbar\omega}^{\varepsilon_F} \frac{n(\varepsilon_n) n(\varepsilon_n + \hbar\omega)}{\hbar\omega} |\langle \phi_q | v_x | \phi_n \rangle|^2 \, d\varepsilon_n. \tag{15A.3.2}$$

Averaging over the positions of the impurities, assumed randomly distributed, gives:

$$\Re e \, \sigma(\omega) = 2\pi e^2 V \hbar \int_{\varepsilon_F - \hbar\omega}^{\varepsilon_F} \frac{n(\varepsilon_n) n(\varepsilon_n + \hbar\omega)}{\hbar\omega} \langle |(v_x)_{qn}|^2 \rangle \, d\varepsilon_n. \tag{15A.3.3}$$

In formula (15A.3.3), $\langle |(v_x)_{qn}|^2 \rangle$ denotes the average of $|\langle \phi_q | v_x | \phi_n \rangle|^2$ over the different configurations of the disordered sample.

If the condition $k_F \ell \gg 1$ is fulfilled, where k_F denotes the Fermi wave vector and ℓ the elastic mean free path of the electrons, the eigenstates of H_0 may be decomposed over a base of plane waves $|\boldsymbol{k}\rangle$, the various plane-wave states forming a wave-packet of width $\sim \ell^{-1}$:

$$|\phi_n\rangle = \sum_{\boldsymbol{k}} a_{\boldsymbol{k}}^n \, |\boldsymbol{k}\rangle, \qquad |\phi_q\rangle = \sum_{\boldsymbol{k}} a_{\boldsymbol{k}}^q \, |\boldsymbol{k}\rangle. \tag{15A.3.4}$$

We set $\varepsilon_n = \hbar^2 k_n^2/2m$ and $\varepsilon_q = \hbar^2 k_q^2/2m$ (with $k_n\ell \gg 1$, $k_q\ell \gg 1$). We assume that, as an effect of disorder, the amplitudes $a_{\boldsymbol{k}}^n$ may be modelized by independent Gaussian random variables of zero mean and of variance approximately given by a Lorentzian[4] of width ℓ^{-1}:

$$\langle a_{\boldsymbol{k}}^{n*} a_{\boldsymbol{k}'}^q \rangle \simeq \delta_{nq}\delta_{\boldsymbol{k}\boldsymbol{k}'} \frac{\pi}{\ell k_n^2 V} \frac{1}{(k-k_n)^2 + (4\ell^2)^{-1}}. \tag{15A.3.5}$$

The normalization of $|\phi_n\rangle$ is thus guaranteed on average.[5] On account of the Gaussian character of the amplitudes, we can deduce from formulas (15A.3.4) and (15A.3.5) the expression for the quantity $\langle |(v_x)_{qn}|^2 \rangle$ for $q \neq n$:

$$\langle |(v_x)_{qn}|^2 \rangle = \sum_{\boldsymbol{k}} \sum_{\boldsymbol{k}'} \langle a_{\boldsymbol{k}}^{n*} \frac{\hbar k_x}{m} a_{\boldsymbol{k}}^q a_{\boldsymbol{k}'}^{q*} \frac{\hbar k'_x}{m} a_{\boldsymbol{k}'}^n \rangle, \qquad q \neq n. \tag{15A.3.6}$$

According to formula (15A.3.5), only the terms with $\boldsymbol{k} = \boldsymbol{k}'$ contribute to the above average. This gives:

$$\langle |(v_x)_{qn}|^2 \rangle = \frac{\pi^2}{\ell^2 k_n^2 k_q^2 V^2} \sum_{\boldsymbol{k}} \frac{\hbar^2 k^2/3m^2}{\left[(k-k_n)^2 + (4\ell^2)^{-1}\right]\left[(k-k_q)^2 + (4\ell^2)^{-1}\right]}, \tag{15A.3.7}$$

that is:[6]

$$\langle |(v_x)_{qn}|^2 \rangle \simeq \frac{2\pi\hbar^2\ell}{3Vm^2} \frac{1}{1 + (k_q - k_n)^2\ell^2}. \tag{15A.3.8}$$

Then, coming back to formula (15A.3.3), we obtain:

$$\Re e\, \sigma(\omega) = \frac{4\pi^2 e^2 \hbar^3 \ell}{3m^2} \int_{\varepsilon_F - \hbar\omega}^{\varepsilon_F} \frac{n(\varepsilon_n) n(\varepsilon_n + \hbar\omega)}{\hbar\omega} \frac{1}{1 + (k_q - k_n)^2\ell^2}\, d\varepsilon_n. \tag{15A.3.9}$$

In the angular frequency range $\hbar\omega \ll \varepsilon_F$, we have:

$$1 + (k_q - k_n)^2\ell^2 \simeq 1 + \ell^2 \frac{m^2\omega^2}{\hbar^2 k_F^2}, \tag{15A.3.10}$$

that is:

$$1 + (k_q - k_n)^2\ell^2 \simeq 1 + \omega^2\tau^2(\varepsilon_F), \tag{15A.3.11}$$

---

4 We have set $k = |\boldsymbol{k}|$.

5 We have:

$$\langle \phi_n | \phi_n \rangle = \sum_{\boldsymbol{k}} |a_{\boldsymbol{k}}^n|^2,$$

and, on average, on account of formula (15A.3.5) and of the hypothesis $k_n\ell \ll 1$:

$$\langle\langle \phi_n | \phi_n \rangle\rangle \simeq \frac{V}{(2\pi)^3} \frac{\pi}{\ell k_n^2 V} k_n^2 \int_0^\infty \frac{4\pi}{(k-k_n)^2 + (4\ell^2)^{-1}}\, dk, \qquad k_n\ell \gg 1.$$

This insures that $\langle\langle \phi_n | \phi_n \rangle\rangle = 1$.

6 The result (15A.3.8) may for instance be obtained by residue integration.

where $\tau(\varepsilon_F) = \ell m/\hbar k_F$ is the relaxation time at the Fermi level. The real part of the conductivity thus reads approximately:

$$\Re e\, \sigma(\omega) \simeq \frac{4\pi^2 e^2 \hbar^3 \ell}{3m^2} n^2(\varepsilon_F) \frac{1}{1+\omega^2\tau^2(\varepsilon_F)}. \tag{15A.3.12}$$

To continue the calculation, we need an expression for the density of states in the disordered metal. We will assume that the density of states is not very different from that of a free electron gas $n(\varepsilon) = \pi^{-2} 2^{-1/2} (m/\hbar^2)^{3/2} \varepsilon^{1/2}$. We then get, on account of the relation $n = k_F^3/3\pi^2$, where n denotes the number of electrons per unit volume:

$$\boxed{\Re e\, \sigma(\omega) \simeq \frac{ne^2\tau(\varepsilon_F)}{m} \frac{1}{1+\omega^2\tau^2(\varepsilon_F)}.} \tag{15A.3.13}$$

This is the result, for a harmonic field at the angular frequency ω, of the classical Drude model.

The above method can also be applied to a one-dimensional sample, provided that it is macroscopic. To carry out the calculation, we have to take into account the appropriate modifications in the expressions for $\langle |(v_x)_{qn}|^2 \rangle$ and $n(\varepsilon_F)$, as well as in the relation between n and k_F.

4. Conductance of a mesoscopic system: Landauer's approach

4.1. Problems arising from the Kubo formulation in a system of small dimensions

When the dimensions of the sample are small enough so that the electronic levels do not form a continuum (mesoscopic system), the previous method cannot be applied. To compute $\Re e\, \sigma(\omega)$ in the framework of the Kubo theory, it is then necessary to come back to formula (15A.2.11), this latter expression for $\Re e\, \sigma(\omega)$ involving a double summation over states which can be viewed as discrete. Formula (15A.2.11), as applied to a finite isolated conductor, leads to a vanishing static conductivity: $\Re e\, \sigma(\omega = 0) = 0$. This result expresses the fact that a finite isolated system, having a discrete energy spectrum, does not absorb energy from the electromagnetic field.

We thus cannot obtain in this way the finite static conductivity of a mesoscopic metallic system.

4.2. Landauer's approach

However, the expression (15A.2.11) for $\Re e\, \sigma(\omega)$ refers to a quantity measured by placing the conducting sample devoid of contacts in an electromagnetic cavity, and measuring the supplementary absorption due to its presence.

As shown by R. Landauer as early as 1957, it is possible to define and to measure in a different way the static conductivity, for instance by putting the conducting sample in contact with a source of current. In Landauer's approach, we consider a disordered conductor crossed through by a flux of incident electrons encountering a barrier of obstacles of length L due to disorder. We assume that, on both sides of these obstacles, there exists a free space without disorder. We denote by R the probability of reflection over the barrier: the relative density of particles in the free space at the left of the barrier is thus $1 + R$ (in this space the incident and reflected current do coexist), whereas the relative density at the right of the barrier is $T = 1 - R$ (only the transmitted current exists in this space) (Fig. 15A.1).

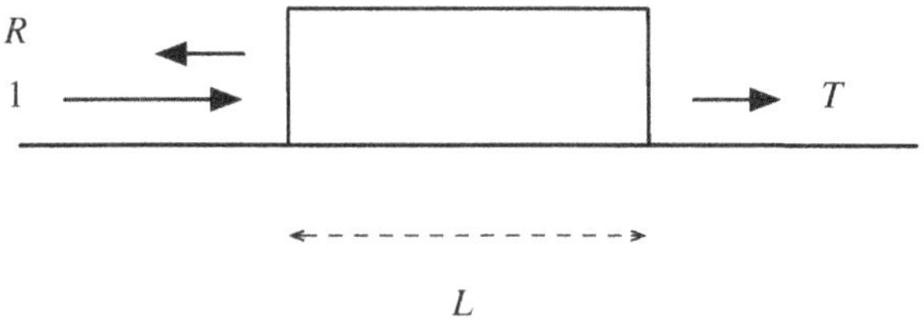

Fig. 15A.1 Landauer's geometry.

The gradient of relative density through the barrier is equal to $-2R/L$. At zero temperature, assuming for the sake of simplicity that all the electrons have the same velocity, equal to the Fermi velocity v_F, we write the current associated with this gradient as $J = v_F(1 - R)$. The current J is a diffusion current. The coefficient D of Fick's law is given by:

$$D = \frac{v_F L}{2} \frac{1 - R}{R}. \tag{15A.4.1}$$

The conductance of the sample follows via the fluctuation-dissipation theorem, written in the form of the relation:

$$\frac{D}{\mu_D} = \frac{n}{e} \frac{\partial \mu}{\partial n}\bigg|_T \tag{15A.4.2}$$

between the diffusion coefficient and the drift mobility (n denotes the number of electrons per unit length). At $T = 0$, we have:

$$\frac{\partial \mu}{\partial n}\bigg|_{T=0} = \frac{1}{2n(\varepsilon_F)}, \tag{15A.4.3}$$

where $n(\varepsilon_F)$ is here the density of states per unit length and per spin direction at the Fermi level. Taking the value $n(\varepsilon_F) = (\pi \hbar v_F)^{-1}$ corresponding to a free electron gas, we obtain for the conductivity $\sigma = ne\mu_D$ the expression:

$$\sigma = \frac{2De^2}{\pi \hbar v_F}. \tag{15A.4.4}$$

The resistivity $\rho = \sigma^{-1}$ is:

$$\rho = \frac{\pi \hbar v_F}{2De^2}. \tag{15A.4.5}$$

On account of the expression (15A.4.1) for D, the resistance $\Omega = \rho L$ of a sample of length L is:

$$\Omega = \frac{\pi\hbar}{e^2}\frac{R}{1-R}. \qquad (15A.4.6)$$

At zero temperature, the conductance $G = \Omega^{-1}$ of the one-dimensional disordered sample, as modelized by a barrier of reflection coefficient R and transmission coefficient $T = 1 - R$, is thus given by the *Landauer formula*:

$$\boxed{G = \frac{e^2}{\pi\hbar}\frac{T}{R}.} \qquad (15A.4.7)$$

5. Addition of quantum resistances in series: localization

A one-dimensional disordered sample can be viewed as a succession of barriers of reflection and transmission coefficients R_i and $T_i = 1 - R_i$.

5.1. Average resistance of a system of two barriers in series

To determine the average resistance of such a succession of barriers, we first consider a system of two barriers in series.

The problem cannot be treated using only the reflection and transmission coefficients (R_1, T_1) and (R_2, T_2) relative to the particle density. Indeed, the interactions of the electrons with fixed impurities, which are responsible for the elastic mean free path ℓ, maintain the phase coherence. This phase coherence is lost as a consequence of random interactions, in particular the electron–phonon and electron–electron interactions. At low temperature, the electron–electron interactions (which do not modify the momentum and thus do not influence ℓ) are mostly responsible for the loss of phase coherence. We are thus led to introduce the coherence length ℓ_ϕ over which the phase coherence of the wave-function of an electron near the Fermi level is lost. If we consider a material for which $\ell_\phi \gg \ell$, the phases of the electron leaving the first barrier, on the one hand, and entering the second one, on the other hand, are perfectly correlated. We therefore must take this coherence into account by using in the calculation probability amplitudes instead of particle densities (Fig. 15A.2).

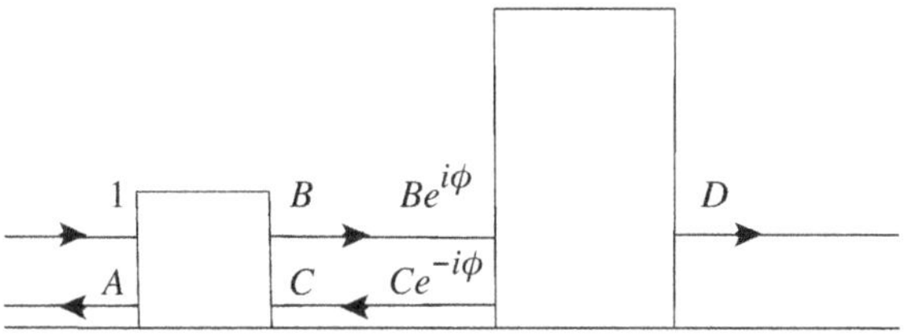

Fig. 15A.2 Reflected and transmitted amplitudes by two barriers in series.

Thus, given an incident wave of unit amplitude, we denote by A the amplitude reflected by the first barrier and by D the amplitude transmitted by the two-barrier set. The wave of amplitude B emerging from the first barrier undergoes a dephasing ϕ before reaching the second barrier. The wave of amplitude C undergoes a similar dephasing between the second barrier and the first one. The continuity equations of the amplitudes entering and leaving the barriers read:

$$\begin{cases} A = r_1 + Ct_1 & B = t_1 + Cr'_1 \\ Ce^{-i\phi} = Be^{i\phi}r_2 & D = Be^{i\phi}t_2. \end{cases} \tag{15A.5.1}$$

In equations (15A.5.1), r_1 and t_1 respectively denote the reflection coefficient of the amplitude entering the first barrier and the transmission coefficient of the amplitude by this barrier, whereas $r'_1 = r_1^*$ denotes the reflection coefficient of the amplitude leaving this barrier. In the same way, r_2 and t_2 denote respectively the reflection coefficient of the amplitude entering the second barrier and the transmission coefficient of this latter.

Equations (15A.5.1) allow us to write A, B, C, D in terms of the parameters of the barriers and of the dephasing. In particular, we have:

$$D = \frac{t_1 t_2 e^{i\phi}}{1 - e^{2i\phi} r_2 r'_1}. \tag{15A.5.2}$$

The transmission coefficient of the two-barrier set is:

$$T^{(2)} = |D|^2 = \frac{|t_1|^2 |t_2|^2}{1 + |r_1|^2 |r_2|^2 - e^{2i\phi} r_2 r_1^* - e^{-2i\phi} r_2^* r_1}, \tag{15A.5.3}$$

that is, with standard notations:

$$T^{(2)} = \frac{T_1 T_2}{1 + R_1 R_2 - 2(R_1 R_2)^{1/2} \cos\theta}. \tag{15A.5.4}$$

In equation (15A.5.4), we have set:

$$\theta = 2\phi + \arg(r_2 r_1^*). \tag{15A.5.5}$$

We then compute the reflection coefficient $R^{(2)} = 1 - T^{(2)}$:

$$R^{(2)} = \frac{R_1 + R_2 - 2(R_1 R_2)^{1/2} \cos\theta}{1 + R_1 R_2 - 2(R_1 R_2)^{1/2} \cos\theta}. \tag{15A.5.6}$$

According to the Landauer formula, the conductance of this obstacle made of two barriers in series is:

$$G^{(2)} = \frac{e^2}{\pi\hbar} \frac{T^{(2)}}{R^{(2)}}, \tag{15A.5.7}$$

that is, from formulas (15A.5.4) and (15A.5.6):

$$G^{(2)} = \frac{e^2}{\pi\hbar}\frac{T_1T_2}{R_1 + R_2 - 2(R_1R_2)^{1/2}\cos\theta}. \tag{15A.5.8}$$

The dephasing ϕ depends on the distance separating the barriers. We now consider a statistical ensemble of samples made up of the two above described barriers, separated by distances varying from one sample to another. We denote by $\langle\ldots\rangle$ the average over this ensemble. We can assume that ϕ is uniformly distributed over the interval $(0, 2\pi)$. Formula (15A.5.5) yields in this case $\langle\cos\theta\rangle = 0$. We then deduce from formula (15A.5.8) the average resistance:

$$\left\langle\frac{1}{G^{(2)}}\right\rangle = \frac{\pi\hbar}{e^2}\frac{R_1 + R_2}{(1 - R_1)(1 - R_2)}. \tag{15A.5.9}$$

Interestingly enough, formula (15A.5.9) shows that Ohm's law of addition of resistances in series is not valid in general[7] (remember, however, that we are carrying out a zero-temperature calculation).

In view of the calculation of the average resistance of a system of N barriers in series, it is convenient to rewrite formula (15A.5.9) in the form:

$$\left\langle\frac{1}{G^{(2)}}\right\rangle = \frac{\pi\hbar}{e^2}\left(\frac{R_1}{1 - R_1} + \frac{R_2}{1 - R_2} + 2\frac{R_1}{1 - R_1}\frac{R_2}{1 - R_2}\right). \tag{15A.5.10}$$

5.2. Average resistance of a disordered conductor: localization

We want now to establish how the average resistance of a one-dimensional disordered conductor modelized by a series of N barriers of the preceding type varies with N. We introduce the dimensionless conductance g, such that $G = (\pi\hbar)^{-1}e^2g$, and the dimensionless resistance[8] $\rho = g^{-1}$.

The addition law (15A.5.10) allows us to compute the average resistance $\langle\rho^{(2)}\rangle$ of a set of two systems of resistances ρ_1 and ρ_2 in series:

$$\langle\rho^{(2)}\rangle = \rho_1 + \rho_2 + 2\rho_1\rho_2. \tag{15A.5.11}$$

Averaging over the parameters related to each system, assumed statistically independent of one another, gives:[9]

$$\langle\rho^{(2)}\rangle = \langle\rho_1\rangle + \langle\rho_2\rangle + 2\langle\rho_1\rangle\langle\rho_2\rangle. \tag{15A.5.12}$$

---

[7] Ohm's law would indeed give, for the resistance of the two-barrier set (independently of the value of ϕ):

$$\frac{1}{G^{(2)}} = \frac{\pi\hbar}{e^2}\left(\frac{R_1}{1 - R_1} + \frac{R_2}{1 - R_2}\right).$$

In the limit of small resistances, that is, for small reflection coefficients ($R_i \ll 1$), Ohm's law identifies with the exact result (15A.5.9).

[8] Note that ρ does not represent here the resistivity, but rather the inverse of the dimensionless conductance g.

[9] To work with simple notations, we keep the same designation for the average resistance either before or after the supplementary averaging over the parameters related to each system.

Thus, considering an obstacle composed of $N-1$ barriers in series with a supplementary barrier, each barrier having an average resistance $\langle\rho\rangle$, we can show that the average resistance $\langle\rho^{(N)}\rangle$ of a conductor of N barriers is:

$$\langle\rho^{(N)}\rangle = \frac{1}{2}\big[(1+2\langle\rho\rangle)^N - 1\big]. \tag{15A.5.13}$$

For $\langle\rho\rangle \ll 1$, we have approximately:

$$\langle\rho^{(N)}\rangle \simeq \frac{1}{2}\big(e^{2\langle\rho\rangle N} - 1\big). \tag{15A.5.14}$$

This exponential increase with N of the average resistance and the ensuing absence of conduction constitute the localization phenomenon in one dimension. For a sample of length L, if d denotes the average distance between barriers, we have $N = Ld^{-1}$ and formula (15A.5.14) reads:

$$\langle\rho^{(N)}\rangle \simeq \frac{1}{2}\big(e^{2\alpha L} - 1\big), \tag{15A.5.15}$$

where $\alpha = \langle\rho\rangle d^{-1}$.

5.3. Scaling variable

The previous approach of localization, which relies on the calculation of the average resistance, is not entirely satisfactory. Indeed, the probability density of the resistances is a broad law, so that the above-found dependence on N of the average resistance is not enough to assess localization.

In such a context, it is interesting to look for a quantity playing the role of a scaling variable, that is, a variable behaving in an extensive (additive) way, with an average increasing linearly with N. The quantity $\log(1+\rho)$ can be seen to possess this property. For each barrier, we have $1+\rho = T^{-1}$, and thus $\log(1+\rho) = -\log T$. The quantity $-\log T$, which plays the role of an absorption coefficient,[10] behaves additively. Indeed, coming back to the obstacle composed of two barriers, we get from formula (15A.5.4), by averaging over the dephasing[11] ϕ, an additive law:

$$\langle\log T^{(2)}\rangle = \log T_1 + \log T_2. \tag{15A.5.16}$$

Formula (15A.5.16) allows us to show that $\log(1+\rho)$ is a relevant scaling variable for this problem. We indeed get, for the conductor of N barriers considered precedently, after averaging over the parameters related to each barrier:

$$\big\langle\log(1+\rho^{(N)})\big\rangle = N\big\langle\log(1+\rho)\big\rangle. \tag{15A.5.17}$$

---

[10] Indeed, the ratio of the intensity transmitted after passage through a sample of length L and of absorption coefficient K to the incident intensity is $T = e^{-KL}$.

[11] We make use of the formula:

$$\int_0^{2\pi} \log(a + b\cos\theta)\, d\theta = \pi \log\frac{1}{2}\Big[a + (a^2 - b^2)^{1/2}\Big].$$

Formula (15A.5.17) implies that the typical resistance of a sample of length L follows the scaling law $\log(1+\rho) = \alpha L$, from which we deduce:

$$\rho = e^{\alpha L} - 1. \tag{15A.5.18}$$

The inverse localization length α acts as the scaling variable. Formula (15A.5.18) shows that, for $\alpha L \gg 1$, the typical resistance of a sample of length L increases exponentially with L ($\rho \simeq e^{\alpha L}$), whereas, for $\alpha L \ll 1$, we recover a 'classical' additive resistance ($\rho \simeq \alpha L$).

Bibliography

É. AKKERMANS and G. MONTAMBAUX, *Mesoscopic physics of electrons and photons*, Cambridge University Press, Cambridge, 2007.

N.W. ASHCROFT and N.D. MERMIN, *Solid state physics*, Holt-Saunders, Philadelphia, 1976.

Y. IMRY, *Introduction to mesoscopic physics*, Oxford University Press, Oxford, second edition, 2002.

L.D. LANDAU and E.M. LIFSHITZ, *Electrodynamics of continuous media*, Butterworth-Heinemann, Oxford, second edition, 1984.

N.F. MOTT, *Conduction in non-crystalline materials*, Clarendon Press, Oxford, 1987.

References

R. LANDAUER, Spatial variation of currents and fields due to localized scatterers in metallic conduction, *IBM J. Res. Dev.* **1**, 223 (1957); Electrical resistance of disordered one-dimensional lattices, *Phil. Mag.* **21**, 863 (1970).

D.A. GREENWOOD, The Boltzmann equation in the theory of electrical conduction in metals, *Proc. Phys. Soc. London* **71**, 585 (1958).

D.J. THOULESS, Relation between the Kubo–Greenwood formula and the Boltzmann equation for electrical conductivity, *Phil. Mag.* **32**, 877 (1975).

P.W. ANDERSON, D.J. THOULESS, E. ABRAHAMS, and D.S. FISHER, New method for a scaling theory of localization, *Phys. Rev. B* **22**, 3519 (1980).

Y. IMRY and R. LANDAUER, Conductance viewed as transmission, *Rev. Mod. Phys.*, Centennial Issue, **71**, S306 (1999).

Chapter 16

Thermal transport coefficients

In a fluid, the gradients of density, of mean velocity, or of temperature give rise to 'thermal forces' internal to the system under study. Since the corresponding perturbations are not described by a Hamiltonian, it is not a priori possible to obtain the response to these forces or the associated generalized susceptibilities by directly applying the linear response theory, since this formalism was originally developed for the responses to the mechanical perturbations described by a Hamiltonian.

It is however generally accepted that we can write, for the linear responses to thermal forces, expressions analogous to those of the linear responses to mechanical forces, in other words, that there are Green–Kubo formulas that allow us to express the thermal transport coefficients in terms of equilibrium correlation functions of the appropriate currents.

In this chapter, we present two methods allowing us to derive Green–Kubo formulas for thermal transport coefficients. The first one, referred to as the 'indirect Kubo method', is applied to the case of a conductor in which impurities or phonons give rise to a resistive behavior in the presence of an applied electric field. Equilibrating the diffusion current and the drift current, we are able to deduce the expression for the diffusion tensor from that for the electrical conductivity tensor. The second method is applied to the case of a fluid. It relies on the expression for the entropy production, with which we associate an equivalent perturbation 'Hamiltonian'. The knowledge of this latter then allows us to apply formally the linear response theory. We can get then the thermal conductivity and the viscosity coefficient of the fluid.

1. The indirect Kubo method

The transport coefficients associated with non-mechanical perturbations cannot be obtained directly by the Kubo theory, since this formalism relies on the existence of the Hamiltonian describing the perturbation. It is nevertheless possible to derive Green–Kubo formulas for these transport coefficients, for instance by using the *indirect Kubo method*. We will illustrate the principle of this procedure in the case of the diffusion tensor in a conducting material. The diffusion tensor is determined from the electrical conductivity tensor, which itself is computed via the Kubo–Nakano formula.

1.1. Stating the problem

Consider a conducting material in which electrons of charge e evolve in the presence of an external potential $\phi(\boldsymbol{r},t)$ (and of an electric field $\boldsymbol{E}(\boldsymbol{r},t) = -\nabla\phi(\boldsymbol{r},t)$). The charge density and electric current density operators are respectively:

$$\rho(\boldsymbol{r},t) = e\sum_i \delta[\boldsymbol{r}-\boldsymbol{r}_i(t)], \tag{16.1.1}$$

and:

$$\boldsymbol{J}(\boldsymbol{r},t) = \frac{e}{2}\sum_i \Big\{\boldsymbol{v}_i(t)\delta\big[\boldsymbol{r}-\boldsymbol{r}_i(t)\big] + \delta\big[\boldsymbol{r}-\boldsymbol{r}_i(t)\big]\boldsymbol{v}_i(t)\Big\}. \tag{16.1.2}$$

In formulas (16.1.1) and (16.1.2), the $\{\boldsymbol{r}_i\}$'s and the $\{\boldsymbol{v}_i = \dot{\boldsymbol{r}}_i\}$'s are the position and velocity operators of the different electrons. The charge and electric current densities are related by the continuity equation:

$$\dot\rho(\boldsymbol{r},t) + \nabla.\boldsymbol{J}(\boldsymbol{r},t) = 0. \tag{16.1.3}$$

We are interested in the local mean values $\langle\rho(\boldsymbol{r},t)\rangle$ and $\langle\boldsymbol{J}(\boldsymbol{r},t)\rangle$. In the linear response regime, the average electric current density is related to the electric field and to the density gradient through the non-local and retarded phenomenological constitutive equation:

$$\begin{aligned}\langle J_\alpha(\boldsymbol{r},t)\rangle = &\sum_\beta \int d\boldsymbol{r}'\int dt'\,\tilde\sigma_{\alpha\beta}(\boldsymbol{r}-\boldsymbol{r}',t-t')E_\beta(\boldsymbol{r}',t')\\ &-\sum_\beta \int d\boldsymbol{r}'\int dt'\,\tilde D_{\alpha\beta}(\boldsymbol{r}-\boldsymbol{r}',t-t')\nabla_\beta\langle\rho(\boldsymbol{r}',t')\rangle,\end{aligned} \tag{16.1.4}$$

in which $\underline{\tilde\sigma}(\boldsymbol{r},t)$ and $\underline{\tilde D}(\boldsymbol{r},t)$ are respectively the electrical conductivity and diffusion tensors.

As usual, we define the spatial and temporal Fourier transforms of the various quantities involved by formulas of the type:[1]

$$A(\boldsymbol{q},\omega) = \int d\boldsymbol{r}\int A(\boldsymbol{r},t)e^{i(\omega t-\boldsymbol{q}.\boldsymbol{r})}\,dt. \tag{16.1.5}$$

[1] We use the same notation $A(.,.)$ for a quantity $A(\boldsymbol{r},t)$ and its spatial Fourier transform $A(\boldsymbol{q},t)$, as well as for its spatial and temporal Fourier transform $A(\boldsymbol{q},\omega)$.

By Fourier transforming equations (16.1.3) and (16.1.4), we see that $\langle\rho(\boldsymbol{q},\omega)\rangle$ and $\phi(\boldsymbol{q},\omega)$ are related by:

$$\langle\rho(\boldsymbol{q},\omega)\rangle = -\frac{\sum\limits_{\alpha,\beta}\sigma_{\alpha\beta}(\boldsymbol{q},\omega)q_\alpha q_\beta}{-i\omega+\sum\limits_{\alpha,\beta}D_{\alpha\beta}(\boldsymbol{q},\omega)q_\alpha q_\beta}\,\phi(\boldsymbol{q},\omega). \tag{16.1.6}$$

This gives us the following linear relation between the Fourier transforms of the average electric current density and the electric field:

$$\langle J_\alpha(\boldsymbol{q},\omega)\rangle = \sum_\beta\left[\sigma_{\alpha\beta}(\boldsymbol{q},\omega) - D_{\alpha\beta}(\boldsymbol{q},\omega)\frac{\sum\limits_{\alpha',\beta'}\sigma_{\alpha'\beta'}(\boldsymbol{q},\omega)q_{\alpha'}q_{\beta'}}{-i\omega+\sum\limits_{\alpha',\beta'}D_{\alpha'\beta'}(\boldsymbol{q},\omega)q_{\alpha'}q_{\beta'}}\right]E_\beta(\boldsymbol{q},\omega). \tag{16.1.7}$$

1.2. Properties in static and homogeneous regime

Let us study the properties of this system of charges in the static and homogeneous limit in which both ω and $\boldsymbol{q}$ tend towards zero. Two cases can be distinguished, depending on the respective order of the limits $\omega\to 0$ and $\boldsymbol{q}\to 0$. We will examine them successively.

- *The 'rapid' case*

The wave vector and the angular frequency both tend towards zero, but $\boldsymbol{q}$ tends towards zero first. More precisely, we assume $\omega \gg D_{\alpha\beta}q_\alpha q_\beta$. In this case, the system of charges does not have enough time to adjust to the spatial variation of the potential and it remains homogeneous. As shown by formula (16.1.7), we can write approximately at small $\boldsymbol{q}$:

$$\langle J_\alpha(\boldsymbol{q},\omega)\rangle \simeq \sum_\beta \sigma_{\alpha\beta}(\boldsymbol{q}=0,\omega)E_\beta(\boldsymbol{q},\omega). \tag{16.1.8}$$

The Kubo–Nakano formula for $\sigma_{\alpha\beta}(\boldsymbol{q}=0,\omega)$ reads:

$$\sigma_{\alpha\beta}(\boldsymbol{q}=0,\omega) = \frac{1}{V}\lim_{\epsilon\to 0^+}\int_0^\infty dt\, e^{(i\omega-\epsilon)t}\int_0^\beta \langle J_\beta(\boldsymbol{q}=0,-i\hbar\lambda)J_\alpha(\boldsymbol{q}=0,t)\rangle\, d\lambda \tag{16.1.9}$$

(V is the volume of the sample). Introducing the current density operator of the homogeneous system, $\boldsymbol{J}(t) = V^{-1}\int \boldsymbol{J}(\boldsymbol{r},t)\,d\boldsymbol{r} = V^{-1}\boldsymbol{J}(\boldsymbol{q}=0,t)$, we can rewrite formula (16.1.9) as:

$$\sigma_{\alpha\beta}(\boldsymbol{q}=0,\omega) = V\lim_{\epsilon\to 0^+}\int_0^\infty dt\, e^{(i\omega-\epsilon)t}\int_0^\beta \langle J_\beta(-i\hbar\lambda)J_\alpha(t)\rangle\, d\lambda. \tag{16.1.10}$$

Taking then the limit $\omega\to 0$, we obtain the static conductivity tensor:

$$\sigma_{\alpha\beta} = V\lim_{\epsilon\to 0^+}\int_0^\infty dt\, e^{-\epsilon t}\int_0^\beta \langle J_\beta(-i\hbar\lambda)J_\alpha(t)\rangle\, d\lambda. \tag{16.1.11}$$

• *The 'slow' case*

The wave vector and the angular frequency both tend towards zero, but ω tends towards zero first. Otherwise stated, we assume $\omega \ll D_{\alpha\beta}q_\alpha q_\beta$. No current can flow, since we are in the presence of a perfectly defined static potential. The system of charges is thus in thermal equilibrium and the current vanishes:

$$\langle J_\alpha(\boldsymbol{q}, \omega = 0)\rangle = 0. \qquad (16.1.12)$$

As shown by formula (16.1.7), we then have, in the limit $\omega \to 0$:

$$\sigma_{\alpha\beta} = D_{\alpha\beta}\frac{\sum_{\alpha',\beta'}\sigma_{\alpha'\beta'}q_{\alpha'}q_{\beta'}}{\sum_{\alpha',\beta'}D_{\alpha'\beta'}q_{\alpha'}q_{\beta'}}. \qquad (16.1.13)$$

Formula (16.1.6) then reads:

$$\big\langle\rho(\boldsymbol{q}, \omega = 0)\big\rangle = -\frac{\sum_{\alpha',\beta'}\sigma_{\alpha'\beta'}q_{\alpha'}q_{\beta'}}{\sum_{\alpha',\beta'}D_{\alpha'\beta'}q_{\alpha'}q_{\beta'}}\phi(\boldsymbol{q}, \omega = 0). \qquad (16.1.14)$$

The relation (16.1.13) may only be satisfied for arbitrary $\boldsymbol{q}$ if there is a proportionality relation between $\sigma_{\alpha\beta}(\boldsymbol{q}, \omega = 0)$ and $D_{\alpha\beta}(\boldsymbol{q}, \omega = 0)$ of the form:

$$\sigma_{\alpha\beta}(\boldsymbol{q}, \omega = 0) = aD_{\alpha\beta}(\boldsymbol{q}, \omega = 0), \qquad (16.1.15)$$

where a is a $\boldsymbol{q}$-independent constant. Formula (16.1.14) shows in addition that a may be deduced from the relation between $\langle\rho(\boldsymbol{q}, \omega = 0)\rangle$ and $\phi(\boldsymbol{q}, \omega = 0)$, which reads, on account of formula (16.1.15):

$$\big\langle\rho(\boldsymbol{q}, \omega = 0)\big\rangle = -a\phi(\boldsymbol{q}, \omega = 0). \qquad (16.1.16)$$

After the proportionality constant a has been determined, formula (16.1.15) allows us to deduce $D_{\alpha\beta}$ from $\sigma_{\alpha\beta}$.

1.3. Relation between the conductivity and diffusion tensors

Formula (16.1.16) gives a prescription for determining a. The quantity $\langle\rho(\boldsymbol{q}, \omega = 0)\rangle$ is the charge density of a system in equilibrium in an external potential of Fourier coefficient $\phi(\boldsymbol{q}, \omega = 0)$. The constant a can thus be obtained from the equilibrium properties of this system.

When the system of charge density $\rho(\boldsymbol{r})$ is submitted to a potential $\phi(\boldsymbol{r}, t)$, the corresponding perturbation Hamiltonian reads:

$$H_1(t) = \int \phi(\boldsymbol{r}, t)\rho(\boldsymbol{r})\,d\boldsymbol{r}. \qquad (16.1.17)$$

Formulas (16.1.16) and (16.1.17) show that the proportionality constant a between $\sigma_{\alpha\beta}$ and $D_{\alpha\beta}$ represents a static susceptibility of the type $\chi_{BA}(\omega = 0)$, in which the operators A and B are both identical to the charge density $\rho(\boldsymbol{r}) = en(\boldsymbol{r})$. More precisely, we have:

$$a = e^2 \lim_{\boldsymbol{q}\to 0} \chi_{nn}(\boldsymbol{q}). \tag{16.1.18}$$

The susceptibility $\chi_{nn}(\boldsymbol{q} = 0)$ may be obtained from the thermodynamic sum rule:

$$\chi_{nn}(\boldsymbol{q} = 0) = V\beta\langle(\Delta n)^2\rangle. \tag{16.1.19}$$

According to the theory of equilibrium thermodynamic fluctuations, we have:

$$\langle(\Delta n)^2\rangle = \frac{kT}{V}\frac{\partial n}{\partial \mu}\bigg|_T, \tag{16.1.20}$$

where μ denotes the chemical potential of the charge carriers. We deduce from formulas (16.1.19) and (16.1.20) the expression for $\chi_{nn}(\boldsymbol{q} = 0)$:

$$\chi_{nn}(\boldsymbol{q} = 0) = \frac{\partial n}{\partial \mu}\bigg|_T. \tag{16.1.21}$$

We then have, according to formula (16.1.18):

$$a = e^2 \left(\frac{\partial \mu}{\partial n}\bigg|_T\right)^{-1}. \tag{16.1.22}$$

We deduce from equations (16.1.15) and (16.1.22) the following relation between the components of the same indices of the tensors $\underline{\sigma}$ and $\underline{D}$ in uniform and static regime:

$$\boxed{\frac{D_{\alpha\beta}}{\sigma_{\alpha\beta}} = \frac{1}{e^2}\frac{\partial \mu}{\partial n}\bigg|_T.} \tag{16.1.23}$$

In the case of a non-degenerate electron gas, we have $(\partial\mu/\partial n)_T = kT/n$, and formula (16.1.23) is the usual Einstein relation.

As a general rule, formula (16.1.23) allows us to determine the diffusion tensor given the conductivity tensor, which itself is computed using the Kubo–Nakano formula (16.1.11). The whole procedure is, for this reason, qualified as the 'indirect Kubo method'.[2]

[2] The indirect Kubo method may be used to derive the Green–Kubo formulas associated with other thermal transport coefficients, such as the thermal conductivity and the viscosity coefficient of a gas or a liquid.

2. The source of entropy and the equivalent 'Hamiltonian'

It turns out to be possible to describe a system submitted to a thermal perturbation in a formally Hamiltonian way, introducing an equivalent perturbation 'Hamiltonian' $H_1(t)$ determined from the entropy production.

To study the relation between the equivalent 'Hamiltonian' $H_1(t)$ and the source of entropy, let us first consider the case of a mechanical perturbation, namely the example of a conductor at uniform temperature submitted to an external potential $\phi(\boldsymbol{r},t)$. This perturbation is described by the Hamiltonian $H_1(t)$ given by formula (16.1.17). Defining the derivative:

$$\dot{H}_1 = \int \phi(\boldsymbol{r},t)\dot{\rho}(\boldsymbol{r})\,d\boldsymbol{r}, \tag{16.2.1}$$

where $\dot{\rho}(\boldsymbol{r})$ corresponds to the unperturbed evolution of $\rho(\boldsymbol{r})$, we verify (using integration by parts) the relation:

$$\dot{H}_1 = -\int \boldsymbol{E}(\boldsymbol{r},t).\boldsymbol{J}(\boldsymbol{r})\,d\boldsymbol{r}. \tag{16.2.2}$$

The entropy source, related to the Joule effect, reads in this case:

$$\sigma_S = \frac{1}{T}\boldsymbol{E}(\boldsymbol{r},t).\boldsymbol{J}(\boldsymbol{r}). \tag{16.2.3}$$

The comparison of formulas (16.2.2) and (16.2.3) shows that $\dot{H}_1/T$ identifies with the opposite of the entropy production within the system:

$$\boxed{\dot{H}_1 = -T\int \sigma_S\,d\boldsymbol{r}.} \tag{16.2.4}$$

We now intend to generalize this relation in order to be able to define $\dot{H}_1$ in situations for which we cannot write a perturbation Hamiltonian, such as heat conduction in a fluid. Such an approach of the calculation of the thermal transport coefficients can, in this sense, be qualified as mechanical.[3]

2.1. Entropy production due to heat conduction

Consider a pure fluid in which, at equilibrium, the temperature and the chemical potential are fixed and respectively equal to T_0 and μ_0. In the presence of a flow, the fluid possesses a global mean velocity $\boldsymbol{u}_0$. We assume that this is not the case here ($\boldsymbol{u}_0 = 0$). We are interested in a situation in which the fluid departs from equilibrium,

[3] Note that this does not signify that we have a detailed microscopic knowledge of the mechanical forces governing the considered processes.

but remains however in local equilibrium. The temperature, the chemical potential, and the local mean velocity become respectively:

$$\begin{cases} T(\boldsymbol{r},t) = T_0 + \delta T(\boldsymbol{r},t) \\ \mu(\boldsymbol{r},t) = \mu_0 + \delta\mu(\boldsymbol{r},t) \\ \boldsymbol{u}(\boldsymbol{r},t) = \delta\boldsymbol{u}(\boldsymbol{r},t). \end{cases} \tag{16.2.5}$$

The irreversible processes generated by this departure from equilibrium give rise to a production of entropy within the fluid. In a pure fluid, there are no diffusive fluxes. Besides, for the sake of simplicity, we do not take into account here the viscous dissipative effects. The source of entropy is thus solely related to heat transport. It reads:

$$\boxed{\sigma_S = \boldsymbol{J}_Q.\nabla\left(\frac{1}{T}\right),} \tag{16.2.6}$$

where $\boldsymbol{J}_Q$ is the heat flux corresponding to the transport by thermal conduction.

2.2. Thermal perturbation

Let us thus consider the out-of-equilibrium fluid and let us introduce the 'Hamiltonian':

$$H_1(t) = \int h_1(\boldsymbol{r},t)\, d\boldsymbol{r}, \tag{16.2.7}$$

with:

$$h_1(\boldsymbol{r},t) = \frac{\delta T(\boldsymbol{r},t)}{T}\Big[\varepsilon(\boldsymbol{r},t) - \frac{\varepsilon+\mathcal{P}}{n} n(\boldsymbol{r},t)\Big]. \tag{16.2.8}$$

In formula (16.2.8), $n(\boldsymbol{r},t)$ and $\varepsilon(\boldsymbol{r},t)$ denote respectively the local densities of particles and energy, whereas n, ε, $\mathcal{P}$ are the global thermodynamic equilibrium values of the particle density, the energy density, and the pressure.[4] It can be shown that the quantity $\varepsilon(\boldsymbol{r},t) - \big[(\varepsilon+\mathcal{P})/n\big]n(\boldsymbol{r},t)$ represents a local density of thermal energy. Indeed, at fixed number of particles, we have the thermodynamic relation:

$$TdS = dE + \mathcal{P}dV. \tag{16.2.9}$$

If $dN = 0$, we have $dV/V = -dn/n$ and $dE = d(\varepsilon V) = Vd\varepsilon + \varepsilon dV = V[d\varepsilon - (\varepsilon/n)dn]$. We then deduce from the relation (16.2.9) the equality:

$$\frac{T}{V}dS = d\varepsilon - \frac{\varepsilon+\mathcal{P}}{n}dn, \tag{16.2.10}$$

which allows us to interpret $[\varepsilon(\boldsymbol{r},t) - (\varepsilon+\mathcal{P})/n]n(\boldsymbol{r},t)$ as a local density of thermal energy.

[4] For short, the temperature T_0 is from now on denoted by T.

2.3. Relation between $\dot{H}_1$ and σ_S

In order to apply the linear response theory to the system perturbed by $H_1(t)$, we first note that this operator is of the general form:

$$H_1 = -\int a(\boldsymbol{r},t)A(\boldsymbol{r})\,d\boldsymbol{r}, \tag{16.2.11}$$

where $a(\boldsymbol{r},t)$ denotes an applied field coupled to a physical quantity $A(\boldsymbol{r})$. In the case of a thermal perturbation, formula (16.2.8) shows that the quantity coupled to the applied field $a(\boldsymbol{r},t) = -T^{-1}\delta T(\boldsymbol{r},t)$ is the local density of thermal energy.

The derivative $\dot{H}_1$ is given by:

$$\dot{H}_1 = -\int a(\boldsymbol{r},t)\dot{A}(\boldsymbol{r})\,d\boldsymbol{r}. \tag{16.2.12}$$

The evolution of the local density of thermal energy can be deduced from the evolution of $n(\boldsymbol{r},t)$ and $\varepsilon(\boldsymbol{r},t)$, as governed by the hydrodynamic equations:

$$\left\{\begin{aligned} &\frac{\partial n(\boldsymbol{r},t)}{\partial t} + \frac{1}{m}\nabla.\boldsymbol{g}(\boldsymbol{r},t) = 0 \\ &\frac{\partial \varepsilon(\boldsymbol{r},t)}{\partial t} + \nabla.\boldsymbol{J}_E(\boldsymbol{r},t) = 0, \end{aligned}\right. \tag{16.2.13}$$

in which $\boldsymbol{g}(\boldsymbol{r},t)$ denotes the local density of kinetic momentum and $\boldsymbol{J}_E(\boldsymbol{r},t)$ the energy flux. In its linearized form, valid for fluctuations of small amplitude, the kinetic momentum density reads:

$$\boldsymbol{g}(\boldsymbol{r},t) = mn\boldsymbol{u}(\boldsymbol{r},t). \tag{16.2.14}$$

As for the linearized energy flux, it reads:[5]

$$\boldsymbol{J}_E(\boldsymbol{r},t) = (\varepsilon + \mathcal{P})\boldsymbol{u}(\boldsymbol{r},t) + \boldsymbol{J}_Q. \tag{16.2.15}$$

To examine the relevance of the expression (16.2.8) proposed for $h_1(\boldsymbol{r},t)$, we have to check that the derivative $\dot{H}_1$ deduced from it effectively verifies the relation (16.2.4) (the entropy source σ_S being given by formula (16.2.6)). Coming back to the definition (16.2.12) of $\dot{H}_1$, and making use of the evolution equations (16.2.13), we obtain:

$$\dot{H}_1 = -\int \frac{\delta T(\boldsymbol{r},t)}{T}\nabla.\Big(\boldsymbol{J}_E - \frac{\varepsilon+\mathcal{P}}{mn}\boldsymbol{g}\Big)\,d\boldsymbol{r}, \tag{16.2.16}$$

[5] The use of the hydrodynamic equations assumes slow variations in space and in time of the various quantities involved. The hydrodynamic regime is defined by the inequalities:

$$\omega\tau \ll 1, \qquad q\ell \ll 1,$$

in which ω and $\boldsymbol{q}$ denote an angular frequency and a wave vector typical of the perturbations imposed on the medium, τ the collision time, and ℓ the mean free path.

that is:

$$\dot{H}_1 = -\int \frac{\delta T(\boldsymbol{r},t)}{T}(\nabla.\boldsymbol{J}_Q)\,d\boldsymbol{r}. \tag{16.2.17}$$

The integral on the right-hand side of equation (16.2.17) can be recast in the following form:

$$\int \delta T(\boldsymbol{r},t)\big[\nabla.\boldsymbol{J}_Q(\boldsymbol{r})\big]\,d\boldsymbol{r} = \int \nabla.\big[\delta T(\boldsymbol{r},t)\boldsymbol{J}_Q(\boldsymbol{r})\big]\,d\boldsymbol{r} - \int \boldsymbol{J}_Q(\boldsymbol{r}).\nabla\delta T(\boldsymbol{r},t)\,d\boldsymbol{r}. \tag{16.2.18}$$

The first term on the right-hand side of equation (16.2.18) may be transformed into a surface integral of the flux $\delta T(\boldsymbol{r},t)\boldsymbol{J}_Q$, and then cancelled owing to a convenient choice of boundary conditions. This gives:

$$\dot{H}_1 = -T\int \boldsymbol{J}_Q.\nabla(\frac{1}{T})\,d\boldsymbol{r}. \tag{16.2.19}$$

Comparing formulas (16.2.19) and (16.2.6), we actually verify the relation (16.2.4) between $\dot{H}_1$ and σ_S. This justifies a posteriori the introduction of the operator $H_1(t)$ defined by formulas (16.2.7) and (16.2.8) as an equivalent perturbation 'Hamiltonian'.

2.4. The Green–Kubo formula for the thermal conductivity

Consider a system inside which a temperature gradient is applied along the direction β. We are looking for the average value of the heat current along the direction α. Having identified the equivalent Hamiltonian $H_1(t)$, we can apply the linear response theory with $\dot{A}(\boldsymbol{r}) = J_{Q\beta}(\boldsymbol{r})$ and $B(\boldsymbol{r}) = J_{Q\alpha}(\boldsymbol{r})$. The applied field is $-T^{-1}\nabla_\beta\delta T(\boldsymbol{r},t)$. To compute $\langle J_{Q\alpha}(\boldsymbol{r},t)\rangle$, we write a non-local and retarded linear response relation:

$$\big\langle J_{Q\alpha}(\boldsymbol{r},t)\big\rangle = -\frac{1}{T}\int d\boldsymbol{r}'\int_{-\infty}^{\infty}\tilde{\chi}_{BA}(\boldsymbol{r}-\boldsymbol{r}',t-t')\nabla_\beta\delta T(\boldsymbol{r}',t')\,dt'. \tag{16.2.20}$$

Expressing the response function $\tilde{\chi}_{BA}(\boldsymbol{r}-\boldsymbol{r}',t-t')$ with the aid of the canonical Kubo correlation function[6] gives:

$$\big\langle J_{Q\alpha}(\boldsymbol{r},t)\big\rangle = -\frac{1}{T}\int d\boldsymbol{r}'\int_{-\infty}^{t}dt'\int_0^\beta \big\langle J_{Q\beta}(\boldsymbol{r}',-i\hbar\lambda)J_{Q\alpha}(\boldsymbol{r},t-t')\big\rangle\nabla_\beta\delta T(\boldsymbol{r}',t')\,d\lambda. \tag{16.2.21}$$

By Fourier transformation of equation (16.2.21), we get:

$$\big\langle J_{Q\alpha}(\boldsymbol{q},\omega)\big\rangle = -\kappa_{\alpha\beta}(\boldsymbol{q},\omega)[\nabla_\beta\delta T](\boldsymbol{q},\omega). \tag{16.2.22}$$

[6] The derivation of the thermal conductivity tensor may be carried out in a classical framework if we are interested in the case of a gas or a liquid in the ordinary sense. However, the quantum formulation of the Green–Kubo formula leaves open the possibility of applying it in other situations, for instance to get the thermal conductivity tensor of a degenerate electron gas (metal).

The components $\kappa_{\alpha\beta}(\boldsymbol{q},\omega)$ of the thermal conductivity tensor $\underline{\kappa}(\boldsymbol{q},\omega)$ are given by:[7]

$$\boxed{\kappa_{\alpha\beta}(\boldsymbol{q},\omega)=\frac{1}{VT}\lim_{\epsilon\to 0^+}\int_0^\infty dt\, e^{(i\omega-\epsilon)t}\int_0^\beta \langle J_{Q\beta}(-\boldsymbol{q},-i\hbar\lambda)J_{Q\alpha}(\boldsymbol{q},t)\rangle\, d\lambda.}\tag{16.2.23}$$

The Green–Kubo formula (16.2.23) is an expression for the thermal conductivity in terms of an equilibrium correlation function of the spatial Fourier transforms of the heat current densities.

In an isotropic fluid, the thermal conductivity tensor is diagonal: $\kappa_{\alpha\beta}(\boldsymbol{q},\omega)=\kappa(\boldsymbol{q},\omega)\delta_{\alpha\beta}$. We simply have (when the classical description suffices):

$$\kappa(\boldsymbol{q},\omega)=\frac{1}{kT^2V}\lim_{\epsilon\to 0^+}\int_0^\infty dt\, e^{(i\omega-\epsilon)t}\langle J_Q(-\boldsymbol{q},0)J_Q(\boldsymbol{q},t)\rangle.\tag{16.2.24}$$

[7] The calculations are similar to those carried out to derive the Kubo–Nakano formula for $\sigma_{\alpha\beta}(\boldsymbol{q},\omega)$.

Bibliography

J.-P. HANSEN and I.R. McDONALD, *Theory of simple liquids*, Academic Press, London, third edition, 2006.

R. KUBO, M. TODA, and N. HASHITSUME, *Statistical physics II: nonequilibrium statistical mechanics*, Springer-Verlag, Berlin, second edition, 1991.

L.D. LANDAU and E.M. LIFSHITZ, *Fluid mechanics*, Butterworth-Heinemann, Oxford, second edition, 1987.

References

M.S. GREEN, Markoff random processes and the statistical mechanics of time-dependent phenomena, *J. Chem. Phys.* **20**, 1281 (1952).

M.S. GREEN, Markoff random processes and the statistical mechanics of time-dependent phenomena. II. Irreversible processes in fluids, *J. Chem. Phys.* **22**, 398 (1954).

R. KUBO, Statistical-mechanical theory of irreversible processes. II. Response to thermal disturbance, *J. Phys. Soc. Japan* **12**, 1203 (1957).

L.P. KADANOFF and P.C. MARTIN, Hydrodynamic equations and correlation functions, *Ann. Phys.* **24**, 419 (1963).

J.M. LUTTINGER, Theory of thermal transport coefficients, *Phys. Rev.* **135**, A1505 (1964).

R. ZWANZIG, Time-correlation functions and transport coefficients in statistical mechanics, *Ann. Rev. Phys. Chem.* **16**, 67 (1965).

Supplement 16A

Diffusive light waves

1. Diffusive light transport

This supplement deals with the transport of light waves in the presence of multiple scatterers. The transport of these waves in a random medium takes place, as a first approximation, in a diffusive (thus, non-propagative) way. From the transport properties of light waves, we can get information about the dynamics of the scatterers present in the medium.

The diffusion equation is a classical equation which neglects the interference effects associated with wave propagation. At this level of description, there is no difference between the diffusion of particles and that of the wave intensity. In the presence of absorption, the diffusion equation for the local intensity $I(\boldsymbol{r}, t)$ reads:

$$\frac{\partial I(\boldsymbol{r},t)}{\partial t} = D\nabla^2 I(\boldsymbol{r},t) - D\kappa^2 I(\boldsymbol{r},t) + S(\boldsymbol{r},t), \qquad (16A.1.1)$$

where D denotes the diffusion coefficient of light intensity, and $S(\boldsymbol{r}, t)$ a source term possibly present in the medium. The second term on the right-hand side of equation (16A.1.1) accounts for absorption, the absorption length being $L_{\text{abs}} = \kappa^{-1}$.

We now assume that there is no source term ($S(\boldsymbol{r}, t) = 0$), and we take the initial condition $I(\boldsymbol{r}, t = 0) = I_0\delta(\boldsymbol{r})$. Considering for the sake of simplicity an infinite medium, we introduce the spatial Fourier transform[1] of $I(\boldsymbol{r}, t)$,

$$I(\boldsymbol{q},t) = \int I(\boldsymbol{r},t)e^{-i\boldsymbol{q}.\boldsymbol{r}}\,d\boldsymbol{r}, \qquad (16A.1.2)$$

and then the Fourier–Laplace transform $I(\boldsymbol{q}, z)$ of $I(\boldsymbol{q}, t)$, defined for a complex argument z of positive imaginary part:

$$I(\boldsymbol{q},z) = \int_0^\infty I(\boldsymbol{q},t)e^{izt}\,dt, \qquad \Im m\, z > 0. \qquad (16A.1.3)$$

[1] We use the same notation $I(.,.)$ for the intensity $I(\boldsymbol{r}, t)$, its spatial Fourier transform $I(\boldsymbol{q}, t)$, and the Fourier–Laplace transform of this latter quantity, $I(\boldsymbol{q}, z)$.

Equation (16A.1.1) yields:

$$I(\boldsymbol{q}, z) = \frac{I_0}{-iz + Dq^2 + D\kappa^2}. \tag{16A.1.4}$$

At $\kappa = 0$, $I(\boldsymbol{q}, z)$ has a pole, designated as the *diffusion pole*, situated in the lower complex half-plane, and whose affix $z_0(\kappa = 0) = -iDq^2$ vanishes as $\boldsymbol{q} \to 0$. In the presence of absorption, $I(\boldsymbol{q}, z)$ still has a pole in the lower complex half-plane, of affix:

$$z_0(\kappa) = -iD(q^2 + \kappa^2). \tag{16A.1.5}$$

To compute $I(\boldsymbol{q}, t)$, we use the inverse Fourier–Laplace transformation:

$$I(\boldsymbol{q}, t) = \frac{I_0}{2\pi} \int_C \frac{e^{-izt}}{-iz + D(q^2 + \kappa^2)} \, dz. \tag{16A.1.6}$$

In equation (16A.1.6), the integration contour C is a parallel to the abscissa's axis of ordinate strictly higher than that of $z_0(\kappa)$. Applying the residue theorem gives:[2]

$$I(\boldsymbol{q}, t) = I_0 e^{-D(q^2 + \kappa^2)t}, \qquad t > 0. \tag{16A.1.7}$$

The intensity $I(\boldsymbol{r}, t)$ follows by inverse Fourier transformation:

$$I(\boldsymbol{r}, t) = I_0 e^{-D\kappa^2 t} \frac{1}{(2\pi)^3} \int e^{-Dq^2 t} e^{i\boldsymbol{q}.\boldsymbol{r}} \, d\boldsymbol{q}. \tag{16A.1.8}$$

This gives:

$$I(\boldsymbol{r}, t) = I_0 \, (4\pi Dt)^{-3/2} \exp\Big(-\frac{r^2}{4Dt}\Big) \exp(-D\kappa^2 t), \qquad t > 0. \tag{16A.1.9}$$

2. Diffusion coefficient of light intensity

2.1. The radiative transfer equation

Our aim is to obtain a microscopic expression for the diffusion coefficient of light intensity. To this end, we start from the balance equation playing the role of a 'Boltzmann equation' for the transport of light energy in a medium containing multiple scatterers. This is the *radiative transfer equation*[3] obeyed by the specific intensity $\mathcal{I}(\boldsymbol{r}, \boldsymbol{n}, t)$:

$$\boxed{\tau \frac{\partial}{\partial t} \mathcal{I}(\boldsymbol{r}, \boldsymbol{n}, t) + \ell \boldsymbol{n}.\nabla_{\boldsymbol{r}} \mathcal{I}(\boldsymbol{r}, \boldsymbol{n}, t) = \frac{1}{4\pi} \int p(\boldsymbol{n}, \boldsymbol{n}') \mathcal{I}(\boldsymbol{r}, \boldsymbol{n}', t) \, d\Omega' - \mathcal{I}(\boldsymbol{r}, \boldsymbol{n}, t).} \tag{16A.2.1}$$

[2] We can also directly solve the differential equation obeyed by $I(\boldsymbol{q}, t)$, which reads:

$$\frac{\partial I(\boldsymbol{q}, t)}{\partial t} = -D(q^2 + \kappa^2) I(\boldsymbol{q}, t), \qquad I(\boldsymbol{q}, t) = I_0.$$

This ends in the result (16A.1.7) without recourse to the Fourier–Laplace transformation. However, the use of this transformation is highly recommended in the more complex situations in which coupled variables are involved (an example of such a case is provided in Supplement 16B).

[3] The study of the diffusion of light waves began in astrophysics, with a view to understanding how the radiation emitted at the center of stars is affected by passing through an interstellar cloud. The radiative transfer equation was first introduced and used in this context.

The definition of $\mathcal{I}(\boldsymbol{r},\boldsymbol{n},t)$ is as follows. The radiation energy contained in an interval of angular frequencies $(\omega, \omega + d\omega)$, transported during the time interval dt through a surface element $d\sigma$ centered at the point $\boldsymbol{r}$ in directions belonging to a solid angle $d\Omega$ centered around the direction $\boldsymbol{n} = (\theta, \phi)$, is:

$$dE = \mathcal{I}(\boldsymbol{r},\boldsymbol{n},t)\cos\theta \, d\omega d\sigma d\Omega dt, \tag{16A.2.2}$$

where θ denotes the angle between $\boldsymbol{n}$ and the normal to the surface element $d\sigma$ (Fig. 16A.1).

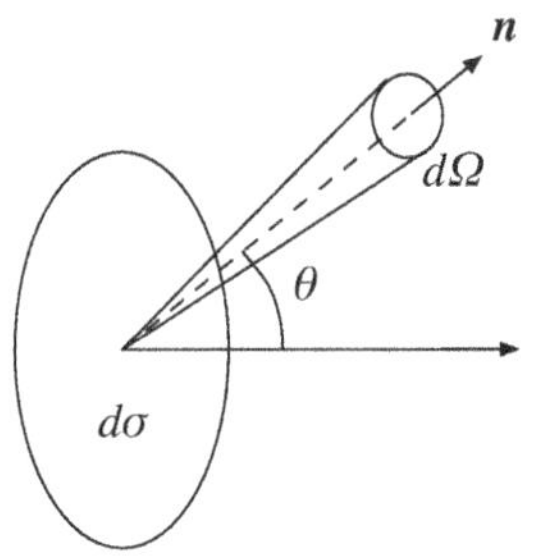

Fig. 16A.1 Geometry for the definition of the specific intensity.

In equation (16A.2.1), $p(\boldsymbol{n},\boldsymbol{n}')(d\Omega'/4\pi)(d\Omega/4\pi)$ is the fraction of the radiation intensity entering a cone of solid angle $d\Omega'$ around the incident direction $\boldsymbol{n}'$, and leaving a cone of solid angle $d\Omega$ around the scattered direction $\boldsymbol{n}$. For a medium containing spherically symmetric scatterers, $p(\boldsymbol{n},\boldsymbol{n}')$ is a function of $\boldsymbol{n}'.\boldsymbol{n} = \cos\Theta$, where Θ is the scattering angle (that is, the angle between the incident and scattered directions). The time τ is the average time separating two successive collision events, and the length ℓ is the corresponding mean free path. Both parameters are assumed to be constant.

2.2. Evolution of the local densities of radiation and current

The local density of radiation $I(\boldsymbol{r},t)$ and the local density of current $\boldsymbol{J}(\boldsymbol{r},t)$ are defined in terms of the specific intensity by the formulas:

$$I(\boldsymbol{r},t) = \int \mathcal{I}(\boldsymbol{r},\boldsymbol{n},t)\, d\Omega, \qquad \boldsymbol{J}(\boldsymbol{r},t) = \frac{\ell}{\tau}\int \boldsymbol{n}\mathcal{I}(\boldsymbol{r},\boldsymbol{n},t)\, d\Omega. \tag{16A.2.3}$$

To derive the evolution equation of $I(\boldsymbol{r},t)$, we integrate the radiative transfer equation over Ω. This gives the relation:[4]

$$\frac{\partial I(\boldsymbol{r},t)}{\partial t} + \nabla.\boldsymbol{J}(\boldsymbol{r},t) = -\frac{1-a}{\tau} I(\boldsymbol{r},t). \tag{16A.2.4}$$

In formula (16A.2.4), we have introduced the *albedo* (that is, the 'whiteness') of the scatterer:

$$a = \frac{1}{4\pi}\int p(\boldsymbol{n},\boldsymbol{n}')\, d\Omega. \tag{16A.2.5}$$

[4] In the following, when no confusion can arise, $\nabla_{\boldsymbol{r}}$ will simply be denoted by ∇.

The albedo is a positive quantity, less than 1 or equal to 1, depending on whether the scattering takes place with or without absorption of light energy. Equation (16A.2.4) is a conservation equation for $I(\boldsymbol{r},t)$ only if $a=1$ (which is indeed the case when there is no absorption).

Similarly, multiplying both sides of equation (16A.2.1) by $\boldsymbol{n}$ and integrating over Ω, we obtain the evolution equation of $\boldsymbol{J}(\boldsymbol{r},t)$:

$$\frac{\tau^2}{\ell}\frac{\partial \boldsymbol{J}(\boldsymbol{r},t)}{\partial t}+\frac{\tau}{\ell}\big(1-\langle\cos\Theta\rangle\big)\boldsymbol{J}(\boldsymbol{r},t)=-\ell\int \boldsymbol{n}(\boldsymbol{n}.\nabla_{\boldsymbol{r}})\mathcal{I}(\boldsymbol{r},\boldsymbol{n},t)\,d\Omega. \tag{16A.2.6}$$

In equation (16A.2.6), we have set:

$$\frac{1}{4\pi}\int p(\boldsymbol{n},\boldsymbol{n}')\boldsymbol{n}\,d\Omega=\boldsymbol{n}'\langle\cos\Theta\rangle, \tag{16A.2.7}$$

where the quantity $\langle\cos\Theta\rangle$ is defined by:

$$\langle\cos\Theta\rangle=\langle\boldsymbol{n}.\boldsymbol{n}'\rangle=\frac{1}{4\pi}\int p(\boldsymbol{n},\boldsymbol{n}')\boldsymbol{n}.\boldsymbol{n}'\,d\Omega. \tag{16A.2.8}$$

Equations (16.2.4) and (16.2.6) do not constitute a closed system of equations for $I(\boldsymbol{r},t)$ and $\boldsymbol{J}(\boldsymbol{r},t)$, since the term $\int \boldsymbol{n}(\boldsymbol{n}.\nabla_{\boldsymbol{r}})\mathcal{I}(\boldsymbol{r},\boldsymbol{n},t)\,d\Omega$ involved on the right-hand side of equation (16.2.6) cannot be expressed in terms of $I(\boldsymbol{r},t)$ and $\boldsymbol{J}(\boldsymbol{r},t)$ only.

However, when the distribution of the light intensity is almost isotropic, we can replace the specific intensity $\mathcal{I}(\boldsymbol{r},\boldsymbol{n},t)$ by $I(\boldsymbol{r},t)/4\pi$ in the term $\int \boldsymbol{n}(\boldsymbol{n}.\nabla_{\boldsymbol{r}})\mathcal{I}(\boldsymbol{r},\boldsymbol{n},t)\,d\Omega$. This approximation allows us to obtain a closed system of equations for $I(\boldsymbol{r},t)$ and $\boldsymbol{J}(\boldsymbol{r},t)$, equation (16.2.6) being replaced by the following approximate equation:[5]

$$\frac{\tau^2}{\ell}\frac{\partial \boldsymbol{J}(\boldsymbol{r},t)}{\partial t}+\frac{\tau}{\ell}\big(1-\langle\cos\Theta\rangle\big)\boldsymbol{J}(\boldsymbol{r},t)=-\frac{\ell}{3}\nabla I(\boldsymbol{r},t). \tag{16A.2.9}$$

2.3. Resolution of the coupled equations for $I(\boldsymbol{r},t)$ and $\boldsymbol{J}(\boldsymbol{r},t)$

Our aim is to solve the system of coupled equations (16.2.4) and (16.2.9) with the initial conditions $I(\boldsymbol{r},t=0)=I_0\delta(\boldsymbol{r})$ and $\boldsymbol{J}(\boldsymbol{r},t=0)=0$.

To this end, we introduce the quantity $I(\boldsymbol{q},z)$, defined from $I(\boldsymbol{r},t)$ by formulas (16.1.2) and (16.1.3), and the quantity $\boldsymbol{J}(\boldsymbol{q},z)$, defined in a similar way from $\boldsymbol{J}(\boldsymbol{r},t)$. We can deduce from equations (16.2.4) and (16.2.9) two coupled equations for $I(\boldsymbol{q},z)$ and $\boldsymbol{J}(\boldsymbol{q},z)$:

$$\begin{cases} -izI(\boldsymbol{q},z)+i\boldsymbol{q}.\boldsymbol{J}(\boldsymbol{q},z)=-\dfrac{1-a}{\tau}I(\boldsymbol{q},z)+I_0 \\[2ex] -\dfrac{\tau^2}{\ell}iz\boldsymbol{J}(\boldsymbol{q},z)+\dfrac{\tau}{\ell}\big(1-\langle\cos\Theta\rangle\big)\boldsymbol{J}(\boldsymbol{q},z)=-\dfrac{\ell}{3}i\boldsymbol{q}I(\boldsymbol{q},z). \end{cases} \tag{16A.2.10}$$

---

[5] Equation (16.2.9) is obtained by writing approximately the term on the right-hand side of equation (16.2.6) as $-(\ell/4\pi)\int \boldsymbol{n}(\boldsymbol{n}.\nabla_{\boldsymbol{r}})I(\boldsymbol{r},t)\,d\Omega=-(\ell/3)\nabla_{\boldsymbol{r}}I(\boldsymbol{r},t)$.

In the 'hydrodynamic' limit of slow evolutions as characterized by the inequality $|z|\tau \ll 1$, we get from the system of equations (16.2.10):

$$\boldsymbol{J}(\boldsymbol{q}, z) \simeq -\frac{\ell^2}{3\tau\big(1 - \langle\cos\Theta\rangle\big)}\, i\boldsymbol{q} I(\boldsymbol{q}, z), \tag{16A.2.11}$$

and, consequently:

$$I(\boldsymbol{q}, z) \simeq \frac{I_0}{-iz + q^2 \dfrac{\ell^2}{3\tau(1 - \langle\cos\Theta\rangle)} + \dfrac{1-a}{\tau}}. \tag{16A.2.12}$$

Comparing formula (16.2.12) with formula (16.1.4) shows that the intensity $I(\boldsymbol{r}, t)$ obeys, in the hydrodynamic limit, a diffusion equation with absorption. We identify the diffusion coefficient of light intensity,

$$D = \frac{\ell^2}{3\tau\big(1 - \langle\cos\Theta\rangle\big)}, \tag{16A.2.13}$$

as well as the absorption length:

$$L_{\text{abs}} = \ell\big[3(1-a)(1 - \langle\cos\Theta\rangle)\big]^{-1/2}. \tag{16A.2.14}$$

The diffusion coefficient of light intensity can be written in the form:

$$D = \frac{1}{3} v \ell_{\text{tr}}, \tag{16A.2.15}$$

where:

$$\ell_{\text{tr}} = \frac{\ell}{1 - \langle\cos\Theta\rangle} \tag{16A.2.16}$$

and:

$$v = \frac{\ell}{\tau} \tag{16A.2.17}$$

are respectively the transport mean free path and the *transport velocity*.

3. Diffusive wave spectroscopy

We will now study an application of diffusive light waves to the optical measurement of the dynamical properties of scatterers situated in materials which otherwise would be transparent. Many commonly encountered fluids are actually suspensions. When the concentration of the solid suspended particles is high and their refractive indexes sufficiently different from that of the liquid, the medium appears as opaque because of the strong scattering by these particles, even if the pure liquid is transparent. Traditional optical measurements are thus very difficult to carry out, except in the extremely dilute case.

A technique of optical measurement introduced in 1988 by several groups, in particular by D.J. Pine and D.A. Weitz, allows us to determine the dynamical properties of the suspended particles. Because of the thermal agitation of the molecules of the liquid, the suspended particles move like Brownian particles. The light they scatter depends randomly on time. The dynamical information about the suspension can be deduced from an analysis of the temporal correlation function of the intensity of the scattered light. The diffusive character of the light waves plays a crucial role in the interpretation of the experimental results, hence the name of *diffusive wave spectroscopy* (DWS) given to this technique.

3.1. Principle of the method

Let $I(t) = \phi(t)\phi^*(t)$ be the intensity at time t of the light scattered at a given point $\boldsymbol{r}$, the (point) light source being assumed situated at $\boldsymbol{r} = 0$. The quantity $\phi(t)$ is the amplitude of the electric field of the wave. We set:

$$\frac{\langle I(t)I(0)\rangle}{\langle I\rangle^2} = 1 + g_2(t), \tag{16A.3.1}$$

where the symbol $\langle\ldots\rangle$ denotes the configuration average over the positions of the scatterers. The function $g_2(t)$ is the autocorrelation function of light intensity. As the concentration of scatterers is high, we can write approximately:[6]

$$\langle I(t)I(0)\rangle = \langle\phi(t)\phi^*(t)\rangle\langle\phi(0)\phi^*(0)\rangle + \langle\phi(t)\phi^*(0)\rangle\langle\phi^*(t)\phi(0)\rangle. \tag{16A.3.2}$$

Using formula (16A.3.2), we obtain for the autocorrelation of light intensity the *Siegert formula*:

$$g_2(t) = |g_1(t)|^2, \qquad g_1(t) = \frac{\langle\phi(t)\phi^*(0)\rangle}{\langle|\phi|^2\rangle}. \tag{16A.3.3}$$

We set:

$$\phi(t) = \sum_{n=1}^{\infty} \phi^{(n)}(t), \tag{16A.3.4}$$

where $\phi^{(n)}(t)$ denotes the amplitude of the field of the wave associated with the trajectories in which the light undergoes n scattering events, assumed elastic, between the source and the measurement point. We can write:

$$\phi^{(n)}(t) = \phi^{(n)}(0)\prod_{i=1}^{n} \exp\{-i\boldsymbol{q}_i.[\boldsymbol{r}_i(t) - \boldsymbol{r}_i(0)]\}, \tag{16A.3.5}$$

where the $\{\boldsymbol{r}_i(t)\}$'s represent the positions of the scatterers at time t whereas the $\{\boldsymbol{q}_i\}$'s are the associated scattering wave vectors. We have:

$$|\boldsymbol{q}_i| = 2q\sin\frac{\Theta_i}{2}, \tag{16A.3.6}$$

[6] To justify formula (16A.3.2), we assume that, as a consequence of the high concentration of scatterers, the amplitude $\phi(t)$ may be considered as a Gaussian random process.

where Θ_i is the scattering angle for the ith scattering and q the modulus of the light wave vector. From the decomposition (16A.3.4) of $\phi(t)$ we get the following decomposition of $g_1(t)$,

$$g_1(t) = \sum_{n=1}^{\infty} g_1^{(n)}(t), \tag{16A.3.7}$$

with, according to formula (16A.3.5):

$$g_1^{(n)}(t) = \frac{\langle |\phi^{(n)}(0)|^2 \rangle}{\langle |\phi|^2 \rangle} \Big\langle \prod_{i=1}^{n} \exp\{-i\boldsymbol{q}_i.[\boldsymbol{r}_i(t) - \boldsymbol{r}_i(0)]\} \Big\rangle. \tag{16A.3.8}$$

The quantity:

$$P(n) = \frac{\langle |\phi^{(n)}(0)|^2 \rangle}{\langle |\phi|^2 \rangle} \tag{16A.3.9}$$

is the probability for a trajectory terminating at the measurement point to undergo n scattering events. Besides $P(n)$, which concerns the diffusive properties of light waves, the expression (16A.3.8) for $g_1^{(n)}(t)$ involves a configuration average relative to the suspended particles. This latter average brings into play, on the one hand, an average over the positions $\boldsymbol{r}_i$ of the scatterers, and, on the other hand, an average over the possible directions of the wave vectors $\boldsymbol{q}_i$. Both sets of random variables will be, in first approximation, treated as independent.

3.2. Configuration average relative to the suspended particles

The scatterers being considered as Brownian particles of diffusion coefficient D_p, the probability distribution of $\boldsymbol{r}_i(t) - \boldsymbol{r}_i(0)$ is the Gaussian law characteristic of diffusion:

$$p\big[\boldsymbol{r}_i(t) - \boldsymbol{r}_i(0)\big] = (4\pi D_p t)^{-3/2} \exp\Big(-\frac{|\boldsymbol{r}_i(t) - \boldsymbol{r}_i(0)|^2}{4D_p t}\Big). \tag{16A.3.10}$$

The average over the positions of the various scatterers once carried out, we get, on account of formula (16A.3.6):

$$g_1^{(n)}(t) = P(n) \langle \prod_{i=1}^{n} \exp(-4D_p q^2 t \sin^2 \frac{\Theta_i}{2}) \rangle. \tag{16A.3.11}$$

In formula (16A.3.11), it remains to average over the scattering angles. We can set[7] $t_0 = (D_p q^2)^{-1}$. At lowest order in t/t_0, the average over the scattering angles may be approximated by:

$$\langle \prod_{i=1}^{n} \exp(-4D_p q^2 t \sin^2 \frac{\Theta_i}{2}) \rangle \simeq \exp\Big[-2n\frac{t}{t_0}(1 - \langle \cos\Theta_i \rangle)\Big]. \tag{16A.3.12}$$

---

[7] Physically, t_0 represents the time necessary for a suspended particle to diffuse over a distance of the order of the light wavelength.

As a consequence we get, in terms of the mean free path ℓ and the transport mean free path ℓ_{tr} defined by formula (16A.2.16):

$$g_1^{(n)}(t) \simeq P(n) \exp\Big(-2n\frac{t}{t_0}\frac{\ell}{\ell_{tr}}\Big) \tag{16A.3.13}$$

and:

$$g_1(t) \simeq \sum_{n=1}^{\infty} P(n) \exp\Big(-2n\frac{t}{t_0}\frac{\ell}{\ell_{tr}}\Big). \tag{16A.3.14}$$

3.3. Computation of $P(n)$

The length of a light trajectory undergoing n scattering events is $n\ell = s$. We now choose as variable, instead of the discrete variable n, the continuous variable s of probability density $p(s)$. Since we have:

$$P(n) = p(s)\, ds, \tag{16A.3.15}$$

we can write, instead of formula (16A.3.14):

$$g_1(t) \simeq \int_0^{\infty} p(s) \exp\Big(-2\frac{t}{t_0}\frac{s}{\ell_{tr}}\Big)\, ds. \tag{16A.3.16}$$

The exact detailed calculation of $p(s)$ is intricate. We can show that this quantity is dominated by an exponential factor $\exp(-r^2/4D\tau_{ph})$, resulting from the diffusive character of light waves ($\tau_{ph} = s/v$ is the photon transit time and $D = v\ell_{tr}/3$ the diffusion coefficient of the waves). More precisely, we can show that the density $p(s)$ is of the form:

$$p(s) = \phi(s) \exp\Big(-\frac{3r^2}{4s\ell_{tr}}\Big), \tag{16A.3.17}$$

where $\phi(s)$ is a function decreasing like a power law.

3.4. Autocorrelation function of light intensity

Replacing $p(s)$ by its expression (16A.3.17) in the formula (16A.3.16) for $g_1(t)$, gives:

$$g_1(t) \simeq \int_0^{\infty} \phi(s) \exp\Big(-\frac{3r^2}{4s\ell_{tr}}\Big) \exp\Big(-2\frac{t}{t_0}\frac{s}{\ell_{tr}}\Big)\, ds. \tag{16A.3.18}$$

We can deduce from formula (16A.3.18) the leading behavior of $g_1(t)$. The argument of the exponential involved in the integrand has indeed a maximum, situated at the point s_0 where the derivative of the argument vanishes:

$$\frac{3r^2}{4\ell_{tr}}\frac{1}{s_0^2} - 2\frac{t}{t_0}\frac{1}{\ell_{tr}} = 0. \tag{16A.3.19}$$

We have $s_0 = r\,(3t_0/8t)^{1/2}$. Since this maximum takes place in an exponent, its position determines the behavior of $g_1(t)$. This gives, approximately:

$$g_1(t) \simeq \exp\left[-\frac{r}{\ell_{\text{tr}}}\left(\frac{6t}{t_0}\right)^{1/2}\right], \tag{16A.3.20}$$

which yields for the light intensity autocorrelation function the behavior:

$$g_2(t) \simeq \exp\left[-2\frac{r}{\ell_{\text{tr}}}\left(\frac{6t}{t_0}\right)^{1/2}\right]. \tag{16A.3.21}$$

From the measurement of $g_2(t)$ in various experimental configurations, we can deduce ℓ_{tr} and t_0, and thus in particular the diffusion coefficient D_p of the suspended particles.

Bibliography

S. Chandrasekhar, *Radiative transfer*, Clarendon Press, Oxford, 1950. Reprinted, Dover Publications, New York, 1960.

A. Ishimaru, *Wave propagation and scattering in random media*, Vol. 1: *Single scattering and transport theory*; Vol. 2: *Multiple scattering and transport theory*, Academic Press, New York, 1978. Reissued, IEEE Press/Oxford University Press Classic Reissue, 1997.

P. Sheng, *Introduction to wave scattering, localization, and mesoscopic phenomena*, Springer-Verlag, Berlin, second edition, 2006.

References

D.J. Pine, D.A. Weitz, P.M. Chaikin, and E. Herbolzheimer, Diffusing-wave spectroscopy, *Phys. Rev. Lett.* **60**, 1134 (1988).

M.C.W. van Rossum and Th. M. Nieuwenhuizen, Multiple scattering of classical waves: microscopy, mesoscopy, and diffusion, *Rev. Mod. Phys.* **71**, 313 (1999).

Supplement 16B

Light scattering by a fluid

1. Introduction

When a beam of monochromatic light passes through a transparent, dense fluid, some part of the light is scattered, since the density of the medium is not uniform. Now, 'frozen' (that is, time-independent) density fluctuations cannot exist in a fluid. The light scattered by the density fluctuations of a fluid thus exhibits a spectrum of angular frequencies characteristic of the time dependence of the density fluctuations.

We aim here to show how we can determine this spectrum in the case of a *normal* fluid, that is, a fluid isotropic, made of spherically symmetric molecules, not charged, and not superfluid. We are interested in the hydrodynamic regime which prevails when the system, after many collisions, has reached a state of local thermodynamic equilibrium. If ω and $\boldsymbol{q}$ denote respectively an angular frequency and a wave vector typical of the perturbations imposed to the medium,[1] the hydrodynamic regime is that of the excitations of low angular frequency and large wavelength, as pictured by the inequalities:

$$\omega\tau \ll 1, \qquad q\ell \ll 1, \tag{16B.1.1}$$

where τ represents the collision time and ℓ the mean free path.[2]

In this regime, the fluid is fully described by the local values of the thermodynamic quantities, whose evolution is determined by the hydrodynamic equations. When the amplitude of the fluctuations is not too high, these latter equations may be linearized.

2. Linearized hydrodynamic equations

Two types of parameters are involved in the hydrodynamic equations, the thermodynamic derivatives, on the one hand, and the transport coefficients, on the other hand.

---

[1] See Section 5.

[2] In a dense liquid, the validity domain of the hydrodynamic regime is very large. Indeed, the mean free path is of the order of the range of the intermolecular forces, that is, a few Å, and the collision time, which we can estimate using similar arguments, is also very small, of the order of 10^{-12} s (except at the very low temperatures where liquids freeze). In a gas, the validity range of the hydrodynamic expressions is comparatively much smaller, since ℓ and τ are much larger than in a liquid.

2.1. Conservation equations

We denote respectively by $n(\boldsymbol{r},t)$, $\boldsymbol{g}(\boldsymbol{r},t)$, and $\varepsilon(\boldsymbol{r},t)$ the local densities of particles, kinetic momentum, and energy. The associated conservation equations are of the general form:[3]

$$\begin{cases} \dfrac{\partial n(\boldsymbol{r},t)}{\partial t} + \dfrac{1}{m}\nabla.\boldsymbol{g}(\boldsymbol{r},t) = 0 \\[2ex] \dfrac{\partial \boldsymbol{g}(\boldsymbol{r},t)}{\partial t} + \nabla.\underline{\Pi}(\boldsymbol{r},t) = 0 \\[2ex] \dfrac{\partial \varepsilon(\boldsymbol{r},t)}{\partial t} + \nabla.\boldsymbol{J}_E(\boldsymbol{r},t) = 0. \end{cases} \tag{16B.2.1}$$

In equations (16B.2.1), $\underline{\Pi}(\boldsymbol{r},t)$ denotes the kinetic momentum flux tensor and $\boldsymbol{J}_E(\boldsymbol{r},t)$ the energy flux, whereas m is the mass of one of the fluid's molecules.

2.2. Linearized constitutive equations

To study the hydrodynamic fluctuations, we have to supplement equations (16B.2.1) by macroscopic constitutive equations giving the expressions for the fluxes.

In their linearized form, the kinetic momentum density (proportional to the particle flux), the kinetic momentum flux tensor (which, once linearized, is identical to the pressure tensor), and the energy flux respectively read:

$$\begin{cases} \boldsymbol{g}(\boldsymbol{r},t) = mn\boldsymbol{u}(\boldsymbol{r},t) \\[2ex] \Pi_{ij}(\boldsymbol{r},t) = \delta_{ij}\mathcal{P}(\boldsymbol{r},t) - \eta\Big(\dfrac{\partial u_j}{\partial x_i} + \dfrac{\partial u_i}{\partial x_j} - \dfrac{2}{3}\delta_{ij}\dfrac{\partial u_l}{\partial x_l}\Big) \\[2ex] \boldsymbol{J}_E(\boldsymbol{r},t) = (\varepsilon + \mathcal{P})\boldsymbol{u}(\boldsymbol{r},t) - \kappa\nabla T(\boldsymbol{r},t). \end{cases} \tag{16B.2.2}$$

In equations (16B.2.2), $\boldsymbol{u}(\boldsymbol{r},t)$ and $\mathcal{P}(\boldsymbol{r},t)$ are respectively the local mean velocity and the local mean pressure, whereas η and κ denote the viscosity coefficient and the thermal conductivity of the fluid. The quantities n, ε, and $\mathcal{P}$ are the thermodynamic equilibrium values of the particle density, the energy density, and the pressure.

The set formed by the conservation equations (16B.2.1) together with the linearized constitutive equations (16B.2.2) constitutes a closed system of equations,[4] from which we deduce the linearized hydrodynamic equations:

$$\begin{cases} \dfrac{\partial n(\boldsymbol{r},t)}{\partial t} + \dfrac{1}{m}\nabla.\boldsymbol{g}(\boldsymbol{r},t) = 0 \\[2ex] \dfrac{\partial \boldsymbol{g}(\boldsymbol{r},t)}{\partial t} + \nabla\mathcal{P}(\boldsymbol{r},t) - \dfrac{\eta}{3mn}\nabla\big[\nabla.\boldsymbol{g}(\boldsymbol{r},t)\big] - \dfrac{\eta}{mn}\nabla^2\boldsymbol{g}(\boldsymbol{r},t) = 0 \\[2ex] \dfrac{\partial \varepsilon(\boldsymbol{r},t)}{\partial t} + \dfrac{\varepsilon+\mathcal{P}}{mn}\nabla.\boldsymbol{g}(\boldsymbol{r},t) - \kappa\nabla^2 T(\boldsymbol{r},t) = 0. \end{cases} \tag{16B.2.3}$$

---

[3] We can interpret equations (16B.2.1) either as the evolution equations of the concerned physical quantities or as the evolution equations of their fluctuations with respect to their global equilibrium values.

[4] We have to take into account the fact that the various thermodynamic quantities are not independent from one another.

To simplify the study of hydrodynamic fluctuations, we first get rid of the boundary conditions by considering an infinite medium, which allows us to carry out a spatial Fourier transformation. Then the remaining initial conditions problem can be solved by using the Fourier–Laplace transformation.

3. Transverse fluctuations

We write the kinetic momentum density as the sum of a longitudinal component and a transverse one,

$$\boldsymbol{g}(\boldsymbol{r},t) = \boldsymbol{g}_L(\boldsymbol{r},t) + \boldsymbol{g}_T(\boldsymbol{r},t), \tag{16B.3.1}$$

with:

$$\nabla \times \boldsymbol{g}_L = 0, \qquad \nabla.\boldsymbol{g}_T = 0. \tag{16B.3.2}$$

We introduce the spatial Fourier transform[5] $\boldsymbol{g}(\boldsymbol{q},t)$ of $\boldsymbol{g}(\boldsymbol{r},t)$, defined by:

$$\boldsymbol{g}(\boldsymbol{q},t) = \int \boldsymbol{g}(\boldsymbol{r},t) e^{-i\boldsymbol{q}.\boldsymbol{r}}\, d\boldsymbol{r}, \tag{16B.3.3}$$

as well as the Fourier transforms $\boldsymbol{g}_L(\boldsymbol{q},t)$ and $\boldsymbol{g}_T(\boldsymbol{q},t)$, defined in a similar way from $\boldsymbol{g}_L(\boldsymbol{r},t)$ and $\boldsymbol{g}_T(\boldsymbol{r},t)$. Note that $\boldsymbol{g}_L(\boldsymbol{q},t)$ and $\boldsymbol{g}_T(\boldsymbol{q},t)$ are respectively parallel and perpendicular to $\boldsymbol{q}$.

3.1. Evolution of the density of transverse kinetic momentum

First, we verify, using the system of equations (16B.2.3), that the component $\boldsymbol{g}_T(\boldsymbol{r},t)$ of the kinetic momentum density is decoupled from the other hydrodynamic variables and that it obeys the closed evolution equation:

$$\boxed{\frac{\partial \boldsymbol{g}_T(\boldsymbol{r},t)}{\partial t} - \frac{\eta}{mn}\nabla^2 \boldsymbol{g}_T(\boldsymbol{r},t) = 0.} \tag{16B.3.4}$$

Although $\boldsymbol{g}_T(\boldsymbol{r},t)$ is not, in general, the quantity in which we are primarily interested,[6] its evolution equation, decoupled from those of the other hydrodynamic variables, is simpler to analyze. Equation (16B.3.4) is a diffusion equation involving a *transverse diffusion coefficient*, related to the viscosity of the fluid and also called the *kinematic viscosity*:

$$\boxed{D_T = \frac{\eta}{mn}.} \tag{16B.3.5}$$

After spatial Fourier transformation, equation (16B.3.4) becomes:

$$\frac{\partial \boldsymbol{g}_T(\boldsymbol{q},t)}{\partial t} + D_T q^2 \boldsymbol{g}_T(\boldsymbol{q},t) = 0. \tag{16B.3.6}$$

---

[5] We use the same notation $\boldsymbol{g}(.,t)$ for the density of kinetic momentum $\boldsymbol{g}(\boldsymbol{r},t)$ and its spatial Fourier transform $\boldsymbol{g}(\boldsymbol{q},t)$.

[6] However, in an incompressible fluid, the fluctuations of kinetic momentum are purely transverse.

3.2. Slow variables

Equation (16B.3.6) displays the fact that the spatial Fourier components of large wavelength of the conserved quantity $\boldsymbol{g}_T(\boldsymbol{r},t)$ are slow variables.

This can be shown by using a simple physical argument. Consider a fluctuation $\boldsymbol{g}_T(\boldsymbol{r},t)$ of wavelength λ. To make it relax, that is, fade away, the molecules of the liquid have to diffuse over a distance of order λ. The longer this distance, the larger the corresponding relaxation time. Indeed, the particles must diffuse over a length $\lambda \sim 1/q$, which, according to equation (16B.3.6), necessitates a time $\sim (D_T q^2)^{-1}$, tending towards infinity with λ. This feature is general: in all systems in which conserved extensive variables exist, the local densities of these variables evolve slowly at large wavelengths.

3.3. Resolution of the diffusion equation

We assume that at time $t = 0$ a fluctuation $\boldsymbol{g}_T(\boldsymbol{q}, t = 0)$ of the density of transverse kinetic momentum exists in the fluid. We are looking for the fluctuation $\boldsymbol{g}_T(\boldsymbol{q},t)$ at a time $t > 0$. This initial conditions problem can be solved by introducing the Fourier–Laplace transform $\boldsymbol{g}_T(\boldsymbol{q},z)$ of $\boldsymbol{g}_T(\boldsymbol{q},t)$, defined by:[7]

$$\boldsymbol{g}_T(\boldsymbol{q},z) = \int_0^\infty \boldsymbol{g}_T(\boldsymbol{q},t)e^{izt}\,dt, \qquad \Im m\, z > 0. \tag{16B.3.7}$$

Equation (16B.3.6) yields the expression for $\boldsymbol{g}_T(\boldsymbol{q},z)$:

$$\boldsymbol{g}_T(\boldsymbol{q},z) = \frac{\boldsymbol{g}_T(\boldsymbol{q},t=0)}{-iz + D_T q^2}. \tag{16B.3.8}$$

Owing to the diffusion process, $\boldsymbol{g}_T(\boldsymbol{q},z)$ exhibits a diffusion pole of affix $z_0 = -iD_T q^2$. We compute $\boldsymbol{g}_T(\boldsymbol{q},t)$ by inverting the Fourier–Laplace transformation:

$$\boldsymbol{g}_T(\boldsymbol{q},t) = \frac{\boldsymbol{g}_T(\boldsymbol{q},t=0)}{2\pi}\int_C \frac{e^{-izt}}{-iz + D_T q^2}\,dz. \tag{16B.3.9}$$

In equation (16B.3.9), the integration contour C is a parallel to the abscissa's axis of strictly positive ordinate. Applying the residue theorem gives:[8]

$$\boldsymbol{g}_T(\boldsymbol{q},t) = \boldsymbol{g}_T(\boldsymbol{q},t=0)e^{-D_T q^2 t}, \qquad t > 0. \tag{16B.3.10}$$

If the initial fluctuation is localized at the origin, that is, if:

$$\boldsymbol{g}_T(\boldsymbol{r},t=0) = \boldsymbol{g}_T\delta(\boldsymbol{r}), \tag{16B.3.11}$$

---

[7] We use the same notation $\boldsymbol{g}(.,.)$ for $\boldsymbol{g}(\boldsymbol{q},t)$ and its Fourier–Laplace transform $\boldsymbol{g}(\boldsymbol{q},z)$.

[8] We can also directly solve equation (16B.3.6) and obtain the result (16B.3.10) without using the Fourier–Laplace transformation. However, we choose to use this method, here and in the rest of this chapter, since it allows us to treat in a relatively simple way the case of longitudinal fluctuations in which coupled variables are involved (see Section 4).

we have:

$$g_T(\mathbf{q}, t=0) = g_T, \tag{16B.3.12}$$

and, at any time $t > 0$:

$$g_T(\mathbf{q}, t) = g_T e^{-D_T q^2 t}. \tag{16B.3.13}$$

Equation (16B.3.13) yields finally:

$$\boxed{g_T(\mathbf{r}, t) = g_T (4\pi D_T t)^{-3/2} \exp\Big(-\frac{r^2}{4D_T t}\Big), \qquad t > 0.} \tag{16B.3.14}$$

The spreading of $\mathbf{g}(\mathbf{r}, t)$ is Gaussian, which is characteristic of a diffusion process.

4. Longitudinal fluctuations

The local density of particles, the longitudinal component of the local density of kinetic momentum, and the local density of energy obey the following coupled evolution equations:

$$\begin{cases} \dfrac{\partial n(\mathbf{r}, t)}{\partial t} + \dfrac{1}{m} \nabla . \mathbf{g}_L(\mathbf{r}, t) = 0 \\ \dfrac{\partial \mathbf{g}_L(\mathbf{r}, t)}{\partial t} + \nabla \mathcal{P}(\mathbf{r}, t) - \dfrac{\eta}{3mn} \nabla \big[\nabla . \mathbf{g}_L(\mathbf{r}, t)\big] - \dfrac{\eta}{mn} \nabla^2 \mathbf{g}_L(\mathbf{r}, t) = 0 \\ \dfrac{\partial \varepsilon(\mathbf{r}, t)}{\partial t} + \dfrac{\varepsilon + \mathcal{P}}{mn} \nabla . \mathbf{g}_L(\mathbf{r}, t) - \kappa \nabla^2 T(\mathbf{r}, t) = 0. \end{cases} \tag{16B.4.1}$$

We choose here as thermodynamic variables the local density of particles and the local temperature. Once $\mathbf{g}_L(\mathbf{r}, t)$ is eliminated, the problem reduces to the resolution of two coupled equations for $n(\mathbf{r}, t)$ and $T(\mathbf{r}, t)$.

4.1. The coupled equations for the density of particles and the temperature

First, taking the divergence of the evolution equation of $\mathbf{g}_L(\mathbf{r}, t)$ (the second of equations (16B.4.1)), and taking into account the evolution equation of $n(\mathbf{r}, t)$ (the first of equations (16B.4.1)), we get the equation:

$$\Big(\frac{\partial^2}{\partial t^2} - D_L \frac{\partial}{\partial t} \nabla^2\Big) n(\mathbf{r}, t) - \frac{1}{m} \nabla^2 \mathcal{P}(\mathbf{r}, t) = 0, \tag{16B.4.2}$$

where:

$$\boxed{D_L = \frac{4\eta}{3mn}} \tag{16B.4.3}$$

is the *longitudinal diffusion coefficient* (or *longitudinal kinematic viscosity*). We deduce from equation (16B.4.2), after some transformations, the following equation:

$$\left[\frac{\partial^2}{\partial t^2} - \left(D_L\frac{\partial}{\partial t} + \frac{1}{m}\frac{\partial \mathcal{P}}{\partial n}\bigg|_T\right)\nabla^2\right] n(\boldsymbol{r},t) = \frac{1}{m}\frac{\partial \mathcal{P}}{\partial T}\bigg|_n \nabla^2 T(\boldsymbol{r},t). \tag{16B.4.4}$$

As previously, we are considering an infinite medium. We take the spatial Fourier transform of equation (16B.4.4). Then, after a Fourier–Laplace transformation with respect to time, we are left with an initial conditions problem:

$$\left(-z^2 - izD_Lq^2 + \frac{c_s^2q^2}{\gamma}\right)n(\boldsymbol{q},z) + \frac{q^2}{m}\frac{\partial \mathcal{P}}{\partial T}\bigg|_n T(\boldsymbol{q},z) = (-iz + D_Lq^2)n(\boldsymbol{q},t=0) + \dot{n}(\boldsymbol{q},t=0). \tag{16B.4.5}$$

In equation (16B.4.5), we have introduced the squared adiabatic sound velocity in the fluid $c_s^2 = (\gamma/m)(\partial\mathcal{P}/\partial n)_T$ (γ is the ratio of the specific heats at constant pressure and constant volume). A priori, the initial conditions term on the right-hand side of equation (16B.4.5) involves both $n(\boldsymbol{q},t=0)$ and $\dot{n}(\boldsymbol{q},t=0)$. It is always possible to choose $\dot{n}(\boldsymbol{q},t=0) = 0$ (for instance by assuming that $\boldsymbol{g}(\boldsymbol{q},t=0)$ is purely transverse). Equation (16B.4.5) then reads:[9]

$$\boxed{\left(-z^2 - izD_Lq^2 + \frac{c_s^2q^2}{\gamma}\right)n(\boldsymbol{q},z) + \frac{q^2}{m}\frac{\partial \mathcal{P}}{\partial T}\bigg|_n T(\boldsymbol{q},z) = (-iz + D_Lq^2)n(\boldsymbol{q},0).} \tag{16B.4.6}$$

We now have to take into account the evolution equation of $\varepsilon(\boldsymbol{r},t)$. Eliminating $\boldsymbol{g}_L(\boldsymbol{r},t)$ between the first and third of equations (16B.4.1), we get the following equation:

$$\frac{\partial}{\partial t}\left[\varepsilon(\boldsymbol{r},t) - \frac{\varepsilon + P}{n}n(\boldsymbol{r},t)\right] - \kappa\nabla^2 T(\boldsymbol{r},t) = 0. \tag{16B.4.7}$$

In terms of the chosen thermodynamic variables $n(\boldsymbol{r},t)$ and $T(\boldsymbol{r},t)$, the local density of thermal energy $\varepsilon(\boldsymbol{r},t) - \big[(\varepsilon + \mathcal{P})/n\big]n(\boldsymbol{r},t)$ reads:

$$\varepsilon(\boldsymbol{r},t) - \frac{\varepsilon + \mathcal{P}}{n}n(\boldsymbol{r},t) = \frac{T}{V}\frac{\partial S}{\partial n}\bigg|_T n(\boldsymbol{r},t) + nc_vT(\boldsymbol{r},t), \tag{16B.4.8}$$

where c_v denotes the specific heat at constant volume per particle. Importing expression (16B.4.8) for the local density of thermal energy into equation (16B.4.7), gives:

$$\frac{T}{V}\frac{\partial S}{\partial n}\bigg|_T \frac{\partial n(\boldsymbol{r},t)}{\partial t} + nc_v\frac{\partial T(\boldsymbol{r},t)}{\partial t} - \kappa\nabla^2 T(\boldsymbol{r},t) = 0, \tag{16B.4.9}$$

[9] In the absence of ambiguity, the quantity $n(\boldsymbol{q},t=0)$ will simply be written $n(\boldsymbol{q},0)$.

which also reads:

$$\Big(\frac{\partial}{\partial t}-\frac{\kappa}{nc_v}\nabla^2\Big)T(\boldsymbol{r},t)+\frac{T}{nc_v}\frac{1}{V}\frac{\partial S}{\partial n}\Big|_T\frac{\partial n(\boldsymbol{r},t)}{\partial t}=0. \tag{16B.4.10}$$

Setting:

$$a=\frac{\kappa}{nc_v}, \tag{16B.4.11}$$

we obtain, after some transformations, the equation:

$$\Big(\frac{\partial}{\partial t}-a\nabla^2\Big)T(\boldsymbol{r},t)+\frac{T}{c_v}\frac{\partial s}{\partial n}\Big|_T\frac{\partial n(\boldsymbol{r},t)}{\partial t}=0, \tag{16B.4.12}$$

in which $s=S/N$ denotes the entropy per particle.

After spatial Fourier transformation and Fourier–Laplace transformation with respect to time, we get from equation (16B.4.12) the equation:

$$(-iz+aq^2)T(\boldsymbol{q},z)-iz\frac{T}{c_v}\frac{\partial s}{\partial n}\Big|_T n(\boldsymbol{q},z)=\frac{T}{c_v}\frac{\partial s}{\partial n}\Big|_T n(\boldsymbol{q},t=0)+T(\boldsymbol{q},t=0). \tag{16B.4.13}$$

The initial conditions term on the right-hand side of equation (16B.4.13) involves both[10] $n(\boldsymbol{q},t=0)$ and $T(\boldsymbol{q},t=0)$. There are no instantaneous correlations between the density fluctuations and the temperature. To calculate the spectrum of density fluctuations, it is thus not necessary to include the term in $T(\boldsymbol{q},0)$ in the expression for $n(\boldsymbol{q},z)$. Otherwise stated, we can make $T(\boldsymbol{q},0)=0$. Equation (16B.4.13) then reduces to:

$$\boxed{(-iz+aq^2)T(\boldsymbol{q},z)-iz\frac{T}{c_v}\frac{\partial s}{\partial n}\Big|_T n(\boldsymbol{q},z)=\frac{T}{c_v}\frac{\partial s}{\partial n}\Big|_T n(\boldsymbol{q},0).} \tag{16B.4.14}$$

4.2. Resolution of the coupled equations for $n(\boldsymbol{q},z)$ and $T(\boldsymbol{q},z)$

The linear system to be solved is that of the two coupled equations (16B.4.6) and (16B.4.14) for $n(\boldsymbol{q},z)$ and $T(\boldsymbol{q},z)$. In matrix form, it reads:

$$\begin{pmatrix} -z^2-izD_Lq^2+\dfrac{c_s^2q^2}{\gamma} & \dfrac{q^2}{m}\dfrac{\partial\mathcal{P}}{\partial T}\Big|_n \\ -iz\dfrac{T}{c_v}\dfrac{\partial s}{\partial n}\Big|_T & -iz+aq^2 \end{pmatrix}\begin{pmatrix} n(\boldsymbol{q},z) \\ T(\boldsymbol{q},z)\end{pmatrix}=\begin{pmatrix} -iz+D_Lq^2 \\ \dfrac{T}{c_v}\dfrac{\partial s}{\partial n}\Big|_T\end{pmatrix} n(\boldsymbol{q},0). \tag{16B.4.15}$$

The determinant of the matrix to be inverted is:

$$\mathrm{Det}=(-iz+aq^2)\Big(-z^2-izD_Lq^2+\frac{c_s^2q^2}{\gamma}\Big)+iz\frac{q^2}{m}\frac{\partial\mathcal{P}}{\partial T}\Big|_n\frac{T}{c_v}\frac{\partial s}{\partial n}\Big|_T. \tag{16B.4.16}$$

[10] When no misinterpretation is possible, these quantities will be respectively denoted by $n(\boldsymbol{q},0)$ and $T(\boldsymbol{q},0)$.

From the thermodynamic identity:

$$-\frac{T}{mc_v}\frac{\partial s}{\partial n}\bigg|_T\frac{\partial \mathcal{P}}{\partial T}\bigg|_n=\frac{\gamma-1}{\gamma}c_s^2, \tag{16B.4.17}$$

established in footnote below,[11] we deduce the following expression for the determinant (16B.4.16):

$$\text{Det}=i\left[(z+iaq^2)\Big(z^2+izD_Lq^2-\frac{c_s^2q^2}{\gamma}\Big)-zc_s^2q^2\frac{\gamma-1}{\gamma}\right]. \tag{16B.4.18}$$

The solution of the linear system (16B.4.15) is, as for the particle density, represented here by $n(\boldsymbol{q},z)$:

$$n(\boldsymbol{q},z)=in(\boldsymbol{q},t=0)\frac{(z+iaq^2)(z+iD_Lq^2)-c_s^2q^2\dfrac{\gamma-1}{\gamma}}{(z+iaq^2)\Big(z^2+izD_Lq^2-\dfrac{c_s^2q^2}{\gamma}\Big)-zc_s^2q^2\dfrac{\gamma-1}{\gamma}}. \tag{16B.4.19}$$

4.3. Calculation of $n(\boldsymbol{q},t)$

The processes associated with the regression of a density fluctuation may be identified through a study of the character of the poles of $n(\boldsymbol{q},z)$. To determine these poles, we have to solve the cubic equation:

$$(z+iaq^2)\Big(z^2+izD_Lq^2-\frac{c_s^2q^2}{\gamma}\Big)-zc_s^2q^2\,\frac{\gamma-1}{\gamma}=0, \tag{16B.4.20}$$

[11] Coming back to the expressions $\gamma=c_p/c_v$ for the ratio of specific heats and $c_s^2=(\gamma/m)(\partial\mathcal{P}/\partial n)_T$ for the squared sound velocity, we write the thermodynamic relation (16B.4.17) in the form:

$$c_p-c_v=-T\frac{(\partial s/\partial n)_T(\partial\mathcal{P}/\partial T)_n}{(\partial\mathcal{P}/\partial n)_T},$$

or, introducing the volume $V=nN$:

$$c_p-c_v=-T\frac{(\partial s/\partial V)_T(\partial\mathcal{P}/\partial T)_V}{(\partial\mathcal{P}/\partial V)_T}.$$

Given the Maxwell relation $(\partial S/\partial V)_T=(\partial\mathcal{P}/\partial T)_V$, we have, in order to establish identity (16B.4.17), to demonstrate the formula:

$$c_p-c_v=-\frac{T}{N}\frac{[(\partial\mathcal{P}/\partial T)_V]^2}{(\partial\mathcal{P}/\partial V)_T}.$$

This may be rewritten as:

$$N(c_p-c_v)=\frac{TV\alpha^2}{\chi_T},$$

where $\alpha=(1/V)(\partial V/\partial T)_P$ is the dilatation coefficient at constant pressure of the fluid and $\chi_T=-(1/V)(\partial V/\partial P)_T$ its isothermal compressibility. We recognize in the above formula a standard thermodynamic identity.

which we will do in the limit of small wave vectors.

For further purposes, let us note that equation (16B.4.20) can be rewritten as:

$$(z+iaq^2)(z^2+izD_Lq^2-c_s^2q^2)=(-iaq^2)c_s^2q^2\frac{\gamma-1}{\gamma}. \tag{16B.4.21}$$

• *Poles in the decoupled case*

If we consider the temperature as decoupled from the density, which would formally be expressed by $\gamma = 1$ (see identity (16B.4.17)), the equation giving the poles of $n(\boldsymbol{q}, z)$ simply reads:

$$(z+iaq^2)(z^2+izD_Lq^2-c_s^2q^2)=0. \tag{16B.4.22}$$

At order q^2, the poles of $n(\boldsymbol{q}, z)$ are, in this decoupling approximation:

$$\begin{cases} z_0^{(0)} = -iaq^2 \\ z_{1,2}^{(0)} = \pm c_s q - \dfrac{i}{2}D_L q^2. \end{cases} \tag{16B.4.23}$$

The pole $z_0^{(0)}$, which has the character of a diffusion pole, is called the *heat pole*, whereas the poles $z_{1,2}^{(0)}$ are called the *sound poles*.

• *Poles in weak coupling*

In the presence of a weak coupling,[12] the heat pole z_0 may be looked for in the form $z_0^{(0)} + \Delta z_0$. Equation (16B.4.20) then reads:

$$\Delta z_0\left[\left(z_0^{(0)}+\Delta z_0\right)^2+iD_Lq^2(z_0^{(0)}+\Delta z_0)-\frac{c_s^2q^2}{\gamma}\right]=(z_0^{(0)}+\Delta z_0)c_s^2q^2\frac{\gamma-1}{\gamma}. \tag{16B.4.24}$$

At first order in Δz_0, we get:

$$\Delta z_0\left[\left(z_0^{(0)}\right)^2+iD_Lq^2z_0^{(0)}-c_s^2q^2\right]=z_0^{(0)}c_s^2q^2\frac{\gamma-1}{\gamma}. \tag{16B.4.25}$$

In the limit of small wave vectors, Δz_0 is of order q^2:

$$\Delta z_0 = -z_0^{(0)}\frac{\gamma-1}{\gamma}. \tag{16B.4.26}$$

The displaced heat pole is thus:

$$z_0=-iaq^2+iaq^2\frac{\gamma-1}{\gamma}=-\frac{ia}{\gamma}q^2=-iD_{\rm th}q^2, \tag{16B.4.27}$$

where we have introduced the thermal diffusion coefficient $D_{\rm th} = a/\gamma = \kappa/nc_p$ (c_p is the specific heat at constant pressure per particle). A similar calculation can be carried

[12] This is realized in a liquid, in which case $\gamma \simeq 1$.

out for the sound poles $z_{1,2}$, which we look for in the form $z_{1,2}^{(0)} + \Delta z_{1,2}$. In the limit of small wave vectors, $\Delta z_{1,2}$ is of order q^2:

$$\Delta z_{1,2} = -\frac{i}{2} a q^2 \frac{\gamma - 1}{\gamma}. \tag{16B.4.28}$$

The displaced sound poles are thus:

$$z_{1,2} = \pm c_s q - \frac{i}{2} D_L q^2 - \frac{i}{2} a q^2 \frac{\gamma - 1}{\gamma} = \pm c_s q - i\Gamma q^2, \tag{16B.4.29}$$

where:

$$\Gamma = \frac{D_L}{2} + \frac{a}{2}\frac{\gamma - 1}{\gamma} \tag{16B.4.30}$$

is the *sound attenuation coefficient*. All three poles z_0 and $z_{1,2}$ lie in the lower complex half-plane.

Let us now come back to formula (16B.4.19). At the order at which the poles have been calculated, we can write the denominator in the form of the following product:

$$(z + iD_{\text{th}} q^2)(z - c_s q + i\Gamma q^2)(z + c_s q + i\Gamma q^2). \tag{16B.4.31}$$

Decomposing $n(\boldsymbol{q}, z)/n(\boldsymbol{q}, t = 0)$ into partial fractions and expressing the residues at lowest order in q gives:

$$\frac{n(\boldsymbol{q}, z)}{n(\boldsymbol{q}, t = 0)} = \frac{\gamma - 1}{\gamma} \frac{i}{z + iD_{\text{th}} q^2} + \frac{i}{2\gamma}\left(\frac{1}{z - c_s q + i\Gamma q^2} + \frac{1}{z + c_s q + i\Gamma q^2}\right). \tag{16B.4.32}$$

Inverting the Fourier–Laplace transformation, we deduce from formula (16B.4.32) the expression for $n(\boldsymbol{q}, t)$:

$$\boxed{\frac{n(\boldsymbol{q}, t)}{n(\boldsymbol{q}, 0)} = \frac{\gamma - 1}{\gamma} e^{-D_{\text{th}} q^2 t} + \frac{1}{\gamma} e^{-\Gamma q^2 t} \cos c_s q t, \qquad t > 0.} \tag{16B.4.33}$$

4.4. Regression of the density fluctuations

The heat pole of $n(\boldsymbol{q}, z)$, given at order q^2 by formula (16B.4.27), is purely imaginary. It corresponds to a fluctuation which regresses without propagating, its lifetime being determined by the thermal diffusion coefficient.

The sound poles of $n(\boldsymbol{q}, z)$, given at order q^2 by formula (16B.4.29), possess a real part equal to $\pm c_s q$ and an imaginary part proportional to the sound attenuation coefficient. They correspond to a fluctuation which propagates within the fluid at the sound velocity, and whose amplitude regresses due to viscosity and thermal conduction effects. The thermal damping of sound waves is weak when $\gamma \simeq 1$, as shown by the expression (16B.4.30) for Γ.

5. Dynamical structure factor

The intensity of scattered light is proportional to the dynamical structure factor $S(\boldsymbol{q},\omega)$, defined as the Fourier transform of $\langle n(\boldsymbol{q},t)n(-\boldsymbol{q},0)\rangle$:

$$S(\boldsymbol{q},\omega) = \int_{-\infty}^{\infty} \langle n(\boldsymbol{q},t)n(-\boldsymbol{q},0)\rangle e^{i\omega t}\, dt. \tag{16B.5.1}$$

In formula (16B.5.1), ω and $\boldsymbol{q}$ denote respectively the changes of angular frequency and of wave vector of light due to scattering.

5.1. Calculation of $S(\boldsymbol{q},\omega)$

For $t > 0$, $n(\boldsymbol{q},t)$ is given by formula (16B.4.33). This gives:

$$\big\langle n(\boldsymbol{q},t)n(-\boldsymbol{q},0)\big\rangle = \big\langle n(\boldsymbol{q},0)n(-\boldsymbol{q},0)\big\rangle\left(\frac{\gamma-1}{\gamma}e^{-D_{\rm th}q^2t} + \frac{1}{\gamma}e^{-\Gamma q^2 t}\cos c_s qt\right), \qquad t>0. \tag{16B.5.2}$$

To obtain $\langle n(\boldsymbol{q},t)n(-\boldsymbol{q},0)\rangle$ for $t < 0$, we make use of the time-translational invariance property:

$$\big\langle n(\boldsymbol{q},t)n(-\boldsymbol{q},0)\big\rangle = \big\langle n(\boldsymbol{q},0)n(-\boldsymbol{q},-t)\big\rangle. \tag{16B.5.3}$$

The quantity $n(-\boldsymbol{q},-t)$ can be obtained from formula (16B.4.33) by changing $\boldsymbol{q}$ into $-\boldsymbol{q}$ and t into $-t$:

$$\frac{n(-\boldsymbol{q},-t)}{n(-\boldsymbol{q},0)} = \frac{\gamma-1}{\gamma}\,e^{D_{\rm th}q^2t} + \frac{1}{\gamma}\,e^{\Gamma q^2 t}\cos c_s qt, \qquad t<0. \tag{16B.5.4}$$

This gives:

$$\big\langle n(\boldsymbol{q},t)n(-\boldsymbol{q},0)\big\rangle = \big\langle n(\boldsymbol{q},0)n(-\boldsymbol{q},0)\big\rangle\left(\frac{\gamma-1}{\gamma}e^{D_{\rm th}q^2t} + \frac{1}{\gamma}e^{\Gamma q^2 t}\cos c_s qt\right), \qquad t<0. \tag{16B.5.5}$$

The dynamical structure factor (formula (16B.5.1)) is thus given by:

$$S(\boldsymbol{q},\omega) = S(\boldsymbol{q})2\,\Re e\left\{\int_0^{\infty}\left(\frac{\gamma-1}{\gamma}e^{-D_{\rm th}q^2t} + \frac{1}{\gamma}e^{-\Gamma q^2 t}\cos c_s qt\right)e^{i\omega t}\,dt\right\}, \tag{16B.5.6}$$

where we have set $S(\boldsymbol{q}) = \langle n(\boldsymbol{q},0)n(-\boldsymbol{q},0)\rangle$. After integration we get:

$$\frac{S(\boldsymbol{q},\omega)}{S(\boldsymbol{q})} = \frac{\gamma-1}{\gamma}\,\frac{2D_{\rm th}q^2}{\omega^2+(D_{\rm th}q^2)^2} + \frac{1}{\gamma}\left[\frac{\Gamma q^2}{(\omega+c_sq)^2+(\Gamma q^2)^2} + \frac{\Gamma q^2}{(\omega-c_sq)^2+(\Gamma q^2)^2}\right]. \tag{16B.5.7}$$

5.2. Spectrum of scattered light

As shown by formula (16B.5.7), the spectrum of the density fluctuations of a normal fluid has three Lorentzian components: the *Rayleigh line*, centered at $\omega = 0$ and of width $2D_{\rm th}q^2$, and the two *Brillouin lines*, centered at $\omega = \pm c_s q$ and of width $2\Gamma q^2$. The two components centered at $\omega = \pm c_s q$ correspond to propagating modes (analogous to the modes associated with the longitudinal acoustic phonons in a solid), whereas the line centered at $\omega = 0$ represents the thermal mode which regresses without propagating.

The total integrated intensity of the Rayleigh line is:

$$\mathcal{I}_R = S(\boldsymbol{q})\frac{\gamma-1}{\gamma}\int_{-\infty}^{\infty}\frac{2D_{\text{th}}q^2}{\omega^2+\left(D_{\text{th}}q^2\right)^2}\,d\omega. \tag{16B.5.8}$$

This gives:

$$\mathcal{I}_R = 2\pi S(\boldsymbol{q})\frac{\gamma-1}{\gamma}. \tag{16B.5.9}$$

The integrated intensity of each one of the Brillouin lines is:

$$\mathcal{I}_B = 2\pi S(\boldsymbol{q})\frac{1}{2\gamma}. \tag{16B.5.10}$$

We therefore have:

$$\mathcal{I}_R + 2\mathcal{I}_B = 2\pi S(\boldsymbol{q}). \tag{16B.5.11}$$

The ratio:

$$\boxed{\frac{\mathcal{I}_R}{2\mathcal{I}_B} = \gamma - 1} \tag{16B.5.12}$$

is called the *Landau–Placzek ratio* (L. Landau and G. Placzek, 1934).

When we come near the critical point, $c_p \to \infty$ and the scattered intensity lies mostly in the Rayleigh line, whose width $\sim D_{\text{th}}q^2$ then approaches zero. Far from the critical point, the Brillouin lines dominate. From the position and the width of the Brillouin lines, we can deduce the sound velocity and attenuation coefficient. From the Rayleigh line, we can determine the thermal diffusion coefficient. Finally, from the Landau–Placzek ratio, we can obtain the specific heats ratio.

Bibliography

B.J. BERNE and R. PECORA, *Dynamic light scattering. With applications to chemistry, biology, and physics*, Wiley, New York, 1976. Reprinted, Dover Publications, New York, 2000.

P.M. CHAIKIN and T.C. LUBENSKY, *Principles of condensed matter physics*, Cambridge University Press, Cambridge, 1995.

D. FORSTER, *Hydrodynamic fluctuations, broken symmetries, and correlation functions*, Westview Press, Boulder, 1995.

J.-P. HANSEN and I.R. MCDONALD, *Theory of simple liquids*, Academic Press, London, third edition, 2006.

L.D. LANDAU and E.M. LIFSHITZ, *Statistical physics*, Butterworth-Heinemann, Oxford, third edition, 1980.

References

L.P. KADANOFF and P.C. MARTIN, Hydrodynamic equations and correlation functions, *Ann. Phys.* **24**, 419 (1963).

R.D. MOUNTAIN, Spectral distribution of scattered light in a simple fluid, *Rev. Mod. Phys.* **38**, 205 (1966).

Index

The manufacturer's authorised representative in the EU for product safety is Oxford University Press España S.A. of El Parque Empresarial San Fernando de Henares, Avenida de Castilla, 2 - 28830 Madrid (www.oup.es/en or product.safety@oup.com). OUP España S.A. also acts as importer into Spain of products made by the manufacturer.
Printed and bound by CPI Group (UK) Ltd, Croydon, CR0 4YY
03/07/2025
01910257-0002